T0254888

Algebraic Approaches to Nuclear Structure

Contemporary Concepts in Physics

Volume 1, *Particle Physics and Introduction to Field Theory* (Revised and Updated First Edition) **T.D. Lee**

Volume 2, *Particle Physics: The Quest for the Substance of Substance*
L.B. Okun

Volume 3, *Gauge Fields and Strings*
A.M. Polyakov

Volume 4, *Nonlinear Physics: From the Pendulum to Turbulence and Chaos*
R.Z. Sagdeev, D.A. Usikov and G.M. Zaslavsky

Volume 5, *Particle Physics and Inflationary Cosmology*
Andrei Linde

Volume 6, *Algebraic Approaches to Nuclear Structure: Interacting Boson and Fermion Models*
Richard F. Casten, Pertti O. Lipas, David D. Warner, Takaharu Otsuka, Kris Heyde and Jerry P. Draayer, edited by **Richard F. Casten**

Volume 7, *Simple Models of Complex Nuclei: The Shell Model and Interacting Boson Model*
Igal Talmi

This book is part of a series. The publisher will accept continuation orders which may be canceled at any time and which provide for automatic billing and shipping of each title in the series upon publication. Please write for details.

Algebraic Approaches to Nuclear Structure

Interacting Boson and Fermion Models

by
Richard F. Casten
Pertti O. Lipas
David D. Warner
Takaharu Otsuka
Kris Heyde
Jerry P. Draayer

Edited by
Richard F. Casten
Brookhaven National Laboratory
Upton, New York, USA

CRC Press is an imprint of the
Taylor & Francis Group, an **informa** business

First published 1993 by Harwood Academic Publishers

Published 2019 by CRC Press
Taylor & Francis Group
6000 Broken Sound Parkway NW, Suite 300
Boca Raton, FL 33487-2742

ISBN 13: 978-3-7186-0538-5 (pbk)

Visit the Taylor & Francis Web site at
http://www.taylorandfrancis.com

and the CRC Press Web site at
http://www.crcpress.com

Library of Congress Cataloging-in-Publication Data

Algebraic approaches to nuclear structure : interacting boson and fermion models / by Richard F. Casten . . . [et al.] ; edited by Richard F. Casten.
p. cm.—(Contemporary concepts in physics ; v. 6)
Includes bibliographical references and index.
ISBN 3-7186-0537-6 (hard).—ISBN 3-7186-0538-4 (soft)
1. Nuclear structure—Mathematical models. 2. Interacting boson models. 3. Fermions—Mathematical models. I. Casten, R. II. Series.
QC793.3.S8A38 1993
539.7′4′015118—dc20

92-26649
CIP

Contributing Authors

Richard F. Casten, Physics Department, Brookhaven National Laboratory, Upton, New York 11973, USA

Jerry P. Draayer, Department of Physics and Astronomy, Louisiana State University, Baton Rouge, Louisiana 70803-4001, USA

Kris Heyde, Institute for Nuclear Physics, Proeftuinstraat 86, B-9000 Gent, Belgium

Pertti O. Lipas, Department of Physics, University of Jyväskylä, SF-40351 Jyväskylä, Finland

Takaharu Otsuka, Faculty of Science, Department of Physics, University of Tokyo, 7-3-1 Hongo, Bunkyo-ku, Tokyo 113, Japan

David D. Warner, Daresbury Laboratory, Warrington, Cheshire WA4 4AD, UK

Contents

Preface to the Series

The series of volumes, *Contemporary Concepts in Physics*, is addressed to the professional physicist and to the serious graduate student of physics. The subjects to be covered will include those at the forefront of current research. It is anticipated that the various volumes in the series will be rigorous and complete in their treatment, supplying the intellectual tools necessary for the appreciation of the present status of the areas under consideration and providing the framework upon which future developments may be based.

Preface

Algebraic, symmetry-based, models of nuclear structure and phenomenology have been in existence, in one form or another, since the early days of nuclear physics. In the 1950s Elliott and coworkers expanded the $SU(3)$ model and applied it to a number of light nuclei. The prominence of algebraic approaches to nuclear structure has grown dramatically since the proposal in the mid-1970s of a model known as the Interacting Boson Approximation (IBA) or Interacting Boson Model (IBM). It is probably fair to say that this model, with its dynamical symmetries, its straightforward treatment of nuclear shape or phase transitional regions, and its simplicity, has helped foster a true renaissance of interest in nuclear structure and allied areas. The successes of the model acted as an encouragement to the ongoing parallel development of fermionic algebraic models which are at once more microscopic but more complex and which have recently made substantial progress through both formal and computational advances. Nowadays, this corpus of algebraic models is available to attack a truly astounding variety of important issues in nuclear structure ranging from light nuclei to the actinides and with implications beyond nuclear structure in the realms of nuclear reactions and scattering and in astrophysics.

Now that the main development phase of these models has passed, and they are entering the standard lexicon of theoretical approaches to nuclear structure, it seems an appropriate occasion for a textbook devoted to algebraic models and their applications. The present volume aspires to fulfill this need. Of course, there is no dearth of literature, review articles and, indeed, other monographs on these models (many of these are cited in the bibliographies to the individual chapters). Yet a simple, but thorough, pedagogic approach, starting from the most elementary ideas and building up to the most recent results of advanced theories, seems a worthwhile addition to the literature.

This book is designed for a graduate level treatment and assumes only very basic knowledge of quantum mechanics, nuclear structure and phenomenology. Most of the requisite background and ideas are explained as needed. The volume starts with a global survey of the data concerning nuclear structure and of existing theoretical approaches (microscopic and geometric) to understanding and correlating this data. With this background, the fundamental group theoretical basis of algebraic models of nuclear structure is introduced in a simple step-by-step way in chapter 2, which forms the theoretical backbone to the rest of the text. Subsequent chapters (3–6) confront the IBM with experiment (for nuclei with even numbers of protons and neutrons), discuss the microscopic foundations of the model, an important and natural extension to it (the IBM-2), its application to nuclei with an odd number of nucleons, and a number of important extensions to the model that expand its purview substantially. Finally, chapter 7 is an introduction to the wealth of fermionic algebraic models that treat the individual nucleons in the nucleus, in contrast to the ansatz of the IBM that these nucleons can be treated, for many purposes, in pairs acting effectively as bosons. A more detailed guide to the text will be found at the end of chapter 1.

Each chapter is individually authored, by experts who are active researchers and acknowledged leaders in the field. Every effort has been made to make the presentations consistent, but individual predilections and the varied nature of the subject matter will naturally (and desirably) lead to differences in style, depth of coverage and level of complexity. Each chapter is intended to be as self-contained as possible, although at the same time each builds on (and cross-references to) other (generally earlier) chapters. The notation throughout is as consistent as reasonable faithfulness to the literature will permit. Where major differences occur, each author has taken the responsibility to explicitly note them (the primary example of this occurs in chapter 6).

This work has truly been a collaborative endeavor, and the extensive interaction between the authors has been mutually beneficial and symbiotic. We hope the reader will see in this coauthored project a coherent, unified whole.

Acknowledgements from individual authors are provided at the end of each chapter. Here, in my editorial role, I would just like to add a few. I am, first of all, extremely grateful to all my coauthors for their dedication to this project and for their (general) timeliness. Each chapter has gone through an iterative process, and I am grateful to the authors for their good-spirited willingness to indulge this process. I am also very grateful to Akito Arima and Franco Iachello, the coinventors of the IBM, for their seminal work and for innumerable discussions, interactions and encouragement over years of exciting research in this field. Their contributions to nuclear structure are lasting and deep. I am grateful to Herman Feshbach, editor of this series, for asking me to produce this book and for encouragement in the lengthy interim prior to its completion. Finally, I am, as usual, extremely grateful to Jackie Mooney for her expertise, professionalism and friendliness, often during hectic periods, without whose help neither this book nor much of my research output would ever have existed.

I am pleased to also thank Brookhaven National Laboratory and, in particular, its physics department, for the opportunity to pursue this project, and the United States Department of Energy for support under contract DE-AC02-76CH00016.

Richard F. Casten

CHAPTER 1

Introduction

RICHARD F. CASTEN

1.1. ORIENTATION

Atomic nuclei are composed of protons and neutrons orbiting a common center of mass and interacting primarily via the attractive short-range strong interaction. The data pertaining to nuclear excitations, transition rates, and other phenomena are extraordinarily rich and have spawned a comparable richness of nuclear models designed to account for various aspects of this phenomenology. These models differ in their assumptions and approaches or methodology and are characterized by varying degrees of microscopy: some are essentially macroscopic, describing primarily collective excitations of atomic nuclei in terms of nuclear shapes (e.g., rotations of deformed ellipsoids, vibrations) while others focus on the specific orbits of individual nucleons and the interactions amongst these nucleons. Generally speaking, these models are not contradictory, but rather complementary approaches to describing different manifestations of the behavior of nuclei. They suffice as an interim stage in the search for an all-encompassing theory that is not only general but, equally important, tractable for all nuclei.

The topic of this book is primarily the study of collective excitations of atomic nuclei as viewed and interpreted in terms of a class of models known as algebraic. This term, used to distinguish such models from microscopic (e.g., shell model) approaches or geometric (e.g., Bohr–Mottelson) models, refers to the exploitation of powerful, yet conceptually simple, group theoretical methods and to an outlook inspired by the search for symmetries underlying the behavior of nuclei. The emphasis throughout this volume will be on the vast realm of nuclear structure (nuclear reactions will be discussed for the most part only insofar as they illuminate structure), and, especially, on the collective excitations characteristically observed in the majority of nuclei at rather low excitation energies.

On the one hand, this emphasis on low-energy nuclear structure restricts the book's purview to only a small fraction of the totality encompassed by the phrase "nuclear physics." On the other, it is the area that, historically, has been the central focus of the field, that today remains one of its very active and exciting areas, with new discoveries and insights emerging at an impressive pace for a field that has been active so long, and that is the basis and background for the rest of the field. Nuclear reactions, the use of nuclear processes to study fundamental conservation laws, astrophysics, or applied aspects of nuclear physics, to name just a few areas, depend to a good extent for their progress on a knowledge of the basic structure of the nuclei whose properties they exploit. For example, the calculation of astrophysical nucleosynthesis in explosive stellar events depends on a knowledge of the structure of known nuclei and on sufficient confidence in nuclear structure models to extrapolate to the properties of unknown nuclei.

Though the first sophisticated algebraic approach to nuclear structure, the so-called Elliott $SU(3)$ scheme, focused on light nuclei in the s-d shell (roughly $A = 16$–30), most recent efforts in the field have centered on medium mass and heavy nuclei and this will be the emphasis here as well. The primary focus of this book, aside from its elucidation of algebraic techniques that have wide applicability, will be a model known as the Interacting Boson Model (or IBM) and its offshoots and extensions. This model is "bosonic" in the sense that its fundamental building blocks are theoretical constructs which can be viewed as correlated pairs of fermions that carry integer spins and can be treated as bosons.

Despite the emphasis on the IBM, however, an important complementary part of the text is an introduction to algebraic fermionic models (Chapter 7): such models (e.g., the Elliott scheme) both pre-date the IBM and carry an independent and important perspective which equally benefits from the powers of group theory. Space constraints prohibit a detailed discussion of these approaches, but it is hoped that the introduction to them here will provide the reader a basis for further study.

This work is the joint result of a collaboration among the editor and five experts in various aspects of the application of algebraic models and group theoretic techniques to nuclear structure. Each chapter attempts a broad overview of its subject matter and hence plays the role of a short monograph of its sub-field. The chapters are written individually although considerable effort has been made to make them compatible in notation and approach and complementary in coverage. Nevertheless, the reader will note different styles, different degrees of formality and abstractness and somewhat different outlooks, as befits a field that is rich both in empirical phenomenology and in theoretical perspectives.

The contents of the chapters will be summarized at the end of this Introduction. Suffice it to say here that, after these initial pages, which offer a general background on nuclear structure, and a brief introduction to the models discussed later, the text proceeds with a presentation of the algebraic (group theoretic) techniques applied in these models, to a confrontation of the principal model to be discussed (the IBM) with the data, followed by a discussion of its microscopic basis, of important extensions to it and, finally, to an introduction to the large arena of complementary algebraic approaches that, unlike the bosonic IBM, focus directly on the fermions (protons and neutrons) that actually comprise the nucleus.

The purpose of the present introduction is to provide a context in which to place the rest of the volume. We will try here to outline, in briefest form, the major model approaches to nuclear structure so as to give a point of reference for comparison with the approach and predictions of algebraic models. We will also attempt a cursory overview of some of the data, phenomenology and systematic properties of atomic nuclei that these, or other, collective models must attempt to account for. In neither aspect can this chapter really broach the field. Indeed, in every sense, this book can never stand alone. The reader is strongly urged to read any (or several) of the excellent general textbooks on nuclear structure that exist. Examples which the editor highly recommends are those by de Shalit and Talmi (1963) and the recent text by Heyde (1991) for the shell model, the works by de Shalit and Feshbach (1974) for a general overview of models, the work by Bohr and Mottelson (1969, 1975) for geometric collective models, and the books by Lane (1964) and by Ring and Schuck (1980) for the microscopic approach to nuclear many-body theory. These and other textbooks and general articles are listed in the bibliography to this chapter.

1.2. NUCLEAR STRUCTURE

To a good approximation, certainly adequate for this work, the nucleus is composed of protons and neutrons which interact by a short range, primarily attractive, nuclear (or strong) force. In addition, the protons experience the repulsive Coulomb interaction. These nucleons (Z protons and N neutrons in a nucleis ${}^{A}_{Z}X_{N}$ where $A = Z + N$) orbit the nuclear center of mass in a mean field which they themselves create. In most treatments, a relatively simple, analytic, form of the central mean field (for example, a spherical harmonic oscillator or a Wood-Saxon potential or their deformed counterparts) is used and the remainder of the internucleonic force is treated as a residual interaction. Though these residual interactions are absolutely crucial to nuclear structure and its evolution with N, Z and A (as we shall abundantly see), the basic concept of individual particle orbits in a central potential remains useful and a valid conceptual approximation. The enigmatic fact of virtually undisturbed nucleon motion in a nuclear medium that is essentially filled with protons and neutrons is one of the many critical consequences of the Pauli Principle. Here, its effect is to render collisions of all except the surface or outermost nucleons ineffective since the interactions are not sufficiently strong to excite the colliding nucleons beyond the filled levels.

The individual particle orbits are quantal, of course, and characterized by principal, orbital and total angular momentum quantum numbers, n, l, j, as well as by a magnetic quantum number, m, that is often dropped in the notation when not contextually important. These orbits are identified by the notation nl_j, as for example $1d_{3/2}$ or $2g_{7/2}$ where s, p, d, f, g, . . . denote the orbital angular momenta $l = 0, 1, 2, 3, 4, \ldots$ (if not specified, it is to be understood that all angular momenta are given in units of $\hbar$ in this book).

It is a rather general consequence of motion in a central field that a "shell" structure develops: that is, the energies depend on n and l in such a way that the levels occur in groupings, corresponding, roughly, to orbits at similar radii. Qualitatively, an increase in either n or l corresponds to increased kinetic energy. Hence, two levels, one with higher n but lower l than the other, may have the same or similar energies. The particular orbits that form a group depend on the form of the central field but, for important cases [simple harmonic oscillator (SHO), a square well], orbits differing by $\Delta l = -2 \cdot \Delta n$ lie close or are degenerate. As an example, in either of these potentials, the $2s_{1/2}$ orbit is degenerate with the $1d_{3/2}$ or $1d_{5/2}$ orbits. These degeneracies are shown on the left in Fig. 1.1 which shows single particle energies in several simple potentials.

Empirically, nuclei with certain "magic" numbers of protons or neutrons enjoy special properties such as high binding energy of the "last" nucleon relative to nuclei with $Z = Z_{magic} + 1$ or $N = N_{magic} + 1$, high lying first excited states, small quadrupole moments, low neutron capture cross sections, and so on. (The main magic numbers are 2, 8, 20, (40), 50, 82, 126.) These features characterize a "closed shell" configuration, and the major task confronting the originators of the independent particle nuclear shell model [Mayer (1950), and Haxel, Jensen and Seuss (1950)] was to devise a shell model Hamiltonian that produced groupings of orbits leading to these major shell closures. The key ingredient was the incorporation of a spin-orbit force (of the form $\bar{l} \cdot \bar{s}$ where s is the nucleonic spin, $(1/2)\hbar$), as well as an attractive l^2 component, in addition to the simple harmonic oscillator. This gives a Hamiltonian that is often written

$$H = H_{SHO} + Cl^2 + D\bar{l} \cdot \bar{s} \qquad (1.1)$$

where the signs of C and D are such that high j orbits are lowered relative to the simple harmonic oscillator and the $j = l + \frac{1}{2}$ orbit is lowered in energy relative to the orbit with $j = l - \frac{1}{2}$.

This Hamiltonian leads to a set of "shell model" orbits, each of which can contain $(2j + 1)$ particles, whose energies are shifted relative to those of H_{SHO} alone so that energy gaps appear after each of the above magic numbers. However, it is important to stress that the mean central field giving these levels arises from the interactions of the nucleons themselves, and it is therefore dependent on the number of neutrons and protons. Hence, the single particle energy levels, and, in particular, even their relative positions, change with N, Z and A: the typical "textbook" shell model diagram is only a "snapshot" for a given mass region. Nevertheless, it is useful, as a point of reference, to give such a diagram, as shown in Fig. 1.1.

In the magnified section on the right, we illustrate and emphasize the above point that important details can change with mass by showing the single particle energies for the $N = 50$–82 shell as experimentally determined for ^{91}Zr, where this shell, and the $1g_{9/2}$

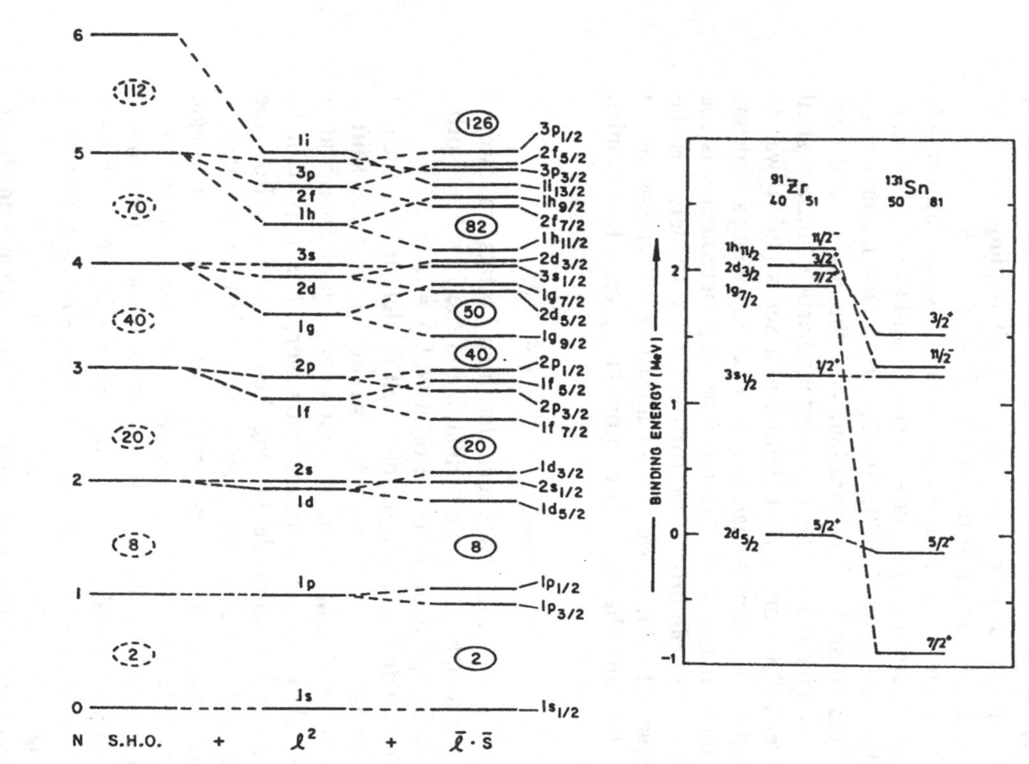

FIGURE 1.1. *Left: Single-particle energy levels for a simple harmonic oscillator* (S.H.O.), *a modified oscillator with l^2 term, and a realistic shell model potential with l^2 and $l \cdot s$ terms.*

Right: Neutron single-particle energies in ^{91}Zr *and* ^{131}Sn *showing the variations that can occur, even in relative positions, as a function of N and Z. The principal change, the lowering of the $\nu g_{7/2}$ level, is primarily due to the filling of the proton $\pi g_{9/2}$ orbit between Z = 40 and 50 since the $\nu g_{7/2}$ and $\pi g_{9/2}$ orbits have high spatial overlap and hence a strong attractive residual p-n interaction. [Right side from Heyde (1988).]*

proton shell, is essentially empty, and for ^{131}Sn, where both are basically full. We note that, not only the energies, but even the order changes: the qualitative reproduction of such changes, primarily but not solely due to the "valence" interactions of the outermost protons with the outermost neutrons, remains, incidentally, one of the challenges of microscopic nuclear theory.

Focusing on the main part of Fig. 1.1, the critical element that achieves the desired shell structure and correct magic numbers is, for nuclei beyond N, Z = 40, the descent, into the next lower oscillator shell, of a single orbit with the highest l and j (hence the largest l^2 and $l \cdot s$ interactions), from the next shell higher. Since the parities of the "normal" orbits in each oscillator shell alternate from shell to shell this special "intruder" orbit has opposite parity to the others in whose midst it now lies. Hence, it is often called the non-normal, abnormal or unique parity orbit. It plays a fundamental role in nuclear structure, not only because it allows the magic numbers to be reproduced but because of other special properties: the two most important of these are its high j (which leads to such effects as enhanced Coriolis interactions and, thereby, to phenomena, which the reader may have encountered, such as backbending and rotation alignment) and the fact that, since its neighboring levels are of opposite parity, residual interactions involving this orbit are weak and its manifestations in actual nuclei are usually quite pure. Its presence also presents conceptual difficulties in certain algebraic approaches [$SU(3)$] which in turn have led to the extremely important development of the concepts of pseudo-spin and pseudo-$SU(3)$ which will be discussed in several places in this book. (See, for example, Chapters 5 and 7.)

Nuclei with a single nucleon outside a closed shell (as either a particle or hole) should be ideal candidates for a shell model description and indeed this is where the model works best. Figure 1.2 shows an example, the low lying states of ^{209}Bi, which, though not pure single particle in character, can be closely correlated with the orbits just above 82 in Fig. 1.1.

When there is more than one "valence" particle, residual interactions come into play and can "mix" different shell model configurations: for example, a nucleus with three valence nucleons in the 50–82 shell could have low lying states with all three nucleons in either the $2d_{5/2}$ or $1g_{7/2}$ orbits or with one in one of these orbits and two in the other. These configurations are denoted, symbolically, as $(2d_{5/2})^3J$, $(1g_{7/2})^3J$, $[(2d_{5/2})^2J_1\ (g_{7/2})]J$ or $[(1g_{7/2})^2J_1\ (2d_{5/2})]J$, where

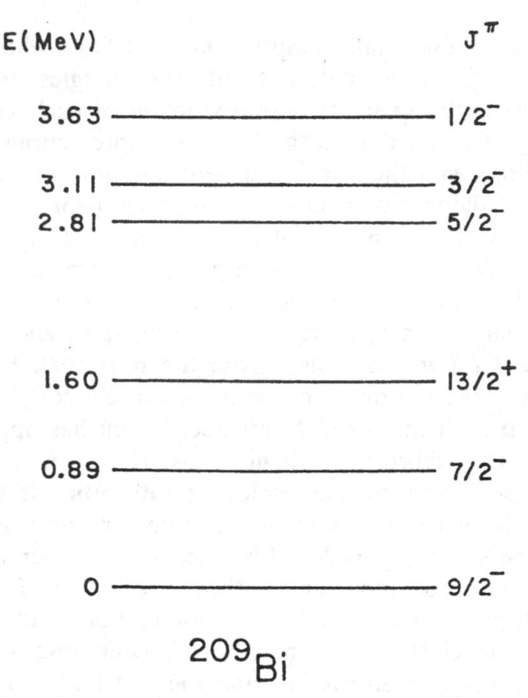

FIGURE 1.2. *Low-lying single-particle neutron levels of* ^{209}Bi *corresponding to orbits in the shell between N = 82 and 126.*

the superscript indicates the number of particles in the orbit (lack of a superscript implies a single particle), J_1 is the angular momentum for the intermediate coupling of two of the particles, and J is the final total angular momentum. The rules of angular momentum coupling will often allow more than one configuration for a given J. Hence, residual interactions will often lead to "real" states defined by linear combinations of these configurations. Fortunately, a well-known empirical fact (which is also a feature of most short-range attractive residual interactions) is that pairs of nucleons in a given orbit tend to couple so that the lowest energy state has spin and parity $J^\pi = 0^+$ (see, below, Fig. 1.17). Hence, the lowest state of three identical nucleons in the $1g_{7/2}$ orbit $((1g_{7/2})^3)$ should have $J^\pi = 7/2^+$, since two of these nucleons will couple to 0^+ so the total angular momentum is given by the third. This fact allows the regime of applicability of the simple shell model to be enormously extended.

Nevertheless, whenever there are more than a few nucleons of a given type, and almost always when there are nucleons of both types (protons and neutrons) outside closed shells, a full multi-particle shell model calculation (diagonalization) incorporating a residual interaction (often expressed as a semi-empirical set of two-body matrix elements) must be performed. However, even for a few valence nucleons (of both types) the number of states of a given angular momentum can easily reach 10^6–10^{12}, and the dimensions of such calculations become prohibitive.

At the same time, it is an empirical feature of such nuclei that "collective" behavior is manifest. This can be illustrated in many ways. Perhaps the most dramatic is in terms of $B(\mathrm{E2}{:}0_1^+ \rightarrow 2_1^+)$ values in even-even nuclei (nuclei with even numbers of protons and neutrons). These $B(\mathrm{E2})$ values are proportional to squares of electric quadrupole E2 transition matrix elements between the first excited 2^+ state and the 0^+ ground state. Figure 1.3 shows their systematics across the nuclear chart in units of $B(\mathrm{E2})_{\mathrm{sp}}$ which is an estimate of the "single particle" value, that is, the value expected if the transition were due to a change of orbit of a single proton. Clearly in several regions, there are enormous enhancements pointing to coherent behavior. Identification of the nuclei with particularly large $B(\mathrm{E2}{:}0_1^+ \rightarrow 2_1^+)$ values shows that they occur wherever there are many valence protons and many valence neutrons, that is, near mid-shell for both kinds of particles. This is the classic regime of collectivity. (We shall see why this is shortly.) These two facts (large shell model dimensions and empirical collective behavior) have motivated, over the last four decades, a wide variety of alternate models, most of which are of collective and "geometric" character: that is, they envisage the nuclear excitations in terms of a classical picture of excitation modes of a non-spherical nucleus, such as rotational flows, vibrations, oscillations, and the like, involving many nucleons in a coherent way.

It is easy to see why this is a relevant picture. We have just noted that regions far from closed shells display strong collective behavior and that this is primarily due to residual interactions among the valence nucleons which accumulate as additional valence nucleons are added past a closed shell (e.g., past a doubly magic nucleus such as $^{208}\mathrm{Pb}$ or $^{132}\mathrm{Sn}$). We also noted that a given state no longer consists of a particle (or particles) in a *single* shell model orbit: rather the wave function consists of a linear combination including several different j-configurations. But, since these different j values have different sets of magnetic substates ($m = -j, -j$

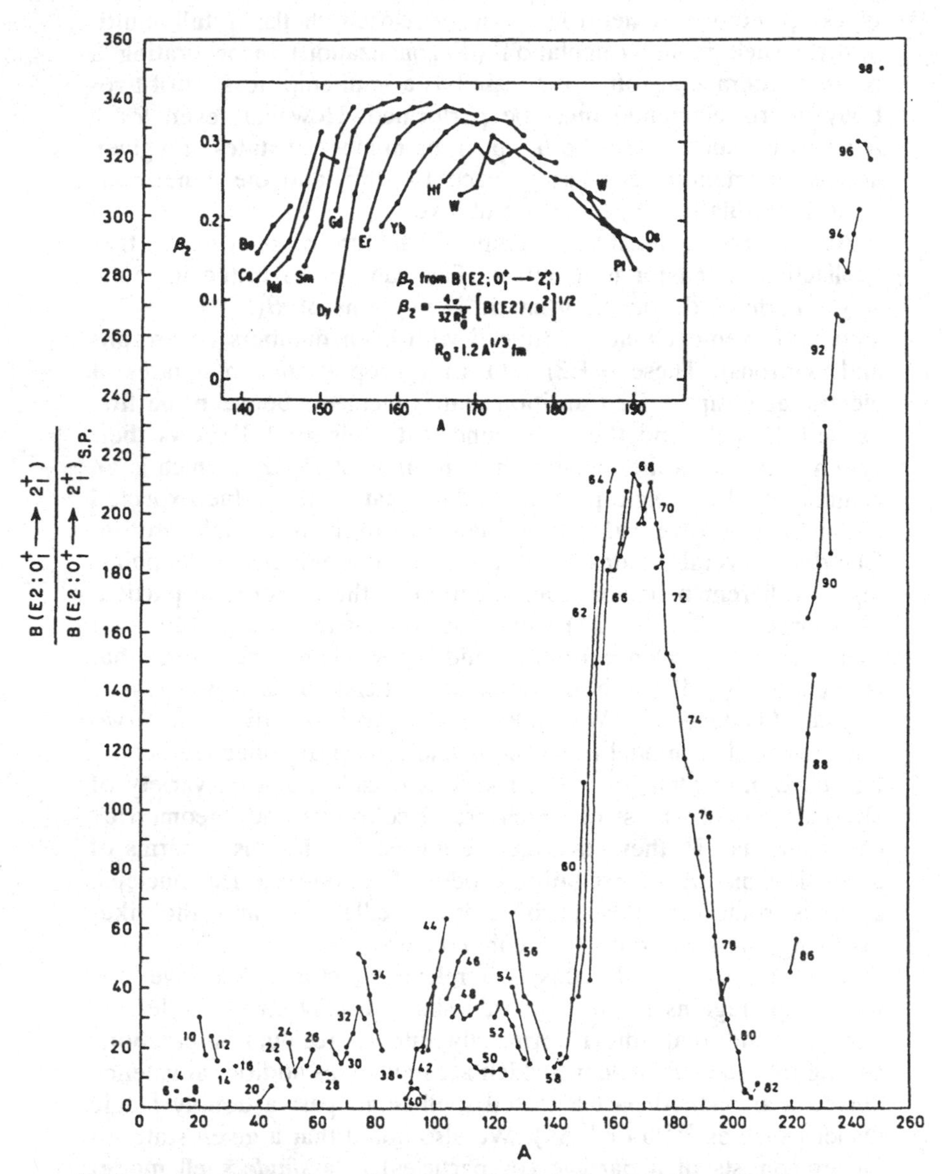

FIGURE 1.3. *B(E2:$0_1^+ \rightarrow 2_1^+$) values (in "single particle" units) for even-even nuclei. From Bohr and Mottelson (1975). The inset shows values of the quadrupole deformation β_2 for the rare earth nuclei extracted from such B(E2) values. From Casten (1990).*

+ 1, . . . , j − 1, j), the net result *must* be an unequal distribution of m-states in the resultant wave function. This, however, is tantamount to deformation since a preference for greater probability density of particles in a particular direction obviously destroys the sphericity. We thus have a simple rationale for the widespread existence of non-spherical nuclei and for the applicability of geometrical models embodying static or dynamic excursions from sphericity.

Most collective model approaches are expressed in terms of nuclear shape variables such as those measuring excursions away from spherical symmetry embodying various types of "deformation." These excursions can be either dynamic oscillations or static shapes. In most of these models the magnitudes of the $B(\text{E2}:0_1^+ \rightarrow 2_1^+)$ values discussed above are closely related to the magnitude of these deformations, and model dependent estimates of the latter can often be extracted from measured $B(\text{E2})$ values. This is illustrated in the inset to Fig. 1.3 which shows values of the quadrupole ellipsoidal deformation parameter β_2 in the rare earth region extracted from the square roots of these $B(\text{E2})$ values. [Note: Unfortunately, there are several notations for deformation. The most common are β, ϵ, and δ, which are within a few percent of each other, for typical deformed nuclei: they measure the elongation of the nucleus or of the potential in the z direction. One can write, for example, $R_z = R_0(1 + \beta Y_{20})$ where R_0 is the equivalent spherical radius.] The inset in Fig. 1.3 is also useful in illustrating, empirically, the development of collectivity and deformation, and the "transition" regions between near-magic nuclei and well-deformed nuclei. Finally, the inset reveals another interesting phenomenon, namely the "saturation" of collectivity or deformation near midshell where additional valence nucleons no longer increase the collectivity. This feature can also be explained in terms of the dependence of the valence p-n interaction on the spatial overlaps of valence proton and neutron orbits but further discussion of this subject is beyond the scope of the present chapter.

A characteristic of deformation and collectivity, besides large $B(\text{E2}:0_1^+ \rightarrow 2_1^+)$ values, is the very low energy of the first 2^+ state $[E(2_1^+)]$ of an even-even nucleus. In near-magic nuclei such a state must involve the breaking of a pair of nucleons coupled to $J = 0^+$: but this requires, typically, 1–2 MeV. Indeed, this is the origin of the familiar "pairing gap" in even-even nuclei: single-particle excitations cannot occur much lower (although collective ones can). In heavy deformed nuclei, $E(2_1^+)$ typically lies around 100 keV.

Thus the spherical-deformed transition region entails a systematic and often dramatic drop in 2_1^+ (and other) energies. [See the discussion later in this chapter of transitional nuclei (Figs. 1.15 and 1.16).]

Deformation sets in, in different regions, under different conditions (different numbers of valence particles) and can arise gradually or suddenly as a function, say, of neutron number. Figure 1.4 illustrates some data for two such transition regions and shows both the general behavior and the simplicity or complexity that can characterize them. (Note that, in the $A \sim 130$ region on the right, the "valence" neutrons are "holes", counted as the number of vacancies below the magic number $N = 82$.)

Our view of these transition regions can be considerably simplified (and unified from one region to another) by exploiting the recognition that the primary controlling factor in the development of collectivity and deformation is *valence* nucleon interactions and, among these, especially those between protons and neutrons. Indeed, in nuclei that are singly magic (nuclei in which one type of nucleon has a magic number, such as the Sn isotopes), $E(2_1^+)$ is nearly constant (at ~1.2 MeV) regardless of the number of valence neutrons, whereas in doubly non-magic nuclei $E(2_1^+)$ rapidly drops (see Fig. 1.4). Inspection of the simpler $A \sim 130$ transition region of Fig. 1.4 (right side) shows, in addition, that $E(2_1^+)$ is lower, for a given number of valence neutrons, the greater the number of

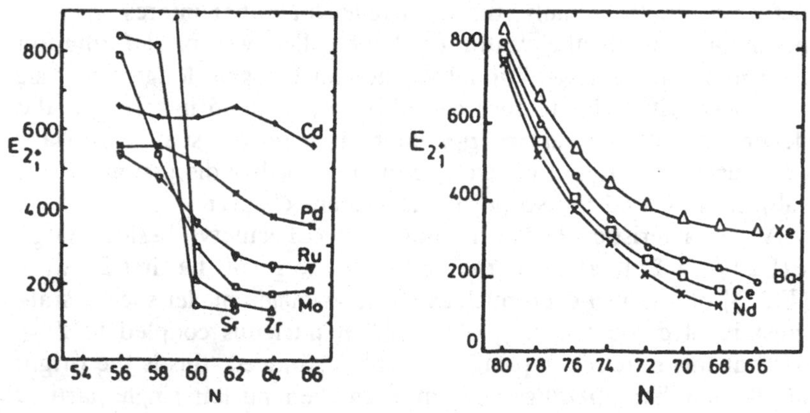

FIGURE 1.4. *Systematics of $E(2_1^+)$ in the $A = 100$ (left) and $A = 130$ (right) mass regions.*

valence protons. These persistent features of nuclear phenomenology give a clue to an alternative way of looking at the systematic behavior of nuclear structure as N and Z change. If we assume (of course, this is only an approximation) that all p-n interactions are equal in strength in a given region, the integrated p-n strength should be proportional to N_pN_n, the product of the number of valence protons times the number of valence neutrons. If properties of transition regions are plotted against N_pN_n, as illustrated in Fig. 1.5 for the two regions shown in Fig. 1.4, the complexity is dramatically reduced and the data fall on a single smooth curve. Moreover, different regions behave very similarly, in contrast to their behavior in "normal" plots against N, Z or A. The N_pN_n scheme or the use of the related P-factor, defined as $P = N_pN_n/(N_p + N_n)$, can correlate the systematic properties and much of the phenomenology of medium mass and heavy nuclei in a very simple formulation with microscopic underpinnings based on the central importance of the residual p-n interaction in nuclear structure.

We are now in a position to summarize the gross categories of nuclear structure throughout the Periodic Table, at least in a qualitative way. Figure 1.6 shows a schematic nuclear chart indicating the magic numbers (horizontal and vertical lines) and the zones where collectivity and deformation should be dominant (circled mid-shell areas). The intermediate nuclei have transitional character and, being therefore complex, are often the best testing grounds of nuclear models. In these introductory remarks, we will describe

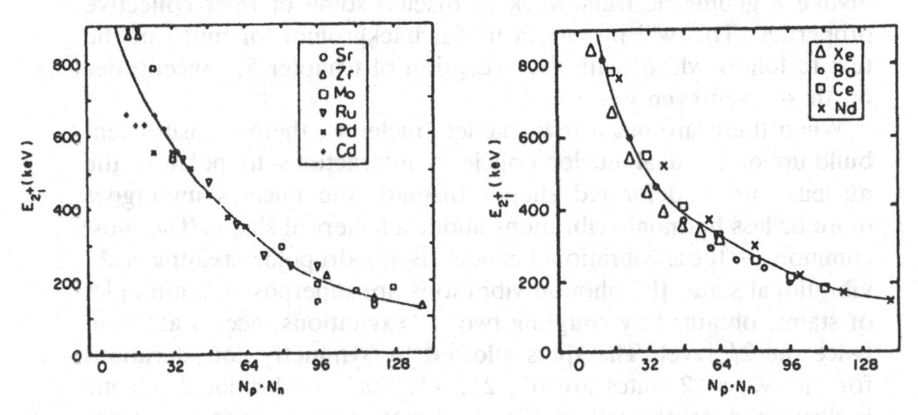

FIGURE 1.5. *Same data as in Fig. 1.4 except plotted against N_pN_n. Based on Casten (1985).*

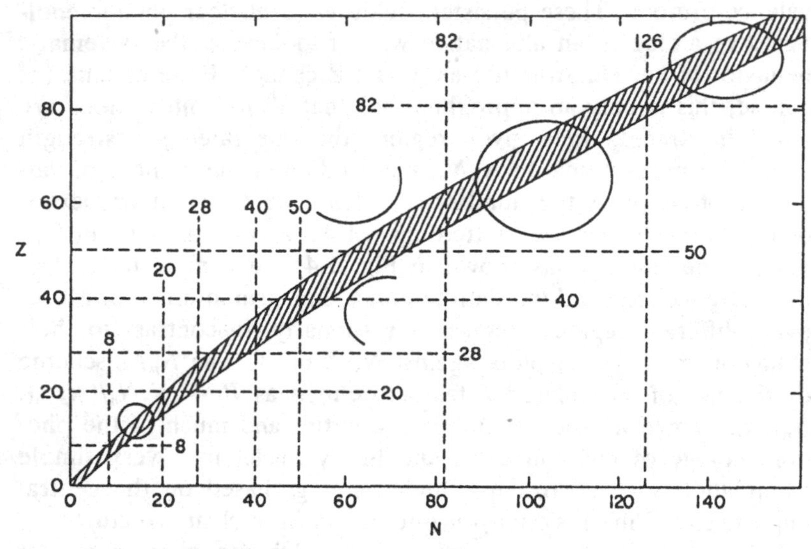

FIGURE 1.6. *Schematic chart of the nuclides showing the locus of the valley of stability (shaded area), the magic numbers (vertical and horizontal lines), and deformed regions (circles or circular segments). From Casten (1990).*

some aspects of collective behavior in a qualitative way to illustrate the wealth of structural possibilities and some of their empirical manifestations. The text throughout the rest of this volume focuses on these phenomena in much greater detail.

To proceed we first consider even-even nuclei, and we will invoke a geometric framework to discuss some of their collective properties. This will provide a useful background for much of the text to follow which, with the exception of Chapter 5, concentrates on these even-even nuclei.

When there are but a few valence nucleons, there is insufficient build-up of residual nucleon-nucleon interactions to polarize the nucleus into a deformed shape. Instead, the nucleus undergoes more or less harmonic vibrations about a spherical shape. The most common of these vibrational modes is quadrupole, creating a 2^+ vibrational state. If 2-phonon vibrations are superposed, a multiplet of states, obtained by coupling two 2^+ excitations, occurs at about twice the 2_1^+ level. The spins allowed by symmetry considerations for the $N_{ph} = 2$ states are 0^+, 2^+, 4^+. Such a vibrational scheme is illustrated on the left in Fig. 1.7. Note that it gives an energy ratio $E(4_1^+)/E(2_1^+) = 2.0$. Indeed, this is one characteristic empirical

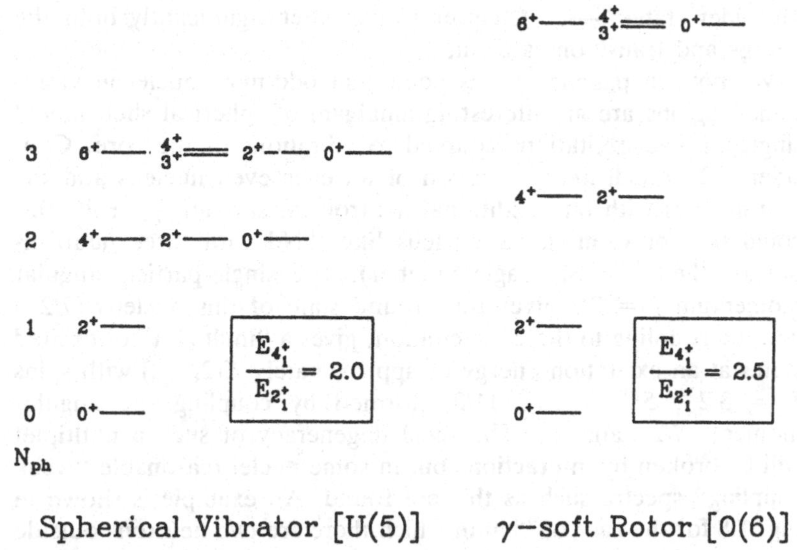

3 6^+ 4^+ 3^+ 2^+ 0^+ | 6^+ 4^+ 3^+ 0^+
2 4^+ 2^+ 0^+ | 4^+ 2^+
1 2^+ | 2^+
0 0^+ | 0^+
N_{ph}

$\frac{E_{4_1^+}}{E_{2_1^+}} = 2.0$ | $\frac{E_{4_1^+}}{E_{2_1^+}} = 2.5$

Spherical Vibrator [U(5)] γ−soft Rotor [O(6)]

FIGURE 1.7. *Illustration of the low-lying "phonon" levels of the harmonic spherical vibrator model (left)—which is closely related to the U(5) symmetry of the* IBM*—and of the γ-soft-rotor (right)—related to the* IBM *O(6) symmetry. In the vibrator model the level energies are proportional to the phonon number. In the γ-soft case, the energies follow a* $\tau(\tau + 3)$ *law where* $\tau = I_{max}/2$ *or half the maximum J value in each multiplet.*

signatures of this type of motion. In contrast, simple shell model configurations typically have $E(4_1^+)/E(2_1^+) \sim 1.6$. [This is illustrated below in Fig. 1.17 which shows examples of empirical level schemes of nuclei with two valence nucleons, as well as the predicted sequences for configurations of the type $|j^2J\rangle$.] Rotational motion (see below) is characterized by $E(4_1^+)/E(2_1^+) = 3.33$. In principle, though rarely observed, a 3-phonon vibrational quintuplet, 0^+, 2^+, 3^+, 4^+, 6^+, can occur still higher in energy at $\sim 3E(2_1^+)$. An interesting example of a near-harmonic nucleus, ^{118}Cd, will be illustrated in Chapter 3.

There are characteristic E2 properties and selection rules for the decay of these vibrational excitations. The $2_1^+ \rightarrow 0_1^+$ transition should be moderately collective. Decays of multi-phonon states are governed by the rules $\Delta N_{ph} = \pm 1$. In addition, the 2-phonon → 1-phonon B(E2) values should be twice the $B(\mathrm{E2}:2_1^+ \rightarrow 0_1^+)$ value. Anharmonicities in the oscillator motion, as well as mixing of phonon states, or mixing with other degress of freedom (such as "intruder states"—see Chapter 6) can alter significantly both the energy and transition rate rules.

We note in passing at this point that odd-mass nuclei in vibrational regions are an interesting amalgam of spherical shell model single-particle excitations coupled to vibrations of the core. Consider a 2^+ quadrupole vibration of an even-even nucleus and the odd nucleus with one additional neutron in, say, an $f_{7/2}$ orbit (this could be, for example, a nucleus like ^{145}Nd with three neutrons outside the $N = 82$ magic number). The single-particle angular momentum $j = 7/2$ gives the ground state of this nucleus ($7/2^-$) and, by coupling to the 2^+ excitation, gives a "multiplet" of excited states at an excitation energy of approximately $E(2^+_{1\,\mathrm{even}})$ with spins $J = 3/2^-, 5/2^-, \ldots, 11/2^-$ formed by coupling the angular momenta $7/2^-$ and 2^+. The ideal degeneracy of such a multiplet will be broken by interactions but in some nuclei reasonable "weak coupling" spectra such as this are found. An example is shown in Fig. 1.8 for the case of ^{93}Nb in which there are two neutrons outside the $N = 50$ closed shell and a single valence proton, in a $g_{9/2}$ orbit, outside the $Z = 40$ magic number.

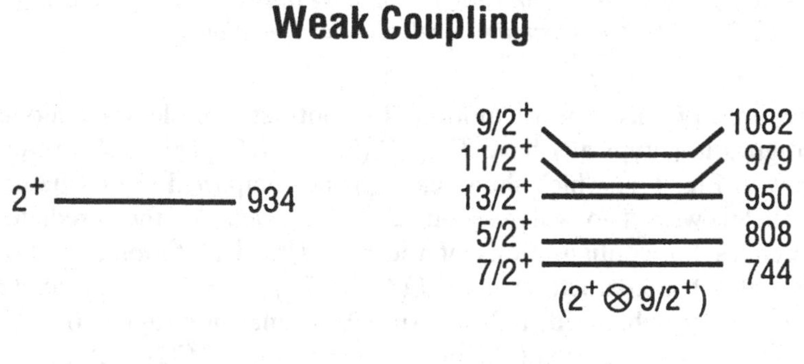

FIGURE 1.8. *Example of weak coupling in which the selected positive parity levels of* ^{93}Nb *are viewed in terms of the coupling of a* $9/2^+$ *single particle to the* 2^+ *excitation of the* ^{92}Zr *core (shown on the left).*

Even-even nuclei with sufficient build up of (mostly) valence nucleon interactions can develop a stable quadrupole deformation as opposed to vibrational excursions (into deformed territory) about an equilibrium spherical shape.

The collective excitations we deal with are primarily but not always (see the end of Chapter 4) those occurring relatively low in energy and below the pairing gap. There are two principal types of collective motions, although a rigorous separation between them is only an approximation: rotational and vibrational (or intrinsic). The energy scale for rotations is roughly an order of magnitude smaller than that of vibrations. Rotation is possible (quantum mechanically) because the nucleus is no longer spherically symmetric. One can picture the collective levels of a deformed even-even nucleus in terms of various intrinsic vibrations on top of which a rotational *band* of states is superposed. Each vibration is characterized by an intrinsic angular momentum (e.g., 2, for quadrupole vibrations) and the projection K of that angular momentum on the nuclear symmetry axis. Built on each vibration is a rotational sequence or band comprised of states $J = K, K + 1, K + 2, \ldots$, except for $K = 0$ modes in which $J^\pi = 0^+, 2^+, 4^+, \ldots$ for positive parity (e.g., quadrupole) excitations or $J^\pi = 1^-, 3^-, 5^-, \ldots$ for negative parity (e.g., octupole) excitations.

Due to pairing, the ground state (no vibrations) of an even-even nucleus has $J^\pi = 0^+$ and hence a rotational band $0^+, 2^+, 4^+, 6^+, \ldots$, with energies following the eigenvalue expression of the quantum symmetrical top rotor, $E(J) = (\hbar^2/2I)J(J + 1)$ where I is the moment of inertia. (More complex formulas exist to account for perturbations to this ideal picture of pure rotation in an axially symmetric nucleus.) As alluded to above, this formula gives a characteristic ratio of energy levels, $E(4_1^+)/E(2_1^+) = 3.33$, that differs significantly from the phonon or spherical vibrator value of 2.0 and which can be used to help distinguish between different structures empirically.

The systematics of $E(4_1^+)/E(2_1^+)$ values for the transitional and deformed rare earth region are shown in Fig. 1.9. One notes some values near 2, then rather rapid transition regions in which this ratio grows quickly toward the asymptotic rotational value of 3.33. In the middle of the deformed rare-earth region, the closeness of the empirical ratios to this value is remarkable and confirms the basic correctness of the rotational picture.

The best known and most commonly observed vibrational exci-

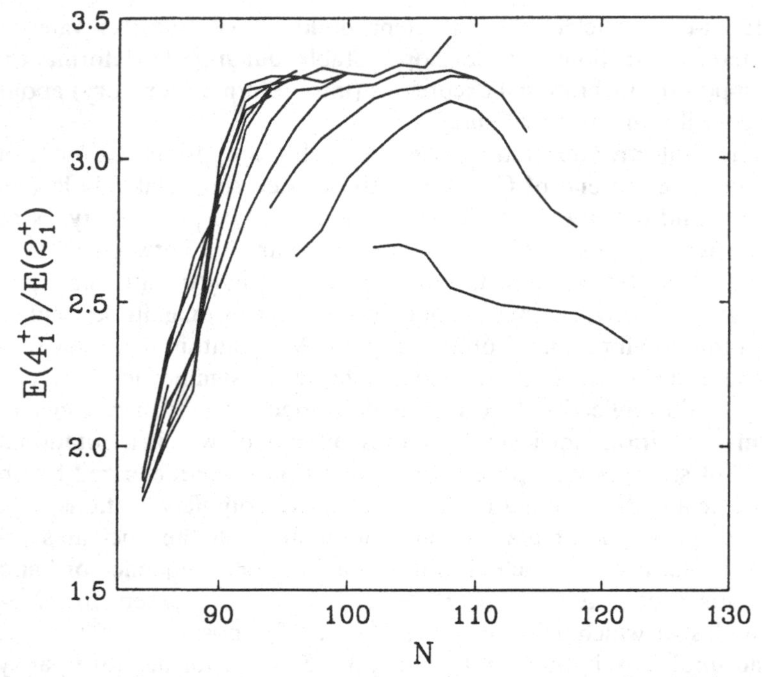

FIGURE 1.9. *$E(4_1^+)/E(2_1^+)$ ratios for the rare-earth transitional and deformed region. The lines connect isotopes of a given element. The rather sharp vibrator → rotor transition around $N = 90$ is evident at the left as well as the asymptotic limiting rotor value of the $4_1^+/2_1^+$ ratio for well deformed nuclear near mid-shell. The nuclei with ratios near 2.5 at the right are the γ-soft (or O(6)-like)* Pt *nuclei.*

tations in deformed nuclei are the so-called β and γ quadrupole vibrations, with $K = 0$ and 2, respectively. The β mode is a vibration along the symmetry axis (a vibration in the deformation variable β) while the γ mode can be viewed semi-classically as an alternate squashing of the prolate quadrupole equilibrium shape in the two directions perpendicular to the symmetry axis. In the standard geometrical model these are independent vibrational modes: microscopically, for those familiar with such approaches as the RPA, they are created by the characteristic operators r^2Y_{20} and $r^2Y_{2\pm2}$, respectively. In actual deformed nuclei low-lying $K = 0, 2$ modes are nearly always seen. The $K = 2$ excitations seem to exhibit the properties expected for the γ mode while the real nature of the $K = 0$ excitations is still unresolved. The latter excitations exhibit some properties expected for β vibrations but also show others (such as collective transitions to the γ bands) that violate the harmonic geometrical model picture of this mode. The IBM (see Chapter 3) accounts rather well for some of these properties. It may be that the lowest empirical $K = 0^+$ modes in deformed nuclei have characteristics of both a true β mode as well as of a 2-phonon $K = 0^+$ double γ mode (and perhaps other components as well). These states remain a key challenge to modern nuclear structure models. In any case, for convenience, we will continue to use the traditional label β to refer to the lowest $K = 0^+$ intrinsic excitation.

Figure 1.10 shows a typical level scheme for a well-studied deformed even-even nucleus, ^{168}Er, with ground, β, and γ vibrations, several octupole modes (octupole vibrations, with intrinsic angular momentum 3, can occur in $K = 0^-, 1^-, 2^-$ or 3^- realizations) and other intrinsic excitations representing 2-particle or, better, 2-quasiparticle, states. This level scheme is essentially complete up through the levels shown.

Figure 1.11 shows the systematics of some of these vibrations across the well-studied rare-earth region. Note the rather different energy systematics of β and γ modes and the characteristic way in which the order of the octupole bands inverts from the beginning of this region where $K = 0^-, 1^-$ lie lowest to the end where $K = 2^-$ is low lying.

In the geometric collective model there are characteristic E2 selection rules for the decay of quadrupole vibrations of deformed nuclei. They are obtained from the fact that a one-body transition operator can create or destroy a single phonon (vibration). Hence, both β and γ bands should show collective transitions to the ground state band; but transitions between β and γ vibrational states are expected to be forbidden. The empirical situation for $\gamma \rightarrow g$, $\beta \rightarrow g$, and intra-band rotational transitions ($g \rightarrow g$, $2_1^+ \rightarrow 0_1^+$) is summarized, again for the rare earths, in Fig. 1.12. There are three very obvious characteristics. Rotational B(E2) values are dominant, at least an order of magnitude stronger than vibrational ones. Secondly, $\gamma \rightarrow g$ B(E2) values are much stronger than for $\beta \rightarrow g$ transitions. Thirdly, $g \rightarrow g$ and $\gamma \rightarrow g$ matrix elements are extremely stable and smoothly varying, as befits collective states that entail the participation of many particles in different orbits and, which, hence, should not vary much with Z and N. In contrast, $\beta \rightarrow g$ transitions are completely erratic. Finally, we mention another very important experimental fact, not illustrated in Fig. 1.12, which is at variance with the simple geometrical model just

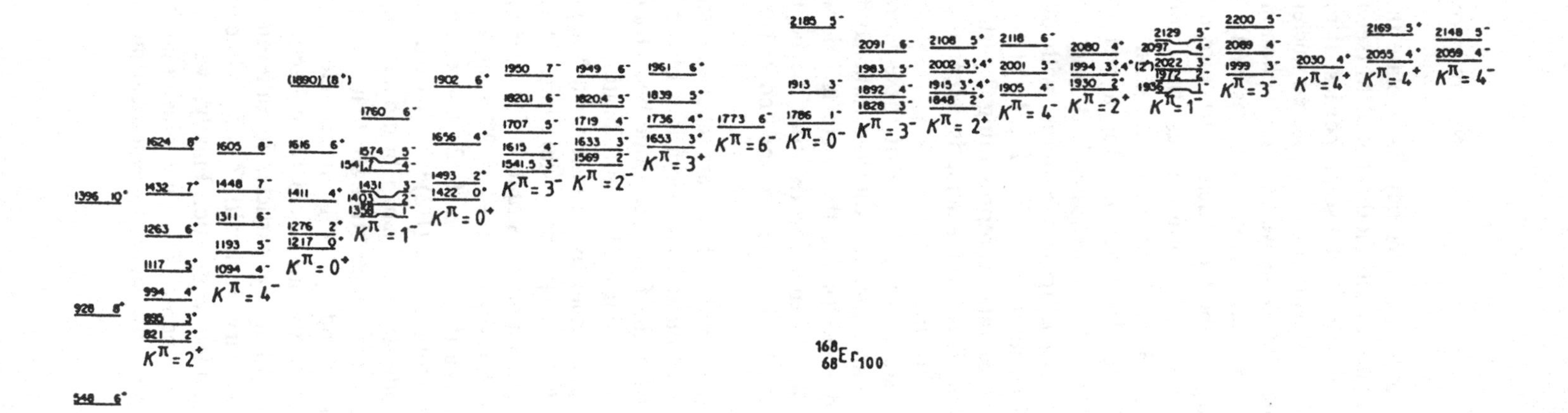

FIGURE 1.10. *Energy levels of the well-studied deformed nucleus* ^{168}Er. *From Warner, Casten, and Davidson (1980). Energy levels, J^π values and the K^π values for each rotational band are indicated. From Davidson* et al. *(1981).*

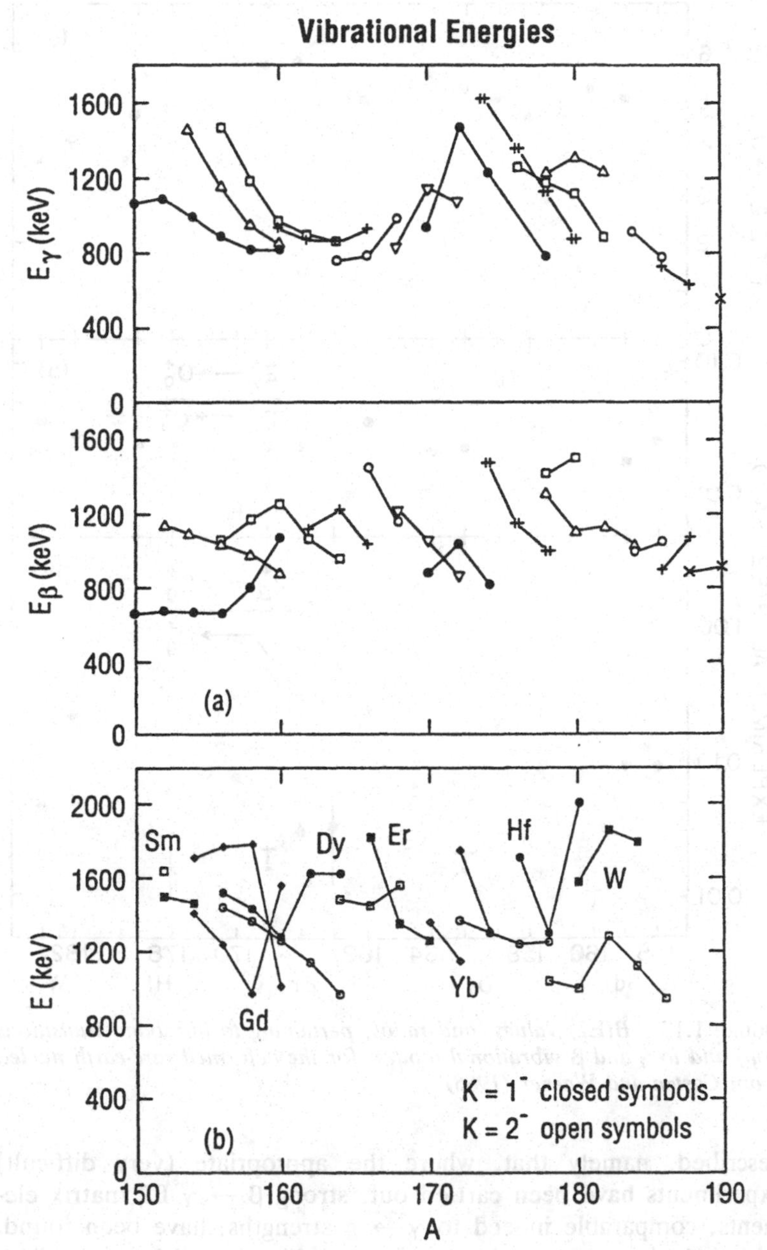

FIGURE 1.11. *Systematics of vibrational energies in the rare earth region for β, γ, and K = 1⁻, 2⁻ octupole excitations.*

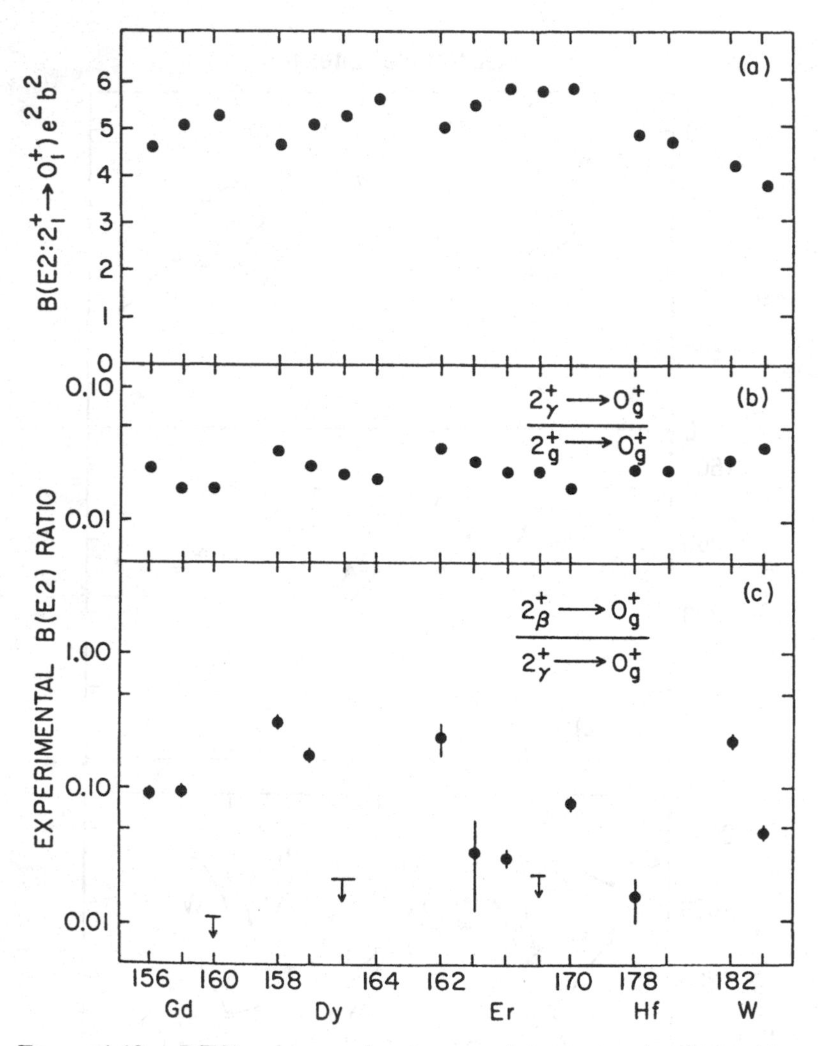

FIGURE 1.12. *B*(E2) *values, and ratios, pertaining to intraband transitions (top) and to* γ *and* β *vibrational modes, for the deformed rare-earth nuclei. From Casten and Warner (1988).*

described, namely that, where the appropriate (very difficult) experiments have been carried out, strong $\beta \rightarrow \gamma$ E2 matrix elements, comparable indeed to $\gamma \rightarrow g$ strengths, have been found. Although the structural implications of this are still being studied, it is worth noting here that collective $\beta \rightarrow \gamma$ transitions are in fact an *inherent feature* of IBM calculations for deformed nuclei. Indeed, this is one of the most noteworthy ways in which the IBM differs in deformed nuclei from traditional geometrical models. Moreover, in the IBM, $\beta \rightarrow \gamma$ and $\gamma \rightarrow g$ $B(E2)$ values are comparable to each other and much stronger than $\beta \rightarrow g$ $B(E2)$ values. Such predictions, which reflect the empirical situation, were historically important in establishing the IBM as a viable model.

A final feature of E2 transitions in deformed nuclei worth noting is the so-called Alaga rules. Ratios of $B(E2)$ values from a given initial state to two final states which themselves belong to the *same* band are given purely by "geometrical" factors independent of the microscopic structure of the vibrations: this latter "intrinsic" part of the matrix element, which does depend on the structure, cancels out in the ratio. Deviation from these ratios arise from mixing of one excitation into another. This mixing is sometimes referred to as rotation-vibration coupling. These deviations from the Alaga rules are systematic: they increase with spin as one goes "up" a band and they favor transitions that *in*crease the spin (e.g., $2_\gamma^+ \rightarrow 4_g^+$) as opposed to spin decreasing transitions (e.g., $2_\gamma^+ \rightarrow 0_g^+$). They are largest near the edges of the deformed region where the energy scales of rotational and vibrational motion are less distinct, and are smallest near mid-shell. Even though the mixing amplitudes are small (typically <0.1), the effect on transition rates is large since, in effect, the admixed amplitude is for a highly collective *intra*-band matrix element. Table 3.3 of Chapter 3 shows some of these branching ratios, compared to the Alaga rules, for the nucleus ^{168}Er. Such deviations are often parameterized in terms of a 2-band mixing parameter called Z_γ (Z_β) for $\gamma - g$ ($\beta - g$) mixing. Figure 3.23 of Chapter 3 illustrates the systematics just cited. Note the clear minimum near mid-shell (middle of the figure).

It is incumbent upon any successful model of deformed even-even nuclei to predict or at least be compatible with all these energy and transition rate features that we have discussed in the preceding paragraphs. Indeed, they are the prime data used to test various model descriptions of deformed nuclei.

Now that we have encountered the concept of deformation and some of the principle collective manifestations of even-even nuclei, in particular, the rotational motion, the seemingly very complicated level schemes of odd mass deformed nuclei are easy to understand and disentangle. We first consider the motion of the individual nucleons in a deformed field, and then the superposition of rotation.

Of course, if the nucleus is deformed, the energies of individual particle orbits will not be those of the spherical shell model (though the wave functions can be written as linear combinations of them). The shell model energies shown in Fig. 1.1 are independent of the magnetic substate occupied by a particle since the potential or field is spherically symmetric. This changes when a deformed field sets in, and does so in a characteristic way that is easy to visualize. Consider the most common case of a prolate quadrupole deformed ellipsoid (an American football shape). The major symmetry axis (the z axis, so-called) is the long one (through the length of the football). A particle orbiting the long way around the prolate shape (an "equatorial" orbit) is, on average, closer to the bulk of the nuclear matter than a particle orbiting at the same radius around the "waist" (a "polar" orbit). Hence, since the nuclear force is short-ranged and attractive, the particle in an equatorial orbit is favored, that is, it has lower energy. The degree of lowering grows with the deformation. It is now useful to introduce a new quantum number, namely Ω, which is the projection of the odd particle angular momentum on the symmetry (z) axis. In axially symmetric nuclei, the rotation has no projection on this axis and hence K, the projection of the total angular momentum on the z-axis, is the same as Ω. For convenience, we use the label K below. K has values $K = \frac{1}{2}, \frac{3}{2}, \frac{5}{2}, \ldots, j$. It is illustrated in the inset at the upper right of Fig. 1.13 (along with a schematic illustration of a prolate deformed nucleus). The equatorial orbit has j pointing nearly normal to the z axis and hence small K ($K = \frac{1}{2}$). The low K orbits are lowered the most for a given j. Systematic application of this idea allows one to construct a *deformed* shell model diagram of energy levels as a function of deformation, that is, a "Nilsson" type diagram, by starting with a set of spherical single-particle energies and considering how their degeneracy breaks, as a function of deformation, according to the value of K. A portion of such a Nilsson diagram is shown in Fig. 1.13. Each single particle level breaks up into several: each is now only twofold degenerate (the particle can orbit clockwise or the opposite), instead of $(2j + 1)$ degenerate. One determines the state of the last particle simply by filling orbits sequentially. Almost invariably, the pairing assumption is invoked so that the ground state J^π of any odd mass deformed nucleus is given by $J^\pi = K^\pi$ of the last orbit occupied. Excited "intrinsic" states (that is, those involving a different basic particle configuration) are formed by exciting that "last" particle or by cre-

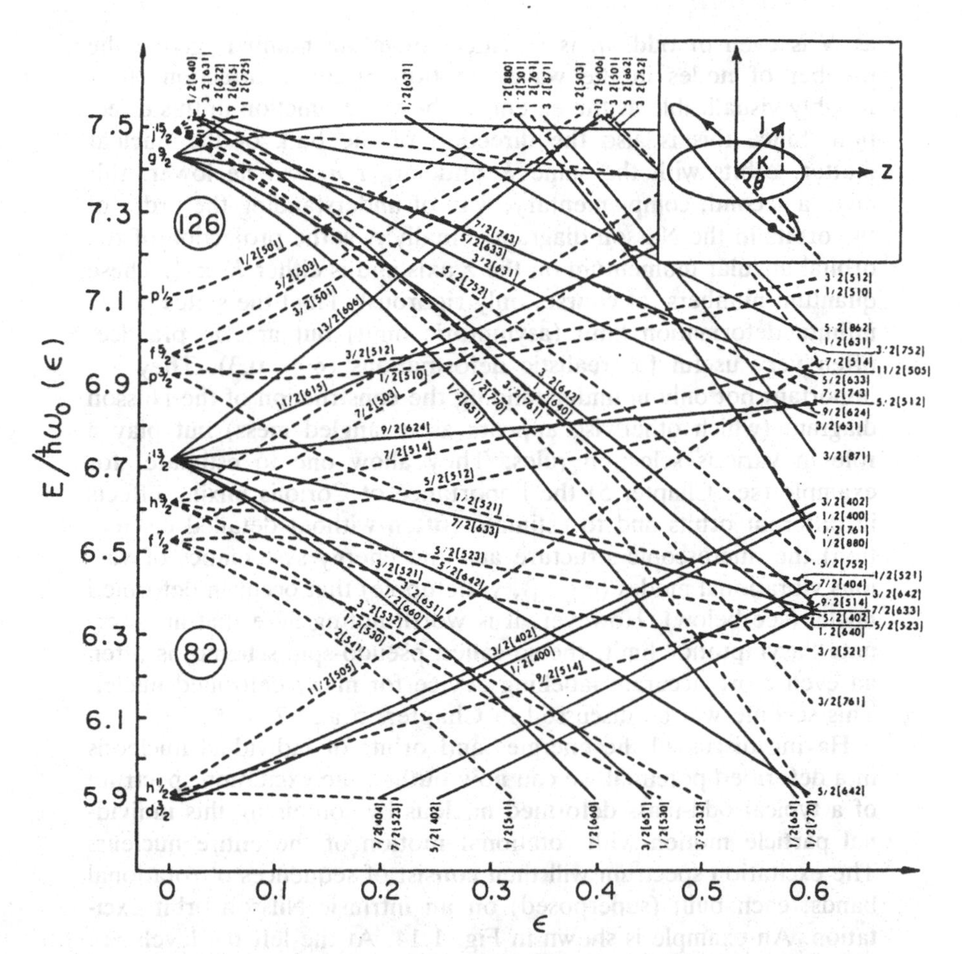

FIGURE 1.13. *Typical example of a Nilsson diagram, applicable to neutrons in the rare earth region (A ~ 150–190). Based on Gustafson* et al. *(1967). The inset at the extreme upper right illustrates an axially symmetric quadrupole deformed prolate nucleus, the orbit of a single particle with its angular momentum j and the projection K of that angular momentum on the symmetry axis z.*

ating a "hole" state, that is, by elevating a particle from an occupied orbit to the "ground state" orbit, creating a $J^{\pi} = 0^{+}$ pair and leaving a half occupied orbit below the Fermi surface.

The quantum numbers on the Nilsson diagram are $K^{\pi}[Nn_z\Lambda]$. We have discussed K. N is the principal quantum number giving the major oscillator shell. The parity is positive or negative according as N is even or odd. n_z is a critical quantum number, giving the number of nodes in the wave function in the z direction. It is roughly visualizable as the extent of the wave function in this direction. Since this is also the direction of the bulk of the nuclear matter, orbits with the same K and larger n_z will lie lower: this gives a second, complementary, way of understanding the order of the orbits in the Nilsson diagram. Finally Λ is the projection of the orbital angular momentum on the z axis and is either $K \pm \frac{1}{2}$. These quantum numbers, of course, only rigorously label the states in the infinite deformation limit (asymptotic limit) but are, in practice, often very useful for realistic deformations ($\delta \sim 0.3$). They are important not only in understanding the construction of the Nilsson diagram (which otherwise appears as a tangled mess) but play a role in various selection rules. They allow one to estimate, for example (see Chapter 5) the importance of Coriolis mixing effects in different orbits and to estimate (often without detailed calculation) the microscopic structure and the energy systematics of various vibrational modes (e.g., β, γ, octupole) that occur in deformed nuclei (see below). However, it is worth noting here that an alternate "asymptotic" limit, the so-called pseudo-spin scheme, is often an even more accurate labelling device for many deformed nuclei. This scheme will be discussed in Chapters 5 and 7.

Having discussed the energies and orbits of individual nucleons in a deformed potential we can now outline the excitation spectrum of a typical odd-mass deformed nucleus by combining this individual particle motion with rotational motion of the entire nucleus. The excitation spectrum will then consist of sequences of rotational bands, each built (superposed) on an intrinsic Nilsson orbit excitation. An example is shown in Fig. 1.14. At the left the levels are simply shown in order of excitation energy and clearly present a complex problem of interpretation. The powerful organizing ability of the Nilsson model, in conjunction with the intimately linked idea of rotational behavior, is evident when the levels are sorted into sequences of rotational bands on the right. One feature to note in the latter is the irregular spacings of bands with $K = \frac{1}{2}$: this feature is a well-known consequence of the Coriolis interaction. These Coriolis effects are largest for the unique parity orbits and can completely upset not only the $J(J + 1)$ rotational patterns but even the monotonic order in J of the rotational states in $K = \frac{1}{2}$ bands. An example is the Nilsson orbit $\frac{1}{2}^+[660]$ from the $i_{13/2}$ spherical orbit. Indeed, in some cases of moderately deformed or transitional

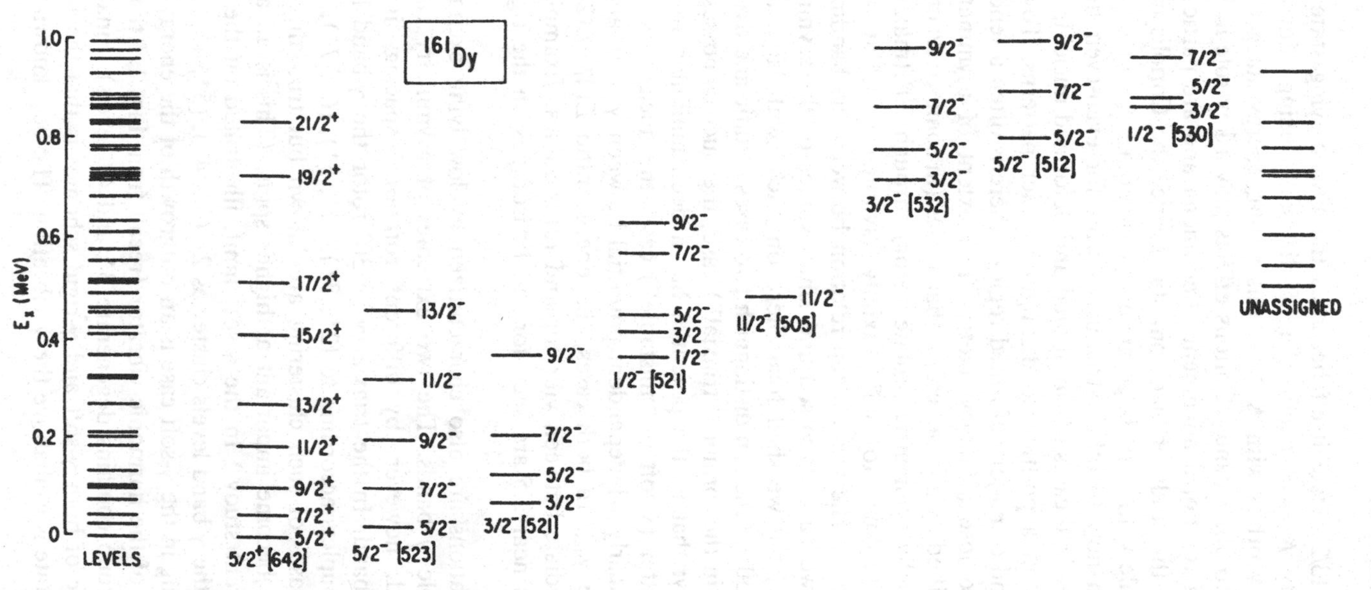

FIGURE 1.14. *Level scheme of* ^{161}Dy. *Left: All levels. Right: Same levels arranged into rotational bands according to the intrinsic Nilsson excitations on which they are built. From Casten (1990).*

nuclei, the $13/2^+$ state itself becomes the lowest lying state of a unique parity $K = \frac{1}{2}$ band. Via off-diagonal mixing with other unique parity orbits with $K > \frac{1}{2}$, this strong energy staggering is propagated to other bands. Coriolis effects play a key role in back-bending and rotational alignment phenomena and are particularly pervasive in the study of high spin states since the Coriolis matrix elements scale with J for large angular momenta.

We return now to other structural forms for even-even nuclei. With the idealized cases of vibrational and deformed nuclei in hand we are now in a position to distinguish other patterns. Towards the heavy end of major deformed regions there is often a tendency for nuclei to lose axial asymmetry. The extent of asymmetry is usually specified by a parameter γ that varies (in some conventions) from 0° (axially symmetric prolate rotor) through 30° (maximally asymmetric rotor) to 60° (axially symmetric oblate—disk shaped—rotor). There is a deep relation between the bandmixing (Z_γ) which we have discussed and small values of the asymmetry parameter γ, but we shall focus here on nuclei with large asymmetry ($\gamma \sim 30°$). One can distinguish two cases—static and dynamic asymmetry: in the former ("triaxial") case, the nucleus possesses a large fixed γ: that is, the potential has a deep minimum for that γ. In the latter (γ-soft or "unstable") case, the nucleus oscillates freely in a nearly γ-independent potential between $\gamma = 0°$ and 60° with a large γ_{rms}. In both cases, the energy ratio $E(4_1^+)/E(2_1^+) \sim 2.5$, intermediate between vibrator and rotor values. Examples of $4_1^+/2_1^+$ ratios near 2.5 are seen for the Pt isotopes on the right in Fig. 1.9.

Both situations are also characterized by low lying excitations that resemble γ bands. The two extremes of asymmetry can be distinguished, however, by different rotational spacings in this "quasi-" γ band. In the triaxial $\gamma \sim 30°$ rotor the γ-band levels cluster in couplets according to $(2^+, 3^+)$, $(4^+, 5^+)$, $(6^+, 7^+)$, ... No actual cases of such clustering are known but triaxiality may nevertheless become important at higher spins. (This is an active area of current study.) In the γ-soft limit, illustrated on the right in Fig. 1.7, the γ band levels cluster as 2^+ $(3^+, 4^+)$, $(5^+, 6^+)$, ... The clustering in the γ-soft case is an outgrowth of the energy level associations of the harmonic vibrator (recall from Fig. 1.7 that the 3-phonon group contains degenerate 3^+ and 4^+ levels). Finally, a characteristic of both γ-soft and γ-rigid schemes is that the lowest excited 0^+ state is well above the γ vibration. The best known cases of such behavior are in the Pt and Xe–Ba regions. Figure 3.2 of Chapter 3 illustrates ^{196}Pt (shown there since the γ-unstable potential also corresponds to the "$O(6)$"-limit of the IBM, one of that model's three characteristic dynamical symmetries). Further comparison of the γ-rigid and γ-soft coupling schemes is given in the context of Fig. 3.5 of Chapter 3.

Transitional nuclei form a class intermediate between vibrational, or γ soft, and deformed. How many transitional nuclei occur in a given region depends on the rapidity of the phase/shape transition. It can be very rapid (in N or Z) as in the $A = 100$ (see Fig. 1.4) or $A = 150$ regions, or gradual as in the $A = 130$ (see Fig. 1.4) or $A = 190$ regions. It can be simple or, in most cases, quite complex (compare the two parts of Fig. 1.4). Besides Figs. 1.4 and 1.5, we have already seen a couple of other examples of nuclear systematics that exhibit transitional behavior. Figure 1.3 showed brief spans of $B(\mathrm{E2}{:}0_1^+ \rightarrow 2_1^+)$ values on either flank of the rare earth nuclei (near ~150 and 190) and early in the actinides ($A \sim 230$) that reveal extremely rapid changes from weakly to very strongly collective. Other, less distinct regions, are discernible near $A = 100$ and 130. Likewise, Fig. 1.9 showed the development of collectivity and deformation at the start of the rare earth region ($N \sim 90$, $A \sim 150$) reflected in the energy ratio $E(4_1^+)/E(2_1^+)$ which increases from shell model values less than 2.0, through a vibrational region with ratios between 2.0 and 2.5, leading to the asymptotic rotor value of 3.33. The beginning of the Os–Pt–Hg transition region, from rotor through axially asymmetric, towards the spherical Pb region, is also evident near $N \sim 115$, $A \sim 190$.

Figure 1.15 illustrates these changes in the context of an imaginary sequence of nuclei undergoing a transition from spherical shell-model structure characteristic of a "single j" scheme through a vibrational structure with phonon structure, a typical transitional scheme, and terminating in a typical rotational level scheme characteristic of deformed nuclei like ^{168}Er which was illustrated in Fig. 1.10. The most characteristic features of such a transitional region, which should be easily understood from our foregoing discussion, are a rapid decrease in $E(2_1^+)$ as the collectivity increases and a concomitant increase in the ratio $E(4_1^+)/E(2_1^+)$. At the same time, levels of the 2-phonon vibrational multiplet develop into the β and γ vibrational modes of deformed nuclei.

We have noted two principal types of nuclear phase/shape transitional regions, spherical vibrator → deformed rotor, and axially

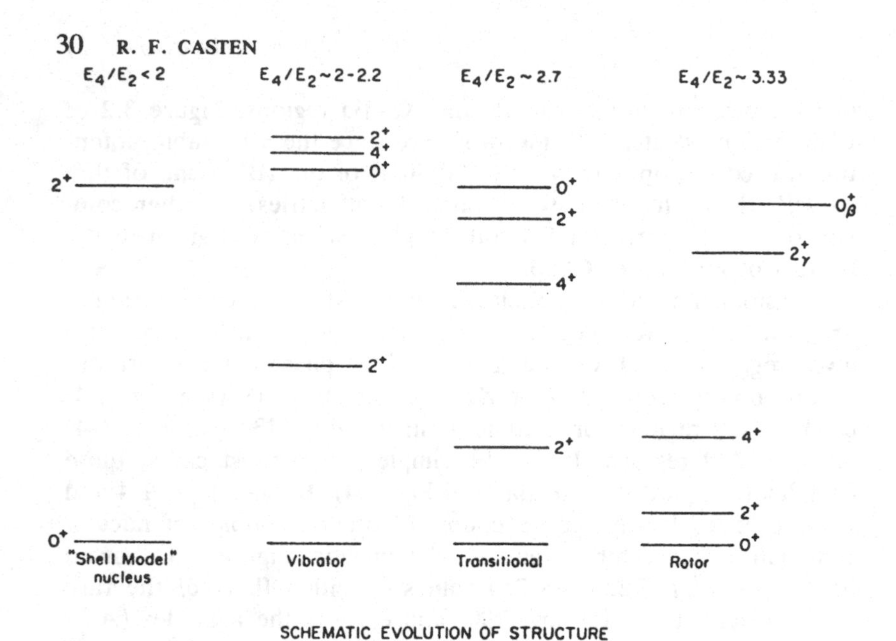

FIGURE 1.15. *A schematic illustration of a typical nuclear transition region involving spherical, vibrational, transitional, and deformed nuclei. Note the rapidly decreasing energy of the 2^+ level as the collectivity increases, and the accompanying increase in the $4_1^+/2_1^+$ energy ratio. The development of the deformed vibrations of β and γ type out of the spherical phonon excitations is also illustrated. Not shown is the sharp increase in $B(\mathrm{E2}{:}0_1^+ \rightarrow 2_1^+)$ values that also would occur in such a region in going from left to right. Based on Casten (1990).*

asymmetric → deformed rotor. In both, the $E(4_1^+)/E(2_1^+)$ ratio decreases [see Fig. 1.15] and $B(\mathrm{E2}{:}2_1^+ \rightarrow 0_1^+)$ values increase. It has therefore been very difficult to distinguish and identify the types of transition regions encountered. Extensive data for energies and $B(\mathrm{E2})$ values involving higher-lying intrinsic states, extending over *sequences* of nuclei, have normally been required to unravel the evolution of structure. Recently, though, this situation has changed with the development of the new signature illustrated in Fig. 1.16. This figure shows that a plot of $B(\mathrm{E2}{:}2_1^+ \rightarrow 0_1^+)$ values, expressed in Weiskopf units (not e^2b^2), and normalized by the mass number A, against the ratio $E(4_1^+)/E(2_1^+)$, follows a universal path for nearly all medium and heavy even–even nuclei ($Z > 30$) with a clear bifurcation into two tracks in transitional nuclei. The upper trajectory corresponds to a vibrator → deformed rotor [$U(5) \rightarrow SU(3)$ in IBM language—see below] phase transition and the lower

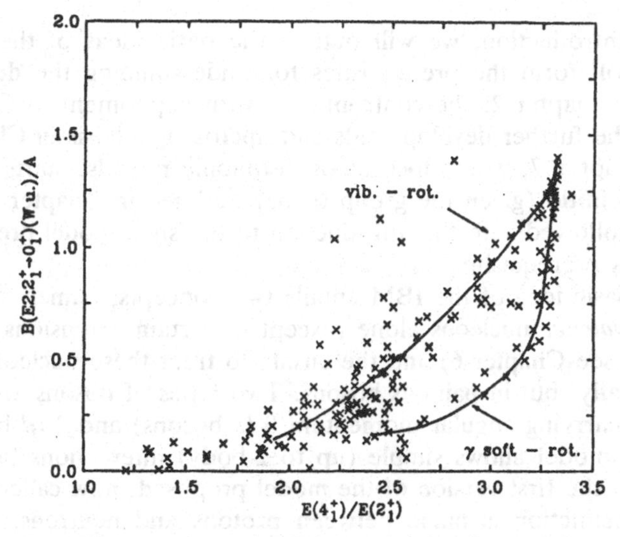

FIGURE 1.16. *Plot of measured $B(\mathrm{E2}{:}2_1^+ \rightarrow 0_1^+)$ values (in W.u.) divided by A against the energy ratio $E(4_1^+)/E(2_1^+)$ for all even–even nuclei with $Z > 30$. The curves are drawn to guide the eye and labelled by the kind of transition region involved. From Casten and Zamfir (private communication, 1992).*

one to an axially symmetric → deformed rotor [$O(6) \rightarrow SU(3)$] transition. Since these two experimental quantities are among the easiest to measure, and since data on only a single transitional nucleus are often sufficient, the classification of phase transitions is greatly facilitated.

While the above models have been remarkably successful, generally over a period of decades, another approach, known as algebraic, which is couched in group theory and emphasizes symmetries, offers an interesting and complementary alternative that has also proved highly successful. It provides the subject matter of this book. Algebraic models, as noted earlier, have been used in nuclear structure almost as long as any other approaches to collectivity (recall the Elliott scheme alluded to earlier) but they attained much more prominence following the proposal of the Interacting Boson Model or IBM, around 1975 (Arima and Iachello, 1975, Janssen, Jolos, and Donau, 1974). Since then, this field of study has virtually exploded with interest and, today, algebraic models, whether of bosonic type like the IBM and its offshoots and extensions, or of fermionic type, such as the pseudo-$SU(3)$, symplectic and other models discussed in Chapter 7, have proliferated. Here, in this introduction, we will outline the basic ideas of the IBM. These will form the prerequisites for understanding the development in Chapter 2, the confrontation with experiment in Chapter 3, and the further developments and microscopic basis in Chapters 4–6. Chapter 7, which focuses on fermionic models, starts essentially ab initio (given the group theory outlined in Chapter 2) and can be followed with the introduction to the shell model presented above as prerequisite.

The basic idea of the IBM entails two concepts, namely a focus on the *valence* nucleons alone (except in certain extensions to the model—see Chapter 6) and the ansatz to treat these nucleons, not individually, but in pairs as bosons. Two types of bosons are envisioned, carrying angular momentum 0 (s bosons) and 2 (d bosons) and the model allows simple (up to 2-body) interactions between them. In the first version of the model proposed, now called IBM-1, no distinction is made between protons and neutrons. In the IBM-2, proton and neutron bosons are distinct. Unless otherwise specified below, we speak in terms of IBM-1: IBM-2 is the subject of Chapter 4. Although the boson concept does not of necessity imply a specific physical picture, it was soon recognized that an appropriate interpretation involved pairs of fermions coupled to angular momentum 0 and 2.

The rationale for implementing just s and d bosons can be phrased in two parallel but intimately linked fashions. The dominance of low lying collectivity in nuclei of quadrupole type clearly motivates the d boson (just as, later, the desire to interpret octupole degrees of freedom led to the introduction of p and f bosons). Similarly, the importance of pairing focuses attention on a monopole (s boson) mode. The other viewpoint looks to the microscopic basis of the IBM in the shell model. Here, short-range attractive interactions dominate. One well known result of such interactions, noted and already used above, is that the lowest state of two identical particles in a given j-state, that is, of the configuration $|j^2 J\rangle$, is 0^+. The next is 2^+, and all the others, up to $J = (2j - 1)$, occur substantially higher. This is illustrated in Fig. 1.17 where 2-particle spectra calculated with a δ-function residual interaction are shown for three j values and compared to the empirical level schemes of nuclei with two valence neutrons predominantly occupying shell model states with those j values. The same result applies in multi-particle configurations $|j^n J\rangle$ for a wide variety of residual interactions [such as those in which "seniority" (see Chapter 4) is a good

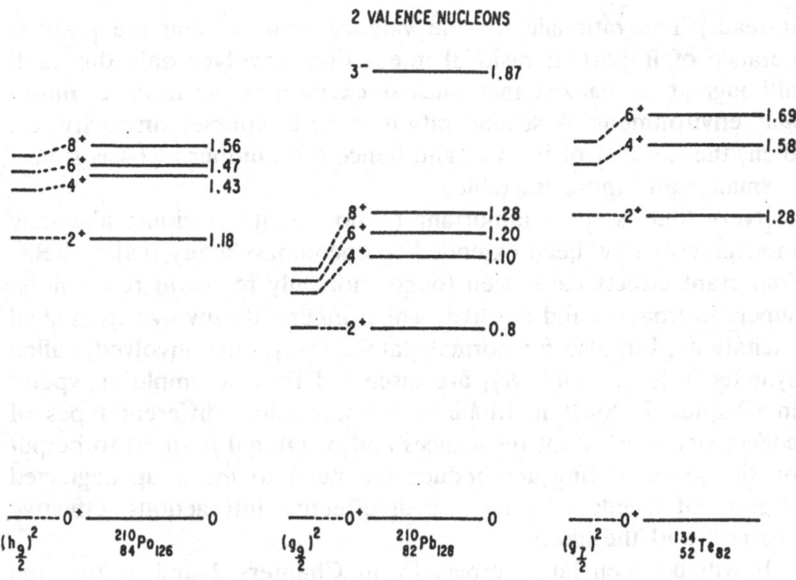

FIGURE 1.17. *Comparison of experimental and calculated low-lying even-spin (yrast) states in three nuclei that have two valence nucleons beyond a closed shell. The orbits used for the two identical nucleons are indicated in each case. The calculations are done with a short-range δ-function residual interaction. Note the lowering of the 0^+ and 2^+ states below the others. Based on Casten (1990).*

quantum number]. And, in many circumstances, similar features are obtained for multi-particle, multi-j configurations $|j_1^{n_1}J_1, j_2^{n_2}J_2, \ldots, J\rangle$. Hence, again, one sees a motivation for invoking 0^+ and 2^+ excitations. Of course, the microscopic justification and foundations of the IBM are more complex than this, and require a mapping to be constructed between the shell model and the IBM-2. (Clearly a shell model basis of the IBM must be couched in terms of a version of the IBM which distinguishes protons and neutrons.) This will be a major focus of Chapter 4.

The valence space aspect of the IBM appears in the boson number, N (not to be confused with neutron number of a nucleus), which is taken in the model as half the total number of *valence* nucleons counting both protons and neutrons, regardless of whether they are particles or holes, to the *nearest* closed shells (magic numbers). As examples, the nuclei ${}^{112}_{46}Pd_{66}$, ${}^{148}_{56}Ba_{92}$, ${}^{168}_{68}Er_{100}$, and ${}^{196}_{78}Pt_{118}$ have $N = 10$, 8, 16, and 6, respectively. (Sometimes it is better to do the counting to major subshell closures such as $Z = 40$ or 64 instead.) The rationale for the valence anatz is that the preponderance of important residual interactions involves only this shell although it is realized that nuclear excitations occur in a "multi-$\hbar\omega$" environment. A second rationale is, of course, simplicity, as, then, the number of bosons, and hence the number of basis states, is smaller and more tractable.

[Nevertheless, it is important to note that fermionic algebraic models have now been extended to encompass many major shells. Important effects have been found, not only for giant resonances, superdeformation and the like, which inherently involve cross shell excitations, but also for normal states. The groups involved, called symplectic [e.g., Sp(3, R)] are discussed from a simple viewpoint in Chapter 7. Such multi-$\hbar\omega$ treatments allow different types of collectivity (e.g., giant resonances and rotational motion) to be put on the same footing and reduce the need to mock up neglected degrees of freedom by the use of effective interactions, effective charges, and the like.]

It will be seen later (especially in Chapters 2 and 3) that the explicit appearance of the boson number in the IBM is extremely important in many of its predictions. As two trivial examples, $B(E2{:}2_1^+ \rightarrow 0_1^+)$ values scale as N for vibrational nuclei in the IBM but as N^2 in rotational nuclei, and bandmixing effects in the latter systematically decrease with N (see Fig. 3.23). Both these examples illustrate that, while the IBM is a phenomenological or macroscopic model, the N-dependence confers on it a microscopic aspect in that, even with constant parameters, the predictions vary in characteristic and automatic ways with N, that is, across a shell. It turns out that these predicted trends are in agreement with the data. Perhaps a more important point is that, were the data different it would have been virtually impossible to alter the IBM predictions accordingly. The finite boson number aspect of the model is crucial to its structure and is essential to many of its predictions which are in accord with the data.

Clearly, the dual limitations, to the valence space, and to s and d bosons, renders the IBM an enormous truncation of the shell model. For example, in ^{154}Sm, just the valence proton and neutron states alone allow the construction of $\sim 3 \times 10^{14}$ 2^+ states. The IBM-1 for ^{152}Sm has twenty-six 2^+ basis states. One hopes, of course, that the s, d ansatz has wisely chosen the most important ones and the success of the model suggests that this is the case.

The s and d bosons of the IBM span a six-dimensional space. This fact leads to the group theoretical side of the model, namely that its structure can be expressed in terms of the group $U(6)$. We will not at all delve into the group theory in these introductory pages since that is the purview of Chapter 2 but suffice it so say that one can decompose such a group into chains of subgroups, that each chain corresponds to a particular geometric picture of the nucleus and that, as a result, the model immediately gives the predictions for such a structure in analytic form. Each of these chains describes what is called a dynamic symmetry. There are three useful dynamic symmetries contained in $U(6)$, namely $U(5)$, $SU(3)$, and $O(6)$.

The first describes a broad class of vibrational models, including both harmonic and highly anharmonic versions. It is generally applicable relatively near closed shells (as with the geometric vibrational model). The low lying levels of a harmonic $U(5)$ scheme were illustrated in Fig. 1.7.

$SU(3)$ represents a deformed, axially symmetric rotor with a potential centered at $\gamma = 0°$. Dynamic excursions in γ are allowed, and hence γ_{rms} is small but finite. It is important to stress that $SU(3)$ is a very specific type of deformed rotor with several key signatures such as degenerate β and γ bands, vanishing $\gamma \rightarrow g$ and $\beta \rightarrow g$ E2 transitions and no γ-g bandmixing. Many of the key features of $SU(3)$ are summarized and discussed in the context of Fig. 3.7. Hence, in its pure form it will seldom be strictly applicable (see Fig. 1.12). As with all the symmetries, though, one of its most appealing uses is as a benchmark: that is, many actual deformed nuclei can be easily and well described in terms of a relatively small amount of $SU(3)$ breaking.

$O(6)$ has the structure of a γ-unstable rotor and was, at the time of its prediction by the IBM, the least well known of the three types of structure. Ironically, since then, several excellent empirical manifestations of $O(6)$ have been found and, at least in its pure form, it is now the best established of the three symmetries. Examples of all three symmetries will be discussed at length in Chapters 2 and 3. Simplified level schemes (for example, for $N = 3$) are shown in Chapter 2 and a typical $O(6)$ level scheme for $N = 6$ is shown in Fig. 3.1 (see also Fig. 1.7 above).

The Hamiltonian for the IBM is expressed in terms of creation and destruction operators for s and d bosons (namely, $s^\dagger$, s, $d^\dagger$, and d). Several different forms have been given, all equivalent but useful in different situations. We give two here, using the notation of Chapters 2 and 3, namely,

$$H = \epsilon_s s^\dagger s + \epsilon_d d^\dagger \cdot \tilde{d} + \frac{1}{2} \sum_{L=0,2,4} c_L (d^\dagger d^\dagger)^{(L)} \cdot (\tilde{d}\tilde{d})^{(L)}$$

$$+ \frac{v_2}{\sqrt{10}} [(d^\dagger d^\dagger)^{(2)} \cdot \tilde{d}s + \text{H.c.}]$$

$$+ \frac{v_0}{2\sqrt{5}} [d^\dagger \cdot d^\dagger ss + \text{H.c.}] + \frac{u_2}{\sqrt{5}} d^\dagger s^\dagger \cdot \tilde{d}s + \frac{u_0}{2} s^\dagger s^\dagger ss, \quad (1.2)$$

$$H = \epsilon'' \hat{n}_d + a_0 P^\dagger P + a_1 \mathbf{J} \cdot \mathbf{J}$$

$$+ a_2 Q \cdot Q + a_3 T_3 \cdot T_3 + a_4 T_4 \cdot T_4, \quad (1.3)$$

where, in eq. (1.2), ϵ_s, ϵ_d, c_L (where $L = 0, 2, 4$), u_0, u_2, v_0, and v_2 are parameters as are ϵ'', a_0, a_1, a_2, a_3, and a_4 in eq. (1.3). The exact definition of these operators and operator combinations will be part of the subject matter of Chapters 2 and 3: we merely present the Hamiltonian here without discussion for purposes of illustration. The form in eq. (1.3), often called the "multipole" form, is the most frequently used because the different terms have an intuitive physical sense. Thus, for example, the $P^\dagger P$ term is called the pairing term (pairing between bosons, not fermions) while $a_2 Q \cdot Q$ is a quadrupole interaction between bosons.

E2 transition rates in the IBM-1 are calculated in terms of the operator Q with

$$T(\text{E2}) = eQ = e[(s^\dagger \tilde{d} + d^\dagger s) + \chi (d^\dagger \tilde{d})^{(2)}] \quad (1.4)$$

where χ is another (extremely important, as we shall see) parameter, and e is a boson effective charge.

The term in $\epsilon'' \hat{n}_d$ in eqs. (1.2) and (1.3) gives an energy proportional to the number of d-bosons. If the d-boson is interpreted (as we did qualitatively above) as a kind of 1-quadrupole phonon excitation in a vibrational picture then this term alone would give a pure vibrator spectrum and ϵ'', the d-boson energy, would be the phonon energy, $E(2_1^+)$. This is the harmonic version of $U(5)$.

Although anharmonic versions include other terms (those in c_L in eq. (1.2)) this illustration exemplifies an extremely important point: each dynamical symmetry of the IBM corresponds to the dominance of a particular term in H. The $Q \cdot Q$ term gives $SU(3)$ (provided χ in Q has the value $-\sqrt{7}/2$) and $O(6)$ results when the $P^{\dagger}P$ term dominates. [An alternate way of getting $O(6)$, described in detail in Chapter 3, is with the $Q \cdot Q$ term alone if χ is set to zero.]

This term-by-term analysis points to another critical feature of the model. Though the Hamiltonian appears complex, single terms determine the structure of the symmetries, and hence transition regions *between* two symmetries can depend only on the *ratio* of the two relevant coefficients. Thus, for example, a vibrator to rotor [$U(5) \rightarrow SU(3)$] transition depends only on the ratio a_2/ϵ'' in eq. (1.2). An $O(6) \rightarrow SU(3)$ phase transition proceeds according to the magnitude of a_2/a_0 (or of χ as this parameter varies between 0 and $-\sqrt{7}/2$)). An $O(6) \rightarrow U(5)$ transition develops according to the magnitude of ϵ''/a_0.

These ideas point to a convenient pedagogical way to represent the IBM, namely in terms of a "symmetry triangle" as illustrated in Fig. 1.18 where each vertex represents a symmetry and each leg is a transitional region which can be calculated as a function of a simple ratio of parameters [χ also has to vary between $U(5)$ and $SU(3)$]. This has the enormously important consequence that complex transition regions are trivial to calculate. Indeed, since most nuclei turn out to be situated, in effect, along the legs of the triangle, one seldom needs to use the full triangle (the full Hamiltonian of eqs. (1.2) or (1.3)). Most calculations entail 1–3 param-

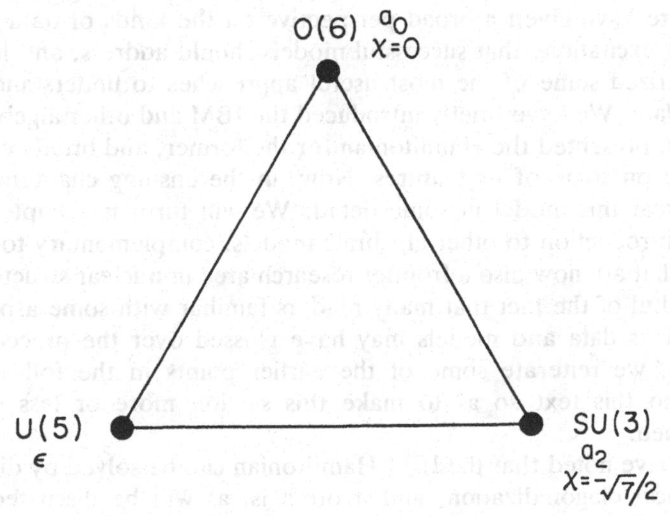

FIGURE 1.18. *The symmetry triangle for the* IBM.

eters only. It is remarkable that a model that can cover such diverse types of nuclear structure can do so this simply. Of course, in itself, that is inconsequential: the test, ultimately, is whether the model works. This will be discussed extensively throughout this volume. But it is nevertheless worth emphasizing its simplicities and the simple relation between the symmetries and transition regions between them.

Nowadays, the IBM has withstood its critical tests and has entered the domain of nuclear structure models, forming, with the shell model and geometric collective models, a triad of complementary approaches. Fermionic models offer the promise of further extending the domain of algebraic approaches and of linking bosonic models like the IBM to their microscopic underpinnings.

With this background, we can conclude this introductory discussion. The next five chapters will discuss the IBM and its extensions in considerable detail but always in a simple, hopefully easily understood, pedagogical way. Chapter 7 will present an analogous introduction to, and discussion of, some of the fermionic algebraic models that are actively pursued today as well. To orient the reader to what follows, we end this chapter with a brief guide to the remainder of the book.

1.3. GUIDE TO THE TEXT

We have discussed some of the basic ideas of low energy nuclear structure have given a broad perspective on the kinds of data and nuclear excitations that successful models should address, and have summarized some of the most useful approaches to understanding these data. We have briefly introduced the IBM and other algebraic models, presented the Hamiltonian for the former, and briefly commented on some of its features. Now, in the ensuing chapters we shall treat this model in some detail. We will turn, in Chapter 7, to an introduction to other algebraic models, complementary to the IBM, that are now also a frontier research area in nuclear structure.

Mindful of the fact that many readers familiar with some aspects of nuclear data and models may have glossed over the preceding section, we reiterate some of the earlier points in the following guide to this text so as to make this section more or less self-contained.

We have noted that the IBM Hamiltonian can be solved by direct numerical diagonalization, and it often is, as will be discussed in Chapter 3. However, one of the most characteristic and appealing features of algebraic models, and in particular of the IBM, is that they lend themselves to an alternate, group theoretical, approach that is both elegant and powerful. Such an approach leads to the recognition of symmetries of the Hamiltonian and the physical systems that it describes and, for these symmetries, gives many results in simple analytic form. Even though most nuclei will not satisfy the exact structures of such symmetries, these symmetries are yet of widespread and deep importance. They have simple geometrical (shape) interpretations, and they act as benchmarks that serve as starting points for the calculation of nuclei that are close in structure to these idealized limits. Moreover, as we shall see, they greatly facilitate the treatment of phase or shape transitional regions in which a sequence of nuclei span a range of structures, as a function of N, Z or A, from those resembling one symmetry to those close to another.

The group theoretical approach is, in fact, the common thread in the entire class of models called algebraic that we treat in this volume. With this mathematical apparatus, a remarkable litany of results can be obtained by simple algebraic techniques, centered around the manipulation of appropriate commutation relations involving particular operators. Although algebraic nuclear models are extremely diverse, even focusing on different entities (bosons or fermions), they share much of the same spirit and the group theory underlying them is based on common techniques.

Indeed, one of the most striking features of group theory is its unity and coherence: exactly the same techniques applicable to the simple groups $O(3)$ and $SU(2)$, which lead to the familiar treatments of angular momentum and isospin, respectively, can be generalized to much more complex, higher dimensional groups. The simple groups therefore provide a pedagogical basis that can be exploited to exhibit the principles and approaches involved. The explicit calculation of results for these simple cases allows the more complex ones to be plausibly motivated by analogy. Such a practical, group-specific, approach, which is the one taken in Chapter 2, is appealing since it avoids the need to introduce group theory by a lengthy series of abstract theorems and lemmas which have little immediate relation with a physical system. Another feature of group theory, which is exploited in Chapter 2, is that the same group can describe quite different physical situations and therefore the results worked out for one can be directly taken over to the other even though the physical application is totally different.

The layout of the rest of this volume is as follows. Following this

Introduction, the group theory of algebraic models in nuclear structure will be discussed in Chapter 2 following the pedagogical approach just outlined. In many ways this chapter is the key to everything that follows. In particular, the simple examples of the group theoretical treatment of isospin and angular momentum in Section 2.3 and some concepts developed in Section 2.4 provide a description for the general group $U(n)$ which in turn provides a basis to understand both the specific group structure $U(6)$ of the IBM and the groups involved in fermion models discussed in Chapter 7. These sections will repay, many times over, as much effort as can be lavished upon them by the reader. Chapter 2 then goes on to elucidate the specific group structure of the IBM, to discuss its dynamical symmetries, and their key properties.

Chapter 3 then turns to the confrontation of this model with the data of low energy nuclear structure. First, empirical manifestations of each of the dynamical symmetries are discussed and assessed. However, since most nuclei do not obey the prescriptions of these limiting symmetries, it is also necessary to consider numerical calculations in which the IBM Hamiltonian is diagonalized. Two common approaches to this, the use of the so-called multipole Hamiltonian and the Consistent Q Formalism (CQF), are discussed and the general predictions of the IBM, especially for deformed and transitional nuclei, are presented and compared to the data. Detailed calculations for particular nuclei (e.g., ^{168}Er) are thoroughly discussed. The geometrical analogues of the IBM and of specific IBM calculations are explored and it is shown how one can exploit the simplicities of the IBM to better understand the geometrical structure of nuclei.

The treatment up to this point will have focused on the IBM-1, the original IBM which does not distinguish between protons and neutrons. Despite the success of this approach (indeed it is often more successful then the presumably more realistic IBM-2 that does distinguish protons and neutrons) there are numerous observables that demand such a distinction (the (p, t) two neutron transfer reaction would be an obvious example). Moreover, any justification of the IBM microscopically (i.e., in terms of the shell model) must be in terms of the IBM-2. This model, therefore, is the focus of Chapter 4. This chapter first outlines a microscopic understanding of the so-called "mapping procedures" whereby one can go from a shell model formulation to the associated IBM-2 treatment. This is followed by a discussion of the IBM-2 itself, its Hamiltonian, some new symmetries and collective modes that emerge, and its confrontation with the data.

The IBM (primarily IBM-1) has been extended to odd-mass nuclei as well, in a model called the Interacting Boson–Fermion Model, or IBFM. This is the subject of Chapter 5. The Hamiltonian and the predictions obtained by coupling a single fermion to the boson structure of the even core are first discussed. Then an interesting aspect arises. Combining the IBM and IBFM, one has a pair of models, couched in compatible language, that can describe even and odd nuclei on the same footing. In certain cases, this leads to a new algebraic approach, centered on the idea of combined Bose–Fermion symmetries (or even supersymmetries). The nature and predictions of these are treated in depth in Chapter 5, along with physical interpretations and a comparison with experiment.

Despite the success of the IBM, it was clear very soon after its development that it had significant limitations. For example, it could not describe high spin states nor nuclei where multiple sets of diverse states (such as the so-called normal and "intruder" levels, with different shapes) appear simultaneously. Various extensions to the basic model were soon proposed. Many are rather obvious. Clearly, the *s-d* boson basis of the IBM can be extended to include $L = 4$ or g bosons. Likewise, negative parity states can be incorporated into the IBM with p ($L = 1$) and/or f ($L = 3$) bosons: this allows octupole correlations or shapes to be considered. Aside from bosons with angular momenta other than 0 or 2, another extension to the IBM is to augment either the Hamiltonian, and/or transition operators, with higher order terms. Finally, "coexistence phenomena" of excitations with differing shapes, or other characteristics, in the same nucleus can be treated by combining, and mixing, two separate subsets of IBM wave functions (with different boson numbers). These and other extensions to the model, and the data relevant to them, are the purview of Chapter 6.

Finally, Chapter 7 focuses on an extremely important and broad subject: that of the rich variety of other algebraic models of nuclear structure. The IBM is neither the first nor the only algebraic model. Indeed, as noted in an earlier section, the famous Elliott description of nuclear rotational behavior in light nuclei in terms of an $SU(3)$ fermion description dates back to the 1950s. Since then, many other fermionic algebraic approaches have proliferated. The interest in algebraic models generated by the IBM has, of course, focused new attention on the opportunities offered by these other approaches, but they have always maintained an active and promising independent existence. Many are based on an extension of the Elliott concept, applicable to heavy nuclei, in terms of the concept of pseudo-spin, and pseudo-$SU(3)$. (This concept has taken on even more interest of late with certain discoveries in superdeformed bands.)

The IBM, pseudo-$SU(3)$, and many other algebraic approaches, share an emphasis on the valence shell, that is, on the shells immediately after the last respective proton and neutron shell closures. However, it is well known that excitations spanning several shells (multi-$\hbar\omega$ excitations) can be important, and a class of algebraic fermionic models has been developed, based on "symplectic groups" (e.g., Sp(3, R)) that seek to incorporate a full "vertical" space. These models are promising and enticing (although they present computational difficulties), because they propose to treat, for example, the collectivity of giant resonances and low lying states on the same footing, and, in the same vein, to calculate, for example, B(E2) values, without effective charges. These and other fermionic algebraic models are introduced and discussed in Chapter 7 where the reader can get at least an initial glimpse of the physics of a very actively pursued and appealing approach to certain issues in nuclear structure.

With the background provided by the following chapters, the reader should be well prepared to follow the literature in this active area of nuclear structure research, to recognize the implications, regarding these models, of existing or new data on nuclear levels and transitions, and to pursue research in this field if he or she chooses.

ACKNOWLEDGEMENTS

I am grateful to the many people who provided comments on my earlier textbook whose approach and philosophy were instrumental in developing this chapter. I would like to thank Jackie Mooney for her invaluable work in helping prepare the final form of this chapter.

This work was supported in part by contract DE-AC02-76CH00016 with the United States Department of Energy.

BIBLIOGRAPHY

A. Arima and F. Iachello, *Phys. Rev. Lett.* **35**, 1069 (1975). The introduction of the IBM.

A. Arima and F. Iachello, *Ann. Phys.* (N.Y.) **99**, 253 (1976). This and the other Annals of Physics references cited here are basic references for the IBM.

A. Arima and F. Iachello, *Ann. Phys.* (N.Y.) **111**, 201 (1978).

A. Arima and F. Iachello, *Ann. Phys.* (N.Y.) **123**, 468 (1979).

A. Arima, T. Otsuka, F. Iachello, and I. Talmi, *Phys. Lett.* **66B**, 205 (1977). Focuses on the IBM-2 and *F*-spin.

D. R. Bes, *Nucl. Phys.* **49**, 544 (1963). RPA for $K = 2^+$ γ vibrations.

D. R. Bes, P. Federman, E. Maqueda, and A. Zuker, *Nucl. Phys.* **65**, 1 (1965). RPA for $K = 0^+$ β vibrations.

A. Bohr and B. Mottelson, *Nuclear Structure*, Vols. I and II (Benjamin, New York, 1969, 1975). One of the classic texts on collective models of nuclear structure.

D. M. Brink, A.F.R. de Toledo Piza, and A. K. Kerman, *Phys. Lett.* **19**, 413 (1965). Anharmonic vibrator.

R. F. Casten, *Nuclear Structure from a Simple Perspective* (Oxford University Press, New York, 1990). A textbook on nuclear structure which focuses on physics concepts and intuitive arguments rather than extensive formalism. Complementary to other standard texts.

R. F. Casten, *Phys. Lett. B* **152**, 145 (1985). The N_pN_n scheme.

R. F. Casten, D. S. Brenner, and P. E. Haustein, *Phys. Rev. Lett.* **58**, 658 (1987). The P-factor.

R. F. Casten and D. D. Warner, *Rev. Mod. Phys.* **60**, 389 (1988). Review article on the IBM. Chapter 3 of this book is based on this work.

W. F. Davidson, D. D. Warner, R. F. Casten, K. Schreckenbach, H. G. Börner, J. Simic, M. Bogdanovic, S. Koicki, W. Gelletly, G. B. Orr, and M. L. Stelts. *J. Phys. G* **7**, 455 (1981).

A. S. Davydov and G. F. Filippov, *Nucl. Phys.* **8**, 237 (1958). The rigid triaxial rotor model.

A. de Shalit and H. Feshbach, *Theoretical Nuclear Physics, Vol. I, Nuclear Structure* (Wiley, New York, 1974).

A. de Shalit and I. Talmi, *Nuclear Shell Theory* (Academic Press, New York, 1963). Classic text on shell model theory. Difficult formalism but essential, especially for the concept of seniority.

J. M. Eisenberg and W. Greiner, *Nuclear Theory*, Vol. 1, 3rd ed. (North Holland, Amsterdam, 1987).

J. P. Elliott, *Proc. Roy. Soc.* **A245**, 128 and 562 (1958). The fermion *SU*(3) model.

P. Federman and S. Pittel, *Phys. Lett.* **69B**, 385 (1977) and **77B**, 29 (1978). Key articles on the effects of the *p-n* interaction in medium and heavy nuclei.

H. Feshbach and F. Iachello, *Phys. Lett.* **45B**, 7 (1973) and *Ann. Phys.*

(N.Y.) **84**, 211 (1974). This pair of articles introduced the acronyms IBA or IBM and were influential progenitors of the model we currently know under the name Interacting Boson Model.

C. J. Gallagher in *Selected Topics in Nuclear Spectroscopy*, ed. B. J. Verhaar (North Holland, Amsterdam, 1964). Excellent early discussion of collective models.

C. Gustafson, I. L. Lamm, B. Nilsson, and S. G. Nilsson, *Ark. Fysik* **36**, 613 (1967).

O. Haxel, J. H. D. Jensen, and H. E. Suess, *Z. Phys.* **128**, 295 (1950). The shell model.

K. Heyde, in *Nuclear Structure of the Zirconium Region*, eds. J. Eberth, R. A. Meyer and K. Sistemich (Springer-Verlag, Berlin, 1988), p. 3.

K. Heyde, P. Van Isacker, M. Waroquier, J. L. Wood, and R. A. Meyer, *Phys. Rep.* **102**, 291 (1983). Intruder states.

K. Heyde, *The Nuclear Shell Model*, Springer Series in Nuclear and Particle Physics (Springer-Verlag, Berlin, 1991).

F. Iachello and A. Arima, *The Interacting Boson Model* (Cambridge University Press, Cambridge, 1987). Discussion and compilation of the concepts and formulas of the IBM.

F. Iachello and I. Talmi, *Rev. Mod. Phys.* **59**, 339 (1987). Review article on the IBM-2.

A. K. Jain, R. K. Sheline, P. C. Sood, and K. Jain, *Rev. Mod. Phys.* **62**, 393 (1990). Update of the classic 1971 article by Bunker and Reich. Compiles all the intrinsic excitations (Nilsson orbits, vibrational and mixed modes) for deformed odd-mass rare-earth and actinide nuclei. Indispensable.

D. R. Janssen, R. V. Jolos, and F. Donau, *Nucl. Phys.* **A224**, 93 (1974). Alternate presentation of the concepts of the IBM.

K. Kumar and M. Baranger, *Nucl. Phys.* **A122**, 273 (1968).

A. M. Lane, *Nuclear Theory* (Benjamin, New York, 1964). Contains reprints of several critical articles concerning the RPA. An excellent reference for pairing theory and the RPA.

C. M. Lederer and S. Shirley, *Table of Isotopes*, 7th Edition (Wiley, New York, 1977). Compact summary of nuclear level schemes. A new edition, due in 1993, is in preparation.

M. G. Mayer, *Phys. Rev.* **78**, 16 (1950). The shell model.

K. Neergaard and P. Vogel, *Nucl. Phys.* **A145**, 33 (1970). Standard early reference on RPA calculations of octupole modes.

S. G Nilsson, *Dan. Mat. Fys. Medd.* **29**, No. 16 (1955). The Nilsson model.

T. Otsuka, A. Arima, and F. Iachello, Nucl. Phys. **A309**, 1 (1978). The OAI boson–fermion mapping procedure.

S. Raman, C. H. Malarkey, W. T. Milner, C. W. Nestor, Jr., and P. H. Stelson, *At. Data and Nucl. Data Tables* **36**, 1 (1987). An extremely useful compilation of $B(E2{:}0_1^+ \rightarrow 2_1^+)$ values.

R. D. Ratna Raju, J. P. Draayer, and K. T. Hecht, *Nucl. Phys.* **A202**, 433 (1973). An excellent discussion of pseudo-$SU(3)$ and related fermion models.

J. P. Draayer in *Understanding the Variety of Nuclear Excitations*, ed. A.

Covello (World Scientific, Singapore, 1991), p. 439. Review of basic ideas of pseudo-$SU(3)$.
L. L. Riedinger, N. R. Johnson, and J. H. Hamilton, *Phys. Rev.* **179**, 1214 (1969). A classic discussion and application of band mixing.
P. Ring and P. Schuck, *The Nuclear Many-Body Problem* (Springer, New York, 1980).
G. Scharff-Goldhaber and J. Weneser, *Phys. Rev.* **98**, 212 (1955). The vibrator model.
J. P. Schiffer and W. W. True, *Rev. Mod. Phys.* **48**, 191 (1976). Discusses residual *p-n* interactions near closed shells. A pioneering paper in this field.
O. Scholten, F. Iachello, and A. Arima, *Ann. Phys.* (N.Y.) **115**, 325 (1978).
P. C. Sood, D. M. Headly, and R. K. Sheline, *At. Data and Nucl. Data Tables* **47**, 89 (1991). Update to Sakai's pioneering compilations. Tabulates the rotational levels of all identified rotational bands in the rare earth region. Indispensable.
F. S. Stephens, *Rev. Mod. Phys.* **47**, 43 (1975). Discusses rotational motion, Coriolis effects, rotation alignment, backbending and related effects.
I. Talmi, *Simple Models of Complex Nuclei: The Shell Model and Interacting Boson Model* (Harwood, New York, in press).
D. D. Warner, R. F. Casten, and W. F. Davidson, *Phys. Rev. Lett.* **45**, 1761 (1980). Presents the (non-CQF) calculations for ^{168}Er.
D. D. Warner and R. F. Casten, *Phys. Rev. Lett* **48**, 1385 (1982). Introduces the CQF.
L. Wilets and M. Jean, *Phys. Rev.* **C102**, 788 (1956). The γ-soft model.
C. L. Wu *et al.*, *Phys. Lett.* **168B**, 313 (1986). Introduces the FDSM.

CHAPTER 2

Group Theory of the IBM and Algebraic Models in General

PERTTI O. LIPAS

2.1. INTRODUCTION: WHY GROUPS?

A beginning graduate student of physics will have used group theory even though he/she may not know it by that name. Apart from the discrete symmetries relating to regularities of crystalline structure, continuous translational and rotational symmetries occur throughout the study of physics. And where there are symmetries, there are related groups—thence group theory.

In the present context of nuclear structure, continuous groups are of paramount importance. The physically and technically most important one is rotational symmetry and the associated group structure. It also serves as the prototype of a number of other symmetries/groups belonging to particular nuclear models. The interacting-boson model (IBM) of Iachello and Arima is our most important area of such group concepts and techniques.

How, specifically, does group theory serve the formulation and application of the IBM? The Hamiltonian of the model contains single-boson energies and boson–boson interactions. One need not think of group theory in writing it down. When it comes to solving the Schrödinger equation, however, some group theory is mandatory even when only numerical solutions are sought: the set of basis states for diagonalization are defined with reference to group-theoretic concepts.

Yet group theory plays also a more direct and impressive role in the IBM. The general Hamiltonian admits of three special cases, referred to as dynamic symmetries, where group theory leads to remarkably simple and useful analytic formulas for level energies and transition rates. These special cases describe basic types of nuclear collective motion: anharmonic vibration of a spherical nucleus, rotation and vibration of a deformed nucleus, and so-called gamma-unstable motion. These collective modes are presented in Chapter 1 as characteristic of the geometric model. Qualitatively, the three special cases of the IBM thus serve to give a vivid physical picture of the meaning of the otherwise abstract algebraic model. Quantitatively, the simple formulas for these cases are often adequate in applications to real nuclei.

Let us quickly review the basic ideas of rotational symmetry or, synonymously, rotational invariance. When the Hamiltonian of a system is the same in all coordinate systems obtained by rotations about the origin, it is rotation invariant, and the physical system has rotational symmetry. Equivalently one can think of rotating the Hamiltonian field in a fixed coordinate system. One speaks of passive and active rotations, respectively, according to these two vantage points. Such an invariant Hamiltonian is also known as a scalar.

The rotations in three-dimensional physical coordinate space form a *group*. The essential defining property is that two consecutive rotations are also a rotation; a given orientation of the axes can be equivalently reached in one step or in several steps.

Angular momentum is intimately related to rotations. The rotational invariance of a physical system means that its total angular momentum is conserved; it has a "good" angular momentum **J**. In quantum mechanics this means that the Hamiltonian commutes with each of the components J_x, J_y, J_z. Infinitesimal rotations are indeed generated by the angular momentum components, and finite rotations are built up from infinitesimal rotations. We touch these points in Section 2.3. For technical details beyond standard quantum mechanics texts, see e.g. the classics by Rose (1957) and de-Shalit and Talmi (1963).

The remarkable properties of quantized angular momentum, particularly its spectrum of integral and half-integral values, can be derived from the commutator relations $[J_x, J_y] = iJ_z$ etc. without reference to a particular physical system. We do this in detail in Section 2.3. This algebraic structure leading to definite physical predictions is the prototype example of group theory in physics. A great variety of more involved commutator algebras have decisively served the theory of nuclear structure in recent years. The IBM is the prime example discussed in this book.

Group theory can be approached at varied levels of formality, ranging from an axiomatic mathematical level to a heuristic level of minimal mathematics. Our present approach is decidedly closer to the latter than to the former. It is directed to the present needs of application and thus opens only a narrow window to the wide subject of group theory.

In Section 2.2 we give a self-contained introduction to the language of creation and annihilation operators, appropriately known as occupation-number representation. It is a locally necessary preliminary because we carry our discussion of group theory in that language. It is also globally called for since the language is used throughout the book.

Section 2.3 is a heuristic introduction to the concepts and techniques of group theory. We build it on the student's previous knowledge, assumed at least at the descriptive level, of isospin and angular momentum. At the risk of redundancy, we give the algebraic derivation of the angular-momentum eigenvalues in complete detail. This is because it serves as a prototype for the determination of quantum numbers for other groups. Within the bounds of the volume, we must much rely on analogs in place of further detailed derivations. Groups designated as $U(2)$, $SU(2)$ and $O(3)$ emerge.

Section 2.4 is an extension and a generalization of Section 2.3. Its subject is the general unitary group $U(n)$: its algebra, quantum numbers, and symmetries expressed through so-called Young tableaux. The concept of group representation, introduced in Section 2.3, is amplified. We revert to $U(2)$ for a detailed study of the symmetries and a complete derivation of the wave functions.

Section 2.5 consists of a rather detailed study of the group $U(6)$ as applied to the IBM. Three subgroup chains are identified and classified in terms of quantum numbers. The Casimir operators (introduced in Section 2.3) relevant to the basic IBM, IBM-1, are presented and their eigenvalues are stated.

Section 2.6 and 2.7 continue the detailed treatment of IBM-1 but at the same time introduce general concepts and techniques. The subject of Section 2.6 is the reduction problem: what representations of a subgroup belong to a given representation of the larger group. Section 2.7 discusses the three dynamic symmetries that emerge from the three subgroup chains. They entail simple predictions for excitation energies and transition rates. The predictions are compared to the properties of real nuclei in Chapter 3.

Finally we sketch in Section 2.8 applications of group theory to the extended nuclear models discussed in Chapters 4–7.

2.2. OCCUPATION-NUMBER REPRESENTATION

The application of group theory to nuclear structure is largely concerned with systems of identical particles. They fall into two categories in nature, *bosons* of integral spin and *fermions* of odd-half-integral spin. While the "true" particles of nuclear-structure physics, nucleons, are fermions, various collective nuclear models avail themselves of bosons. In the "geometric" Bohr–Mottelson collective model the quantized vibrations of the nuclear surface are bosons called phonons; see Chapter 1. In the "algebraic" IBM, so-called s and d bosons are the basic elementary excitations. Microscopically they are related to nucleon pairs, as is discussed at length in Chapter 4.

Bosons and fermions coexist in some models, particularly those dealing with odd-A nuclei. Within the IBM family of models, the interacting-boson–fermion model IBFM is a case in point; see Chapter 5.

Our subsequent discussion of group theory will be largely carried out in terms of *creation* and *annihilation operators* for bosons and fermions. They are the basic operators of what is known as the *occupation-number representation* or "second quantization" (the quotation marks stand to indicate that it is really a misnomer: there is no further quantization beyond the initial step into quantum mechanics).

We now give a self-contained heuristic derivation of the occupation-number representation. The pivotal point is that, as a general fact of nature, states consisting of identical bosons (fermions) must be symmetric (antisymmetric) under exchange of particle labels. We carry the discussion in terms of states of product form. Such states represent noninteracting particles. Wave functions for systems of interacting particles, which is the case of nontrivial physical interest in nuclear structure, can be expressed as linear combinations of such states. Thus the latter can be used as a *basis*, and we suffer no loss of generality.

2.2.1. Bosons

Consider a state of two identical bosons

$$\begin{aligned}\Psi_{ij}(12) &= \sqrt{\tfrac{1}{2}}\,[\psi_i(1)\psi_j(2) + \psi_i(2)\psi_j(1)] \\ &= \sqrt{\tfrac{1}{2}}\,[\psi_i(1)\psi_j(2) + \psi_j(1)\psi_i(2)].\end{aligned} \tag{2.1}$$

The arguments 1 and 2 are shorthand for the coordinates $\mathbf{r}_1$ and $\mathbf{r}_2$ (and possible intrinsic spins); this is the *coordinate representation* of the state, i.e., the usual wave function of basic quantum mechanics. This state is properly symmetrized according to the above requirement. It is also normalized since we assume the *single-particle states* ψ_i orthonormal: $(\psi_i, \psi_j) = \delta_{ij}$. It is a two-particle state where one particle is in single-particle state ψ_i and the other in single-particle state ψ_j, but we cannot say which particle, 1 or 2, is in which state. This *must* in fact be so because of the identity of the particles, and the labels 1 and 2 are a mere notational device for writing down the wave function.

Similarly to eq. (2.1) we write the generic three-boson state as

$$\begin{aligned}\Psi_{ijk}(123) = \frac{1}{\sqrt{3!}}\,[&\psi_i(1)\psi_j(2)\psi_k(3) + \psi_j(1)\psi_k(2)\psi_i(3) \\ &+ \psi_k(1)\psi_i(2)\psi_j(3) + \psi_j(1)\psi_i(2)\psi_k(3) \\ &+ \psi_i(1)\psi_k(2)\psi_j(3) + \psi_k(1)\psi_j(2)\psi_i(3)].\end{aligned} \tag{2.2}$$

In fact we can agree to write the particle labels always in the order 123 . . . and permute the state labels as we have done above. Then we can abbreviate the notation by dispensing altogether with the particle labels. Thus state (2.1) would read simply

$$\Psi_{ij} = \sqrt{\tfrac{1}{2}}\,(\psi_i\psi_j + \psi_j\psi_i). \tag{2.3}$$

Even with omitted particle labels the notation of many-particle states clearly becomes cumbersome with increasing particle numbers. The occupation-number representation comes to the rescue. In it we define a *creation operator* $b_i^\dagger$ which, when operating on the *vacuum* state $|0\rangle$, creates a boson in the single-particle state ψ_i: $\psi_i = b_i^\dagger|0\rangle$. The two- and three-boson states (2.1) and (2.2) become

$$\Psi_{ij} = N_{ij} b_i^\dagger b_j^\dagger |0\rangle, \qquad \Psi_{ijk} = N_{ijk} b_i^\dagger b_j^\dagger b_k^\dagger |0\rangle, \tag{2.4}$$

where N_{ij} and N_{ijk} are normalization constants. In order that the states indeed be symmetric, i.e., $\Psi_{ij} = \Psi_{ji}$ and $\Psi_{ijk} = \Psi_{jik}$ etc., the creation operators must *commute*:

$$[b_i^\dagger, b_j^\dagger] \equiv b_i^\dagger b_j^\dagger - b_j^\dagger b_i^\dagger = 0 \tag{2.5}$$

for all single-particle indices i, j.

The annihilation operators b_i come about when we consider the bra states corresponding to the ket states defined above: $(b_i^\dagger|0\rangle)^\dagger = \langle 0|b_i$. The dagger operation means Hermitian conjugation, which appears as complex conjugation in coordinate representation. In particular let us look at the single-particle orthonormality condition

$$\delta_{ij} = (\psi_i, \psi_j) \equiv \langle 0|b_i b_j^\dagger|0\rangle. \tag{2.6}$$

The Hermitian conjugate b_i of the creation operator $b_i^\dagger$ is the corresponding *annihilation operator*. We can read eq. (2.6) so that b_i annihilates a particle from the state $b_i^\dagger|0\rangle$; thence the name.

By taking the Hermitian conjugate of eq. (2.5) we see that the boson annihilation operators also commute,

$$[b_i, b_j] = 0 \quad \text{for all } i, j. \tag{2.7}$$

There is nothing to annihilate in the vacuum, so we must have $b_i|0\rangle = 0$ for all i. The orthonormality condition (2.6) can now be satisfied if we require the validity of the commutation relation

$$[b_i, b_j^\dagger] = \delta_{ij} \quad \text{for all } i, j. \tag{2.8}$$

Equations (2.5), (2.7) and (2.8) constitute the *boson commutation relations* that can be used as the starting point in an axiomatic presentation of the topic.

We have yet to determine the normalization constants N_{ij} and N_{ijk} of eq. (2.4). The commutation relations show that they are one as long as the indices are different. Likewise we find for n_i bosons in the single-particle state ψ_i the normalized state

$$|n_i\rangle = \frac{1}{\sqrt{n_i!}}(b_i^\dagger)^{n_i}|0\rangle. \tag{2.9}$$

In eq. (2.4) we would thus have $N_{ii} = 1/\sqrt{2}$, $N_{iii} = 1/\sqrt{6}$, $N_{ii,j\neq i} = 1/\sqrt{2}$.

The general structure of the boson states is now clear. Once the single-particle basis $\{\psi_i\}$ has been given, the general boson state reads

$$|n_1 n_2 \cdots n_i \cdots\rangle = N_{n_1 n_2 \cdots n_i \cdots}(b_1^\dagger)^{n_1}(b_2^\dagger)^{n_2} \cdots (b_i^\dagger)^{n_i} \cdots |0\rangle, \tag{2.10}$$

with as many single-particle states as the system contains. For bosons the *occupation numbers* n_i can be anything 0, 1, 2, . . . with the only restriction that $\Sigma_i\, n_i = N$, the total number of particles. This expression literally justifies the name "occupation-number representation." With the general state thus defined through eqs. (2.9) and (2.10), use of the commutators yields two important relations:

$$\begin{aligned} b_i|n_1 n_2 \cdots n_i \cdots\rangle &= \sqrt{n_i}\,|n_1 n_2 \cdots n_i - 1 \cdots\rangle, \\ b_i^\dagger|n_1 n_2 \cdots n_i \cdots\rangle &= \sqrt{n_i + 1}\,|n_1 n_2 \cdots n_i + 1 \cdots\rangle. \end{aligned} \tag{2.11}$$

Furthermore, we hence deduce the single-particle and total *number operators*

$$\hat{n}_i = b_i^\dagger b_i, \qquad \hat{N} = \sum_i \hat{n}_i \tag{2.12}$$

with the eigenvalue equations

$$\begin{aligned} \hat{n}_i|n_1 n_2 \cdots n_i \cdots\rangle &= n_i|n_1 n_2 \cdots n_i \cdots\rangle, \\ \hat{N}|n_1 n_2 \cdots n_i \cdots\rangle &= N|n_1 n_2 \cdots n_i \cdots\rangle, \qquad \sum_i n_i = N. \end{aligned} \tag{2.13}$$

The one element still missing from our formalism is the representation of physical operators such as the Hamiltonian. What is the appearance of one-body and two-body operators for a boson system? The kinetic energy is a typical one-body operator, while the two-body interaction energy is a typical two-body operator. Written for a system of N particles in coordinate representation they have the forms

$$T = \sum_{\alpha=1}^{N} T(\alpha), \qquad V = \sum_{\alpha<\beta}^{N} V(\alpha\beta), \tag{2.14}$$

where we have the *particle* labels α and β to be clearly distinguished from the *state* labels such as i above. We assert that in occupation-number representation these operators become

$$T = \sum_{ij} T_{ij} b_i^\dagger b_j, \qquad V = \tfrac{1}{2} \sum_{ijkl} V_{ijkl} b_j^\dagger b_i^\dagger b_k b_l, \tag{2.15}$$

where we have the single-particle and two-particle matrix elements

$$\begin{aligned} T_{ij} &\equiv (\psi_i(\alpha), T(\alpha)\psi_j(\alpha)), \\ V_{ijkl} &\equiv (\psi_i(\alpha)\psi_j(\beta), V(\alpha\beta)\psi_k(\alpha)\psi_l(\beta)). \end{aligned} \tag{2.16}$$

These relations are proved by showing that they give the same matrix elements between many-particle states in occupation-number representation as the operators (2.14) give in coordinate representation. Easiest is to prove the case $N = 2$ and be satisfied with it. As a technical detail we note that the two-particle matrix element V_{ijkl} present above is to be computed without symmetrizing the two-particle wave functions. *If* one wants to have symmetric two-particle states in the two-particle matrix element, then the factor $\frac{1}{2}$ in front of the sum in eq. (2.15) is replaced by $\frac{1}{4}$. A further remark is that it is customary to employ a basis where the one-body part of the Hamiltonian is diagonal, i.e., of form $\Sigma_i\, \epsilon_i b_i^\dagger b_i$.

We are now finished with the occupation-number representation for bosons. For *fermions* we need not repeat much of the above discussion; the results can be gleaned by inspection leading to appropriate modifications.

2.2.2. Fermions

The one basic difference between bosons and fermions is that the latter are described by *antisymmetric* wave functions. Accordingly the fermion counterparts of eqs. (2.1) and (2.2) are

$$\Phi_{ij} = \sqrt{\tfrac{1}{2}}\, [\phi_i\phi_j - \phi_j\phi_i], \tag{2.17}$$

$$\begin{aligned} \Phi_{ijk} = \frac{1}{\sqrt{3!}} [&\phi_i\phi_j\phi_k + \phi_j\phi_k\phi_i + \phi_k\phi_i\phi_j \\ &- \phi_j\phi_i\phi_k - \phi_i\phi_k\phi_j - \phi_k\phi_j\phi_i], \end{aligned} \tag{2.18}$$

where we have followed the practice, established in eq. (2.3), of omitting the particle labels since they are always put in ascending order. We define the fermion creation operator $a_i^\dagger$ analogously to the boson case and write the one-, two- and three-fermion states

$$\phi_i = a_i^\dagger|0\rangle, \quad \Phi_{ij} = N_{ij}a_i^\dagger a_j^\dagger|0\rangle, \quad \Phi_{ijk} = N_{ijk}a_i^\dagger a_j^\dagger a_k^\dagger|0\rangle. \quad (2.19)$$

The fermion wave functions are endowed with their predicated antisymmetry if we replace the commutators in eqs. (2.5), (2.7) and (2.8) by corresponding *anticommutators*:

$$\{a_i^\dagger, a_j^\dagger\} \equiv a_i^\dagger a_j^\dagger + a_j^\dagger a_i^\dagger = 0, \quad \{a_i, a_j\} = 0, \quad \{a_i, a_j^\dagger\} = \delta_{ij}. \quad (2.20)$$

The Pauli exclusion principle (cf. Chapter 1) is present here: the first anticommutator implies $a_i^\dagger a_i^\dagger = 0$, so at most one fermion may occupy any given single-particle state. As we learned in Subsection 2.2.1, there is no such restriction for bosons. The restriction $n_i = 0, 1$ renders the fermion counterparts of several details of the boson case trivial. Thus the normalization constants of eq. (2.19) are one if the single-particle labels are different, otherwise zero; the fermion equations corresponding to eqs. (2.9)–(2.13) admit only of $n_i = 0$ or 1. The one-body and two-body operators for fermions have the same form as the boson operators (2.15).

2.3. ISOSPIN AND ANGULAR MOMENTUM

Rather than expound generalities about the elements of group theory, we introduce its language and basic concepts by considering the familiar physical constructs of isospin and angular momentum. Although physically very different, it turns out that the two bear an intimate mathematical relationship.

The idea and purpose of this section is not to teach isospin and angular momentum for their own sake. Rather we assume a basic knowledge of them and exploit it as a platform for teaching the precepts of group theory in a concrete context. The subsequent sections lean heavily on the paradigm introduced here. Therefore this section merits a careful study even when parts of it may appear overly familiar.

2.3.1. Isospin: Groups *U*(2) and *SU*(2)

Consider the two-dimensional space spanned by the state vectors

$$|\mathrm{p}\rangle = a_\mathrm{p}^\dagger|0\rangle, \qquad |\mathrm{n}\rangle = a_\mathrm{n}^\dagger|0\rangle, \qquad (2.21)$$

where we only specify the existence of a proton and a neutron, without regard to any further specification of the state. These vectors are orthogonal, $\langle n|p\rangle = 0$, and normalized, $\langle p|p\rangle = 1$, $\langle n|n\rangle = 1$. In our space the general state ψ, also assumed normalized as $(\psi, \psi) \equiv \langle\psi|\psi\rangle = 1$, is a linear combination of $|p\rangle$ and $|n\rangle$:

$$\psi = \langle p|\psi\rangle|p\rangle + \langle n|\psi\rangle|n\rangle. \tag{2.22}$$

The situation, depicted in Fig. 2.1, is completely analogous to that of an elementary *xy* space with unit vectors **i** and **j** and a general unit vector $\mathbf{r}/r \equiv \mathbf{e}_r$. To eq. (2.22) then corresponds $\mathbf{e}_r = (\mathbf{i}\cdot\mathbf{e}_r)\mathbf{i} + (\mathbf{j}\cdot\mathbf{e}_r)\mathbf{j}$.

Now consider rotating the vector ψ to different orientations in the pn space. Rotations preserve length and are therefore *unitary* transformations; in general we include also complex transformations preserving $(\psi, \psi) = 1$. Any two consecutive rotations can obviously be replaced by a single rotation to arrive at the same orientation of ψ. This is the essential defining property of a *group* of transformations; specifically the unitary transformations on ψ are said to form the unitary group in two dimensions, designated as $U(2)$ or U_2.

The two vectors $|p\rangle$ and $|n\rangle$ are said to form the *fundamental representation* of the group $U(2)$. The concept of representation will be amplified in the sequel. To clearly see what we mean by unitary transformations in the present context, let us write down the effect on the basis vectors of an operator U for a particular unitary transformation:

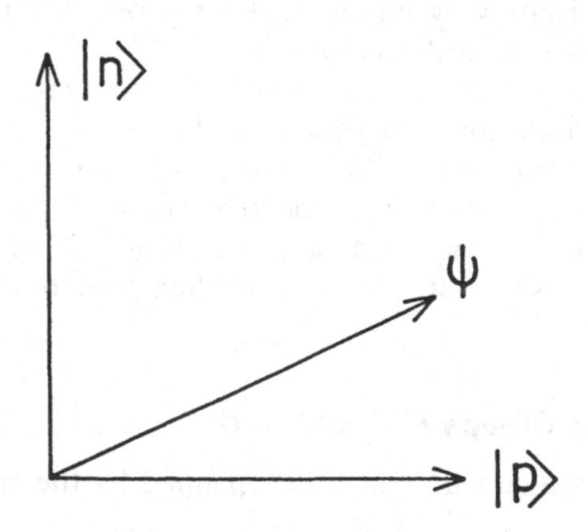

FIGURE 2.1. *Proton–neutron isospin space with general state ψ.*

$$U|\mathrm{p}\rangle = U_{\mathrm{pp}}|\mathrm{p}\rangle + U_{\mathrm{np}}|\mathrm{n}\rangle, \qquad U|\mathrm{n}\rangle = U_{\mathrm{pn}}|\mathrm{p}\rangle + U_{\mathrm{nn}}|\mathrm{n}\rangle. \quad (2.23)$$

The generally complex coefficients $U_{\mathrm{pp}} \equiv \langle \mathrm{p}|U|\mathrm{p}\rangle$ etc. are matrix elements of the operator U. They form a 2×2 matrix which satisfies the unitarity condition $U^\dagger = U^{-1}$. The two equations (2.23) are just eq. (2.22) with $\psi = U|\mathrm{p}\rangle$ and $\psi = U|\mathrm{n}\rangle$, respectively.

The group $U(2)$ is a definite mathematical structure of transformations. In that respect it is incidental that we chose to introduce it in the physical context of proton–neutron isospin. Such a specific environment for the group is known as a chosen *realization* of it.

We now consider the nucleon creation operators $a_\mathrm{p}^\dagger$, $a_\mathrm{n}^\dagger$ appearing in eq. (2.21) and the corresponding annihilation operators a_p, a_n. When we attempt to construct many-nucleon states, the Pauli principle restricts the possible states to be antisymmetric since in the isospin formalism the nucleons are identical fermions. Therefore our operators obey the anticommutation relations (2.20):

$$\begin{aligned}
&\{a_\mathrm{p}^\dagger, a_\mathrm{p}^\dagger\} = \{a_\mathrm{p}^\dagger, a_\mathrm{n}^\dagger\} = \{a_\mathrm{n}^\dagger, a_\mathrm{n}^\dagger\} = 0,\\
&\{a_\mathrm{p}, a_\mathrm{p}\} = \{a_\mathrm{p}, a_\mathrm{n}\} = \{a_\mathrm{n}, a_\mathrm{n}\} = 0, \qquad (2.24)\\
&\{a_\mathrm{p}, a_\mathrm{p}^\dagger\} = 1, \quad \{a_\mathrm{p}, a_\mathrm{n}^\dagger\} = 0, \quad \{a_\mathrm{n}, a_\mathrm{p}^\dagger\} = 0, \quad \{a_\mathrm{n}, a_\mathrm{n}^\dagger\} = 1.
\end{aligned}$$

The next step in our discourse is to form from the creation and annihilation operators the bilinear combinations

$$a_\mathrm{p}^\dagger a_\mathrm{p}, \quad a_\mathrm{p}^\dagger a_\mathrm{n}, \quad a_\mathrm{n}^\dagger a_\mathrm{p}, \quad a_\mathrm{n}^\dagger a_\mathrm{n}. \qquad (2.25)$$

These four operators are clearly linearly independent. Let us now form and compute, with eq. (2.24), all their nontrivial commutators:

$$\begin{aligned}
&[a_\mathrm{p}^\dagger a_\mathrm{p}, a_\mathrm{p}^\dagger a_\mathrm{n}] = a_\mathrm{p}^\dagger a_\mathrm{n}, \quad [a_\mathrm{p}^\dagger a_\mathrm{p}, a_\mathrm{n}^\dagger a_\mathrm{p}] = -a_\mathrm{n}^\dagger a_\mathrm{p},\\
&[a_\mathrm{p}^\dagger a_\mathrm{p}, a_\mathrm{n}^\dagger a_\mathrm{n}] = 0, \quad [a_\mathrm{p}^\dagger a_\mathrm{n}, a_\mathrm{n}^\dagger a_\mathrm{p}] = a_\mathrm{p}^\dagger a_\mathrm{p} - a_\mathrm{n}^\dagger a_\mathrm{n}, \qquad (2.26)\\
&[a_\mathrm{p}^\dagger a_\mathrm{n}, a_\mathrm{n}^\dagger a_\mathrm{n}] = a_\mathrm{p}^\dagger a_\mathrm{n}, \quad [a_\mathrm{n}^\dagger a_\mathrm{p}, a_\mathrm{n}^\dagger a_\mathrm{n}] = -a_\mathrm{n}^\dagger a_\mathrm{p}.
\end{aligned}$$

We note that these commutators have the remarkable property that they are, in general, equal to linear combinations of the operators (2.25) entering the commutators. This state of affairs is described by saying that the four operators (2.25) *close on commutation*. As defined by this property, our four operators are said to form a *Lie algebra*. The particular Lie algebra present here is properly denoted by $u(2)$; however, in physics the distinction between group and Lie algebra is often glossed over so that here both would be designated as $U(2)$. The present group of transformations, as discussed above, is likewise called the *Lie group* $U(2)$. The operators (2.25) are called the *generators* of the group $U(2)$.

To make contact with the normal isospin formalism and to gain further insight in the concepts of group theory, we proceed to cast the above commutator algebra into alternative forms. To that end we define certain linear combinations of the operators (2.25):

$$a_p^\dagger a_n \equiv T_+ \equiv T_1 + iT_2, \qquad a_n^\dagger a_p \equiv T_- \equiv T_1 - iT_2, \qquad (2.27)$$
$$\tfrac{1}{2}(a_p^\dagger a_p - a_n^\dagger a_n) \equiv T_3, \qquad a_p^\dagger a_p + a_n^\dagger a_n \equiv \hat{N}.$$

Thus we have defined two new alternative sets of four linearly independent operators. One set is T_+, $T_- = (T_+)^\dagger$, T_3, $\hat{N}$; the other set is T_1, T_2, T_3, $\hat{N}$, all of which are Hermitian. When we calculate all commutators for the latter set, we find with use of eq. (2.26)

$$[T_1, T_2] = iT_3, \quad [T_2, T_3] = iT_1, \quad [T_3, T_1] = iT_2, \qquad (2.28)$$
$$[\hat{N}, T_1] = 0, \quad [\hat{N}, T_2] = 0, \quad [\hat{N}, T_3] = 0.$$

The first three commutation relations have the same form as the familiar relations for angular-momentum components. Therein lies the formal analogue of isospin to angular momentum. We explore this point in Subsection 2.3.3 below.

The four operators T_1, T_2, T_3, $\hat{N}$ are a set of generators, alternative to eq. (2.25), of the group $U(2)$; yet another alternative set is T_+, T_-, T_3, $\hat{N}$. Equation (2.28) shows that the operator $\hat{N}$ commutes with all generators $T_{1,2,3}$ and $\hat{N}$ of the group. Such an operator is called a *Casimir operator* of the group. This Casimir operator is linear in the generators; generally there are also quadratic etc. Casimir operators.

Not only is $\hat{N}$ the linear Casimir operator of $U(2)$, but we also see from eq. (2.28) that the smaller set of generators T_1, T_2, T_3 close by themselves on commutation and thus form a Lie algebra of their own. This Lie algebra and corresponding group are called $SU(2)$, the S standing for "special." Since the generators of $SU(2)$ are a subset of the generators of $U(2)$, the former is called a *subgroup* of the latter. The operator $\hat{N}$ by itself also forms a trivial $U(1)$ subalgebra of $U(2)$. The relation between the groups is denoted as

$$U(2) = SU(2) \times U(1) \supset SU(2). \qquad (2.29)$$

The *product group* is marked with an ordinary $\times$; the symbol $\otimes$ is also used in the literature but we reserve it for a different purpose (Subsection 2.6.3). In general, a product group means that the generators of one factor group commute with those of the other factor group. The generators of the product group consist of the totality of the generators of the factor groups.

By eq. (2.28) we see that the operator

$$T_1^2 + T_2^2 + T_3^2 \equiv \mathbf{T}\cdot\mathbf{T} \equiv \mathbf{T}^2 \tag{2.30}$$

commutes with $T_{1,2,3}$ (this is also obvious from the angular-momentum analogue). Accordingly it is a Casimir operator of $SU(2)$. It is quadratic in the generators and it is the only Casimir operator of $SU(2)$. It is also the quadratic Casimir operator of $U(2)$ since it follows from eq. (2.28) that $[\mathbf{T}^2, \hat{N}] = 0$. We have here defined an operator $\mathbf{T}$ which is a vector in isospin space with components $T_{1,2,3}$. This vector operator is analogous to the angular-momentum operator $\mathbf{J}$. Physically, of course, isospin and mechanical angular momentum are completely disparate things.

2.3.2. The Spectrum of Angular Momentum

Beyond elementary treatments restricted to the orbital angular momentum of a single particle, the study of angular momentum is carried out in complete generality by formal algebra. Since we anchor our subsequent discussion of other groups on the prototype of angular momentum, we give here the algebraic treatment in detail, although it is given in good texts on quantum mechanics.

The quantal angular-momentum vector operator $\hbar\mathbf{J}$ is defined as a vector whose Cartesian components, referred to axes xyz fixed in space, are Hermitian ($J^\dagger_{x,y,z} = J_{x,y,z}$) and satisfy the commutation relations

$$[J_x, J_y] = iJ_z, \quad [J_y, J_z] = iJ_x, \quad [J_z, J_x] = iJ_y. \tag{2.31}$$

From these follows that

$$\mathbf{J}\cdot\mathbf{J} \equiv \mathbf{J}^2 = J_x^2 + J_y^2 + J_z^2 \tag{2.32}$$

commutes with $J_{x,y,z}$. We choose $\mathbf{J}^2$ and J_z as the two commuting operators for which we set the simultaneous eigenvalue problem:

$$\mathbf{J}^2|\lambda M\rangle = \lambda|\lambda M\rangle, \qquad J_z|\lambda M\rangle = M|\lambda M\rangle, \tag{2.33}$$

where λ and M are the eigenvalues of $\mathbf{J}^2$ and J_z, respectively. As eigenvalues of Hermitian operators they must be real numbers. Hermiticity also implies that the eigenstates, which we take to be normalized, are orthogonal:

$$\langle \lambda M | \lambda' M' \rangle = \delta_{\lambda\lambda'} \delta_{MM'}. \tag{2.34}$$

We now define $J_\pm \equiv J_x \pm iJ_y$ (note that $J_\pm^\dagger = J_\mp$) and record the alternative commutation relations

$$[J_\pm, J_z] = \mp J_\pm, \quad [J_+, J_-] = 2J_z, \quad [J_\pm, \mathbf{J}^2] = 0. \tag{2.35}$$

Operating with these on the eigenvector $|\lambda M\rangle$ we get

$$\begin{aligned} J_z J_\pm |\lambda M\rangle &= (M \pm 1) J_\pm |\lambda M\rangle, \\ \mathbf{J}^2 J_\pm |\lambda M\rangle &= \lambda J_\pm |\lambda M\rangle, \end{aligned} \tag{2.36}$$

whence we conclude that $J_\pm |\lambda M\rangle$ is the eigenvector, yet unnormalized, belonging to the eigenvalue $M \pm 1$ of J_z and to the eigenvalue λ of $\mathbf{J}^2$. We thus write

$$J_\pm |\lambda M\rangle = C_\pm(\lambda M) |\lambda, M \pm 1\rangle \tag{2.37}$$

and determine $C_\pm(\lambda M)$ from normalization and the identity

$$\mathbf{J}^2 = J_\pm J_\mp + J_z^2 \mp J_z: \tag{2.38}$$

$$|C_\pm(\lambda M)|^2 = \langle \lambda M | J_\mp J_\pm | \lambda M \rangle = \lambda - M^2 \mp M. \tag{2.39}$$

With the standard choice of real phases this yields the "raising and lowering relations"

$$J_\pm |\lambda M\rangle = \sqrt{\lambda - M^2 \mp M}\, |\lambda, M \pm 1\rangle. \tag{2.40}$$

We have yet to determine the possible values of λ and M. First, we have from eqs. (2.32) and (2.33)

$$\lambda - M^2 = \langle \lambda M | J_x^2 + J_y^2 | \lambda M \rangle \geq 0 \tag{2.41}$$

because by the Hermiticity of J_x

$$\langle \lambda M | J_x^2 | \lambda M \rangle \equiv (\psi_{\lambda M}, J_x^2 \psi_{\lambda M}) = (J_x \psi_{\lambda M}, J_x \psi_{\lambda M}) \geq 0 \tag{2.42}$$

and similarly for J_y^2. Secondly, eq. (2.39) dictates that

$$\lambda - M^2 \mp M \geq 0. \tag{2.43}$$

For any given value of λ, M will thus have an upper (lower) limit $M_{\max} \geq 0$ ($M_{\min} \leq 0$) that will mark the end of possible states in raising (lowering):

$$J_+|\lambda M_{max}\rangle = 0, \qquad J_-|\lambda M_{min}\rangle = 0. \tag{2.44}$$

The limiting values are solved from

$$\lambda - M^2_{max} - M_{max} = 0, \qquad \lambda - M^2_{min} + M_{min} = 0. \tag{2.45}$$

Since $M_{max} \geq M_{min}$ the only possible pair of solutions is

$$M_{max} = -\frac{1}{2} + \sqrt{\frac{1}{4} + \lambda}, \qquad M_{min} = \frac{1}{2} - \sqrt{\frac{1}{4} + \lambda}. \tag{2.46}$$

We see that $M_{min} = -M_{max}$. Starting with raising and lowering operators from any M, we see that M_{max} and M_{min} must be separated by an integral number of steps of one. It follows that

$$M_{max} - M_{min} = 2M_{max} = -1 + 2\sqrt{\frac{1}{4} + \lambda} = n,$$

$$n = 1, 2, 3, \ldots . \tag{2.47}$$

This yields the possible values of λ and M_{max} as

$$\lambda = \tfrac{1}{2}n(\tfrac{1}{2}n + 1), \qquad M_{max} = \tfrac{1}{2}n. \tag{2.48}$$

We denote $\frac{1}{2}n \equiv J$ as the standard quantum number for the magnitude of the angular momentum. Through our purely algebraic procedure we have thus established the quantum-number spectra

$$\begin{aligned} J &= 0, \tfrac{1}{2}, 1, \tfrac{3}{2}, 2, \ldots; \\ M &= J, J-1, J-2, \ldots, -J+1, -J \quad \text{for each } J. \end{aligned} \tag{2.49}$$

The usual appearance of angular-momentum results is obtained by replacing the eigenvalue λ by $J(J+1)$ and the quantum number λ by J in the above equations. For example, the raising and lowering relations (2.40) then become

$$\begin{aligned} J_\pm|JM\rangle &= \sqrt{J(J+1) - M^2 \mp M}\,|J, M \pm 1\rangle \\ &= \sqrt{(J \pm M + 1)(J \mp M)}\,|J, M \pm 1\rangle. \end{aligned} \tag{2.50}$$

2.3.3. Angular Momentum and Rotations: Group O(3)

Angular momentum is intimately associated with rotations in ordinary three-dimensional space. Without delving into details, we state the unitary rotation operator $R(\mathbf{n}, \Theta)$ as follows:

$$\psi' = R(\mathbf{n}, \Theta)\psi = \exp(-i\Theta\mathbf{n}\cdot\mathbf{J})\psi. \tag{2.51}$$

Here a general wave function ψ is rotated, in a fixed coordinate system, about a direction specified by the unit vector $\mathbf{n}$ through an angle Θ, which results in a new wave function ψ'. The operator $\mathbf{J}$ is the *total* angular momentum of the system. Each rotation (2.51) is an *element* of the group consisting of all possible rotations; the group thus has an infinite number of elements and it is a continuous group.

If the rotation (2.51) occurs about the z axis through an infinitesimal angle ϵ, we have

$$\psi' = (1 - i\epsilon J_z)\psi; \tag{2.52}$$

similar expressions obtain for infinitesimal rotations about the x and y axes. This role of the angular-momentum operators $J_{x,y,z}$ is described by calling them *generators* of spatial rotations. This usage coincides with the group-theoretical concept of generator introduced in Subsection 2.3.1. Three-dimensional rotations clearly have the group property that consecutive rotations can be replaced by a single rotation leading to the same orientation of the system.

Alternative to the description (2.51), spatial rotations of a *vector* can be represented by three-by-three *orthogonal* matrices. For example, when vector $\mathbf{r}$ is subjected to the rotation (2.52), the result can be written as

$$\begin{pmatrix} x' \\ y' \\ z' \end{pmatrix} = \begin{pmatrix} 1 & -\epsilon & 0 \\ \epsilon & 1 & 0 \\ 0 & 0 & 1 \end{pmatrix} \begin{pmatrix} x \\ y \\ z \end{pmatrix}. \tag{2.53}$$

The three-dimensional rotation group is therefore called $O(3)$. When emphasizing that only proper rotations are intended and reflections are excluded, the notation is made more precise: $SO(3)$ or $O^+(3)$. The notation $R(3)$ also occurs in the literature. The group of proper rotations is of course a subgroup of the complete $O(3)$, i.e., $O(3) \supset SO(3)$. In what follows we say simply $O(3)$, as is customary in the IBM literature. We also note that rotations restricted to two dimensions form a subgroup of $O(3)$: $O(3) \supset O(2)$.

Comparing eqs. (2.28) and (2.31) we see that the isospin components $T_{1,2,3}$ obey the same commutation relations as the angular-momentum components $J_{x,y,z}$. Thus they have the same Lie algebra. Yet the group we associated with isospin was $SU(2)$ while that associated with angular momentum is $O(3)$. The groups are in fact essentially the same. Such similarity is called *isomorphism*, and we denote it in the present case as $O(3) \approx SU(2)$. (We ignore the distinction between isomorphism and homomorphism, discussed in the literature, since it is of no consequence for our present purposes.)

2.3.4. Representations of Groups

The basis of angular-momentum eigenstates $|JM\rangle$ provides *representations* of the group $O(3)$. Our first encounter with the notion of representation was in Subsection 2.3.1, where we had the $U(2)$ representation consisting of the basis states $|\mathrm{p}\rangle$ and $|\mathrm{n}\rangle$. A representation can thus be conceived in terms of a particular set of basis states.

Alternatively, a representation of a group can be construed in terms of the matrices of the generators. The connection to the concept of a representation expressed in terms of basis states is clear because the matrices are constructed with the same basis states. In the present case of the rotation group $O(3)$, the generators are $J_{x,y,z}$, or equivalently $J_\pm$ and J_z. Equations (2.50) and (2.33) yield the matrices according to

$$\begin{aligned} \langle J'M'|J_\pm|JM\rangle &= \sqrt{(J \pm M + 1)(J \mp M)}\delta_{J'J}\delta_{M',M\pm 1}, \\ \langle J'M'|J_z|JM\rangle &= M\delta_{J'J}\delta_{M'M}. \end{aligned} \tag{2.54}$$

The matrix elements do not connect states with different values of J. If we then use a basis with a single J, we have an *irreducible* representation, often called *irrep* for short in the literature. The dimension of an $O(3)$ irrep is $2J + 1$ since there are that many M values, eq. (2.49).

As a concrete sample, the matrices (2.54) for $J = 1$ are

$$\begin{aligned} (J_+) &= \begin{pmatrix} 0 & \sqrt{2} & 0 \\ 0 & 0 & \sqrt{2} \\ 0 & 0 & 0 \end{pmatrix}, \quad (J_-) = \begin{pmatrix} 0 & 0 & 0 \\ \sqrt{2} & 0 & 0 \\ 0 & \sqrt{2} & 0 \end{pmatrix}, \\ (J_z) &= \begin{pmatrix} 1 & 0 & 0 \\ 0 & 0 & 0 \\ 0 & 0 & -1 \end{pmatrix}. \end{aligned} \tag{2.55}$$

This is the three-dimensional irrep of $O(3)$, or equivalently of

$SU(2)$. [It is also the fundamental representation of $O(3)$.] Likewise for $J = 0$ we have a (trivial) one-dimensional irrep, for $J = \frac{1}{2}$ a two-dimensional irrep (Pauli spin matrices), and so forth. If we had elected a set of basis states with more than one J value, the matrices would split into disconnected blocks, one for each J; that would be a reducible representation. In practice we talk mostly about irreps and often call them just representations.

Our formulation of isospin in Subsection 2.3.1 gives for one nucleon a two-dimensional representation with $T = \frac{1}{2}$; the states $|TM_T\rangle$ are $|\mathrm{p}\rangle = |\frac{1}{2}\frac{1}{2}\rangle$ and $|\mathrm{n}\rangle = |\frac{1}{2}, -\frac{1}{2}\rangle$. In order to have states containing more than one nucleon of each kind, we must include in the single-particle quantum numbers not only isospin but all spatial quantum numbers as well. Only then can we have representations higher than $T = \frac{1}{2}$. Let k stand for a complete set of the spatial quantum numbers. The operators (2.27) are then dressed to the form $T_+ = \Sigma_k\, a^\dagger_{\mathrm{p}k} a_{\mathrm{n}k}$ etc., and they are still seen to obey the $U(2)$ Lie algebra (2.28). Isospin representations $T = 1, \frac{3}{2}, \ldots$ thus become possible. Their form is identical to the angular-momentum representations $J = 1, \frac{3}{2}, \ldots$; e.g., the $T = 1$ matrix representation is given by eq. (2.55).

Representation matrices are also defined for finite rotations. The matrix elements are then

$$\langle J'M'|R|JM\rangle \equiv D^J_{M'M}, \tag{2.56}$$

where the rotation operator R of eq. (2.51) is usually parametrized in terms of the Euler angles α, β, γ. The complex conjugates $D^{J*}_{MK}(\alpha\beta\gamma)$ play an important direct physical role as rotational wave functions $|JMK\rangle$ where K is the angular-momentum component along the symmetry axis of the rotating body. The quantum rotor will be discussed in Chapter 7.

The preceding discussion demonstrates how a given basis defines a matrix representation of the generators or elements of a group. In our $O(3)$ example the only Casimir operator is the quadratic one $\mathbf{J}^2$, and its quantum number J completely labels each representation. Within a given J representation the quantum number M then labels the basis states. In fact we can generally discuss representations without actually constructing matrices since the basis states with their quantum numbers carry the complete information.

In the general case a group has several commuting generators, whose quantum numbers (like M) label the basis states. The number of commuting generators is called the *rank* r of the group.

There are then also the same number r of commuting Casimir operators, whose quantum numbers (like J) serve to label the representations. The structure of the basis states is accordingly $|J_1 J_2 \cdots J_r, M_1 M_2 \cdots M_r\rangle$. The J_i are *representation labels* and the M_i *basis-state labels* applicable within a representation.

The group $O(3) \approx SU(2)$ thus has rank 1 and indeed just one Casimir operator. The group $U(2)$ has two commuting generators $\hat{N}$ and T_3, whence $r = 2$, and two commuting Casimir operators $\hat{N}$ and $\mathbf{T}^2$. Their eigenvalue equations are

$$\begin{aligned} \hat{N}|NT, NM_T\rangle &= N|NT, NM_T\rangle, \\ \mathbf{T}^2|NT, NM_T\rangle &= T(T+1)|NT, NM_T\rangle, \end{aligned} \tag{2.57}$$

where we have omitted spatial quantum numbers. The nucleon number N is written in also as an M-type quantum number because the space is one-dimensional with respect to N; see eq. (2.29). The situation is analogous to two-dimensional angular momentum, group $O(2)$, where M labels the eigenvalue of the Casimir operator J_z and also its "component."

2.3.5. Summary of Group-Theoretic Procedure

From the discussion in this section emerges our practical program for applying group theory to the IBM model. We first write all relevant bilinear operators. Then we form all commutators among them and identify the Lie algebra they constitute. We proceed to identify possible chains of subgroups. The next task is to determine all Casimir operators. Then the simultaneous eigenvalue problem is solved for the Casimir operators and the commuting generators. Their quantum numbers provide the representation and basis-state labels. The algebraic treatment of angular momentum in Subsection 2.3.2 is the prototype of such a solution.

The next question is: what representations of a subgroup belong to a given representation of the larger group? Answering this question is known as the *reduction* of a group with respect to its subgroup. Again our angular-momentum problem serves as a concrete example. The representations of $O(3)$ are labeled by J and those of its subgroup $O(2)$ by $M = J, J-1, \ldots, -J$. Note that the basis-state label M of $O(3)$ is the representation label of $O(2)$.

For a visual impression of the flow of group reduction it is helpful to indicate the representation labels of each group below that group symbol in a subgroup chain. Thus illustrated, our angular-momentum example reads

$$\underset{J}{O(3)} \supset \underset{M}{O(2)} \tag{2.58}$$

and the values of M must be separately listed.

2.4. THE GENERAL UNITARY GROUP $U(n)$

2.4.1. Algebra and Casimir Operators

As an extrapolation of our foregoing $U(2)$ example we now consider the general group $U(n)$. The dimension n designates the number of single-particle states involved. Typically in the context of nuclear structure we could have $n = 2j + 1$ referring to the degenerate substates of a shell-model orbital j (Section 2.8, Chapters 5 and 7); or we could have $n = 6$ referring to the six basic boson states of the IBM model (Section 2.5).

In the case of n single-fermion states the simplest set of generators of $U(n)$ are the bilinear operators $a_i^\dagger a_j$ with $i, j = 1, 2, \ldots, n$ whose factors anticommute according to eq. (2.20). The commutators forming the $U(n)$ Lie algebra are

$$[a_i^\dagger a_j, a_k^\dagger a_l] = \delta_{jk} a_i^\dagger a_l - \delta_{il} a_k^\dagger a_j. \tag{2.59}$$

Any n^2 linearly independent linear combinations of the $a_i^\dagger a_j$ will also close under commutation and provide an alternative set of generators of $U(n)$.

If instead of fermions the particles are bosons, whose commutation relations are given by eqs. (2.5), (2.7) and (2.8), the generators are $b_i^\dagger b_j$ with $i, j = 1, 2, \ldots, n$ and the Lie algebra is

$$[b_i^\dagger b_j, b_k^\dagger b_l] = \delta_{jk} b_i^\dagger b_l - \delta_{il} b_k^\dagger b_j. \tag{2.60}$$

We thus see that bosons and fermions give the *same* Lie algebra $U(n)$, although the single-particle operators commute and anticommute, respectively.

The rank of $U(n)$ is n. This means that there are n commuting operators among the n^2 generators. The simplest set of them for fermions is $a_i^\dagger a_i$, $i = 1, 2, \ldots, n$ and similarly for bosons. Other equivalent sets can be defined as arbitrary linearly independent linear combinations of these n generators. Furthermore, there are n

Casimir operators. By definition they commute with all the generators $a_i^\dagger a_j$, whence they also commute with each other. As can be verified by the defining property and eq. (2.59), a complete set of Casimir operators for fermions is

$$\begin{aligned}
C_{1Un} &= \sum_i a_i^\dagger a_i, \\
C_{2Un} &= \sum_{ij} a_i^\dagger a_j a_j^\dagger a_i, \\
C_{3Un} &= \sum_{ijk} a_i^\dagger a_j a_j^\dagger a_k a_k^\dagger a_i, \\
&\vdots \\
C_{nUn} &= \sum_{ijk\ldots} a_i^\dagger a_j a_j^\dagger a_k a_k^\dagger \cdots a_i.
\end{aligned} \tag{2.61}$$

Because eq. (2.60) is identical in form to eq. (2.59), the $U(n)$ Casimir operators for bosons also have the form (2.61), with boson operators replacing the fermion operators.

In view of their defining property, the Casimir operators of $U(n)$ are not unique. Clearly the ones given by eq. (2.61) can be multiplied by arbitrary constants. Moreover, $aC_{2Un} + bC_{1Un}^2$ with arbitrary constants a and b is an equally legitimate quadratic Casimir operator as C_{2Un}. In general practice, the notation appearing on the left-hand side of eq. (2.61) need not mean literally the right-hand sides.

The linear Casimir operator $C_{1Un} = \Sigma_i\, a_i^\dagger a_i$ is recognized [eq. (2.12)] as the operator $\hat{N}$ for the total number of particles. Moreover, it actually duplicates one of the n commuting generators, since it is possible to choose the latter so as to include $\Sigma_i\, a_i^\dagger a_i$ in the independent linear combinations of the $a_i^\dagger a_i$. In fact, we already recognized this in our detailed example of $U(2)$ in Subsections 2.3.1 and 2.3.4. The same applies of course to bosons.

An important property follows from the fact that all the generators of a group commute with its Casimir operators. The latter define a representation. In other words, the Casimir operators have common eigenstates. Now operating with a generator on such a state will not change the representation labels, only the basis-state labels, in the terminology of Subsection 2.3.4. Equation (2.37) is the basic example of this. The general conclusion regarding matrix elements is that *a generator of a group cannot connect different representations of the group*.

The number operator $\hat{N}$ can be separated from the generators of $U(n)$. The remaining $n^2 - 1$ bilinear operators close on commutation and form the algebra $SU(n)$ of rank $n - 1$: $U(n) = SU(n) \times U(1)$. The $SU(n)$ groups, in particular $SU(2)$ and $SU(3)$, find application in many areas of physics.

2.4.2. Quantum Numbers, Symmetries and Young Tableaux

From the preceding we see that $U(n)$ states are labeled by n quantum numbers of "J type." They are associated with the n Casimir operators and specify a representation of $U(n)$. Furthermore the states carry another n quantum numbers of "M type." They are associated with the n commuting generators and serve to label a basis state within a representation. It is instructive to look back at this point and check that our $U(2)$ example of Section 2.3 conforms to the general properties of $U(n)$ given here.

We note here a complication that occurs for $n \geq 3$. The M-type quantum numbers do not generally suffice for labeling the basis states within a given representation. Additional labels are needed for a complete specification. They can be chosen for convenience as suggested by the physics. In particular such labels may be provided by the representation labels of the subgroups of $U(n)$.

The quantum states we have in mind are many-particle states where specified numbers of particles occupy each single-particle state. For $U(n)$ we have n single-particle states available. The states with good $U(n)$ quantum numbers are states of a definite symmetry pattern under permutation of particle labels.

The complete overall wave function of a system of identical fermions (bosons) can only be antisymmetric (symmetric). At that level $U(n)$ symmetry has little variety. However, wave functions are often constructed from parts belonging to different spaces: coordinate space and spin space in atomic physics; mechanical space and isospin space in the nuclear shell model; mechanical boson space and F-spin space in the proton-neutron IBM (see Chapter 4). Then a variety of symmetry is possible in the single spaces; only the combined wave function must have pure antisymmetry (symmetry) for fermions (bosons).

The symmetry type of a state is given by the set of J-type quantum numbers. To define them, we first define integers $\lambda_1 \geq \lambda_2 \geq$

$\cdots \geq \lambda_n \geq 0$ such that their sum is the total number N of particles in the state. Since N can be associated with the commuting generators instead of the Casimir operators (see above), we can say, for the purpose of the ensuing argument, that the λ_i provide $n - 1$ quantum numbers of the J type. Their precise meaning will be explained below. The symmetry pattern implied by the λ_i is denoted as $[\lambda_1\lambda_2 \cdots \lambda_n]$ and a pictorial rendition of it is known as a *Young pattern*. It consists of rows of boxes, of length $\lambda_1, \lambda_2, \ldots$, placed on top of each other, with the λ_1 row at the top etc. An example is the $U(3)$, $N = 8$ Young pattern [431]:

(2.62)

The still missing last J-type quantum number consists of a definite assignment of particle labels to the eight squares of the pattern (2.62). The rule is to start from the upper left corner with 1 and then have the numbers increase right and down. Three possibilities are

$$\begin{array}{|c|c|c|c|}\hline 1&2&3&4\\\hline 5&6&7\\\cline{1-3} 8\\\cline{1-1}\end{array} \qquad \begin{array}{|c|c|c|c|}\hline 1&2&4&5\\\hline 3&6&7\\\cline{1-3} 8\\\cline{1-1}\end{array} \qquad \begin{array}{|c|c|c|c|}\hline 1&3&4&5\\\hline 2&6&7\\\cline{1-3} 8\\\cline{1-1}\end{array} \tag{2.63}$$

and there are 70 of them altogether. Each of them stands for a definite representation of $U(3)$. The detailed symmetries are complicated, but the leading idea is that the horizontal direction indicates symmetry, the vertical direction antisymmetry; see the simple examples in Subsection 2.4.3. Young patterns with particle labels are known as *Young tableaux*, or, fully specified, Young particle-number tableaux.

We record here the easy but important Young patterns for complete symmetry and complete antisymmetry. Within our example of $n = 3$ and $N = 8$, complete symmetry reads $[800] \equiv [8]$. Within this example we cannot have a completely antisymmetric state, as we will discover below. If N were 3, we could: $[111] \equiv [1^3]$. These simple symmetries have the appearance

$$[8] = \begin{array}{|c|c|c|c|c|c|c|c|}\hline &&&&&&&\\\hline\end{array} \qquad [1^3] = \begin{array}{|c|}\hline \\\hline \\\hline \\\hline\end{array} \tag{2.64}$$

It is also possible to embellish a Young pattern with a single-particle quantum number in each square. The result is a Young quantum-number tableau. The occupation number of each single-particle state can be seen from it. For $U(n)$ there are n of them, and they are the M-type quantum numbers which, apart from generally needed additional labels, complete the specification of the state. Staying within our $U(3)$ example, we have for each representation, such as those displayed in eq. (2.63), quantum-number tableaux

$$\begin{array}{|c|c|c|c|}\hline a & a & a & a \\ \hline b & b & b & \\ \cline{1-3} c & & & \\ \cline{1-1}\end{array} \qquad \begin{array}{|c|c|c|c|}\hline b & b & b & b \\ \hline a & a & a & \\ \cline{1-3} c & & & \\ \cline{1-1}\end{array} \qquad \begin{array}{|c|c|c|c|}\hline c & c & c & c \\ \hline a & a & a & \\ \cline{1-3} b & & & \\ \cline{1-1}\end{array} \tag{2.65}$$

and three more permutations of the single-particle quantum numbers a, b, c. On the analogy to angular momentum, each of these quantum-number tableaux is like a value of M while the fixed particle-number tableau to which they belong is like a particular J. The specific three M-type quantum numbers in the present example are the occupation numbers of the three single-particle states; e.g., for the first tableau $n_a = 4$, $n_b = 3$, $n_c = 1$.

The one rule observed in drawing up the quantum-number tableaux (2.65) is that only *different* quantum numbers may appear in the same column. This is because the vertical direction implies antisymmetry and antisymmetric states do not allow repeated quantum numbers (Pauli principle). This rule restricts the possible occupation numbers, analogously to the restriction $|M| \leq J$, so that, e.g., $n_a = 5$, $n_b = 2$, $n_c = 1$ is not allowed.

The role of the extra quantum numbers can be demonstrated within the present example. Each state (2.65), for a given particle-number tableau, is uniquely defined by its configuration; thus, e.g., the first configuration a^4b^3c cannot be placed in the tableau in any other way. However, the type of state displayed is not general. The same shape can accommodate more configurations $a^k b^n c^m$ such that $k + n + m = N = 8$ (15 in all).

An example is $a^3b^3c^2$. For this configuration there are *two* distinct quantum-number tableaux, with rows aaab, bbc, c and aaac, bbb, c, respectively. They stand for two orthogonal states, so we have here effectively an extra label which sets them apart. We have here used the additional rule, necessary for avoiding double counting, that both horizontally and vertically a is placed before b and c, and b is placed before c.

The particle-number tableau and quantum-number tableau, of

the same shape, together contain a unique definition of a $U(n)$ basis state. The particle-number tableau constitutes the J-type quantum numbers defining the representation. The quantum-number tableau constitutes the M-type quantum numbers (occupation numbers) and the necessary extra labels defining the particular basis state within the representation.

2.4.3. Detailed Example: $U(2)$

The example used so far is too complicated for threshing out the detailed construction of wave functions. We therefore return to the isospin example of Section 2.3, i.e., the group $U(2)$. There we only had the isospin space of basis states $|\mathrm{p}\rangle \equiv \mathrm{p}$ and $|\mathrm{n}\rangle \equiv \mathrm{n}$; the only antisymmetric state is then $2^{-1/2}(\mathrm{pn} - \mathrm{np})$. In real nuclear physics we can have any isospin symmetry together with a mechanical symmetry as long as the complete combined wave function is antisymmetric. Below we treat in detail the $U(2)$ cases $N = 2$ and $N = 3$. Although the isospin quantum number T is incidental to the present argument, we carry it along for completeness and in preparation for the study of analogous F spin in Chapter 4.

For the $N = 2$ system the complete unsymmetrized basis is pp, pn, np, nn. Thence we have the states of good exchange symmetry

$$\mathrm{pp}, \quad \sqrt{\tfrac{1}{2}}\,(\mathrm{pn} + \mathrm{np}), \quad \mathrm{nn}; \qquad \sqrt{\tfrac{1}{2}}\,(\mathrm{pn} - \mathrm{np}). \tag{2.66}$$

These have the triplet–singlet form familiar from spin wave functions. The first three are symmetric, of Young pattern [2] and isospin $T = 1$, with $M_T = +1, 0, -1$, respectively. The last one is antisymmetric, of Young pattern $[1^2]$ and isospin $T = 0$, $M_T = 0$. The respective pairs of quantum-number and particle-number tableaux are

$$\begin{array}{llll} \begin{array}{|c|c|}\hline \mathrm{p} & \mathrm{p} \\ \hline \end{array} & \begin{array}{|c|c|}\hline 1 & 2 \\ \hline \end{array} & \qquad \begin{array}{|c|c|}\hline \mathrm{p} & \mathrm{n} \\ \hline \end{array} & \begin{array}{|c|c|}\hline 1 & 2 \\ \hline \end{array} \\ \begin{array}{|c|c|}\hline \mathrm{n} & \mathrm{n} \\ \hline \end{array} & \begin{array}{|c|c|}\hline 1 & 2 \\ \hline \end{array} & \qquad \begin{array}{|c|}\hline \mathrm{p} \\ \hline \mathrm{n} \\ \hline \end{array} & \begin{array}{|c|}\hline 1 \\ \hline 2 \\ \hline \end{array} \end{array} \tag{2.67}$$

Next we consider $N = 3$. The simplest configurations are p^3 and n^3. Their only possible states are ppp and nnn, both of Young pattern [3] and isospin $T = \frac{3}{2}$, $M_T = \pm\frac{3}{2}$. The quantum-number and particle-number tableaux are

$$\begin{array}{|c|c|c|}\hline \mathrm{p} & \mathrm{p} & \mathrm{p} \\ \hline \end{array} \quad \begin{array}{|c|c|c|}\hline 1 & 2 & 3 \\ \hline \end{array} \qquad \begin{array}{|c|c|c|}\hline \mathrm{n} & \mathrm{n} & \mathrm{n} \\ \hline \end{array} \quad \begin{array}{|c|c|c|}\hline 1 & 2 & 3 \\ \hline \end{array} \tag{2.68}$$

The remaining $N = 3$ configurations p^2n and pn^2 lead to more interesting symmetries. Again the simplest states are fully symmetric. Their Young pattern is [3] and isospin $T = \frac{3}{2}$, $M_T = \pm\frac{1}{2}$ with tableaux

$$\begin{array}{|c|c|c|}\hline \mathrm{p} & \mathrm{p} & \mathrm{n} \\ \hline \end{array} \quad \begin{array}{|c|c|c|}\hline 1 & 2 & 3 \\ \hline \end{array} \qquad \begin{array}{|c|c|c|}\hline \mathrm{n} & \mathrm{n} & \mathrm{p} \\ \hline \end{array} \quad \begin{array}{|c|c|c|}\hline 1 & 2 & 3 \\ \hline \end{array} \tag{2.69}$$

The normalized wave functions are

$$\sqrt{\tfrac{1}{3}}\,(\mathrm{ppn} + \mathrm{pnp} + \mathrm{npp}), \qquad \sqrt{\tfrac{1}{3}}\,(\mathrm{pnn} + \mathrm{nnp} + \mathrm{npn}). \tag{2.70}$$

The next symmetry possibility for p^2n and np^2 is represented by the pattern [21] with tableaux

$$\begin{array}{|c|c|}\hline \mathrm{p} & \mathrm{p} \\ \hline \mathrm{n} \\ \cline{1-1} \end{array} \quad \begin{array}{|c|c|}\hline 1 & 2 \\ \hline 3 \\ \cline{1-1} \end{array} \qquad \begin{array}{|c|c|}\hline \mathrm{n} & \mathrm{n} \\ \hline \mathrm{p} \\ \cline{1-1} \end{array} \quad \begin{array}{|c|c|}\hline 1 & 2 \\ \hline 3 \\ \cline{1-1} \end{array} \tag{2.71}$$

The particle-number tableaux mean that the wave functions are symmetric under exchange of 1 and 2; they have no symmetry with respect to 3. We construct the explicit wave functions by insisting on the $1 \leftrightarrow 2$ symmetry and by making the functions orthogonal to the symmetric states (2.70). We thus find the normalized wave functions

$$\sqrt{\tfrac{1}{6}}\,(2\mathrm{ppn} - \mathrm{pnp} - \mathrm{npp}), \qquad \sqrt{\tfrac{1}{6}}\,(2\mathrm{nnp} - \mathrm{npn} - \mathrm{pnn}). \tag{2.72}$$

These states have isospin $T = \frac{1}{2}$. This can be seen by coupling, with Clebsch–Gordan coefficients, the isospin of particles 1 and 2, $T_{12} = 1$, to the isospin $\frac{1}{2}$ of particle 3 so that the total isospin is $T = \frac{1}{2}$ with component $M_T = \pm\frac{1}{2}$. The result is, to within an overall sign, that given by eq. (2.72). The result justifies the simple algorithm of saying that the p and n on top of each other in the quantum-number tableaux (2.71) "cancel" each other so that only one $T = \frac{1}{2}$ remains.

The second possibility for the pattern [21] is

$$\begin{array}{|c|c|}\hline \mathrm{p} & \mathrm{p} \\ \hline \mathrm{n} \\ \cline{1-1} \end{array} \quad \begin{array}{|c|c|}\hline 1 & 3 \\ \hline 2 \\ \cline{1-1} \end{array} \qquad \begin{array}{|c|c|}\hline \mathrm{n} & \mathrm{n} \\ \hline \mathrm{p} \\ \cline{1-1} \end{array} \quad \begin{array}{|c|c|}\hline 1 & 3 \\ \hline 2 \\ \cline{1-1} \end{array} \tag{2.73}$$

Now the particle-number tableaux mean that these states are antisymmetric under the exchange $1 \leftrightarrow 2$ with no symmetry regarding

3. They are to be orthogonal both to eq. (2.70) and to eq. (2.72), and normalized. The result is

$$\sqrt{\tfrac{1}{2}}\,(\text{pnp} - \text{npp}), \qquad \sqrt{\tfrac{1}{2}}\,(\text{npn} - \text{pnn}). \tag{2.74}$$

By the algorithm conjectured after eq. (2.72), the isospin is again $T = \frac{1}{2}$, $M_T = \pm\frac{1}{2}$. The truth of this can be seen by the trivial coupling of $T_{12} = 0$ to the remaining isospin $\frac{1}{2}$. We note at this juncture that the [21] states of eqs. (2.72) and (2.74) will be useful as prototype examples of the *mixed-symmetry* states of IBM-2, discussed in Chapter 4.

One might expect one further pattern, namely the fully antisymmetric one $[1^3]$. However, with only two basis states p and n signifying the $U(2)$ algebra, it is impossible to construct such a state. This is an example of the general rule stated following eq. (2.65). We can also see that the orthonormal states (2.70), (2.72) and (2.74), three for each configuration, exhaust all possibilities because the unsymmetrized space for each configuration is three-dimensional: ppn, pnp, npn; pnn, nnp, npn. In fact, we can see as a check that for $N = 3$ there are altogether eight unsymmetrized basis states and eight states with good $U(2)$ symmetry: ppp, nnn and the pairs in eqs. (2.70), (2.72) and (2.74).

Our foregoing example does not give an exhaustive algorithm for constructing $U(n)$ wave functions, although it does set forth the principles. For general rules, consult the technical literature. Fortunately, in the present context of nuclear structure, we seldom need to handle complicated cases explicitly.

The unitary transformations of $U(2)$ mix states having the same Young particle-number tableau, i.e., states belonging to the same representation of $U(2)$. It is like the Y_{lm} with fixed l mix under rotations. So the states ppp, nnn, and the pair (2.70) define a four-dimensional representation of $U(2)$; the pair (2.72) a two-dimensional one; and the pair (2.74) another two-dimensional one. It is an instructive exercise to apply eq. (2.23) to, say, the first state (2.74) and find that the result is a linear combination of the two states (2.74). The states belonging to a given Young particle-number tableau are said to form a *tensor representation* of the group.

We conclude this subsection by a subsidiary remark of vital importance. The Young patterns, without regard to quantum numbers, define representations of the discrete group of permutations, the so-called *symmetric group* S_n. For example, [3] is a representation of S_3; it is one-dimensional because it has only one particle-number tableau. Then [21] is a two-dimensional representation of S_3: permutations of the particle numbers 123 mix the two tableaux. In $U(2)$ we could not have $[1^3]$ but with $U(3)$ we would have one such state, and $[1^3]$ is another one-dimensional representation of S_3. The same Young patterns are thus relevant to unitary and symmetric groups.

Permutations connect the various particle-number tableaux belonging to the same Young pattern. Any tableau can be obtained from any other by a suitable linear combination of permutation operators. It follows that if one of the tableaux is impossible for a given set of quantum numbers, the whole pattern is ruled out. This in effect proves the rule we stated for quantum-number tableaux following eq. (2.65).

Altogether a rather sweeping conclusion emerges. Since we are concerned with systems of identical particles, the wave functions must be essentially insensitive to permutations of the particle labels. Therefore a $U(n)$ or an $SU(n)$ wave function is associated with a definite Young pattern rather than with a particular particle-number tableau. This point will be fortified by the following discussion of the construction of the overall wave function.

2.4.4. Combining Spaces into Simple Symmetry

We finally consider the construction of overall states of simple symmetry: antisymmetry for fermions, full symmetry for bosons. We do this in necessary detail within our preceding $U(2)$ example of nuclear isospin. In this case the overall wave function must be antisymmetric.

For overall antisymmetry, a fully symmetric (antisymmetric) isospin function must be multiplied by an antisymmetric (fully symmetric) mechanical function. Accordingly an isospin function of symmetry [2] ($[1^2]$) must be multiplied with a mechanical function of symmetry $[1^2]$ ([2]). Likewise isospin symmetry [3] is to multiply mechanical symmetry $[1^3]$; the converse case does not exist.

For the mixed-symmetry cases [21] the construction is more complicated. As in the simple-symmetry cases, the building blocks are again *conjugate* tableaux. For [21] they are the two different particle-number tableaux belonging to it. The tableaux and wave functions are given in eqs. (2.71)–(2.74).

Now we need explicit mechanical wave functions. They must

have the symmetry structure of eqs. (2.72) and (2.74). We thus assume two mechanical single-particle states and call them s and d. We choose isospin configuration p^2n and mechanical configuration s^2d. The relevant mechanical states are thus

$$\sqrt{\tfrac{1}{6}}\,(2ssd - sds - dss), \qquad \sqrt{\tfrac{1}{2}}\,(sds - dss), \tag{2.75}$$

$1 \leftrightarrow 2$ symmetric and $1 \leftrightarrow 2$ antisymmetric, respectively.

The desired overall antisymmetric wave function is obtained as

$$\begin{aligned}\frac{1}{\sqrt{2}}\Big[&\sqrt{\tfrac{1}{6}}\,(2ssd - sds - dss)\cdot\sqrt{\tfrac{1}{2}}\,(pnp - npp)\\ &- \sqrt{\tfrac{1}{2}}\,(sds - dss)\cdot\sqrt{\tfrac{1}{6}}\,(2ppn - pnp - npp)\Big] \qquad (2.76)\\ &= \sqrt{\tfrac{1}{6}}\,(s_ps_nd_p - s_ns_pd_p + s_nd_ps_p - d_ps_ns_p + d_ps_ps_n - s_pd_ps_n)\\ &\equiv |s_ps_nd_p\rangle_{as},\end{aligned}$$

where $s_ps_nd_p \equiv ssd\cdot pnp$ etc. Our result $|s_ps_nd_p\rangle_{as}$ is normalized and it has precisely the form (2.18). In multiplying out the left-hand side of eq. (2.76), terms appear for the configuration $s_p^2d_n$ but they cancel out, as they should consistent with the Pauli principle.

The construction of a fully symmetric overall wave function for a boson system is similar. Then wave functions of the *same* symmetry are multiplied and the products are added. We conclude our discussion of exchange symmetries by reminding ourselves that only simple final symmetries are allowed; any mixed symmetry in one space must be compensated by the corresponding mixed symmetry in the other space.

2.5 THE GROUP $U(6)$ IN THE IBM

Rather than continue with the general group $U(n)$, we now specialize to the group $U(6)$. The reason is twofold. For one, $U(6)$ is large enough a group so as to display all essential general properties of $U(n)$, beyond those already discussed in Section 2.4. For another, our primary application the IBM has precisely the group structure $U(6)$, as discussed below.

2.5.1 The Operators of the IBM

The elementary excitations of the IBM are single-particle states of identical bosons. Microscopically the bosons are constructed from pairs of like nucleons, as discussed in Chapter 4. However, in the basic version of the model, IBM-1, the bosons are taken as given phenomenological objects. Moreover, IBM-1 even ignores the difference between protons and neutrons.

For the time being, we focus on the bosons of IBM-1; until otherwise indicated, by IBM we mean IBM-1. A single boson has available to it two angular-momentum states, an s state ($J = 0$) and a d state ($J = 2$), the latter with five substates $M = 2, 1, 0, -1, -2$. So there are altogether *six* states available to a boson. They define a six-dimensional space just like the states p(roton) and n(eutron) defined a two-dimensional space in the isospin example of Section 2.3. Unitary transformations, i.e., complex rotations, on a vector in that space form the group $U(6)$.

Some remarks are in order. The group $U(6)$ appears in other physical contexts as well. An example is a shell-model nucleon with $j = \frac{5}{2}$ with its six substates $m = \pm\frac{5}{2}, \pm\frac{3}{2}, \pm\frac{1}{2}$. The present sd-boson context is one of many realizations of the group. Another remark is that the IBM literature speaks of "s and d bosons"; we will say likewise for simplicity. This is to be understood as a figure of speech since the bosons are considered identical and full symmetry is required of their wave function. Our third remark is that extensions of the model (Chapter 6) include states beyond s and d.

The basic operators of the IBM are the creation operators $s^\dagger$ and $d^\dagger_\mu$ with their Hermitian conjugates, i.e., annihilation operators, s and d_μ. They obey the boson commutation relations (2.5), (2.7), and (2.8), where the indices i, j carry s and the five d's on an equal footing. Their bilinear products form the basic set of 36 generators of $U(6)$:

$$s^\dagger s, \quad s^\dagger d_\mu, \quad d^\dagger_\mu s, \quad d^\dagger_\mu d_\nu; \quad \mu, \nu = \pm 2, \pm 1, 0. \tag{2.77}$$

These generators close on commutation according to eq. (2.60).

In preparation for an alternative formulation of the generators, we now shortly study the tensorial properties of our creation and annihilation operators. When operating on the vacuum, $s^\dagger$ gives a state s which has angular momentum $J = 0$. When $d^\dagger_\mu$ operates on the vacuum, the result is an eigenstate d_μ of the angular momentum with $J = 2$, $M = \mu$. The vacuum of the IBM consists of the closed major shells and we denote it by $|\tilde{0}\rangle$; its role in the model is that of a passive background. The basic s and d states are then

$$|\mathrm{s}\rangle = s^\dagger|\tilde{0}\rangle = |J = 0, M = 0\rangle, \qquad (2.78)$$
$$|\mathrm{d}_\mu\rangle = d^\dagger_\mu|\tilde{0}\rangle = |J = 2, M = \mu\rangle.$$

Accordingly the operators $s^\dagger$ and s are scalars, i.e., spherical tensors of degree 0. We now speak of tensors as defined by their angular-momentum contents, or equivalently, by their behavior under rotations; see the Appendix. Compared to the spherical harmonics $Y_{lm}(\theta, \phi)$, which are the prototype spherical tensors, $s^\dagger$ is like Y_{00} $(= 1/\sqrt{4\pi})$ and s like Y^*_{00}. Similarly $d^\dagger_\mu$ is like $Y_{2\mu}$, i.e., a spherical tensor of degree 2 and component μ. The annihilation operator d_μ is consequently like $Y^*_{2\mu} = (-1)^\mu Y_{2,-\mu}$, which is *not* a proper tensor component with indices 2 and μ. Therefore we are led to define a modified d-boson annihilation operator

$$\tilde{d}_\mu \equiv (-1)^\mu d_{-\mu}. \qquad (2.79)$$

This operator does behave like $Y_{2\mu}$, so that $d^\dagger$ and $\tilde{d}$ are the proper tensor creation and annihilation operators, respectively, with components μ.

As pointed out in Subsection 2.4.1, the generators may be expressed as linear combinations of the basic ones (2.77). Particularly useful linear combinations result from angular-momentum coupling into tensor products (Appendix). The first three entries of eq. (2.77) remain unchanged but the fourth is replaced by the tensor product

$$[d^\dagger\tilde{d}]_{lm} \equiv \sum_{\mu\nu} (2\mu 2\nu|lm) d^\dagger_\mu \tilde{d}_\nu, \qquad l = 0, 1, 2, 3, 4, \qquad (2.80)$$

where the $(2\mu 2\nu|lm)$ are Clebsch–Gordan coefficients. Note that it was necessary to define the operator (2.79) in order to be able to do the coupling in eq. (2.80). We note that the numbers of components m for the various l add up to $1 + 3 + 5 + 7 + 9 = 25$. This is the same as the number $5\cdot 5 = 25$ of uncoupled operators $d^\dagger_\mu d_\nu$, which serves as a check of completeness.

2.5.2. Subgroups of *U*(6)

The utility of the coupled form of the generators, eq. (2.80), lies in the fact that commuting subsets are easy to identify among them.

Certain general principles can be stated off hand. We have in fact come across them in the detailed elementary examples of Section 2.3.

Clearly, unitary transformations in an $(n - 1)$-dimensional space are a subgroup of unitary transformations in the full n-dimensional space, and so forth: $U(n) \supset U(n - 1) \supset \cdots$. A special unitary group is always a subgroup of the corresponding full group: $U(n) \supset SU(n)$. An orthogonal group, by definition restricted to real rotations in an n-dimensional space, is a subgroup of the corresponding unitary group of complex transformations: $U(n) \supset O(n)$.

We also note that these three types of group, unitary, special unitary and orthogonal, are the ones of most frequent occurrence in physics and the only ones that will appear in the present context of the IBM. In fact, the remaining regular type of Lie group, the so-called *symplectic group* $Sp(n)$, occurs in fermion models of nuclear structure (Section 2.8, Chapters 5 and 7). Five so-called exceptional groups complete the list of Lie groups.

On the general grounds touched above we may expect whole chains of subgroups of $U(6)$. It turns out that there are *three* such chains of relevance to the IBM. To understand the physical mechanism at work, we must recognize that the IBM Hamiltonian (Chapter 1, Section 2.7) can be expressed in terms of $U(6)$ generators. The Hamiltonian contains one-boson energies and two-boson interactions of the form (2.15). The former are directly $U(6)$ generators, the latter can be commuted into them: $b_j^\dagger b_i^\dagger b_k b_l = b_j^\dagger b_k b_i^\dagger b_l - \delta_{ik} b_j^\dagger b_l$. This property is referred to by saying that the IBM has $U(6)$ *group structure*.

It is well to note that although the IBM Hamiltonian is composed of the $U(6)$ generators, the set of generators do not commute with it. Yet its eigenstates can be labeled with the quantum numbers of $U(6)$ and at least one of its subgroups. This one necessary subgroup symmetry is the ordinary rotational symmetry $O(3)$. In other words, the eigenstates of the Hamiltonian must also be eigenstates of the angular momentum. The labeling of states will become clear as our discussion progresses.

From the above emerges the conclusion that any subgroup chains that could serve the labeling of physical states must contain $O(3)$. We now examine what subalgebras are contained in the $U(6)$ algebra of the 36 generators

$$s^\dagger s, \quad s^\dagger d_\mu, \quad d^\dagger_\mu s, \quad [d^\dagger \tilde{d}]_{lm} \equiv T_{lm}. \tag{2.81}$$

The 25 components of T_{lm} are immediately seen to close on commutation because they separate from the generators containing s terms. The corresponding group is $U(5)$; it is just like the initial $U(6)$ case but with the s degree of freedom omitted. We can proceed and seek further subgroups. On general grounds one would expect $O(5)$ as the next link in the chain. However, we wish to be able to identify the subalgebra among the generators of $U(5)$. This presents us with a general problem to which we now digress.

The criterion of a subalgebra is that it close on commutation. We therefore want to be able to compute commutators of the generators expressed in tensor-product form. For general applicability beyond the present sd case, we now consider boson states $b_{\lambda\mu}$ of an arbitrary multipolarity λ. This means that b_0 = s and b_2 = d. The tensorial creation and annihilation operators are

$$b^\dagger_{\lambda\mu}, \quad \tilde{b}_{\lambda\mu} \equiv (-1)^{\lambda-\mu} b_{\lambda,-\mu}. \tag{2.82}$$

The coupled bilinear operators are $[b^\dagger_\lambda \tilde{b}_{\lambda'}]_{lm}$ and their Lie algebra consists of the commutation relations

$$\begin{aligned}
&[[b^\dagger_{\lambda_1}\tilde{b}_{\lambda_2}]_{l_1m_1}, [b^\dagger_{\lambda_3}\tilde{b}_{\lambda_4}]_{l_2m_2}] \\
&\quad = \sum_{lm} \sqrt{(2l_1+1)(2l_2+1)}(l_1m_1l_2m_2|lm) \\
&\quad \times \left[\delta_{\lambda_2\lambda_3}(-1)^{\lambda_1-\lambda_4+l} \begin{Bmatrix} l_1 & l_2 & l \\ \lambda_4 & \lambda_1 & \lambda_2 \end{Bmatrix} [b^\dagger_{\lambda_1}\tilde{b}_{\lambda_4}]_{lm} \right. \\
&\quad \left. - \delta_{\lambda_1\lambda_4}(-1)^{\lambda_2-\lambda_3+l_1+l_2} \begin{Bmatrix} l_1 & l_2 & l \\ \lambda_3 & \lambda_2 & \lambda_1 \end{Bmatrix} [b^\dagger_{\lambda_3}\tilde{b}_{\lambda_2}]_{lm} \right].
\end{aligned} \tag{2.83}$$

The coefficient $\{\cdots\}$ is a $6j$ symbol of angular-momentum recoupling. Equation (2.83) can be viewed as merely a technical extension of the simple commutation relation (2.60). As long as the space of basis states created by the $b^\dagger_i$ is the same, the Lie algebra and the associated physics remain unchanged. For s and d states we have the $U(6)$ algebra of the basic IBM. When other boson states (g, etc.) are included, the relevant group is larger; see Chapter 6. In eq. (2.83) we have taken care with phases so that the result is *true also for fermion operators*.

When a subalgebra has been found by the commutation test, it remains to identify it. We can accomplish this by counting the generators and then comparing with the known numbers of generators for the various types of group as given in Table 2.1. From Subsection 2.4.1 we already have the results for $U(n)$ and $SU(n)$. We do not prove the result for $O(n)$ but we can see its plausibility from the familiar cases of $O(3)$ and $O(2)$ as discussed in Section 2.3. We also state without proof the result for $Sp(n)$, where only *even* values of n may occur.

It can happen that the count gives two or more answers: e.g., $SU(4)$ and $O(6)$ both have 15 generators. Then the two groups are essentially the same, isomorphic, which we denote as $SU(4) \approx O(6)$ following the discussion at the end of Subsection 2.3.3. In fact, Table 2.1 shows a three-way isomorphism $SU(2) \approx O(3) \approx Sp(2)$.

Applying eq. (2.83) to the generators T_{lm} of $U(5)$, one finds that the seven components T_{3m} and the three components T_{1m} close on commutation. From Table 2.1, we identify the corresponding group of ten generators as $O(5)$. Of those ten, the three operators T_{1m} again close by themselves. Worked out, the commutators are

$$[T_{11}, T_{1,-1}] = -\sqrt{\tfrac{1}{10}}\, T_{10}, \qquad [T_{10}, T_{1,\pm1}] = \pm\sqrt{\tfrac{1}{10}}\, T_{1,\pm1}. \tag{2.84}$$

Expressed in the spherical basis, i.e., as components of a spherical tensor of degree 1 (see Appendix), the angular-momentum components are $J_{\pm1} = \mp\sqrt{\tfrac{1}{2}}(J_x \pm iJ_y)$, $J_0 = J_z$. Their commutation relations are, from eq. (2.31),

$$[J_1, J_{-1}] = -J_0, \qquad [J_0, J_{\pm1}] = \pm J_{\pm1}. \tag{2.85}$$

Since the IBM bosons really describe the mechanical motion in the

TABLE 2.1. *Numbers of generators of common Lie groups.*

Group	Number
$U(n)$	n^2
$SU(n)$	$n^2 - 1$
$O(n)$	$\tfrac{1}{2}n(n-1)$
$Sp(n)$	$\tfrac{1}{2}n(n+1)$

nucleus, the similarity of eqs. (2.84) and (2.85) is more than just a formal coincidence, and we identify

$$J_m = \sqrt{10}\, T_{1m} = \sqrt{10}[d^\dagger \tilde{d}]_{1m} \qquad (2.86)$$

as the angular-momentum operator for the bosons of the IBM.

As we certainly want Hamiltonian eigenstates of good angular momentum, we have indeed established the appropriate subgroup $O(3)$ as the next link in the chain of subgroups. The group $O(4)$, although of course formally a subgroup of $O(5)$, is absent from the chain because it would not contain the *physical* $O(3)$ just identified. And finally of course we have $O(2)$ as a subgroup of $O(3)$. It is the group of rotations about the z axis. However, in the normal case of no external field the energy will be degenerate with respect to the associated quantum number M. Therefore mention of $O(2)$ is often neglected as trivial.

The complete group chain resulting from our foregoing discussion is

$$U(6) \supset U(5) \supset O(5) \supset O(3) \supset O(2). \qquad (2.87)$$

This is the first of three chains of subgroups of $U(6)$ and it is referred to as *Chain I*.

We can now derive the following two chains on the above principles. First we recognize that by forming certain linear combinations of the generators (2.81) we can find other subalgebras than $U(5)$. For the second chain such a subalgebra consists of the nine operators

$$s^\dagger s + \sqrt{5}\, T_{00}, \quad T_{1m}, \quad s^\dagger \tilde{d}_m + d_m^\dagger s \pm \tfrac{1}{2}\sqrt{7}\, T_{2m}. \qquad (2.88)$$

The algebra is $U(3)$. The first operator is the boson number operator:

$$s^\dagger s + \sqrt{5}\, T_{00} = s^\dagger s + \sum_m d_m^\dagger d_m \equiv \hat{n}_s + \hat{n}_d \equiv \hat{N}. \qquad (2.89)$$

As discussed in general terms in Subsection 2.4.1, this operator can be separated from the set of nine, and the remaining eight generate the group $SU(3)$. We elect to include $SU(3)$ rather than $U(3)$ in the chain, because the eigenvalue N of $\hat{N}$ is the same for all states of a given nucleus. Now again the generators of $O(3)$ separate into a subalgebra, given in detail in eq. (2.84), which must be in the chain for conserved angular momentum. We have thus established *Chain II* as

$$U(6) \supset SU(3) \supset O(3) \supset O(2). \tag{2.90}$$

Before proceeding, we note that five of the generators of $SU(3)$,

$$Q_m \equiv s^\dagger \tilde{d}_m + d_m^\dagger s - \tfrac{1}{2}\sqrt{7}\, T_{2m}, \tag{2.91}$$

have the tensorial properties of the electric quadrupole tensor. We have here elected the minus sign in front of T_{2m}. It is the customary IBM choice, which relates to the prolate shape prevalent in the geometric model for deformed nuclei. However, the choice is no restriction on the applicability of the model. The positive sign has found some use in IBM-2 (Chapter 4).

The third chain starts from the observation that the ten generators of $O(5)$, i.e., T_{1m} and T_{3m}, and the five operators $s^\dagger \tilde{d}_m + d_m^\dagger s$ close on commutation. These 15 operators generate the group $O(6)$, as indicated by Table 2.1. By this construction it is clear that $O(5)$ is a subgroup of $O(6)$, and the chain again, and necessarily, ends with $O(3)$ and $O(2)$. Our *Chain III* is then

$$U(6) \supset O(6) \supset O(5) \supset O(3) \supset O(2). \tag{2.92}$$

We have now established the three chains of subgroups of $U(6)$ which contain the spatial rotation group $O(3)$ as a necessary link. These exhaust the possibilities. Alternative to the designations I, II and III, the chains are also referred to by their second link: $U(5)$, $SU(3)$ and $O(6)$, respectively.

The three group chains play a fundamental role in the structure and physical contents of the IBM. They yield a wealth of impressively simple predictions that compare favorably with experiment. Chains I, II, and III, respectively, relate to vibrational, rotational, and so-called γ-soft nuclei. We will touch upon this physical interpretation in Section 2.7 and it is elaborated in Chapter 3.

2.5.3. Casimir Operators

From the general discussion in Section 2.4, we know what Casimir operators are and how they can be found. For the unitary groups eq. (2.60) provides a straightforward algorithm for finding the Casimir operators. For all groups the basic defining property, namely commutation with all generators, is valid and can be used in searching for the Casimir operators.

The rank, i.e., the number of commuting generators, of a group is also the number of its Casimir operators. From Section 2.4 we know immediately that $U(6)$ has six, $U(5)$ five and $SU(3)$ two Casimir operators. We also know from Section 2.3 that $O(3)$ and $O(2)$ both have one Casimir operator. How many do $O(5)$ and $O(6)$ have? We give the answer in general terms in Table 2.2, whence we see that $O(5)$ has two and $O(6)$ three Casimir operators.

We now state the Casimir operators of direct relevance to the IBM. Since the standard Hamiltonian of the model includes only two-body interactions, there will be no explicit need for Casimir operators beyond quadratic. So restricted, and with some arbitrary overall multiplicative constants customary in the IBM literature, the operators are

$$C_{1U6} = \hat{N}, \qquad C_{2U6} = \hat{N}(\hat{N} + 5), \tag{2.93}$$

$$C_{1U5} = \hat{n}_d, \qquad C_{2U5} = \hat{n}_d(\hat{n}_d + 4), \tag{2.94}$$

$$C_{2O5} = 4(T_1 \cdot T_1 + T_3 \cdot T_3) = 2\hat{n}_d(\hat{n}_d + 3) - 2(d^\dagger \cdot d^\dagger)(\tilde{d} \cdot \tilde{d}), \tag{2.95}$$

$$C_{2O3} = 2\mathbf{J} \cdot \mathbf{J}, \tag{2.96}$$

$$C_{2O2} = 2J_z^2, \tag{2.97}$$

$$C_{2SU3} = \tfrac{4}{3}\, Q \cdot Q + \tfrac{1}{2}\, \mathbf{J} \cdot \mathbf{J}, \tag{2.98}$$

$$\begin{aligned} C_{2O6} &= 2(s^\dagger\tilde{d} + d^\dagger s) \cdot (s^\dagger\tilde{d} + d^\dagger s) + C_{2O5} \\ &= 2\hat{N}(\hat{N} + 4) - 8P^\dagger P, \end{aligned} \tag{2.99}$$

where we have defined for later use

TABLE 2.2. *Ranks of common Lie groups.*

Group	Rank
$U(n)$	n
$SU(n)$	$n - 1$
$O(n)$, n even	$\frac{1}{2}n$
$O(n)$, n odd	$\frac{1}{2}(n - 1)$
$Sp(n)$	$\frac{1}{2}n$

$$P^{\dagger} \equiv \tfrac{1}{2}\,(d^{\dagger}\cdot d^{\dagger} - s^{\dagger}s^{\dagger}), \qquad P = (P^{\dagger})^{\dagger} = \tfrac{1}{2}\,(\tilde{d}\cdot\tilde{d} - ss). \quad (2.100)$$

The optional forms in eqs. (2.95) and (2.99) will be called upon in Section 2.6. In eqs. (2.93)–(2.99) we have abbreviated the notation, and made the equations physically more transparent, through the use of eqs. (2.86), (2.89) and (2.91). For brevity we have also used the dot product of tensors (see Appendix 2), defined for two arbitrary tensors A_l and B_l as

$$\begin{aligned} A_l\cdot B_l &\equiv (-1)^l\sqrt{2l+1}[A_lB_l]_{00} \\ &= (-1)^l\sqrt{2l+1}\sum_m (lml, -m|00)A_{lm}B_{l,-m} \qquad (2.101) \\ &= \sum_m (-1)^m A_{lm}B_{l,-m}. \end{aligned}$$

We denote the angular-momentum vector ($l = 1$ tensor) **J** for consistency with Section 2.3 and to distinguish it from the quantum number J. To emphasize the tensorial character, we have retained the dot throughout and refrained from the further abbreviation $T_l\cdot T_l \equiv T_l^2$. Only the group $U(n)$ has a linear Casimir operator.

2.5.4. Quantum Numbers

We learned in the context of the general unitary group in Section 2.4 that a group has the same number of commuting generators and of Casimir operators. This number is the rank r of the group, and it is given in Table 2.2. The commuting $2r$ operators possess simultaneous eigenvalue equations for common eigenstates. In other words, in terminology relating to matrices, the $2r$ operators can be simultaneously diagonalized. Analogously to angular momentum, the r quantum numbers relating to the Casimir operators are of the J type, the r quantum numbers relating to the commuting generators are of the M type. These features are not unique to $U(n)$ but are shared by all Lie groups.

Returning to the discussion of quantum numbers in Section 2.4, we recall that we argued in Subsections 2.4.2 and 2.4.3 that a definite Young tableau, not only a pattern, is required for a complete specification of the $U(n)$ representation. The detailed $U(2)$ example made this view concrete. However, our observation at the end of Subsection 2.4.3 of the role of permutations effectively reversed the earlier view: the Young pattern does specify the representation of $U(n)$. Thus the quantum numbers $[\lambda_1\lambda_2\cdots\lambda_n]$ are the proper and complete set of J-type quantum numbers specifying the representation of $U(n)$. Since $SU(n)$ has one fewer generator and Casimir operator, its corresponding set is $[\lambda_1\lambda_2\cdots\lambda_{n-1}]$.

We are now faced with the task of finding for each group the J-type quantum numbers, which serve to label the representations. These quantum numbers will of course appear in the expressions for the eigenvalues of the Casimir operators. From our knowledge of how many quantum numbers there are, we can sketch the J-type quantum numbers as follows:

$$\begin{aligned} U(6)&: [N_1N_2N_3N_4N_5N_6] \\ U(5)&: [n_1n_2n_3n_4n_5] \\ O(5)&: (\tau_1\tau_2) \\ O(3)&: J \\ O(2)&: M \\ SU(3)&: [f_1f_2] \\ O(6)&: (\sigma_1\sigma_2\sigma_3). \end{aligned} \tag{2.102}$$

Enclosing the multiple quantum numbers in different brackets has a reason. The square brackets used for the unitary groups indicate that their basis states can be described by Young tableaux as introduced in Section 2.4. The orthogonal groups do not admit of this; thence the parentheses. The choice of symbols anticipates standard IBM notation. The complete sets of quantum numbers given in eq. (2.102) tie up with complete sets of Casimir operators, of which only the linear and quadratic ones are spelled out in eqs. (2.93)–(2.99). The quantum numbers appearing in eq. (2.102) are non-negative integers, except that the orthogonal groups admit also of half integers (but of course not for boson systems).

How does one find the quantum numbers? In principle by solving the various eigenvalue equations algebraically; our treatment of angular momentum in Subsection 2.3.2 serves as the prototype of such a process. General algorithms have in fact been derived for finding the eigenvalues of the Casimir operators of unitary, orthogonal and symplectic groups. Algorithms and general results derived with them have been listed by Iachello (1980). Below we list the eigenvalues $\langle C \rangle$ of the linear and quadratic Casimir operators (2.93)–(2.99) as functions of the quantum numbers (2.102):

$$\langle C_{1U6} \rangle = N_1 + N_2 + \cdots + N_6, \tag{2.103}$$

$$\langle C_{2U6} \rangle = N_1(N_1 + 5) + N_2(N_2 + 3) + \cdots + N_6(N_6 - 5), \tag{2.104}$$

$$\langle C_{1U5} \rangle = n_1 + n_2 + \cdots + n_5, \tag{2.105}$$

$$\langle C_{2U5} \rangle = n_1(n_1 + 4) + n_2(n_2 + 2) + \cdots + n_5(n_5 - 4), \tag{2.106}$$

$$\langle C_{2O5} \rangle = 2\tau_1(\tau_1 + 3) + 2\tau_2(\tau_2 + 1), \tag{2.107}$$

$$\langle C_{2O3} \rangle = 2J(J + 1), \tag{2.108}$$

$$\langle C_{2O2} \rangle = 2M^2, \tag{2.109}$$

$$\langle C_{2SU3} \rangle = \tfrac{2}{3}\,(f_1^2 + f_2^2 - f_1 f_2 + 3f_1), \tag{2.110}$$

$$\langle C_{2O6} \rangle = 2\sigma_1(\sigma_1 + 4) + 2\sigma_2(\sigma_2 + 2) + 2\sigma_3^2. \tag{2.111}$$

The same unessential overall factors appear here as in the operators (2.93)–(2.99). In Subsection 2.6.2 we actually derive eqs. (2.107) and (2.111) for the special case relevant to the IBM.

Instead of f_1 and f_2 it is customary to use for $SU(3)$ the Elliott quantum numbers

$$\lambda \equiv f_1 - f_2, \qquad \mu \equiv f_2. \tag{2.112}$$

In an $SU(3)$ Young pattern λ is the "overhang" of the top row and μ is the length of the bottom row. The notation for the representation is $(\lambda\mu)$, for example

$$[f_1 = 3, f_2 = 2] = (\lambda = 1, \mu = 2) = \begin{array}{|c|c|c|}\hline \;&\;&\; \\ \hline \;&\; \\ \cline{1-2}\end{array} \tag{2.113}$$

With the Elliott quantum numbers the eigenvalue (2.110) takes on the form

$$\langle C_{2SU3} \rangle = \tfrac{2}{3}(\lambda^2 + \mu^2 + \lambda\mu + 3\lambda + 3\mu). \tag{2.114}$$

For completeness we also state the eigenvalue of the cubic Casimir operator of $SU(3)$,

$$\langle C_{3SU3} \rangle = \tfrac{2}{9}(\lambda^3 - \mu^3) + \tfrac{1}{3}\lambda\mu(\lambda - \mu) + (\lambda + 2)(2\lambda + \mu). \tag{2.115}$$

Equations (2.114) and (2.115) are a complete statement of the Casimir operators of $SU(3)$. Although the cubic term does not occur in the Hamiltonian of the standard IBM, it is instructive to display it as the other source of the two quantum numbers $(\lambda\mu)$, both of which are essential in the IBM.

2.6. GROUP REDUCTIONS

2.6.1. Symmetric Wave Function and Quantum Numbers

The group-theoretic machinery needed for the IBM has been built up step by step in the foregoing Sections 2.4 and 2.5. Although we chose in Section 2.5 to discuss the group $U(6)$ in terms of IBM boson operators, we kept the presentation on a rather general level transcending the immediate needs of the IBM. We determined those subgroups of $U(6)$ that are relevant to the IBM, but treated their quantum numbers in complete generality.

In particular we allowed unrestricted Young patterns producing the strings of quantum numbers recorded in eq. (2.102) and the eigenvalues of Casimir operators (2.103)–(2.111). A radical simplification comes about when we apply the theory specifically to IBM-1 with its single kind of boson. The wave function can only be completely symmetric, and it is constructed in a single space, unlike the two-space case discussed in Subsection 2.4.4. It follows that the only $U(6)$ representation allowed is the completely symmetric one $[N]$.

We elected to present the results for general Young patterns for two reasons. One of them was to demonstrate that the representations of a group of rank r (Table 2.2) are indeed labeled by that many J-type quantum numbers. The other reason was to prepare the background for IBM-2, where two-rowed Young patterns occur. This is briefly discussed in Subsection 2.8.1 and more extensively in Chapter 4.

In this section we only discuss the completely symmetric representations of $U(6)$ and its subgroups. It is clear from the analogy to $U(6)$ that $U(5)$ can only have the representation $[n_d]$. What happens with $O(5)$, $O(6)$ and $SU(3)$ is not immediately clear. The answer is that $O(5)$ and $O(6)$ only allow representations with a single quantum number, $(\tau_1 \equiv \tau, \tau_2 = 0) \equiv \tau$ and $(\sigma_1 \equiv \sigma, \sigma_2 = 0, \sigma_3 = 0) \equiv \sigma$, respectively. To the contrary, both $SU(3)$ quantum numbers $(\lambda\mu)$ will persist. The familiar angular-momentum quantum numbers J [group $O(3)$] and M [group $O(2)$] behave as usual and cause no problem.

2.6.2. The Reduction Problem: Chains I and III

We are now faced with the problem which representations, or J-type quantum numbers, of a subgroup belong to a given representation of the larger group. This is called the reduction problem. The result is also known as a branching rule. We now address the problem for the three chains of $U(6)$ subgroups established in Section 2.5.

2.6.2.1. Reductions $U(6) \supset U(5)$ and $U(5) \supset O(5)$

In our Chain I of $U(6)$ subgroups, eq. (2.87), we have $U(6) \supset U(5)$ at the outset. What is the reduction of representations of $U(6)$ with respect to $U(5)$? In the IBM context we only have $[N]$ for $U(6)$, N being the total number of bosons, and $[n_d]$ for $U(5)$, n_d being the number of bosons in the d state. We can have all the bosons in the d state, $N = n_d$; all but one boson in the d state, $N - 1 = n_d$; . . . ; no bosons in the d state, $n_d = 0$. The desired reduction is thus

$$n_d = N, N - 1, N - 2, \ldots, 1, 0. \qquad (2.116)$$

The next reduction problem in Chain I is $U(5) \supset O(5)$. In other words, we ask which values of τ can occur for a given value of n_d. The problem can be solved by means of the technique of *boson quasispins*. This is strictly a mathematical device and has no direct bearing on the physical angular momentum of the bosons.

We define the boson-quasispin operators

$$S_+ \equiv \tfrac{1}{2}d^\dagger \cdot d^\dagger, \qquad S_- \equiv S_+^\dagger = \tfrac{1}{2}\tilde{d}\cdot\tilde{d}, \qquad (2.117)$$
$$S_0 \equiv \tfrac{1}{4}(d^\dagger \cdot \tilde{d} + \tilde{d}\cdot d^\dagger) = \tfrac{1}{2}\hat{n}_d + \tfrac{5}{4} = S_0^\dagger.$$

These are found to commute according to

$$[S_\pm, S_0] = \mp S_\pm, \qquad [S_+, S_-] = -2S_0. \qquad (2.118)$$

Apart from the minus sign in front of the 2, the commutation relations have the same form as the angular-momentum relations (2.35). This is a so-called noncompact Lie algebra, denoted as $O(2, 1) \approx SU(1, 1)$ with a distinct kinship to $O(3) \approx SU(2)$. The operator

$$C \equiv -S_+S_- - S_0 + S_0^2 = -S_-S_+ + S_0 + S_0^2 \qquad (2.119)$$

commutes with the three generators and is thus the Casimir operator of the group. We write its eigenvalue as $S(S-1)$ and give the eigenvalue of S_0 the name μ, so that the eigenvalue problem analogous to eq. (2.33) is

$$C|S\mu\rangle = S(S-1)|S\mu\rangle, \qquad S_0|S\mu\rangle = \mu|S\mu\rangle. \tag{2.120}$$

The analogy with angular momentum, Subsection 2.3.2, is rather complete. Corresponding pairs are $C \leftrightarrow \mathbf{J}^2$, $S_0 \leftrightarrow J_z$, $|S\mu\rangle \leftrightarrow |JM\rangle$. The one exception in eq. (2.120) is the ansatz $S(S-1)$ in lieu of $J(J+1)$, reflecting the small difference in the algebras.

In complete analogy to Subsection 2.3.2 we find

$$S_\pm|S\mu\rangle = \sqrt{\mu^2 \pm \mu - S(S-1)}\,|S, \mu \pm 1\rangle. \tag{2.121}$$

From the expression for S_0 in eq. (2.117) we see immediately that

$$\mu = \tfrac{1}{2}n_d + \tfrac{5}{4}, \tag{2.122}$$

so that μ has an absolute minimum $\frac{5}{4}$ and no maximum. For the ladder to cut off at a certain $\mu = \mu_{min}$, i.e., $S_-|S\mu_{min}\rangle = 0$, we must have $S(S-1) \geq \frac{5}{16}$. We may choose S positive, whence $\mu_{min} = S$. The spectrum of μ for a given S is then

$$\mu = S, S+1, S+2, \ldots. \tag{2.123}$$

According to eq. (2.122), to a given $\mu_{min} = S$ corresponds a certain minimum value of n_d; let us give it the name τ. Then we have

$$S = \tfrac{1}{2}\tau + \tfrac{5}{4}. \tag{2.124}$$

Combining eqs. (2.122)–(2.124), we obtain

$$n_d = \tau, \tau + 2, \tau + 4, \ldots. \tag{2.125}$$

Equivalently to the quantum numbers S, μ we may use τ, n_d, so that the quasispin eigenstates are $|S\mu\rangle \equiv |\tau n_d\rangle$.

What is the boson structure of the state $|SS\rangle \equiv |\tau\tau\rangle$? We can see it by writing

$$S_-|SS\rangle = \tfrac{1}{2}\tilde{d}\cdot\tilde{d}|\tau\tau\rangle = \tfrac{1}{2}\sqrt{5}[\tilde{d}\tilde{d}]_{00}|\tau\tau\rangle = 0, \tag{2.126}$$

where the last equality follows from eq. (2.121). This means that the state $|\tau\tau\rangle$, consisting of $n_d = \tau$ bosons in the d state, contains no pairs of bosons coupled to angular momentum zero. More precisely, this defines the state $|\tau\tau\rangle$ as the state of maximal *boson seniority* τ (the notation $v \equiv \tau$ is also used in the literature). In analogy to nucleon seniority, boson seniority is physically understood as the number of bosons that are not coupled in pairs to angular momentum zero.

The algebraic result (2.125) gives the values of n_d that belong to an $SU(1, 1)$ representation labeled by τ. We want to invert the point of view since we are seeking the values of τ that belong to a given fixed n_d, i.e., the reduction $U(5) \supset O(5)$. The result, from eq. (2.125), is

$$\tau = n_d, n_d - 2, n_d - 4, \ldots, 0 \text{ or } 1, \tag{2.127}$$

all values being either even or odd.

We can also find the eigenvalues of the Casimir operator C_{2O5} by the quasispin method. Substituting the quasispin operators (2.117) into the Casimir operator (2.95) and using eq. (2.119), a straightforward calculation yields

$$C_{2O5}|\tau n_d\rangle = 2\tau(\tau + 3)|\tau n_d\rangle. \tag{2.128}$$

This result, valid for the symmetric representation ($\tau 0$) required for identical bosons, agrees with the general result (2.107).

2.6.2.2. The reduction $O(5) \supset O(3)$

The following link is the reduction $O(5) \supset O(3)$. In our present physical terms this boils down to angular-momentum coupling of d bosons and assigning boson seniority τ to the states. We now consider states consisting of d bosons only; bosons in the s state play no role here. The ground state $n_d = 0$ has trivially $\tau = 0$. Likewise the first excited state $n_d = 1$ has $\tau = 1$. Two d bosons can be coupled to the exchange-symmetric states $J = 0$, 2, and 4. Only they are acceptable since the states $J = 1$ and 3 are antisymmetric. The $J = 0$ state has $\tau = 0$, while the $J = 2$ and 4 states have $\tau = 2$. The configuration d^3 yields symmetric states $J = 0, 2, 3, 4, 6$. Of these, the $J = 2$ state is built so that two bosons are coupled to 0, 2, or 4. By definition, the seniority assignment is made on the basis of the simplest possibility so that the $J = 2$ state is assigned $\tau = 1$. The remaining d^3 states have $\tau = 3$.

As we proceed to d^4, the symmetric states are $J = 0, 2^2, 4^2, 5, 6, 8$. As a new feature, there are now *two* states with $J = 2$ and also with $J = 4$. The two states of the same J are distinguished by their seniority. Since a $J = 2$ state can be built so that one pair is coupled to zero, we assign $\tau = 2$ to it; the other $J = 2$ state is then $\tau = 4$. The $J = 0$ state has $\tau = 0$ and the states $J = 5, 6, 8$ have $\tau = 4$.

When detailed wave functions are required, the orthonormal states are constructed through the use of so-called (boson) coefficients of fractional parentage (cfp). We also remark that beyond d^2 one should use the *m*-table technique or Young patterns to find the possible states. The former can be found in textbooks on nuclear physics; the latter will be touched upon in Subsection 2.6.3.

Yet a new feature comes up for configuration d^6. There appear three $J = 6$ states and there are only two possible seniorities, $\tau = 4$ and $\tau = 6$, for them. Since d^4 gives one $J = 6$, one of the three states is $\tau = 4$ and two have $\tau = 6$. So the quantum number τ is not sufficient to label orthogonal states of the same J. An *additional label* is needed. It is not a quantum number associated with some Casimir operator. In fact, it need not be more than a numbering label for arbitrarily orthogonalized states of the same J and τ. The choice made in the IBM, however, does have a physical meaning: the additional label is called ν_Δ (sometimes n_Δ) and it enumerates d-boson triplets of zero angular momentum. Its spectrum is

$$\nu_\Delta = 0, 1, 2, \ldots . \tag{2.129}$$

Thus the two d^6 states with $J = 6$ and $\tau = 6$ are distinguished through $\nu_\Delta = 0$ and $\nu_\Delta = 1$. Further labels beyond ν_Δ will not be needed.

For completeness and consistency the quantum numbers τ and ν_Δ are assigned even when they are not necessary to distinguish states of the same J. There is the following general algorithm for the reduction $O(5) \supset O(3)$. Define the auxiliary quantity

$$\Lambda \equiv \tau - 3\nu_\Delta \geq 0, \tag{2.130}$$

which means the number of d bosons not in pairs or triplets of angular momentum zero. The allowed angular momenta are

$$J = 2\Lambda, 2\Lambda - 2, 2\Lambda - 3, \ldots, \Lambda + 1, \Lambda. \tag{2.131}$$

We note that $2\Lambda - 1$ is missing and $J = 1$ will never occur. For a check, one can reproduce the above examples with this algorithm.

We are now finished with Chain I since the remaining reduction $O(3) \supset O(2)$ is elementary and given in eq. (2.49). Following the form of eq. (2.58), we restate the whole chain (2.87) with the quantum numbers indicated below:

$$\underset{[N]}{U(6)} \supset \underset{n_d}{U(5)} \supset \underset{\tau}{O(5)} \supset_{\nu_\Delta} \underset{J}{O(3)} \supset \underset{M}{O(2)} \tag{2.132}$$

Here we have followed the usual practice of saying $[N]$ rather than just N to emphasize the symmetric nature of the state. The additional quantum number ν_Δ is placed between the groups $O(5)$ and $O(3)$ to stress its intermediary role in the otherwise incomplete reduction.

2.6.2.3. The reduction $U(6) \supset O(6)$

Chain III, given in eq. (2.92), differs from Chain I in that $O(6)$ replaces $U(5)$ as the second largest group. Accordingly we have the new reduction problems $U(6) \supset O(6)$ and $O(6) \supset O(5)$. Let us first address the first one.

The reduction $U(6) \supset O(6)$ goes very similarly to the foregoing treatment of $U(5) \supset O(5)$. The only difference is that s bosons are now included. Instead of the quasispin operators (2.117) we now define

$$\begin{aligned} &S_+ \equiv P^\dagger, \quad S_- \equiv P, \\ &S_0 \equiv \tfrac{1}{4}(d^\dagger \cdot \tilde{d} + \tilde{d} \cdot d^\dagger + s^\dagger s + s s^\dagger) = \tfrac{1}{2}\hat{N} + \tfrac{3}{2}, \end{aligned} \tag{2.133}$$

where we have the operators $P^\dagger$ and P defined in eq. (2.100). These operators satisfy the *same* Lie algebra (2.118) as the pure d-boson quasispins. Equations (2.122) and (2.124) are now replaced by

$$\mu = \tfrac{1}{2}N + \tfrac{3}{2}, \qquad S = \tfrac{1}{2}\sigma + \tfrac{3}{2}, \tag{2.134}$$

where σ is the value of N corresponding to $\mu_{\text{min}} = S$. The eigenstates are thus $|S\mu\rangle \equiv |\sigma N\rangle$, and similarly to eq. (2.125) we have the spectrum

$$N = \sigma, \sigma + 2, \sigma + 4, \ldots. \tag{2.135}$$

What is the physical meaning of the quantum number σ? It extends the concept of boson seniority by including the s degree of freedom in accordance with

$$S_-|SS\rangle = \tfrac{1}{2}(\tilde{d} \cdot \tilde{d} - ss)|\sigma\sigma\rangle = 0. \tag{2.136}$$

This condition is analogous to eq. (2.126). We cannot perceive a more tangible meaning of σ. Inverting eq. (2.135), we have the desired reduction $U(6) \supset O(6)$:

$$\sigma = N, N - 2, N - 4, \ldots, 0 \text{ or } 1, \tag{2.137}$$

the choice depending on whether N is even or odd.

Working out the eigenvalue equation for the Casimir operator C_{2O6} of eq. (2.99) we obtain a result analogous to eq. (2.128):

$$C_{2O6}|\sigma N\rangle = 2\sigma(\sigma + 4)|\sigma N\rangle. \tag{2.138}$$

We have thus derived the symmetric special case ($\sigma 00$) of the general result quoted in eq. (2.111).

2.6.2.4. The reduction $O(6) \supset O(5)$

One step of reduction remains in Chain III: $O(6) \supset O(5)$. In other words, we ask what values of τ belong to a given σ. We can answer the question by inspection and comparison of the reductions to $O(6)$ and $O(5)$. One possible $O(6)$ state is where all bosons associated with the quantum number σ are d bosons; then we have $\sigma = \tau$. Another possibility is that all but one of the bosons associated with σ are d bosons; such a state has $\sigma = \tau + 1$. Continuing similarly we arrive at the reduction $O(6) \supset O(5)$:

$$\tau = \sigma, \sigma - 1, \sigma - 2, \ldots, 1, 0. \tag{2.139}$$

Now that we have established the possible values of the various quantum numbers we restate Chain III, eq. (2.92), complete with quantum numbers:

$$\underset{[N]}{U(6)} \supset \underset{\sigma}{O(6)} \supset \underset{\tau}{O(5)} \supset \underset{\nu_\Delta \quad J}{O(3)} \supset \underset{M}{O(2)} \tag{2.140}$$

2.6.3. The Reduction Problem: Chain II

Chain II of subgroups, eq. (2.90), contains two further reductions: $U(6) \supset SU(3)$ and $SU(3) \supset O(3)$. These reductions are carried out by a building-up process which involves the multiplication of Young patterns.

2.6.3.1. Multiplication of Young Patterns

We first introduce the multiplication of Young patterns through an example whose result is familiar from an elementary context. Consider the angular-momentum coupling of two single-particle wave functions of angular momentum $j = 2$. Such a state has five substates $m = \pm 2, \pm 1, 0$. Unitary transformations in the space spanned by them form the group $U(5)$, or $SU(5)$ if we exclude the particle number from the generators. The analogy to the isospin case of Section 2.3 and to the IBM case of Section 2.5 is clear.

Let the single-particle wave functions be $\psi_{2m}(1)$ and $\psi_{2m}(2)$. The two-particle wave function of angular momentum J, M is

$$\Psi_{JM}(12) = \sum_{m\mu} (2m2\mu|JM)\psi_{2m}(1)\psi_{2\mu}(2). \tag{2.141}$$

If we symmetrize/antisymmetrize this state, we have (unnormalized)

$$\Psi_{JM}(12) \pm \Psi_{JM}(21) = [1 \pm (-1)^J]\Psi_{JM}(12), \tag{2.142}$$

where we have used symmetry properties of the Clebsch–Gordan coefficients. The result means that the symmetric states have $J = 0, 2, 4$ and the antisymmetric states have $J = 1, 3$. This can be represented by multiplying and summing Young patterns:

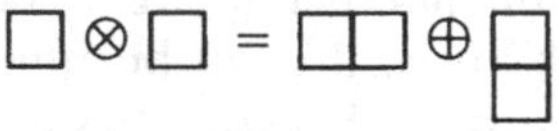

$$\square \otimes \square = \square\square \oplus \begin{array}{c}\square\\ \square\end{array} \tag{2.143}$$

Here each separate box stands for a spin-2 object. Their coupling without regard to symmetry, represented by the product on the left-hand side, gives $J = 0, 1, 2, 3, 4$. On the right-hand side, the horizontal box represents the symmetric states $J = 0, 2, 4$ and the vertical box the antisymmetric states $J = 1, 3$. In the notation $[\lambda_1\lambda_2\cdots]$ of Section 2.4 we can rewrite eq. (2.143) as $[1] \otimes [1] = [2] \oplus [1^2]$. We can continue and form the products $[2] \otimes [1]$ and $[1^2] \otimes [1]$:

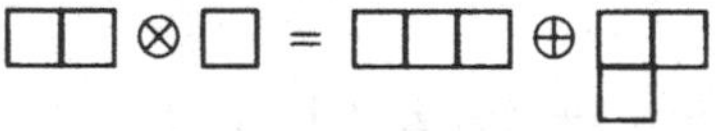

$$\square\square \otimes \square = \square\square\square \oplus \begin{array}{l}\square\square\\ \square\end{array} \tag{2.144}$$

$$\begin{array}{c}\square\\ \square\end{array} \otimes \square = \begin{array}{l}\square\square\\ \square\end{array} \oplus \begin{array}{c}\square\\ \square\\ \square\end{array} \tag{2.145}$$

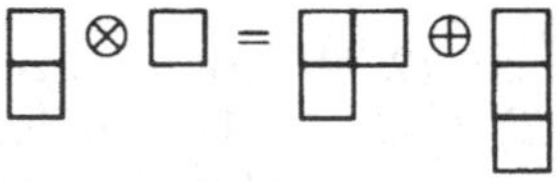

These can be treated as algebraic equations. From

$$[2] \otimes [1] = [3] \oplus [21], \quad [1^2] \otimes [1] = [21] \oplus [1^3] \tag{2.146}$$

we can eliminate [21] and solve for [3]:

$$[3] = [2] \otimes [1] - [1^2] \otimes [1] \oplus [1^3]. \tag{2.147}$$

The state $[1^3]$ gives the same angular momenta as $[1^2]$, i.e. $J = 1, 3$. This is because $[1^5]$ is a "closed shell" with $J = 0$, so that $[1^3]$

can be viewed as two holes in the closed shell and thus equivalent with $[1^2]$. By elementary coupling rules, the first term on the right-hand side gives $J = 0, 1, 2^3, 3^2, 4^2, 5, 6$. The second term gives $J = 1^2, 2^2, 3^2, 4, 5$. Subtracting the second items from the first ones and adding $J = 1, 3$ from the last term, we find $J = 0, 2, 3, 4, 6$ as the angular momenta of [3]. These are the proper J values for the boson configuration d^3 since the state must be completely symmetric. This was one of the results mentioned in Subsection 2.6.2. From either equation (2.146) we can now find also the angular momenta possessed by [21]: $J = 1, 2^2, 3, 4, 5$. It is noteworthy that $J = 1$ can only occur in patterns containing some antisymmetry.

The above examples illustrate the multiplication of Young patterns; the products are called outer products. The full rules, known as Littlewood's rules, needed for more complicated cases can be found in the literature. The products considered here are products of different representations of the group $U(5)$. They also carry angular momentum, or representations of $O(3)$, by the elementary rules of angular-momentum coupling.

2.6.3.2. The reduction $SU(3) \supset O(3)$

We have now prepared the background, and in fact the practical rules, for the reduction $SU(3) \supset O(3)$. The building-up process consists in multiplying representations of $SU(3)$ and identifying angular momenta just as above. Since the group $SU(3)$ is based on *three* single-particle states, we associate angular momentum $j = 1$ with each box. Let us consider the following example of multiplication of $SU(3)$ representations:

$$[31] \otimes [1] = [41] \oplus [32] \oplus [311] \qquad (2.148)$$

The last pattern [311] is not a proper $SU(3)$ pattern because it has more than two rows. Indeed, the filled first column represents a "closed shell" which does not contribute to the quantum numbers. By removing it we gain a proper $SU(3)$ pattern [20]. We can now restate eq. (2.148) in terms of the $SU(3)$ quantum numbers $(\lambda\mu)$ as

$$(21) \otimes (10) = (31) \oplus (12) \oplus (20). \qquad (2.149)$$

In fact, from this example we see the general relation

$$(\lambda\mu) \otimes (10) = (\lambda + 1, \mu) \oplus (\lambda - 1, \mu + 1) \oplus (\lambda, \mu - 1). \quad (2.150)$$

The improper pattern $[1^3]$ encountered above is equivalent to empty, i.e., an $SU(3)$ scalar with quantum numbers (00). Its angular momentum is zero. The representation (10), a single square, has $J = j = 1$. From eq. (2.150) we have

$$(10) \otimes (10) = (20) \oplus (01). \quad (2.151)$$

The left-hand side gives angular momenta 0, 1, and 2. On the right-hand side we have $(01) = [1^2]$; it represents one hole and thus has $J = 1$. It follows that (20) has $J = 0$ and 2. Continuing with (01), we have

$$(01) \otimes (10) = (11) \oplus (00), \quad (2.152)$$

whence we deduce $J = 1, 2$ for (11). One more example is

$$(20) \otimes (10) = (30) \oplus (11), \quad (2.153)$$

yielding $J = 1, 3$ for (30). We remark here that the $J = 1$ states that appear here will be excluded in IBM-1 by the reduction $U(6) \supset SU(3)$, as we see below.

The building-up process can be continued similarly. It has given rise to an algorithm due to Elliott. It consists of two steps. First, define an auxiliary quantity

$$K_0 \equiv \min\{\lambda, \mu\}, \min\{\lambda, \mu\} - 2, \min\{\lambda, \mu\} - 4, \ldots, 0 \text{ or } 1. \quad (2.154)$$

Then the values of J are given by

$$J = \begin{cases} K_0, K_0 + 1, K_0 + 2, \ldots, K_0 + \max\{\lambda, \mu\} & \text{for } K_0 \neq 0, \\ 0, 2, 4, \ldots, \max\{\lambda, \mu\} & \text{for } K_0 = 0. \end{cases} \quad (2.155)$$

The quantity K_0 is related to the K quantum number of the geometric rotational model (see Chapter 1). We will elaborate the point in Section 2.7.

2.6.3.3. The reduction $U(6) \supset SU(3)$

The last remaining reduction within our scheme is $U(6) \supset SU(3)$. In other words we want to know which $SU(3)$ quantum numbers $(\lambda\mu)$ belong to a given symmetric representation $[N]$ of $U(6)$. The idea and technique of the reduction is essentially the same as in the previous case, only somewhat more complicated to carry through. Nevertheless we sketch the procedure and state the algorithm.

The starting point is to note that the fundamental representation [1] of $U(6)$ is six-dimensional as there are six single-particle states. On the other hand, the $SU(3)$ representation (20) is also six-dimensional; its basis states are of the form aa, bb, cc, ab + ba, bc + cb and ca + ac. Analogously to the correspondence (10) and $J = 1$, the $U(6)$ square [1] corresponds to the symmetric pair (20) of $SU(3)$.

The building-up process requires parallel multiplication of $U(6)$ Young patterns and $SU(3)$ Young patterns. To distinguish the patterns of the two groups, we put a spot inside the $U(6)$ squares. The first step of multiplication in $U(6)$ is then

$$\begin{array}{|c|}\hline \cdot \\ \hline\end{array} \otimes \begin{array}{|c|}\hline \cdot \\ \hline\end{array} = \begin{array}{|c|c|}\hline \cdot & \cdot \\ \hline\end{array} \oplus \begin{array}{|c|}\hline \cdot \\ \hline \cdot \\ \hline\end{array} \tag{2.156}$$

The corresponding $SU(3)$ product is

$$\begin{array}{|c|c|}\hline \ & \ \\ \hline\end{array} \otimes \begin{array}{|c|c|}\hline \ & \ \\ \hline\end{array} = \begin{array}{|c|c|c|c|}\hline \ & \ & \ & \ \\ \hline\end{array} \oplus \begin{array}{|c|c|c|}\hline \ & \ & \ \\ \hline \ \\ \cline{1-1}\end{array} \oplus \begin{array}{|c|c|}\hline \ & \ \\ \hline \ & \ \\ \hline\end{array} \tag{2.157}$$

The first pattern (40) on the right-hand side is symmetric. The last pattern (02) is equivalent to the two-hole pattern (20), and thus also symmetric. We associate the symmetric $U(6)$ pattern [2] with these symmetric $SU(3)$ patterns and conclude that the $SU(3)$ representations (40) and (02) are contained in the $U(6)$ representation [2]. The remaining $SU(3)$ representation (21) is contained in the $U(6)$ representation $[1^2]$. This latter $U(6)$ representation does not occur in IBM-1, expressly restricted to $[N]$, which rules out the $J = 1$ state contained in (21); all $J = 1$ states disappear similarly.

Without showing details, we give the results for $N = 3$:

$$[3]:\ (60), (22), (00); \qquad [21]:\ (41), (22), (11). \qquad (2.158)$$

We refrain from further specific examples and state the general result for the *symmetric* $U(6)$ representation $[N]$:

$$\begin{aligned}
(\lambda\mu) = {} & (2N, 0), (2N-4, 2), (2N-8, 4), \ldots, \\
& \begin{cases} (0N) & N \text{ even}, \\ (2, N-1) & N \text{ odd}, \end{cases} \\
& (2N-6, 0), (2N-10, 2), (2N-14, 4), \ldots, \\
& \begin{cases} (2, N-4) & N \text{ even}, \\ (0, N-3) & N \text{ odd}, \end{cases} \\
& (2N-12, 0), (2N-16, 2), (2N-20, 4), \ldots, \\
& \begin{cases} (0, N-6) & N \text{ even}, \\ (2, N-7) & N \text{ odd}, \end{cases} \\
& \cdots \\
& \vdots
\end{aligned} \qquad (2.159)$$

with the understanding that λ and μ may never be negative. We note that all values of λ and μ given by eq. (2.159) are even; odd values can only come from $U(6)$ representations other than $[N]$. As an example of the application of eq. (2.159) we give

$$[4]:\ (80), (42), (04), (20). \qquad (2.160)$$

We have now completed the analysis of the quantum numbers of Chain II, eq. (2.90). The group $SU(3)$ is not fully reducible with respect to $O(3)$. The situation is analogous to the reduction $O(5) \supset O(3)$ as summarized in eqs. (2.132) and (2.140). Instead of the auxiliary quantum number K_0 introduced in eq. (2.154), we use a related quantity K, specified in Section 2.7, as the necessary additional quantum number. Chain II thus complemented reads

$$\underset{[N]}{U(6)} \supset \underset{(\lambda\mu)}{SU(3)} \supset_{K} \underset{J}{O(3)} \supset \underset{M}{O(2)} \qquad (2.161)$$

2.7. DYNAMIC SYMMETRIES

2.7.1. General Background

The idea of a dynamic symmetry is easiest to understand from an elementary physical example. Consider a Hamiltonian

$$H = 2a\mathbf{J}^2 + 2bJ_z^2 = aC_{2O3} + bC_{2O2}, \tag{2.162}$$

where we have substituted the Casimir operators of $O(3)$ and $O(2)$ from eqs. (2.96) and (2.97). The Hamiltonian is substantially like one for the Zeeman effect. Since $[\mathbf{J}^2, H] = 0$ and $[J_z, H] = 0$, J and M are good quantum numbers. Yet the Hamiltonian is not spherically symmetric, or a scalar, since $[J_{x,y}, H] \neq 0$; it *is* symmetric with respect to the z axis. The eigenvalues are

$$E_{JM} = 2aJ(J + 1) + 2bM^2. \tag{2.163}$$

The eigenfunctions are the same angular-momentum eigenstates $|JM\rangle$ that would obtain in the absence of the b term; only then the energies would be degenerate with respect to M. The symmetry properties of our Hamiltonian can be characterized by saying that it has a *dynamic symmetry* associated with the group chain $O(3) \supset O(2)$. This example serves us as the prototype of a dynamic symmetry and it is illustrated in Fig. 2.2. Note that a twofold degeneracy remains for $M > 0$.

Certain basic systems of physics have dynamic symmetries, although they are not often mentioned by that name. Furthermore, these symmetries are essentially classical and need not be considered in terms of quantum mechanics. The Kepler problem, or hydrogen atom, with a potential proportional to $1/r$ has dynamic symmetry $O(4)$ in addition to the normal geometric rotational symmetry. The three-dimensional harmonic oscillator has dynamic symmetry $SU(3)$. Classically these two dynamic symmetries manifest themselves in that the orbital ellipse stays put and does not precess. For basic details see Schiff (1968) and Lipkin (1965); for a complete technical study see Wybourne (1974).

The most fertile realm of dynamic symmetries is in particle physics. The $SU(3)$ symmetry of hadrons built from u, d, and s quarks is a prime example. Dynamic symmetries are also referred to as weakly broken symmetries. The idea is then that the quantum numbers of the pure symmetry remain good and the departure from

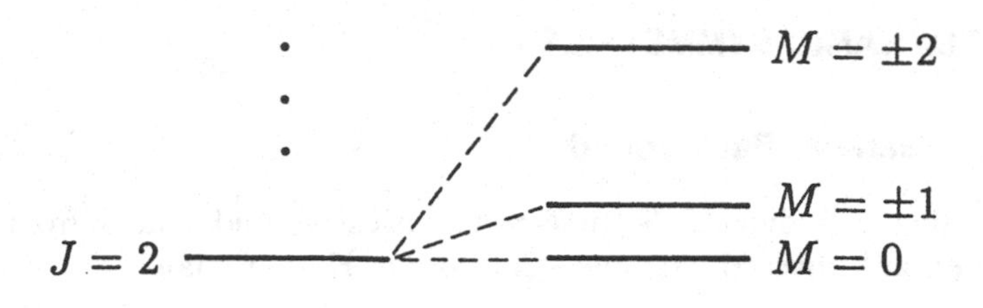

$J = 1$ $M = \pm 1$ $M = 0$

$J = 0$ $M = 0$

$$\begin{array}{ccc} O(3) & \supset & O(2) \\ E_{JM} = 2a\,J(J+1) & + & 2b\,M^2 \end{array}$$

FIGURE 2.2. *The dynamic symmetry O(3) ⊃ O(2) possessed by the Hamiltonian (2.162).*

degeneracy is small at each link of the group chain, as illustrated in Fig. 2.2.

2.7.2. Dynamic Symmetries in the IBM

The IBM Hamiltonian consists of one- and two-body terms of the form (2.15). It is of course a scalar, i.e., the various terms are coupled to angular momentum zero. This is best expressed using tensor products and scalar products of tensors as introduced in eqs. (2.80) and (2.101) and in Appendix 2. Written out, with the constants expressed in a traditional form, the Hamiltonian then reads

$$H = \epsilon_s s^\dagger s + \epsilon_d d^\dagger \cdot \tilde{d} + \frac{1}{2} \sum_{L=0,2,4} c_L [d^\dagger d^\dagger]_L \cdot [\tilde{d}\tilde{d}]_L$$
$$+ \frac{v_2}{\sqrt{10}} ([d^\dagger d^\dagger]_2 \cdot \tilde{d}s + \text{H.c.})$$
$$+ \frac{v_0}{2\sqrt{5}} (d^\dagger \cdot d^\dagger ss + \text{H.c.})$$
$$+ \frac{u_2}{\sqrt{5}} d^\dagger s^\dagger \cdot \tilde{d}s + \frac{1}{2} u_0 s^\dagger s^\dagger ss, \qquad (2.164)$$

where we have used eq. (2.89) to express the one-boson terms. The nine parameters of this Hamiltonian are the two single-boson energies ϵ_s and ϵ_d, and the seven boson–boson interaction strengths c_0, c_2, c_4, v_2, v_0, u_2, u_0. The parameters are assumed real, and H.c. stands for Hermitian conjugate. The operators $s^\dagger s = \hat{n}_s$ and $d^\dagger \cdot \tilde{d} = \hat{n}_d$ count the numbers of bosons in the s and d states, respectively. The c_L terms describe the interactions of two bosons which remain the the d state. Likewise the last term describes the interaction of two bosons remaining in the s state. All of these terms conserve s- and d-boson numbers separately. The remaining terms represent "inelastic scattering" of bosons: one or both bosons change state in the interaction.

The Hamiltonian (2.164) can be rewritten by substituting $N - \hat{n}_d$ for $\hat{n}_s$ since the operator $\hat{N}$ is always diagonal with eigenvalue N, which is the *constant* number of bosons for a given nucleus. The resulting additive constant can be neglected when, as usual, we are interested in excitation energies rather than absolute energies. Then the Hamiltonian becomes

$$H = \epsilon' \hat{n}_d + \frac{1}{2} \sum_L c'_L [d^\dagger d^\dagger]_L \cdot [\tilde{d}\tilde{d}]_L$$
$$+ \frac{v_2}{\sqrt{10}} ([d^\dagger d^\dagger]_2 \cdot \tilde{d}s + \text{H.c.}) \qquad (2.165)$$
$$+ \frac{v_0}{2\sqrt{5}} (d^\dagger \cdot d^\dagger ss + \text{H.c.}),$$

$$\epsilon' \equiv \epsilon_d - \epsilon_s + (N-1)(u_2/\sqrt{5} - u_0), \qquad (2.166)$$
$$c'_L \equiv c_L + u_0 - 2u_2/\sqrt{5}.$$

Appendix 2 contains operator relations needed for this result. Note that the Hamiltonian (2.165) has only six parameters. Another equivalent six-parameter form is the so-called multipole form

$$H = \epsilon'' \hat{n}_d + a_0 P^\dagger P + a_1 \mathbf{J}\cdot\mathbf{J} + a_2 Q\cdot Q + a_3 T_3\cdot T_3 + a_4 T_4\cdot T_4 \qquad (2.167)$$

with the operators given by eqs. (2.81), (2.86), (2.91), and (2.100). The new parameters are linear functions of the old ones; see Appendix 2. Numerical IBM work is commonly based on this form.

A truly remarkable thing is that the six-parameter IBM Hamiltonian can also be expressed in terms of the Casimir operators of the subgroups of $U(6)$, given in eqs. (2.94)–(2.99); the Casimir operators (2.93) of $U(6)$ itself would only contribute constants for a given nucleus with a fixed N. The result is

$$H = \epsilon''' C_{1U5} + \alpha C_{2U5} + \beta C_{2O5} + \gamma C_{2O3} + \delta C_{2SU3} + \eta C_{2O6}, \qquad (2.168)$$

where again the constants are simply related to those of the other forms. Due to the term $2\hat{N}(\hat{N} + 4)$ in eq. (2.99), this Hamiltonian effectively contains a constant term $2\eta N(N + 4)$ as compared to eqs. (2.165) and (2.167).

Note that the Casimir operators of all three subgroup chains are included; only $C_{2O2} = 2J_z^2$ is trivially absent because it is not a scalar. However, interesting and important special cases emerge when only Casimir operators belonging to one of the three subgroup chains, eqs. (2.132), (2.140), (2.161), are endowed with nonzero coefficients. These *special cases* of the general Hamiltonian represent three dynamic symmetries associated with the respective group chains. These special cases are also called $U(5)$, $SU(3)$, and $O(6)$ *limits*.

2.7.2.1. Dynamic symmetry I, $U(5)$

By putting $\delta = 0 = \eta$ in eq. (2.168), or equivalently $a_0 = 0 = a_2$ in eq. (2.167), we obtain a Hamiltonian which contains only Casimir operators belonging to Chain I, also identified by its second link $U(5)$:

$$\begin{aligned} H_{\mathrm{I}} &= \epsilon''' C_{1U5} + \alpha C_{2U5} + \beta C_{2O5} + \gamma C_{2O3} \\ &= \epsilon'' \hat{n}_{\mathrm{d}} + a_1 \mathbf{J} \cdot \mathbf{J} + a_3 T_3 \cdot T_3 + a_4 T_4 \cdot T_4. \end{aligned} \tag{2.169}$$

The good quantum numbers can be read off eq. (2.132), so that we have the eigenvalue equation

$$H_{\mathrm{I}} |[N] n_{\mathrm{d}} \tau \nu_{\Delta} JM\rangle = E_{\mathrm{I}} |[N] n_{\mathrm{d}} \tau \nu_{\Delta} JM\rangle. \tag{2.170}$$

Equations (2.105)–(2.108) give the general eigenvalues of the present Casimir operators. We now apply them to the completely symmetric states of the IBM. Subsections 2.6.1 and 2.6.2, including eq. (2.128), provide ready results, and we obtain

$$E_{\mathrm{I}} = \epsilon''' n_{\mathrm{d}} + \alpha n_{\mathrm{d}}(n_{\mathrm{d}} + 4) + 2\beta\tau(\tau + 3) + 2\gamma J(J + 1). \tag{2.171}$$

The possible values of the quantum numbers we know, in the order of occurrence, from eqs. (2.116), (2.127), (2.129)–(2.131), and (2.49). The quantum number N is the same for all states of a nucleus. It does not directly affect the energy spectrum, but indirectly it does, because it governs the possible values of the other quantum numbers. The quantum number ν_{Δ} has no effect on the energy because it does not appear in any Casimir eigenvalue. Because of spherical symmetry, there is no C_{2O2} in the Hamiltonian and hence no M dependence for the energy.

We may now stop and admire the power of group theory. The Hamiltonian (2.169), albeit a special case of the general IBM Hamiltonian (2.165) or (2.167), is rather complicated. Yet we have managed to find its quantum numbers and eigenvalues by algebraic means, without solving a differential equation or diagonalizing a matrix. The detailed eigenstates are obtained only with further algebraic effort, but various selection rules can be inferred immediately by means of group theory (Subsection 2.7.3). For a sample and flavor of the wave functions, we do give them for the $U(5)$ case in Table 2.4, in connection with our $N = 3$ examples in Subsection 2.7.4.

We gain further insight into the idea of a dynamic symmetry by recalling the elementary example contained in eq. (2.162) and Fig. 2.2. In analogy to it, we now note that the eigenstates would be the same if the Hamiltonian (2.169) only had the first term n_{d}. The energy would then be degenerate with respect to the other active quantum numbers τ and J. As the other terms are added, the degeneracy is lifted, but the eigenstates remain unchanged.

The eigenstates of H_{I} are the states $|[N] n_{\mathrm{d}} \tau \nu_{\Delta} JM\rangle$. Inasmuch as they are generated by a Hermitian operator, they are automatically orthogonal. In such states, however, the meaning of the additional quantum number ν_Δ is not exactly that defined in connection with eq. (2.129), viz., the number of d-boson triplets. We will not belabor the point, but refer to an analogous situation arising in Chain II and discussed below.

The eigenstates $|[N]n_d\tau\nu_\Delta JM\rangle$ form a complete orthogonal basis not only for H_I but also for the general IBM Hamiltonian. While the former is diagonal in that basis, the latter must be diagonalized numerically. Indeed, the standard IBM program PHINT diagonalizes the general Hamiltonian in this basis, known as the $U(5)$ basis. The eigenfunctions of a general Hamiltonian are thus linear combinations of the $|[N]n_d\tau\nu_\Delta JM\rangle$.

Equation (2.171) can be found in differently parametrized forms in the literature. The same holds true for its counterparts giving the energies E_II and E_III below. However, the dependence on the quantum numbers is decisive and it is the same independently of the choice of parametrization.

2.7.2.2. Dynamic symmetry II, *SU*(3)

When we put ϵ''', α, β, and η all equal to zero in the general Hamiltonian (2.168), we are left with Casimir operators belonging to Chain II, eq. (2.161). The same is achieved by leaving only a_1 and a_2 nonzero in eq. (2.167), as can be seen from the explicit expressions (2.96) and (2.98) for the $O(3)$ and $SU(3)$ Casimir operators. The Hamiltonian for the dynamic symmetry associated with Chain II, or its second link $SU(3)$, is thus

$$\begin{aligned} H_\text{II} &= \gamma C_{2O3} + \delta C_{2SU3} \\ &= a_1 \mathbf{J}\cdot\mathbf{J} + a_2 Q\cdot Q. \end{aligned} \tag{2.172}$$

Since we know the eigenvalues of the Casimir operators, eqs. (2.108) and (2.114), we can immediately write down the eigenvalue

$$E_\text{II} = 2\gamma J(J+1) + \tfrac{2}{3}\delta(\lambda^2 + \mu^2 + \lambda\mu + 3\lambda + 3\mu). \tag{2.173}$$

The possible values of the quantum numbers can be read off eqs. (2.159), (2.154), (2.155), and (2.49). The energies are degenerate with respect to K_0, which is not related to any Casimir operator, and the geometric M degeneracy is trivial.

The question of the eigenstates of Chain II requires further examination, as indicated in Subsection 2.6.3. Offhand, the eigenstates would seem to be simply $|[N](\lambda\mu)K_0JM\rangle$, where K_0 is the additional quantum number defined in eq. (2.154). These states, known as the Elliott basis, do form a complete set but they are not orthogonal. The Elliott states as such are thus not appropriate eigenfunctions of the Hermitian operator H_{II}, nor are they a convenient basis for diagonalization.

Orthogonality is restored by an application of the Schmidt scheme. The assignment of quantum numbers K_0 and J for a given $(\lambda\mu)$ is governed by eqs. (2.154) and (2.155). We now define a quantum number K to label orthogonal linear combinations of Elliott states with different values of K_0. The set of K values is taken to be the same as the set of K_0 values. We number them $K_1 < K_2 < K_3 < \cdots$. The orthonormal states are now

$$\begin{aligned}
|[N](\lambda\mu)K_1JM\rangle &= |[N](\lambda\mu)K_{01}JM\rangle, \\
|[N](\lambda\mu)K_2JM\rangle &= c_{21}|[N](\lambda\mu)K_{01}JM\rangle \\
&\quad + c_{22}|[N](\lambda\mu)K_{02}JM\rangle, \\
|[N](\lambda\mu)K_3JM\rangle &= c_{31}|[N](\lambda\mu)K_{01}JM\rangle \\
&\quad + c_{32}|[N](\lambda\mu)K_{02}JM\rangle \\
&\quad + c_{33}|[N](\lambda\mu)K_{03}JM\rangle,
\end{aligned} \tag{2.174}$$

and so forth. The coefficients c_{ij} are determined from the orthonormality requirement

$$\langle [N](\lambda\mu)K_iJM|[N](\lambda\mu)K_jJM\rangle = \delta_{ij}. \tag{2.175}$$

Computing the overlaps between different values of K_0 is quite involved; see Iachello and Arima (1987). The basis of states $|[N](\lambda\mu)KJM\rangle$ is known as the *Vergados basis*. The energy, given by eq. (2.173), remains degenerate with respect to K.

Let us see about the assignment of K values within a given $SU(3)$ representation $(\lambda\mu)$. When a particular J occurs only once, it is assigned to the lowest possible K value, K_1. Note that for J even, the lowest possible K is zero. When a J occurs twice, it is assigned to the two lowest K values, K_1 and K_2; and so forth. Note that the J values resulting from a particular K_0 according to eq. (2.155) are generally not the same as the J values assigned to $K = K_0$. A concrete example will be given in Subsection 2.7.4.

2.7.2.3. Dynamic symmetry III, O(6)

When we put the parameters ϵ''', α, and δ of eq. (2.168) equal to zero, or equivalently ϵ'', a_2, $a_4 = 0$ in eq. (2.167), we are left with

$$\begin{aligned} H_{\text{III}} &= \beta C_{2O5} + \gamma C_{2O3} + \eta C_{2O6} \\ &= a_0 P^\dagger P + a_1 \mathbf{J}\cdot\mathbf{J} + a_3 T_3 \cdot T_3 + 2\eta \hat{N}(\hat{N} + 4). \end{aligned} \tag{2.176}$$

The Hamiltonian H_{III} consists of Casimir operators from Chain III, eqs. (2.95), (2.96), (2.99), and represents the corresponding dynamic symmetry, also referred to by the second link $O(6)$. The eigenvalue equation is, with quantum numbers from eq. (2.140),

$$H_{\text{III}} |[N]\sigma\tau\nu_\Delta JM\rangle = E_{\text{III}} |[N]\sigma\tau\nu_\Delta JM\rangle. \tag{2.177}$$

Equations (2.107), (2.108), and (2.111), or the relevant special cases (2.128) and (2.138), give the eigenvalues of the Casimir operators, whence

$$E_{\text{III}} = 2\beta\tau(\tau + 3) + 2\gamma J(J + 1) + 2\eta\sigma(\sigma + 4). \tag{2.178}$$

Just as in the case of Chain I, the additional quantum number ν_Δ has no effect on the energies. The M degeneracy of course is present as before.

The possible values for the various quantum numbers are found from eqs. (2.137), (2.139), (2.129)–(2.131), and (2.49). The eigenstates $|[N]\sigma\tau\nu_\Delta JM\rangle$ form a complete orthogonal basis when the additional quantum number ν_Δ is properly defined as in the case of Chain I. In fact, any of the three bases spans the complete IBM space and can thus be used as a basis in numerical diagonalization of the general Hamiltonian. The prevalence of the $U(5)$ basis $|[N]n_d\tau\nu_\Delta JM\rangle$ derives only from practical reasons. The bases may be transformed into one another by unitary transformations.

2.7.3. Multipole Operators and Selection Rules

The IBM operators for electromagnetic multipole transitions are composed of the generators (2.77) or (2.80) of $U(6)$. Normally only linear terms are included. Electric quadrupole (E2) transitions are by far the most important ones in the IBM and in other collective models of nuclear structure. Below we discuss them, particularly in relation to the three dynamic symmetries introduced above. While the multipole operators do not change the total boson number N, we note as an aside that the IBM operators for two-nucleon, i.e., one-boson, transfer reactions do change N by one.

The general linear E2 operator of the IBM is the $l = 2$ tensor operator

$$T(\mathrm{E2}) = \alpha_2(s^\dagger\tilde{d} + d^\dagger s) + \beta_2[d^\dagger\tilde{d}]_2 \equiv \alpha_2(s^\dagger\tilde{d} + d^\dagger s + \chi[d^\dagger\tilde{d}]_2), \tag{2.179}$$

where α_2 and β_2, or the conventional χ, are free parameters. We now look at the whole tensor without denoting its components; see Appendix 2.

The important physical quantities calculated with the E2 operator are the reduced transition probability $B(\mathrm{E2}; J \to J')$ and the quadrupole moment $Q(J)$; in the notation we have suppressed dependence on other quantum numbers. These quantities are defined by the expressions

$$B(\mathrm{E2}; J \to J') \equiv \sum_{mM'} |\langle J'M'|T(\mathrm{E2})_m|JM\rangle|^2 = \frac{1}{2J+1}|(J'\|T(\mathrm{E2})\|J)|^2, \tag{2.180}$$

$$Q(J) \equiv \sqrt{\frac{16\pi}{5}}\langle JJ|T(\mathrm{E2})_0|JJ\rangle = \sqrt{\frac{16\pi}{5}}\begin{pmatrix} J & 2 & J \\ -J & 0 & J \end{pmatrix}(J\|T(\mathrm{E2})\|J). \tag{2.181}$$

These definitions are model independent. Their second forms are obtained through the Wigner–Eckart theorem

$$\langle J'M'|T_{lm}|JM\rangle = (-1)^{J'-M'}\begin{pmatrix} J' & l & J \\ -M' & m & M \end{pmatrix}(J'\|T_l\|J), \tag{2.182}$$

where T_l is a general tensor operator, $(\cdots)$ is a $3j$ symbol and $(\|\ \|)$ is a reduced matrix element.

The first term of eq. (2.179) changes one boson between s and d states. The second term only affects d bosons without changing their number. As noted above, the total boson number $N = n_s + n_d$ remains intact. This is a necessary feature for describing transitions in a nucleus whose N is fixed by definition of the model. In addition to the general angular-momentum and parity selection rules for E2 transitions, we then have the specific IBM selection rules

$$\Delta n_s = -\Delta n_d = \pm 1, \qquad \Delta n_s = 0 = \Delta n_d, \tag{2.183}$$

for the first and second term of eq. (2.179), respectively.

Strict adherence to the idea of dynamic symmetries would demand that the multipole operators be generators of the second link of the subgroup chain in question. Thus for Chain I we should use the generators of $U(5)$, i.e., the $T_{lm} = [d^\dagger\tilde{d}]_{lm}$ of eq. (2.81). Consequently the E2 operator for Chain I would be $\beta_2 T_2$, whose selection rule is $\Delta n_d = 0$. Since Chain-I states have n_d as a good quantum number (see Table 2.3 and Fig. 2.3), $\beta_2 T_2$ would give nonvanishing quadrupole moments but no E2 transitions at all. This is just opposite to what we know experimentally about nuclei whose energy levels approximately conform to the $U(5)$ formula (2.171). It is therefore customary to adopt the first term for Chain I,

$$T(\mathrm{E2})_{\mathrm{I}} = \alpha_2(s^\dagger\tilde{d} + d^\dagger s), \tag{2.184}$$

which corresponds to the choice $\chi = 0$ in eq. (2.179). This operator changes not only n_d (and n_s) but also the boson seniority τ by unity, so we have the selection rules

$$\Delta n_d = \pm 1, \qquad \Delta\tau = \pm 1. \tag{2.185}$$

For Chain II we have the $SU(3)$ generator Q of eq. (2.91) as the only possibility for an E2 operator:

$$T(\mathrm{E2})_{\mathrm{II}} = \alpha_2 Q = \alpha_2(s^\dagger\tilde{d} + d^\dagger s - \tfrac{1}{2}\sqrt{7}[d^\dagger\tilde{d}]_2). \tag{2.186}$$

This is clearly a special case of eq. (2.179) with $\chi = -\frac{1}{2}\sqrt{7}$. Since Q is a generator of $SU(3)$, it cannot connect different representations (Subsection 2.4.1); it is like the components of **J** cannot connect states with different values of J. Thence we have the selection rule

$$\Delta(\lambda\mu) = 0. \tag{2.187}$$

For Chain III the only generator of tensor degree $l = 2$ is $s^\dagger\tilde{d} + d^\dagger s$. it follows that the E2 operator is the same as for Chain I,

$$T(\mathrm{E2})_{\mathrm{III}} = \alpha_2(s^\dagger\tilde{d} + d^\dagger s). \tag{2.188}$$

A generator cannot change the $O(6)$ representation. Also, the role of τ is the same as in Chain I. The selection rules are then

$$\Delta\sigma = 0, \qquad \Delta\tau = \pm 1. \tag{2.189}$$

Finally we look briefly at M1 and E0 transitions. The only possible linear M1 operator is

$$T(\mathrm{M1}) = \beta_1[d^\dagger \tilde{d}]_1 = \beta_1 T_1. \tag{2.190}$$

However, according to eq. (2.86), this is proportional to the angular-momentum operator $\mathbf{J}$. Since J is a good quantum number for all nuclear states, it follows that operator (2.190) cannot cause any transitions, only static moments. To describe M1 transitions within the IBM, one must postulate a higher-order operator.

The E0 operator is

$$\begin{aligned} T(\mathrm{E0}) &= \gamma_0 + \alpha_0 s^\dagger s + \beta_0[d^\dagger \tilde{d}]_0 \\ &= \gamma_0 + \alpha_0 \hat{N} + (\beta_0/\sqrt{5} - \alpha_0)\hat{n}_d, \end{aligned} \tag{2.191}$$

where we have used eq. (2.89). For E0 transitions the effective term is the one proportional to $\hat{n}_d$; the rest of the operator amounts to a constant. Now n_d is a good quantum number in Chain I, so there can be no E0 transitions. For Chains II and III there *can* be E0 transitions; yet $\hat{n}_d$ is not a generator and therefore strictly speaking not an acceptable operator.

2.7.4. Examples of the Three Dynamic Symmetries

We have now derived the main properties of the three dynamic symmetries associated with the three chains of subgroups of $U(6)$. For a concrete and simple, yet nontrivial, demonstration we now examine the three cases for boson number $N = 3$. Such a small boson number would be physically best applicable to the $U(5)$ symmetry, less so for $O(6)$, and quite unrealistic for $SU(3)$. Nevertheless, the salient features of the three symmetries are most effectively brought home in terms of such a simple example when worked out in detail. All states are then easily included.

Below we calculate the energy levels from the simple formulas for the three dynamic symmetries, eqs. (2.171), (2.173), and (2.178), respectively. We also include E2 transitions and quadrupole moments in our examples. Qualitatively they are obtained from the selection rules derived in Subsection 2.7.3. Their quantitative calculation is a more difficult matter requiring detailed analysis of the wave functions for the three chains.

For the $U(5)$ case general wave functions are constructed with boson coefficients of fractional parentage (cfp), as already mentioned in Subsection 2.6.2. Consequently it turns out that the E2 matrix elements consist essentially of the same coefficients. However, the full machinery need not be employed in the simple $U(5)$ example below, and we write down the wave functions (Table 2.4). The $O(6)$ case is built up with wave functions expanded in the $U(5)$ basis. The $SU(3)$ case is treated differently and it is quite involved. Fortunately there are computer programs, PHINT by Scholten the best known of them, to handle the needs of practical applications. In this text we mostly content ourselves with stating the results for transition rates. For a detailed analytic treatment one should consult Iachello and Arima's book (1987) and original papers.

The purpose of our simple examples is to reinforce the foregoing theoretical discussion. A thorough and many-faceted study of the predictions of the model and comparison with experiment is the subject of Chapter 3.

2.7.4.1. Dynamic symmetry $U(5)$

The quantum numbers of Chain I are found from eqs. (2.116), (2.127), (2.129)–(2.131). We neglect the trivial quantum number M. The quantum numbers for $N = 3$ are given in Table 2.3. The auxiliary number Λ is included although it does not label the states.

We want to stress that Table 2.3 includes everything for $N = 3$. It is indeed a characteristic feature of the IBM that for a given finite N it has a definite number of states. This is an important

TABLE 2.3. *$U(5)$-chain quantum numbers for $N = 3$.*

n_d	τ	ν_Δ	Λ	J
0	0	0	0	0
1	1	0	1	2
2	2	0	2	4, 2
	0	0	0	0
3	3	0	3	6, 4, 3
		1	0	0
	1	0	1	2

difference from the geometric model where the spectrum goes on to infinity. In fact, the IBM limit $N \to \infty$ corresponds to the geometric model.

The number of states in Table 2.3, not counting the M substates, is seen to be 10. This number is only a property of the $U(6)$ space for $N = 3$, independent of whether or not we have a dynamic symmetry providing the quantum numbers. We can check the number of states by computing the number of ways in which we can put three indistinguishable objects into six boxes. The answer is $8!/3!5! = 56$. The number of substates contained in Table 2.3 is $1 + 5 + 9 + 5 + 1 + 13 + 9 + 7 + 1 + 5 = 56$, which checks with our previous figure.

Not only is the total number of states a function of N only, but for a given N we also get the same J states irrespective of the Hamiltonian. Formally this can be seen from the fact that $U(6) \supset O(3)$. The following examples of the two other dynamic symmetries demonstrate the validity of the assertion in detail. We may draw the general conclusion that, given an N, we can determine the J spectrum via any one of the three chains, no matter what the Hamiltonian is.

The $U(5)$ energy spectrum is given by eq. (2.171). With the four parameters ϵ''', α, β, and γ picked for a qualitative resemblance to experiment, the energy levels are shown in Fig. 2.3. The levels are grouped into the conventional ground, beta, and gamma quasirotational bands, although such bands have little validity well outside the rotational/deformed region of nuclei.

Apart from the finite number of levels and the anharmonicity due to the α, β, and γ terms of eq. (2.171), the spectrum is the same as that of the spherical quadrupole vibrator discussed in Chapter 1. We can clearly discern the "two-phonon triplet" $J = 0, 2, 4$ and the "three-phonon quintuplet" $J = 0, 2, 3, 4, 6$. Even the anharmonicity is not an essential difference because the same anharmonicity can be produced by suitable extra terms in the geometric model. The $U(5)$ dynamic symmetry of the IBM thus corresponds to the geometric anharmonic vibrator.

So far we have not needed explicit wave functions; the quantum numbers of Table 2.3 and the energies were obtained directly from the group-theoretic formalism. However, we also want to calculate the E2 transition rates and for that we do need explicit wave functions, although the selection rules come directly from the group theory (Subsection 2.7.3). The $U(5)$ wave functions can be written

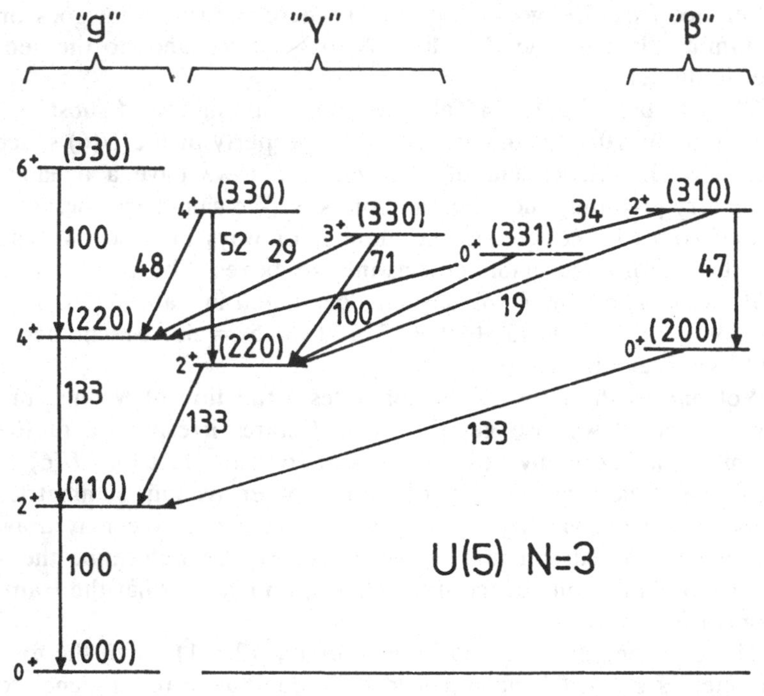

FIGURE 2.3. *Complete U(5) spectrum for N = 3. Quantum numbers* $(n_d \tau v_\Delta)$, *B*(E2) *values normalized to* $2_l^+ \to 0_l^+$, *and quasibands are shown.*

down by inspection for $n_d = 0$, 1, and 2; see eq. (2.9). Even for $n_d = 3$ the form is easy, but due to angular-momentum coupling the normalization constant requires a nontrivial calculation.

In the wave functions we use tensor products $[d^\dagger d^\dagger]_l$, similar to the T_l of eq. (2.81) and generally defined in the Appendix. The vacuum state of no bosons is denoted by $|\tilde{0}\rangle$ as in eq. (2.78). The normalization constants are computed by application of the boson commutators (2.5), (2.7) and (2.8), and some angular-momentum algebra. The resulting wave functions $|[3]n_d \tau v_\Delta JM\rangle$ are given in Table 2.4. The definition of boson seniority τ as the number of bosons not pairwise coupled to zero angular momentum can be concretely seen in the wave functions. Likewise the meaning of v_Δ as the number boson triplets coupled to zero is evident.

The E2 matrix elements for the $U(5)$ case are now calculated with the operator (2.184) and the wave functions of Table 2.4. The reduced matrix element for each transition is found by eq. (2.182)

TABLE 2.4. *U*(5) *wave functions for N* = 3.

n_d	τ	ν_Δ	J	Wave Function
0	0	0	0	$\sqrt{1/6}(s^\dagger)^3\lvert\bar{0}\rangle$
1	1	0	2	$\sqrt{1/2}(s^\dagger)^2 d^\dagger_M\lvert\bar{0}\rangle$
2	2	0	4	$\sqrt{1/2}s^\dagger[d^\dagger d^\dagger]_{4M}\lvert\bar{0}\rangle$
		0	2	$\sqrt{1/2}s^\dagger[d^\dagger d^\dagger]_{2M}\lvert\bar{0}\rangle$
	0	0	0	$\sqrt{1/2}s^\dagger[d^\dagger d^\dagger]_{00}\lvert\bar{0}\rangle$
3	3	0	6	$\sqrt{1/6}[[d^\dagger d^\dagger]_4 d^\dagger]_{6M}\lvert\bar{0}\rangle$
		0	4	$\sqrt{7/22}[[d^\dagger d^\dagger]_2 d^\dagger]_{4M}\lvert\bar{0}\rangle$
		0	3	$\sqrt{7/30}[[d^\dagger d^\dagger]_2 d^\dagger]_{3M}\lvert\bar{0}\rangle$
		1	0	$\sqrt{1/6}[[d^\dagger d^\dagger]_2 d^\dagger]_{00}\lvert\bar{0}\rangle$
	1	0	2	$\sqrt{5/14}[d^\dagger d^\dagger]_{00} d^\dagger_M\lvert\bar{0}\rangle$

and the reduced transition probability B(E2) by eq. (2.180). The results are given in Fig. 2.3 for all transitions allowed by the selection rules (2.185). The B(E2) values given are normalized to $B(\mathrm{E2}; 2_1^+ \to 0_1^+) = 100$.

In the $U(5)$ case the B(E2) values of the IBM are simply related to those of the geometric model. As we indicated above, the geometric limit is realized when $N \to \infty$. The relation is

$$B(\mathrm{E2})_N = (1 - n_d/N)B(\mathrm{E2})_\infty, \tag{2.192}$$

where n_d refers to the final state. A falloff of E2 rates characteristic of the IBM is evident in the three strengths equal to 133; in the geometric model they would be 200.

The quadrupole moments (2.181) vanish for all states in the $U(5)$ case because of the selection rules (2.185); the second term in eq. (2.179) would make them finite.

2.7.4.2. Dynamic symmetry *SU*(3)

For Chain II we find the quantum numbers ($\lambda\mu$), K_0, and J from eqs. (2.159), (2.154), and (2.155). These are the quantum numbers of the Elliott basis, as discussed in Subsection 2.7.2 above. The results for our choice $N = 3$ are stated in Table 2.5. We proceed to go over into the orthogonal Vergados basis (2.174) with the quantum numbers ($\lambda\mu$), K, and J. The result is given in Table 2.6. The difference between Tables 2.5 and 2.6, namely for ($\lambda\mu$) =

TABLE 2.5. *SU(3)-chain quantum numbers for N = 3, Elliott basis.*

$(\lambda\mu)$	K_0	J
(60)	0	0, 2, 4, 6
(22)	2	2, 3, 4
	0	0, 2
(00)	0	0

(22), shows how the Vergados scheme works. We can anticipate from the discussion following Table 2.3 that we have again the same J spectrum as in the $U(5)$ case, $J = 0^3, 2^3, 3, 4^2, 6$. These values of J are indeed the ones present in Tables 2.5 and 2.6.

Figure 2.4 shows the $SU(3)$ energy levels computed from eq. (2.173) with suitable parameters γ and δ. A genuine rotational structure is discernible with ground, beta, and gamma bands in the pattern of the geometric model, where K is the quantum number for the angular-momentum component along the nuclear symmetry axis. The beta and gamma bands are degenerate. This we knew to expect from the discussion of the quantum numbers K_0 and K in Subsection 2.7.2.

The E2 transitions are now calculated with the operator (2.186) and the Vergados wave functions (2.174). As intimated at the beginning of this subsection, the $SU(3)$ case is quite involved as regards the wave functions and computation of matrix elements. Nevertheless, we may observe that since the $U(5)$ basis is a complete basis for a given total boson number N, the $SU(3)$ wave functions for $N = 3$ could be expressed as some linear combinations of the wave functions of Table 2.4. We neglect further consideration of the $SU(3)$ wave functions and only state the results for E2 matrix elements.

The B(E2) values, with strength normalized as in Fig. 2.3, are

TABLE 2.6. *SU(3)-chain quantum numbers for N = 3, Vergados basis.*

$(\lambda\mu)$	K	J
(60)	0	0, 2, 4, 6
(22)	0	0, 2, 4
	2	2, 3
(00)	0	0

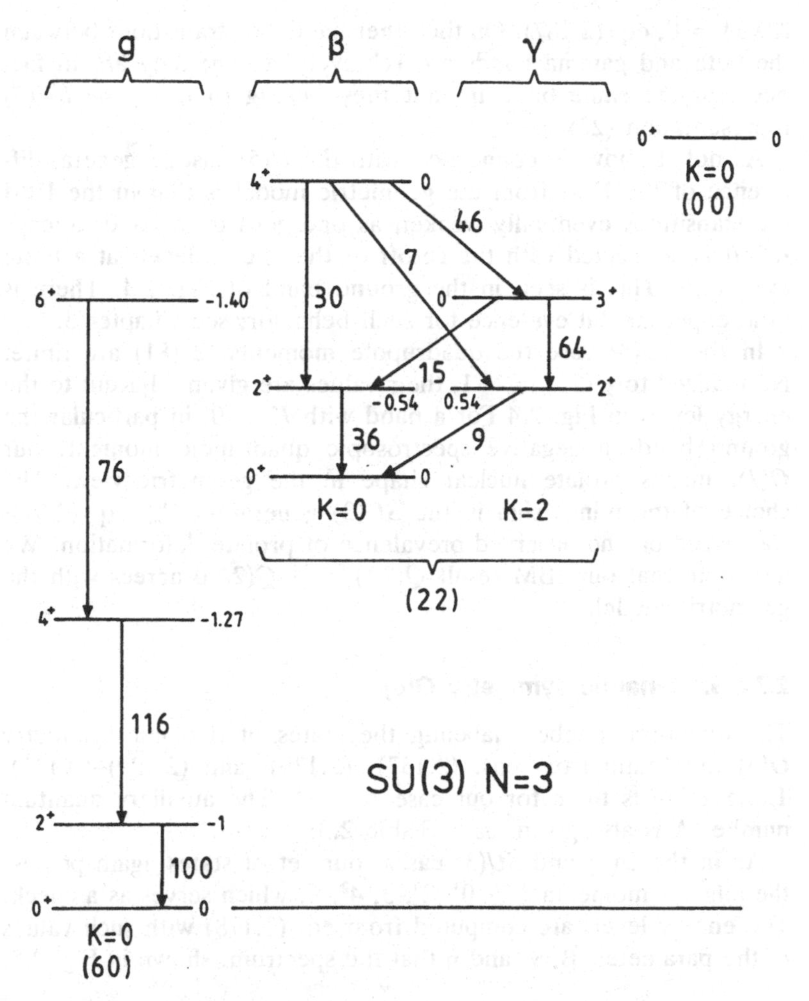

FIGURE 2.4. *Complete SU(3) spectrum for N = 3. The K quantum numbers defining the bands are assigned according to Vergados. Quantum numbers (λμ), B(E2) values normalized to $2_1^+ \rightarrow 0_1^+$, quadrupole moments normalized to $Q(2_1^+) = -1$, and band names are shown.*

shown in Fig. 2.4. When we add to the figure the datum $B(\mathrm{E2}; 2_\beta^+ \leftrightarrow 2_\gamma^+) = 26$, all allowed transitions are included. Their pattern *roughly* conforms to the geometric model: intraband transitions are strong, interband transitions weak. In fact, the interband transitions $\beta \rightarrow g$ and $\gamma \rightarrow g$ are completely forbidden by the selection rule

$\Delta(\lambda\mu) = 0$, eq. (2.187). On the other hand, the transitions between the beta and gamma bands are relatively strong; they are in fact one and the same band in that they belong to the same $SU(3)$ representation (22).

As noted above in connection with the $U(5)$ case, a general difference of the IBM from the geometric model is that in the IBM the transitions eventually weaken as one goes up a band. Such a falloff is connected with the cutoff of the energy levels at a finite excitation. This is seen in the ground band of Fig. 2.4. There is little experimental evidence for such behavior; see Chapter 3.

In the $SU(3)$ case the quadrupole moments (2.181) are finite. Normalized to $Q(2_1^+) = -1$, their values are given adjacent to the energy levels in Fig. 2.4 For a band with $K = 0$, in particular the ground band, a negative spectrosopic quadrupole moment, our $Q(J)$, means prolate nuclear shape in the geometric view. The choice of the minus sign in the $SU(3)$ generators Q_m, eq. (2.91), was based on the observed prevalence of prolate deformation. We also note that our IBM result $Q(2_\gamma^+) = -Q(2_\beta^+)0$ agrees with the geometric model.

2.7.4.3. Dynamic symmetry O(6)

The quantum numbers labeling the states of dynamic symmetry $O(6)$ are found from eqs. (2.137), (2.139), and (2.129)–(2.131). Table 2.7 lists them for our case $N = 3$. The auxiliary quantum number Λ is also given, as in Table 2.3.

As in the $U(5)$ and $SU(3)$ cases, our set of states again possess the angular momenta $J = 0^3, 2^3, 3, 4^2, 6$, which serves as a check. The energy levels are computed from eq. (2.178) with such values of the parameters β, γ, and η that the spectrum, shown in Fig. 2.5,

TABLE 2.7. *$O(6)$-chain quantum numbers for $N = 3$.*

σ	τ	ν_Δ	Λ	J
3	3	0	3	6, 4, 3
		1	0	0
	2	0	2	4, 2
	1	0	1	2
	0	0	0	0
1	1	0	1	2
	0	0	0	0

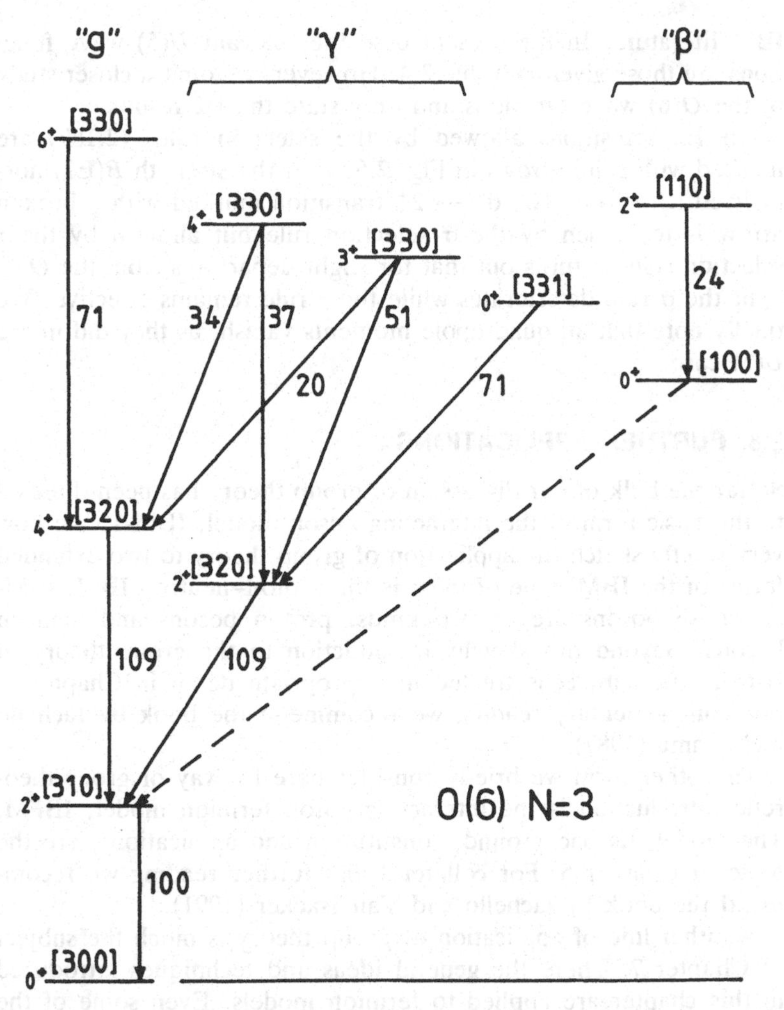

FIGURE 2.5. *Complete O(6) spectrum for N = 3. Quantum numbers $[\sigma\tau\nu_\Delta]$, B(E2) values normalized to $2_1^+ \rightarrow 0_1^+$, and quasibands are shown.*

has a physically realistic appearance. The spectrum is rather like the $U(5)$ spectrum of Fig. 2.3. One distinguishing feature is that the "two-phonon triplet" 0_2^+, 2_2^+, 4_1^+ of $U(5)$ is broken up in $O(6)$ so that the 0^+ member has moved away up.

In the $O(6)$ case the E2 operator is given by eq. (2.188). The wave functions are definite linear combinations of the $U(5)$ wave functions for the same N, and are expressed as such in the detailed

IBM literature. In the present case the relevant $U(5)$ wave functions are those given in Table 2.4. However, we omit a closer study of the $O(6)$ wave functions and only state the E2 results.

All E2 transitions allowed by the selection rules (2.189) are marked with solid arrows in Fig. 2.5, with the strength B(E2) normalized as before. The $0^+ \rightarrow 2^+$ transition marked with a broken arrow is forbidden by the σ selection rule but allowed by the τ selection rule. It turns out that for slight departures from the $O(6)$ limit the σ rule deteriorates while the τ rule remains effective. We finally note that all quadrupole moments vanish, as they did in the $U(5)$ case.

2.8. FURTHER APPLICATIONS

So far the bulk of our discussion of group theory has been directed to the basic form of the interacting-boson model, IBM-1. We now very briefly sketch the application of group theory to two extended forms of the IBM. One of them is the proton–neutron IBM, IBM-2, whose bosons are of two kinds: proton bosons and neutron bosons. Beyond this sketchy introduction to the group theory of IBM-2, the subject is treated in appropriate detail in Chapter 4. For complementary reading we recommend the book by Iachello and Arima (1987).

The other form we briefly consider here by way of group-theoretic introduction is the interacting-boson–fermion model, IBFM. The model, its background, constitution and applications, are the topic of Chapter 5. For collateral and further reading we recommend the book by Iachello and Van Isacker (1991).

Another line of application of group theory is much the subject of Chapter 7. There the general ideas and techniques introduced in this chapter are applied to fermion models. Even some of the fairly detailed material developed here in the context of the IBM will be directly applicable there. The group $SU(3)$ is a case in point. Further preparatory material of relevance to Chapter 7 is presented below in our sketch of fermion groups in the context of the IBFM in Subsection 2.8.2.

2.8.1. Proton–Neutron IBM: IBM-2

The physical idea of the proton–neutron IBM is that one type of boson is built from proton pairs, another type from neutron pairs.

At this point we pay no further attention to the microscopic origin of the bosons but merely take them as given objects. What follows is essentially putting a two-valued index, p for proton bosons and n for neutron bosons, on the operators of Section 2.5. It turns out that the doubling of degrees of freedom much enriches the group-theoretic structure and ensuing physical predictions.

The structure of the IBM-2 Hamiltonian is

$$H = H_{\mathrm{p}} + H_{\mathrm{n}} + H_{\mathrm{pn}}, \tag{2.193}$$

where H_{p} and H_{n} have the structure of the IBM-1 Hamiltonian (see Subsection 2.7.2) and H_{pn} is the interaction between p and n bosons.

Now there are six states available to each kind of boson. The set of 12 states is s_k, $\mathrm{d}_{\mu k}$ with k = p, n and $\mu = \pm 2, \pm 1, 0$. For short we call these states b_{ik}, $i = 1, 2, \ldots, 6$. The Lie algebra of the bilinear operators $b^\dagger_{ik} b_{jk'}$, of the form (2.60), is then $U(12)$. However, it is assumed that not only the total boson number $N = N_{\mathrm{p}} + N_{\mathrm{n}}$ but also its two terms separately are conserved. This means that generators of the type $b^\dagger_{i\mathrm{p}} b_{j\mathrm{n}}$ are excluded, and we have for the model separate algebras $U_{\mathrm{p}}(6)$ and $U_{\mathrm{n}}(6)$. The generators of $U_{\mathrm{p}}(6)$ commute with those of $U_{\mathrm{n}}(6)$. An analogous situation of a product group was encountered in eq. (2.29). The present relationship is denoted as

$$U(12) \supset U_{\mathrm{p}}(6) \times U_{\mathrm{n}}(6). \tag{2.194}$$

The algebra of $U_{\mathrm{p}}(6) \times U_{\mathrm{n}}(6)$, which is the group-theoretic starting point of IBM-2, has a variety of subalgebras considerably richer than plain $U(6)$, the starting point of IBM-1. For one, we can recover the structure of $U(6)$, which is to be expected through the simple device of obliterating the p and n labels. To accomplish this technically, we form from the 36 + 36 = 72 generators of the product group new generators $b^\dagger_{i\mathrm{p}} b_{j\mathrm{p}} + b^\dagger_{i\mathrm{n}} b_{j\mathrm{n}}$. Alternatively we can have for d bosons the coupled form $[d^\dagger_{\mathrm{p}} \tilde{d}_{\mathrm{p}}]_{lm} + [d^\dagger_{\mathrm{n}} \tilde{d}_{\mathrm{n}}]_{lm}$, similarly to eq. (2.80). The new generators, 36 in number, are proton-neutron symmetric. The group they generate has the structure of $U(6)$ and is denoted by $U_{\mathrm{p+n}}(6)$. It is a subgroup of the product group,

$$U_{\mathrm{p}}(6) \times U_{\mathrm{n}}(6) \supset U_{\mathrm{p+n}}(6). \tag{2.195}$$

The symmetric combinations of the generators clearly give rise to subalgebras of symmetric combinations just the same as in IBM-1. The three group chains (2.87), (2.90), and (2.92) are therefore valid with the addition of the subscripts p + n. Chain I, for example, becomes

$$U_{p+n}(6) \supset U_{p+n}(5) \supset O_{p+n}(5) \supset O_{p+n}(3) \supset O_{p+n}(2). \quad (2.196)$$

The situation represented by eq. (2.196) looks very much like the $U(5)$ symmetry of IBM-1. The present case does include IBM-1, but there is more to it. As long as we restrict ourselves to the fully symmetric representation $[N]$ of $U_{p+n}(6)$, we indeed have nothing but IBM-1. Now, however, we have not one but *two* kinds of boson, p and n. This means that we can have Young patterns not only with one but also with two rows, of the form $[N - f, f]$. It follows that the quantum numbers of IBM-1, as given by eq. (2.132) for Chain I, are no longer adequate. Instead we need two quantum numbers for each group as stated in eq. (2.102). There will also be needed not one but two additional quantum numbers like ν_Δ. The outcome is that the IBM-1 chain (2.132) is replaced in IBM-2 by

$$\underset{[N_1 N_2]}{U_{p+n}(6)} \supset \underset{(n_1 n_2)}{U_{p+n}(5)} \supset \underset{(\tau_1 \tau_2)}{O_{p+n}(5)} \underset{(\nu_{\Delta 1} \nu_{\Delta 2})}{\supset} \underset{J}{O_{p+n}(3)} \supset \underset{M}{O_{p+n}(2)} \quad (2.197)$$

Chains II and III similarly acquire more quantum numbers.

An alternative formulation of IBM-2 is analogous to the use of the isospin formalism in dealing with protons and neutrons; see Section 2.3. Then bosons of both types are viewed as identical objects, analogously to nucleons in isospin. There are two distinct spaces: a mechanical space of the six s and d states, and a so-called *F-spin* space of the two states p and n. The Young tableaux of the two spaces are combined into an overall symmetric state. The procedure is analogous to our fermion example of Subsection 2.4.4.

There are a good number of other possible group chains than the familiar three starting with (2.195). One of them is

$$\begin{aligned} &U_p(6) \times U_n(6) \supset U_p(5) \times U_n(5) \supset U_{p+n}(5) \\ &\qquad \supset O_{p+n}(5) \supset O_{p+n}(3) \supset O_{p+n}(2). \end{aligned} \quad (2.198)$$

Here the generators of $U_p(5)$ and $U_n(5)$ are put together into symmetric combinations, which then are the generators of $U_{p+n}(5)$. In any chain starting from $U_p(6) \times U_n(6)$, the p and n labels must meet at the latest in $O_{p+n}(3)$ because good angular momentum is required.

Any group chain is capable of supporting a dynamic symmetry:

the Hamiltonian must consist of its Casimir operators. The IBM-1-like dynamic symmetries, such as $U_{p+n}(5)$ of eq. (2.197), seem to play a more important practical role than the many other dynamic symmetries, such as one based on the chain (2.198).

2.8.2. Boson–Fermion IBM: IBFM

While IBM-1 and IBM-2 are models for doubly even nuclei, the IBFM is their extension to odd-mass and doubly odd nuclei. The IBFM can be based on either IBM-1 or IBM-2. The first alternative is by far the simpler of the two, and we restrict our present remarks to it.

The simplest case is when the model space consists of the IBM-1 space and the space of a single shell-model orbital $(nlj) \equiv j$. The idea is then to have a doubly even core described by the bosons of the IBM, and a single fermion in the orbit j which interacts with the core.

The boson and fermion operators are assumed to commute, i.e., the two are assumed to represent independent degrees of freedom. The microscopic fermion constitution of the bosons is ignored. The group structure of the model is $U_B(6) \times U_F(2j + 1)$. This means that the Hamiltonian is constructed form the generators of this product group, analogously to the IBM-2 case discussed above. The Hamiltonian consists of a boson term, a fermion term and a term for the boson–fermion interaction:

$$H = H_B + H_F + H_{BF}, \tag{2.199}$$

where H_B is the IBM Hamiltonian of Subsection 2.7.2,

$$H_F = \sum_m \epsilon_j a_{jm}^\dagger a_{jm} \tag{2.200}$$

with ϵ_j the single-particle energy, and H_{BF} consists of terms of the form $b^\dagger a^\dagger ba$. In the general multi-orbit case there is a sum over j in eq. (2.200).

As for the physical meaning of the IBFM, we note that it is basically equivalent to the geometric Nilsson model for odd nuclei. This will be elaborated in Chapter 5.

When the generators of the fermion group $U_F(2j + 1)$ are expressed in angular-momentum coupled form, i.e., as tensor products analogous to eq. (2.80), a chain of subalgebras emerges. We recall from the general discussion in Subsection 2.4.1 that removing the number operator

$$\hat{n}_j = \sum_m a^\dagger_{jm} a_{jm} = -\sqrt{2j+1}[a^\dagger_j \tilde{a}_j]_{00}, \tag{2.201}$$

where $\tilde{a}_{jm} \equiv (-1)^{j-m} a_{j,-m}$, from among the generators leaves us with the subalgebra $SU_F(2j+1)$. Further, it follows from eq. (2.83), which is valid also for fermions, that the generators $[a^\dagger_j \tilde{a}_j]_{lm}$ with odd l close on commutation. The number of these generators is $(j+1)(2j+1) = \frac{1}{2}n(n+1)$, and we identify the algebra from Table 2.1 as $Sp_F(2j+1)$. Again the three generators $[a^\dagger_j \tilde{a}_j]_{1m}$ form a subalgebra. Supported by comparison with the boson equations (2.84)–(2.86), we identify the algebra as $O_F(3) \equiv SU_F(2)$ and its generators as proportional to the fermion angular-momentum components J^F_m. Thus we have established the fermion group chain

$$\begin{aligned} U_F(2j+1) \supset SU_F(2j+1) \supset Sp_F(2j+1) \\ \supset O_F(3) \supset O_F(2). \end{aligned} \tag{2.202}$$

One can create dynamic symmetries for the boson–fermion system. According to the general principle learned in Section 2.7, the Hamiltonian must then be composed of the Casimir operators of a group chain. Since total angular momentum must be a good quantum number, such a chain must contain the group $O_{B+F}(3)$. A nontrivial example is the following chain for a $j = \frac{3}{2}$ fermion:

$$\begin{aligned} U_B(6) \times U_F(4) \supset O_B(6) \times SU_F(4) \supset O_{B+F}(6) \\ \supset O_{B+F}(5) \supset O_{B+F}(3) \supset O_{B+F}(2). \end{aligned} \tag{2.203}$$

A crucial ingredient in this chain is that $SU(4)$ is isomorphic to $O(6)$, as we noted in Subsection 2.5.2. Accordingly the boson–fermion fusion can be viewed as $O_B(6) \times O_F(6) \supset O_{B+F}(6)$. It is indeed a general rule that only the *same* boson and fermion groups can be fused. The generators of the B + F group are definite linear combinations of the respective generators of the B and F groups. For example, in the chain (2.203) the generators of $O_{B+F}(3)$ are proportional to the components of the total angular momentum:

$$J^{B+F}_m = J^B_m + J^F_m = \sqrt{10}[d^\dagger \tilde{d}]_{1m} - \sqrt{5}[a^\dagger_{3/2} \tilde{a}_{3/2}]_{1m}. \tag{2.204}$$

In the literature the orthogonal group $O_{B+F}(n)$ is occasionally called $\mathrm{Spin}_{B+F}(n)$.

APPENDIX

A.1. Tensor Operators

The prototype of a spherical tensor is the totality of the spherical harmonics $Y_{lm}(\theta, \phi)$ with fixed l and all values of m: $m = l, l - 1, \ldots, -l$. It is denoted $Y_l(\theta, \phi)$. The most familiar example is Y_1. In terms of x, y, z its components are

$$Y_{1,\pm 1} = \frac{1}{r}\sqrt{\frac{3}{4\pi}}\frac{\mp 1}{\sqrt{2}}(x \pm iy), \qquad Y_{10} = \frac{1}{r}\sqrt{\frac{3}{4\pi}}\, z. \tag{A.1}$$

We thus see that Y_1 is essentially the vector $\mathbf{r}$.

When Y_{lm} plays the role of a wave function, it satisfies eqs. (2.33) and (2.50):

$$J_z Y_{lm} = mY_{lm}, \qquad J_\pm Y_{lm} = \sqrt{(l \pm m + 1)(l \mp m)}\, Y_{l,m\pm 1}. \tag{A.2}$$

However, it is possible that Y_{lm} plays the role of an *operator*. Then it operates on some wave function ψ: $Y_{lm}\psi$. If we now want, e.g., $(J_z Y_{lm})\psi$, we must write

$$\begin{aligned}[J_z, Y_{lm}]\psi &= J_z(Y_{lm}\psi) - Y_{lm}J_z\psi \\ &= (J_z Y_{lm})\psi + Y_{lm}J_z\psi - Y_{lm}J_z\psi = (J_z Y_{lm})\psi,\end{aligned} \tag{A.3}$$

where we have just used the rule for differentiating a product (recall that here $J_z = -i\partial/\partial\phi$). Therefore the left-hand sides of eq. (A.2) are to be replaced by commutators when Y_{lm} is in the position of an operator:

$$\begin{aligned}[J_z, Y_{lm}] &= mY_{lm}, \\ [J_\pm, Y_{lm}] &= \sqrt{(l \pm m + 1)(l \mp m)}\, Y_{l,m\pm 1}.\end{aligned} \tag{A.4}$$

For general information we mention an alternative derivation of eq. (A.4). What happens to operator Y_{lm} under *rotations* is expressed formally with the rotation operator R of eq. (2.51) and its matrix element (2.56):

$$Y'_{lm} = RY_{lm}R^{-1} = \sum_{m'} D^l_{m'm} Y_{lm'}. \tag{A.5}$$

If we now choose to consider an *infinitesimal* rotation of angle ϵ about the z axis, $R = 1 - i\epsilon J_z$ [appears already in eq. (2.52)], the J_z relation of eq. (A.4) can be shown to follow. Similarly the $J_\pm$

relations follow from considering infinitesimal rotations about the x and y axes.

We are now prepared to define a *general* tensor operator T_l, with components T_{lm}, $m = l, l-1, \ldots, -l$, by analogy with Y_l. It need not have anything to do with Y_l except for the defining property that it have the same angular-momentum and rotation properties. Accordingly a tensor operator T_l, of *degree* l, must obey

$$[J_z, T_{lm}] = mT_{lm}, \qquad \text{(A.6)}$$
$$[J_\pm, T_{lm}] = \sqrt{(l \pm m + 1)(l \mp m)}\,T_{l,m\pm 1}$$

or equivalently, with reference to finite rotations,

$$T'_{lm} = RT_{lm}R^{-1} = \sum_{m'} D^l_{m'm}T_{lm'}. \qquad \text{(A.7)}$$

Operationally, the important property of tensor operators is that they obey the coupling rules of angular momentum. Eigenfunctions of two particles, say ψ_{jm} and $\phi_{j'm'}$, are coupled with Clebsch–Gordan coefficients to a state of good total angular momentum JM according to

$$\sum_{mm'} (jmj'm'|JM)\psi_{jm}\phi_{j'm'} \equiv [\psi_j\phi_{j'}]_{JM}. \qquad \text{(A.8)}$$

This structure in fact defines a *tensor product*: the tensor product of degree L of two tensor operators T_l and S_λ is $[T_lS_\lambda]_L$ whose components are

$$[T_lS_\lambda]_{LM} = \sum_{m\mu} (lm\lambda\mu|LM)T_{lm}S_{\lambda\mu}. \qquad \text{(A.9)}$$

A special case of the tensor product is the *scalar* or *dot product* of two tensors, necessarily of the same degree:

$$T_l\cdot S_l \equiv (-1)^l\sqrt{2l+1}[T_lS_l]_{00} = \sum_m (-1)^m T_{lm}S_{l,-m}. \qquad \text{(A.10)}$$

For ordinary vectors, with $l = 1$, the definition coincides with the usual dot product: $T_1\cdot S_1 = \mathbf{T}\cdot\mathbf{S}$.

The spherical tensors discussed above are tensors with respect to the group $O(3)$. Tensors can be defined with respect to other groups as well. The assignment of quantum numbers for the group chains in Section 2.6 actually serves this purpose as well. Thus, for example, the d-boson creation operator $d^\dagger_M$ carries the quantum numbers $n_d = 1$, $\tau = 1$, $\nu_\Delta = 0$, $J = 2$, and M. These are the quantum numbers of the state $d_M^\dagger|\tilde{0}\rangle$, and it is clear that the quantum numbers are solely due to the operator $d_M^\dagger$. Accordingly we can say that $d_M^\dagger$ transforms as a tensor with the above labels under the groups $U(5) \supset O(5) \supset O(3) \supset O(2)$. This can be denoted as

$$d_M^\dagger = \mathcal{R}^{[1,1,0,2,M]}. \tag{A.11}$$

A.2. Recoupling of Operators

The IBM-1 Hamiltonian appears in various forms in Subsection 2.7.2. We show here their connection.

The nontrivial relations involve d bosons only. The Hamiltonians contain terms with four d-boson operators in two basic forms. They are the normal-ordered form $d^\dagger d^\dagger dd$ and the multipole form $d^\dagger dd^\dagger d$. We adopt the abbreviations

$$T_l^2 \equiv T_l \cdot T_l \equiv [d^\dagger \tilde{d}]_l \cdot [d^\dagger \tilde{d}]_l, \qquad l = 0, 1, 2, 3, 4; \tag{A.12}$$

$$D_L \equiv [d^\dagger d^\dagger]_L \cdot [\tilde{d}\tilde{d}]_L, \qquad L = 0, 2, 4. \tag{A.13}$$

The reason why only $L = 0, 2, 4$ occur in D_L is that these are the only possible angular momenta for a state of two d bosons, which is necessarily symmetric; see Subsection 2.6.3. Instead of T_0 and T_1 it is customary to use the operators for d-boson number and angular momentum:

$$\hat{n}_d = \sqrt{5}\, T_0, \qquad \mathbf{J} = \sqrt{10}\, T_1. \tag{A.14}$$

Through standard formulas for angular-momentum recoupling, one can derive the relation

$$T_l^2 = (2l + 1) \sum_L \begin{Bmatrix} 2 & 2 & L \\ 2 & 2 & l \end{Bmatrix} D_L + \frac{2l + 1}{5} \hat{n}_d, \tag{A.15}$$

where the last term results from commutation.

Direct application of eq. (A.15) yields

$$5T_0^2 = \hat{n}_d^2 = \hat{n}_d + D_0 + D_2 + D_4, \tag{A.16}$$

$$10T_1^2 = \mathbf{J}^2 = 6\hat{n}_d - 6D_0 - 3D_2 + 4D_4, \tag{A.17}$$

$$T_2^2 = \hat{n}_d + D_0 - \tfrac{3}{14} D_2 + \tfrac{2}{7} D_4, \tag{A.18}$$

$$T_3^2 = \tfrac{7}{5}\hat{n}_d - \tfrac{7}{5} D_0 + \tfrac{4}{5} D_2 + \tfrac{1}{10} D_4, \tag{A.19}$$

$$T_4^2 = \tfrac{9}{5}\hat{n}_d + \tfrac{9}{5} D_0 + \tfrac{18}{35} D_2 + \tfrac{1}{70} D_4. \tag{A.20}$$

Since there are only three operators D_L, there can only be three linearly independent operators T_l^2. The two dependent ones can be solved from eqs. (A.16)–(A.20). When $\mathbf{J}^2$, T_3^2, and T_4^2 are picked as the independent ones, the other two are given as

$$5T_0^2 = \hat{n}_d^2 = -5\hat{n}_d + \tfrac{2}{9}\mathbf{J}^2 + \tfrac{5}{6}T_3^2 + \tfrac{35}{18}T_4^2, \qquad \text{(A.21)}$$

$$T_2^2 = \tfrac{7}{90}\mathbf{J}^2 - \tfrac{1}{3}T_3^2 + \tfrac{5}{9}T_4^2. \qquad \text{(A.22)}$$

Then we have the inverse relations

$$D_0 = -\tfrac{2}{5}\hat{n}_d + \tfrac{1}{225}\mathbf{J}^2 - \tfrac{7}{30}T_3^2 + \tfrac{7}{18}T_4^2, \qquad \text{(A.23)}$$

$$D_2 = -2\hat{n}_d - \tfrac{1}{45}\mathbf{J}^2 + \tfrac{17}{21}T_3^2 + \tfrac{5}{9}T_4^2, \qquad \text{(A.24)}$$

$$D_4 = -\tfrac{18}{5}\hat{n}_d + \tfrac{6}{25}\mathbf{J}^2 + \tfrac{9}{35}T_3^2 + T_4^2. \qquad \text{(A.25)}$$

The above operator relations allow all transformations relating the Hamiltonians (2.164), (2.165), (2.167), and (2.168). As an example, with direct relevance to the discussion of dynamic symmetries in Subsection 2.7.2, we give the parameters of eq. (2.167) as functions of the parameters of eq. (2.168):

$$\begin{aligned}
\epsilon'' &= \epsilon''' - \alpha, \\
a_0 &= -8\eta, \\
a_1 &= \tfrac{2}{9}\alpha + \tfrac{2}{5}\beta + 2\gamma + \tfrac{1}{2}\delta, \\
a_2 &= \tfrac{4}{3}\delta, \\
a_3 &= \tfrac{5}{6}\alpha + 4\beta, \\
a_4 &= \tfrac{35}{18}\alpha.
\end{aligned} \qquad \text{(A.26)}$$

For a complete glossary of such relations, see Lipas (1983).

ACKNOWLEDGEMENTS

I would like to thank Piet Van Isacker and Jerry Draayer for a number of very helpful comments.

BIBLIOGRAPHY

A. Bohr and B. R. Mottelson, *Nuclear Structure* (Benjamin, New York,

1969), Vol. 1, Appendix 1C. In this monumental general reference on nuclear structure, this is a compact, educational, and practical introduction to group theory.

A. de-Shalit and I. Talmi, *Nuclear Shell Theory* (Academic, New York, 1963). *The* book on angular momentum and tensor techniques.

J. P. Elliott and P. G. Dawber, *Symmetry in Physics* (Macmillan, London, 1979). A general text on group theory, in two volumes. The needs of the nuclear shell model are well observed.

M. Hamermesh, *Group Theory and Its Application to Physical Problems* (Addison-Wesley, Reading, MA, 1962). A classic on group theory in physics. Because of its wide coverage, it does not answer very directly for our present needs.

F. Iachello, in *Nuclear Spectroscopy* (Springer-Verlag, Berlin, 1980), eds. G. F. Bertsch and D. Kurath, Chap. V. Lecture notes on group theory directed to the IBM.

F. Iachello and A. Arima, *The Interacting Boson Model* (Cambridge University Press, Cambridge, 1987). The definitive statement of the IBM for doubly even nuclei. A comprehensive collection of equations with relatively little text.

F. Iachello and P. Van Isacker, *The Interacting Boson–Fermion Model* (Cambridge University Press, Cambridge, 1991). Comprehensive treatment of the IBFM in the vein of the previous book by Iachello and Arima.

P. O. Lipas, *Progr. Part. Nucl. Phys.* **9**, 511 (1983). Contains a glossary of parameter relations in IBM-1.

H. Lipkin, *Lie Groups for Pedestrians* (North-Holland, Amsterdam, 1965). This book has set the model for our presentation of group theory: extrapolating and generalizing the formalism of angular momentum, with minimal formal mathematics.

E. Merzbacher, *Quantum Mechanics* (Wiley, New York, 1970), 2nd ed. A solid standard text that provides the necessary background in general quantum mechanics, including angular momentum and rotations.

J. C. Parikh, *Group Symmetries in Nuclear Structure* (Plenum, New York, 1978). Theory and applications presented in considerable detail.

M. E. Rose, *Elementary Theory of Angular Momentum* (Wiley, New York, 1957). A classic particularly strong on rotations. Notationally somewhat dated.

L. I. Schiff, *Quantum Mechanics* (McGraw-Hill Kogakusha, Tokyo, 1968) 3rd ed., Sec. 30. Presents dynamic symmetries of the hydrogen atom and the harmonic oscillator.

M. Weissbluth, *Atoms and Molecules* (Academic, New York, 1978). An advanced text with emphasis on angular momentum and group theory.

B. G. Wybourne, *Classical Groups for Physicists* (Wiley, New York, 1974). A detailed exposition of Lie groups, followed by physical applications.

CHAPTER 3

Empirical Tests of the IBM-1

RICHARD F. CASTEN and DAVID D. WARNER

INTRODUCTION

The previous chapter introduced and outlined the group theory of the IBM and other algebraic approaches to nuclear structure. It summarized the group chains and the three dynamical symmetries of the IBM-1 group $U(6)$ and briefly discussed the properties (eigenvalues, quantum numbers, and E2 electromagnetic selection rules) that characterize each symmetry. In this chapter, we will discuss the confrontation of the IBM-1 with the data.

First, we shall consider the symmetries themselves, where analytic expressions and simple selection rules pertain and make the comparisons particularly easy. Most nuclei, however, do not manifest one of the symmetries. They are intermediate between these idealized limits, and their treatment in the IBM-1 requires a diagonalization of the IBM-1 Hamiltonian. Despite this added complexity, the model is still remarkably simple, in part because the limiting symmetries themselves act as convenient benchmarks, and in part because of the very small number of parameters and relatively few basis states in the model. As will be extensively discussed in Chapter 4, the IBM-1 is, in effect, an extreme truncation of the shell model that seeks to isolate collective degrees of freedom (primarily quadrupole). Indeed, the model has been described as "Janus-like," with two faces, one looking towards the shell model for its microscopic underpinning and rationalization and the other towards collective models and the empirical data base of information on the collective modes of medium and heavy nuclei ($A \geq 60$).

After treating the dynamical symmetries per se, we will therefore turn to numerical IBM calculations reflecting varying degrees of symmetry breaking. We shall focus extensively on deformed nuclei, which comprise a very large, and important subset of medium and heavy mass nuclei, and then on the critical transition regions where sequences of nuclei span structures ranging from one symmetry towards another: these phase or shape transition regions are often the most sensitive testing grounds for nuclear models. The discussion will utilize two complementary approaches to numerical treatments of the IBM, one based on the full "multipole" Hamiltonian (eq. (2.167)) and another on a truncated version of this, the so-called Consistent Q Formalism (or CQF).

We begin our treatment with the $O(6)$ symmetry which, historically, played a very important role in the recognition and acceptance of the IBM. While the $SU(5)$ and $SU(3)$ symmetries describe nuclei with easily recognizable vibrational and rotational features, the $O(6)$ limit revealed a third basic form of collective motion, namely, that of a deformed γ unstable oscillator, which has now been shown to occur in many different regions of the nuclear chart. The existence of its characteristic structure had not been recognized previously in actual nuclei nor had its link to the potential of Wilets and Jean been realized.

3.1. THE O(6) SYMMETRY

The level structure and E2 selection rules for the $O(6)$ limit have been discussed earlier. The first nucleus to be identified as manifesting the $O(6)$ symmetry was ^{196}Pt. The eigenvalue expression for the $O(6)$ limit contains 3 parameters denoted η, β, and γ in eq. (2.178). This equation is frequently encountered in the literature with renamed parameters as

$$E_{O(6)} = A\sigma(\sigma + 4) + B\tau(\tau + 3) + CJ)(J + 1). \qquad (3.1)$$

Clearly, A, B and C are just twice η, β, and γ. Figure 3.1 shows a typical spectrum of states in $O(6)$, for $N = 6$. It is similar to Fig. 2.5 but, because of the larger N, shows the multiplet structure and familial relationships among levels more clearly.

The parameters A, B and C have been chosen in Fig. 3.1 to

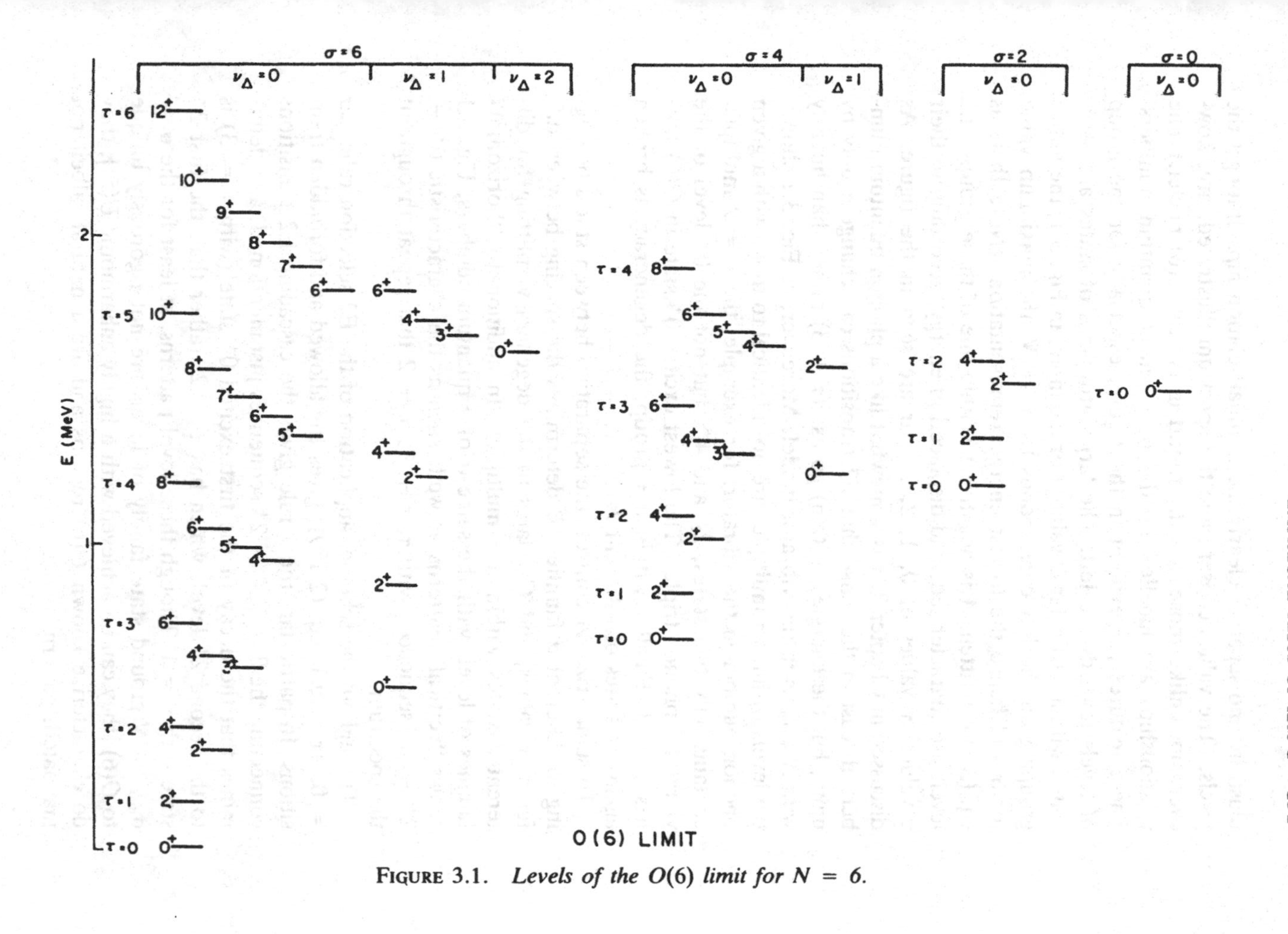

FIGURE 3.1. *Levels of the O(6) limit for N = 6.*

illustrate and separate clearly the familial relationships amongst the levels. (The values chosen, and the spectrum illustrated, are, however, not unlike those actually found in typical $O(6)$ nuclei.) The relationships are labelled by the σ, τ and J quantum numbers. These features are evident in the figure, especially for the group of levels towards the left. The large groupings of states are associated with a particular σ value. For example, in Fig. 3.1, the entire grouping on the left corresponds to $\sigma = N$, the maximum value for σ and hence the lowest energy representation. (Note that, as eq. (3.1) is written, A is negative.) Within one of these groups, the levels are characterized, and displaced in energy, according to their τ values. τ values of 0, 1, 2, . . . σ are seen in the figure. As discussed in Chapter 2, τ is somewhat like a phonon quantum number, at least in the sense that E2 transitions can change τ only by unity, but the energies in $O(6)$ go as $\tau(\tau + 3)$ rather than linearly with N_{phonon} as in the vibrator model. Moreover, as Fig. 3.1 shows, the levels within a τ multiplet are not identical to those with a given phonon number in the vibrator: for example, the $\tau = 2$ multiplet contains only two states, 2_2^+ and 4_1^+, but not the 0^+ level of the vibrator 2-phonon triplet. The lowest excited 0^+ state in $O(6)$ thus has $\tau = 3$. Finally, within a τ group, the degeneracy is broken again for states of different J.

To summarize, A controls the separation between states belonging to different σ families, B determines the spacing between different τ values, and C relates to the degeneracy splitting for different J states within a τ multiplet. In addition to "horizontal" families of levels with the same σ or τ quantum numbers, Fig. 3.1 shows "vertical" patterns as well, such as the characteristic $0^+ - 2^+ - 2^+$ sequences with τ, $\tau + 1$, $\tau + 2$ that repeat throughout the spectrum.

Throughout the scheme, application of the E2 selection rules $\Delta\sigma = 0$, $\Delta\tau = \pm 1$ (eq. (2.189)) gives the allowed and forbidden transitions. In particular, the τ rule gives the cascading E2 transitions connecting the $0^+ - 2^+ - 2^+$ sequences just mentioned, and determines that the decay of the first excited 0^+ state (with $\tau = 3$) is to the second 2^+ level, which has $\tau = 2$, rather than the first 2^+ state with $\tau = 1$. Though these level patterns, at least for the $\sigma = \sigma_{\max} = N$ ground state family of levels are not rigorously unique to $O(6)$ [they can be achieved with a highly anharmonic $U(5)$], they do characterize known $O(6)$ regions and are a natural, albeit not mandated, pattern.

Figure 3.2 compares the $O(6)$ scheme with the empirical level scheme of ^{196}Pt which was the first $O(6)$ nucleus identified. The figure shows, first, that there is a 1–1 correspondence between predicted and empirical levels. This extends through the complete multiplet of levels with $\tau = 3$ and includes levels of two $\sigma < N$ families as well. This latter result is still unique among known $O(6)$-like nuclei. Three examples of the characteristic $O(6)$ sequences of $0^+ - 2^+ - 2^+$ levels with cascading E2 transitions are evident in Fig. 3.2, namely for the quasi-ground band, for the levels related to the 0^+ ($\tau = 3$) level, and for the lowest levels of the $\sigma = N - 2 = 4$ representation.

As noted, the $O(6)$ limit is analogous to the geometric model of a γ-unstable rotor (the Wilets–Jean model), and thus has the typical 2^+, (3^+, 4^+), (5^+, 6^+) . . . γ-band energy staggering of a γ-soft potential. This is opposite to the staggering in a rigidly asymmetric rotor (Davydov model), and this is in fact one of the few practical empirical distinctions between these two types of axially asymmetric rotors. The levels of ^{196}Pt exhibit the same level couplets as the $O(6)$ limit but the staggering, as discussed in more detail in Chapter 6, is less extreme than predicted by a completely γ-flat potential. We shall return to this point below.

It will be noted that, despite the 1–1 correspondence of theoretical and empirical levels, there are also some notable energy discrepancies. Of particular note is the large predicted spacing between high τ states (due to the $\tau(\tau + 3)$ dependence) which is contrary to experiment. Also in the $\tau = 3$ multiplet, both $E(6_1^+)$ and $E(0_2^+)$ are greater than $E(3_1^+)$ empirically, in contrast to the monotonic dependence of energies on spin within a τ multiplet embodied in $O(6)$. These discrepancies will be discussed shortly and their resolution illustrated in Chapter 6: they imply a very small γ dependence (a few %) superposed on the extreme γ-flat potential of $O(6)$.

Despite the correspondence of theoretical and experimental levels in Fig. 3.2, the mere presence of a 1–1 level correspondence, in the absence of other data, does not necessarily imply a structural identity and, hence, we now turn to more crucial structure indicators, namely E2 branching ratios and absolute B(E2) values. In Fig. 3.2, the relative B(E2) values are indicated on the transition arrows and compared with the predictions of the $O(6)$ limit. All transitions allowed by the $O(6)$ selection rules are observed and represent strong branches and all forbidden transitions are either

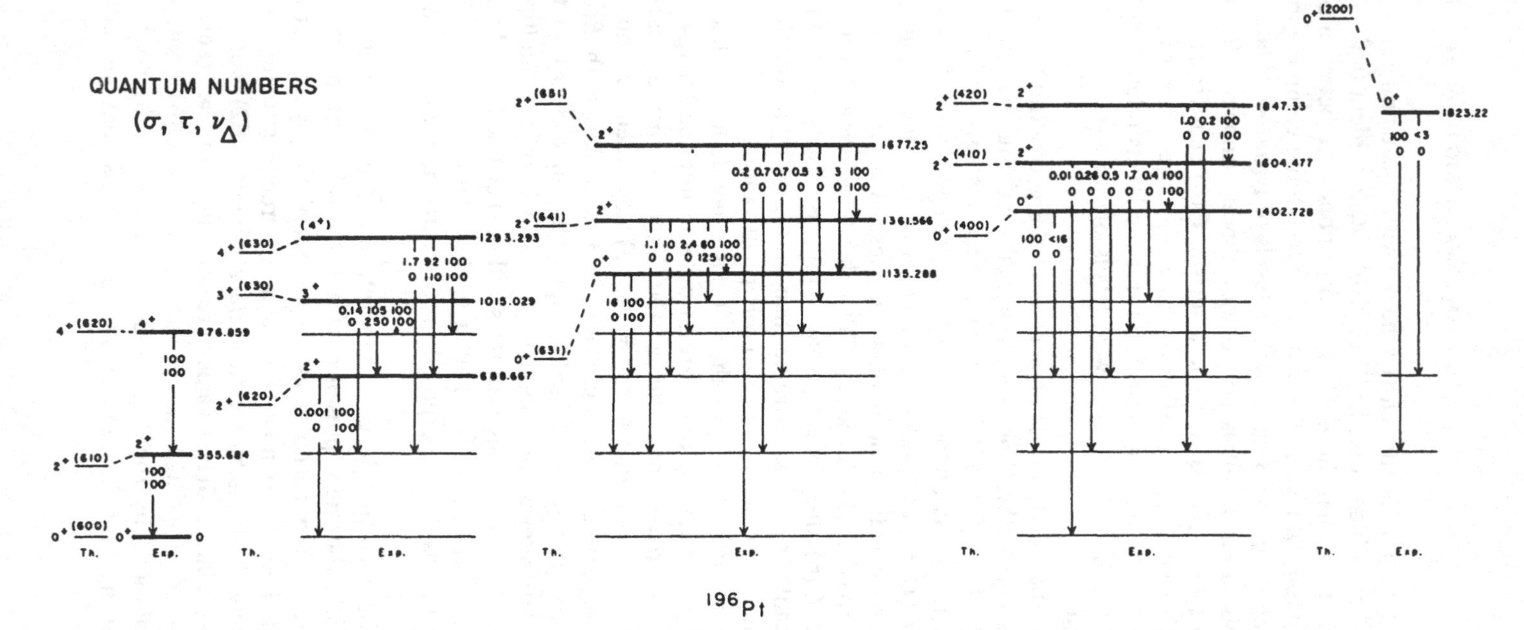

FIGURE 3.2. *Comparison of the predictions of the O(6) limit with the empirical level scheme for* ^{196}Pt. *The upper numbers on the transition arrows are the empirical relative B*(E2) *values, and the lower set gives the O(6) predictions. From Cizewski et al.* (1978).

weak or unobserved. In addition, many of the detailed branching ratios are close to the predicted values.

A comment on the decay of the 0^+ bandheads of the $\sigma < N$ families is useful. According to the $\Delta\sigma = 0$ selection rule (eq. (2.189)) these levels cannot, in principle, decay by E2 transitions. Of course, they must decay, namely by a small amount of symmetry breaking. The $O(6)$ wave functions can be expressed (as is commonly done) in a $U(5)$ basis, and generally consist of several terms. The corresponding E2 transition matrix elements in $O(6)$ are then linear combinations of $U(5)$ matrix elements and the σ and τ selection rules arise in different ways. The σ selection rule occurs because of an exact cancellation of the many non-zero terms contributing to the total E2 matrix element. This rule is therefore more easily broken than the τ selection rule for which each individual term vanishes and whose breaking requires new terms in the Hamiltonian or E2 operator (beyond those necessary for the symmetry). Thus, the 0^+ bandheads (with $\tau = 0$) of the $\sigma < N$ groups would be expected to decay to the 2_1^+, $\tau = 1$, member of the ground state band rather than the 2_2^+, $\tau = 2$, state. This is indeed the case as seen in Fig. 3.2. Though these transitions violate the σ selection rule, the symmetry breaking is quite small: the critical $0_3^+ \to 2_1^+$ transition rate has recently been measured with the result that $B(\mathrm{E2}:0_3^+(\sigma = 4) \to 2_1^+(\sigma = 6)) < 0.034e^2b^2$. That is, it is hindered by at least an order of magnitude relative to allowed $O(6)$ transitions, thus confirming the usefulness of the σ quantum number classification.

It has been mentioned that a highly anharmonic $U(5)$ spectrum (e.g., with 2-phonon levels lying above some 3-phonon states) can simulate $O(6)$. However, there is little danger of confusing the two schemes if absolute $B(\mathrm{E2})$ values, such as the $B(\mathrm{E2}:0_3^+(\sigma = N - 2) \to 2_1^+(\sigma = N))$ value just mentioned, are measured or if E2 branching ratios of higher lying states (multi-phonon states in $U(5)$ and $\sigma < N$ states in $O(6)$) are measured. This is exemplified in Fig. 3.3 for the decay of one of the higher 2^+ levels in ^{196}Pt where the agreement with the $O(6)$ scheme and disagreement with $U(5)$ is readily apparent.

Another extensive region of near-$O(6)$ nuclei occurs in the Ba and Xe isotopes near $A = 130$. In its isotopic and isotonic extent, this region is in fact a better example of $O(6)$ than the Pt isotopes. In addition, a number of $\tau = 4$ and several $\tau = 5$ states are identified. On the other hand, few if any $\sigma < N$ levels are firmly

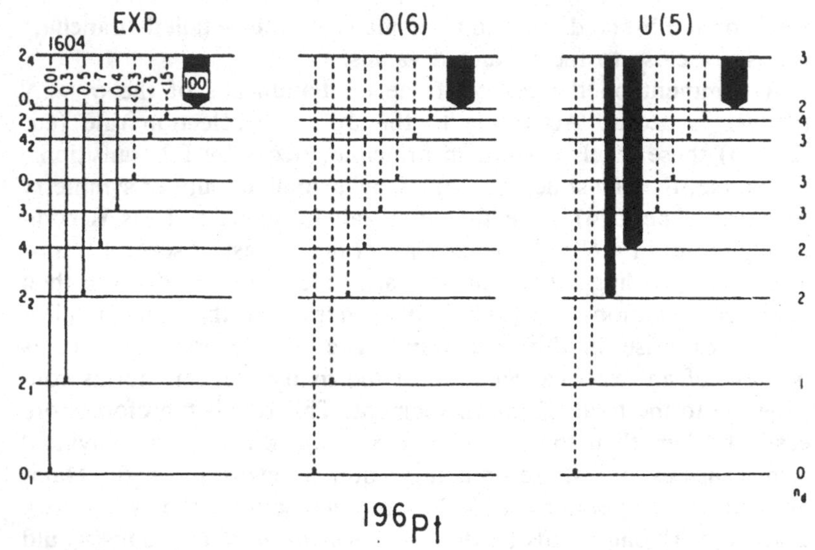

FIGURE 3.3. *Comparison of O(6) and U(5) predictions. Branching ratios for the decay of a σ < N level in* ^{196}Pt *compared with O(6) and U(5) predictions. The widths of the transition arrows are relative B(E2) values. Forbidden or experimentally unobserved transitions are dashed. From Casten and Cizewski (1987).*

assigned. A summary of the comparison of these nuclei with the $O(6)$ limit is shown in Fig. 3.4.

One much discussed problem with the $O(6)$ description of ^{196}Pt has been the observation of finite quadrupole moments in that nucleus. The strict $O(6)$ limit predicts vanishing moments because these diagonal matrix elements conserve τ and therefore violate the $\Delta\tau = \pm 1$ selection rule. The quadrupole moment problem in ^{196}Pt, and the observation of weak but finite strengths for some forbidden $\Delta\tau = 0, 2$ transitions both in ^{196}Pt and the $A = 130$ region, can be partly ameliorated by adding a term $\chi(d^\dagger \tilde{d})^{(2)}$ to the $O(6)$ T(E2) operator of eq. (2.188) giving T(E2) the form of eq. (2.179). Since the two terms in T(E2) connect different basis states, the predictions for the quadrupole moments and for τ-forbidden transitions are completely independent of those for allowed transitions: therefore the introduction of small but finite χ values does not disturb the already good agreement for the latter. Rather good accord for a large number of $\Delta\tau = 0, 2$ transitions has been obtained in this way for ^{134}Ba, ^{126}Xe, and to a slightly lesser extent, for ^{196}Pt. Some of these results are shown in Table 3.1, for ^{126}Xe.

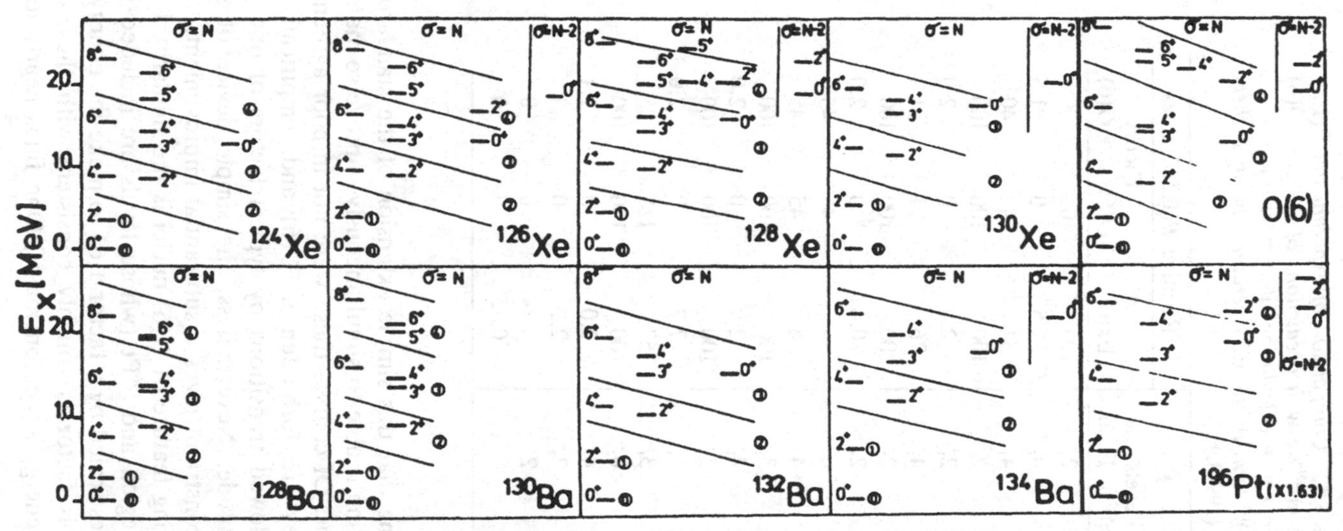

FIGURE 3.4. *Comparison of the low-lying levels of several* Xe *and* Ba *nuclei with* $O(6)$ *limit and also with a scaled level scheme for* ^{196}Pt. *The* $O(6)$ *parameters are* A = 67 keV, B = 70 keV, *and* C = 10 keV *for* N = 6. *From Casten, von Brentano and Haque* (1985).

TABLE 3.1. *Comparison of experimental B(E2) values in* ^{126}Xe *with predictions of the pure O(6) limit and with calculations incorporating a* $\chi(d^{\dagger}\tilde{d})^{(2)}$ *term in the* E2 *operator. (From Lieberz et al. (1990).)*

Levels		Relative B(E2) Values		
			Theory	
Initial	Final	Exp.	$O(6)$	$O(6)_x$
2_2^+	2_1^+	100	100	100
	0_1^+	1.5	0	1.5
3_1^+	4_1^+	34	40	40
	2_2^+	100	100	100
	2_1^+	2.0	0	2.0
4_2^+	4_1^+	76	91	91
	2_2^+	100	100	100
	2_1^+	0.4	0	2.0
5_1^+	6_1^+	75	45	45
	4_2^+	76	45	45
	3_1^+	100	100	100
	4_1^+	2.9	0	2.0
0_2^+	2_2^+	100	100	100
	2_1^+	7.7	0	2.0
2_3^+	3_1^+	67	125	125
	0_2^+	100	100	100
	4_1^+	2.0	0	1.1
	2_2^+	2.2	0	3.5
	2_1^+	0.14	0	0
	0_1^+	0.13	0	0

It is interesting that this simple extension of the standard (χ = 0) $O(6)$ description is able to simultaneously improve so many transition predictions. Of course, these were not in poor agreement to begin with: they are forbidden in $O(6)$ and, empirically, were smaller than allowed transitions by typical factors of one to two orders of magnitude. Nevertheless, the simple device of slightly relaxing the χ constraint gives a substantial improvement.

Another striking feature of $O(6)$ nuclei is the similarity between the A = 130 region and ^{196}Pt (whose levels are included in Fig. 3.4, with energies scaled by a factor 1.63 to make the energy scales comparable: This factor is roughly consistent with the expected dependence of energy scales on mass.) One finds nearly identical parameters in the two regions. The ratio η/β (or A/B) is also particularly interesting. It can be extracted from the empirical spacings by using the relation

$$\frac{\eta}{\beta} = \frac{A}{B} = \frac{E(0^+)(\sigma = N-2)}{E(2_2^+)(\tau = 2) - E(2_1^+)(\tau = 1)} \cdot \frac{3}{2(N+1)} \quad (3.2)$$

which follows from eq. (2.178). The η/β values for ^{196}Pt and the $A = 130$ region are listed in Table 3.2. Recalling that this ratio may take on any value whatsoever in the $O(6)$ symmetry, it is remarkable that the empirical values are so similar for nuclei far separated in mass and occupying different major shells. Their closeness to unity will be discussed below as it is a natural outcome of the CQF: it has a deep relation to the scales of degeneracy splitting in the $O(6)$ and $U(5)$ steps in the group chain characterizing this symmetry.

The similarity of the $A = 130$ and Pt regions runs very deep, extending even to the discrepancies with the $O(6)$ limit. As noted earlier, the energies of states within a τ multiplet must be monotonic with J. Thus, if $E(4_2^+) < E(6_1^+)$ (as is common) then $E(0_2^+)$ must be $< E(3_1^+)$. In both Pt and the $A = 130$ region this prediction is violated. Moreover, in both regions the energy staggering in the γ band is less extreme than predicted by the $O(6)$ limit (if the parameters are fixed from other spacings that are insensitive to the details of the potential in the γ degree of freedom). Finally, the smaller than predicted spacing between high τ states (e.g., the $\tau = 3 - 4 - 5$, $0^+ - 2^+ - 2^+$ sequence in Fig. 3.2) is also observed in the $A = 130$ region.

The observed energy staggering suggests a simple way to improve the agreement between the model and experiment by slightly extending the IBM Hamiltonian. To see this, consider two familiar geometric models of axial asymmetry, the Davydov model in which the potential is γ-rigid (i.e., with a deep minimum at a specific γ) and the Wilets–Jean model of complete γ-flatness ($V(\gamma)$ = con-

TABLE 3.2. *Empirical and theoretical values of the ratio η/β or A/B. See eq. (3.2).*

	CQF[a]	^{196}Pt	^{134}Ba	^{130}Xe	^{128}Xe	^{126}Xe
η/β or A/B	1.0	0.90	0.96	0.86	0.76	0.67

[a]This ratio may take any value whatsoever in the general $O(6)$ limit of the IBM, but is always unity in the consistent-Q formalism (CQF).

stant) which is the geometric counterpart to $O(6)$. In the Davydov picture, $\gamma = \gamma_{rms}$ can be freely chosen. In the Wilets–Jean model γ_{rms} is 30° since the shape oscillates uniformly from 0° to 60°. The energy levels of these two models are shown in Fig. 3.5 in which, for convenience, the Davydov results for $\gamma = 30°$ are separated out. The figure highlights the difference in energy staggering of the γ-band in the two models of axial asymmetry. Since the observed staggering in $O(6)$-like nuclei is intermediate between these extremes, it suggests that better agreement might be obtained with a potential with at least some γ-dependence whilst retaining the property that $\gamma_{rms} = 30°$. This can be done conveniently by extending the IBM to incorporate a particular higher order term which is cubic in d boson creation and destruction operators and which, geometrically, introduces a minimum in the potential at $\gamma = 30°$. Such an approach is appealing since it does not change the $\gamma_{rms} =$ 30° value characteristic of the $O(6)$ limit but only relaxes the extreme γ-independence of the $O(6)$ limit. In effect, it reduces the extent of zero point motion around $\gamma = 30°$. Such modifications to the original IBM will be discussed in Chapter 6. Suffice it to say here that the calculated spectra are substantially improved: indeed not only the γ-band staggering, but all of the discrepancies in ener-

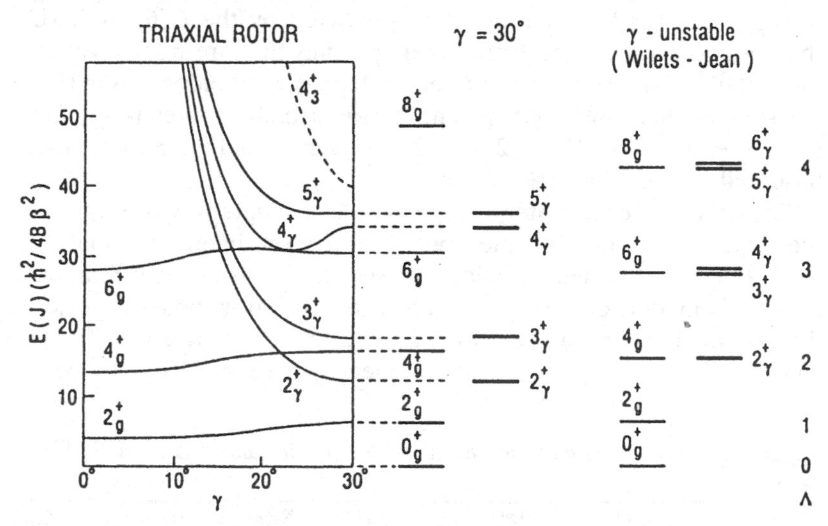

FIGURE 3.5. *Comparison of triaxial (or Davydov) rigid γ rotor and γ-soft models. The Davydov results for $\gamma = 30°$ are shown explicitly in the middle for comparison with the γ-unstable, or Wilets–Jean, model. The γ-band levels are shown as thicker lines. From Zamfir and Casten* (1991).

gies noted earlier are ameliorated. Interestingly, the degree of triaxiality needed in the potential is extremely small. At $\gamma = 30°$, where the effect is a maximum, it amounts to only a 3–5% change. Thus the potentials for these nuclei, while not rigorously γ-independent, are nevertheless extremely γ-soft.

3.2. THE *U*(5) SYMMETRY

The *U*(5) symmetry offers a description of vibrational nuclei that encompasses an extremely broad range of anharmonicities. Here, however, the discussion of the *U*(5) symmetry will be limited to the question of whether or not nuclei have been observed that are relatively harmonic expressions of *U*(5) resembling the traditional vibrational spectra. Such nuclei are characterized by levels corresponding to one, two and multi-phonon structure arranged in reasonably compact multiplets, each multiplet centered at an energy proportional to the number of phonons characterizing it.

Since the energy spacings in typical vibrational-like nuclei lying close to closed shells are larger than those of deformed nuclei, there is a difficulty in identifying unperturbed two-phonon and three-phonon levels. The 2_1^+ level in such nuclei is typically at 500–800 keV and thus the two and three-phonon states will be between 1 and 2.5 MeV, which is well into the region of particle-hole or two quasiparticle excitations. Moreover, in regions where vibrational character is expected, so-called intruder states, which can be described in terms of excitations crossing a major shell gap, may descend into the low lying region.

The Cd nuclei have long been considered archetypical vibrational nuclei. However, intruder levels (see Chapter 6) have been discovered and studied in considerable detail in several Cd isotopes. Indeed, at excitation energies near the supposedly 2-phonon triplet states, a set of five levels is now known in several Cd nuclei, with strong interconnecting *B*(E2) values. It is possible that two of these states are, in parentage, intruders, but other interpretations in terms of mixed F-spin proton-neutron symmetry (see Chapter 4) have been proposed as well.

As a consequence of the proton-neutron interaction, the intruder levels are expected to descend in energy toward mid-shell (^{114}Cd) and rise thereafter. Thus, the lighter or heavier Cd isotopes might be expected to display relatively higher lying intruder states and a more intact phonon spectrum. This may have been confirmed in experiments on ^{118}Cd whose level scheme is shown in Fig. 3.6. Here an isolated 0^+ state occurs at 1615 keV, rather far from the other groupings of levels. The other levels seem to represent a nearly degenerate two-phonon triplet and five levels which comprise candidates for the three-phonon quintuplet. (In this view, the 1615 keV

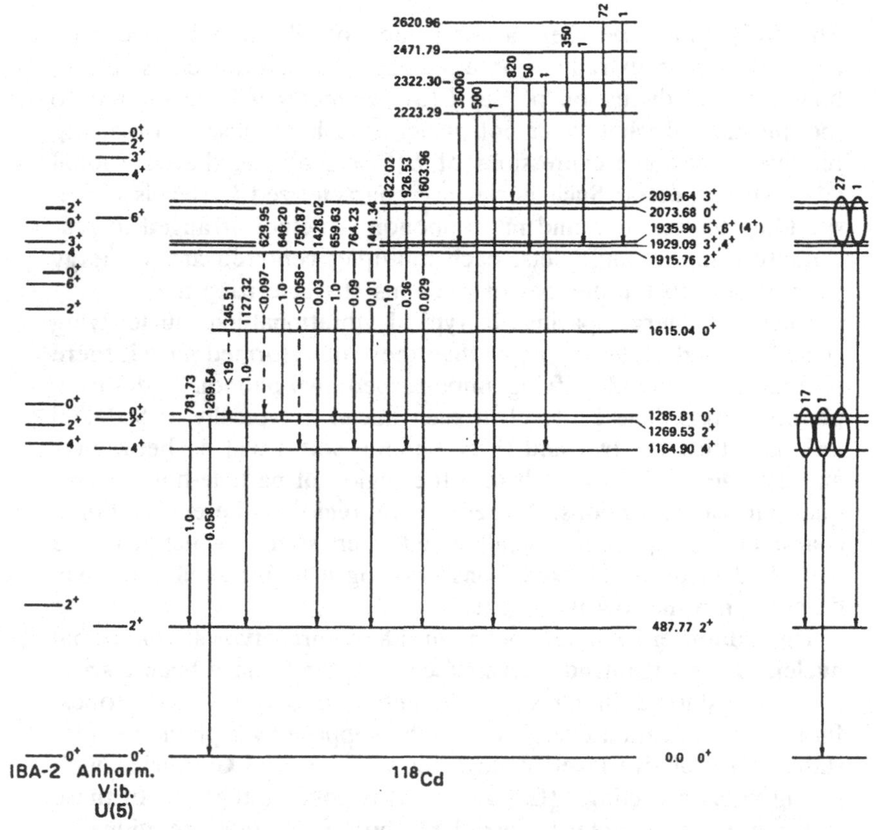

FIGURE 3.6. *Low-lying levels of* ^{118}Cd. *The numbers on the transition arrows are relative* B(E2) *values. On the far right are the averaged relative empirical* E2 *branching ratios for transitions changing the phonon number by one and two. The* 0^+ *level at 1615* keV *has been assigned to the intruder configuration. On the left is a comparison with the* $U(5)$ *limit of the* IBM, *which gives the same predictions as the anharmonic vibrator model of Brink, de Toledo Piza and Kerman* (1965). IBM-2 *calculations are also shown: the short horizontal bars indicate levels with calculated dominant intruder amplitudes. From Aprahamian et al.* (1987).

level would be the intruder.) The data allow multiple spin choices for two of these five levels, but, in each case, one of these is consistent with the J^π values required of the three-phonon quintuplet.

The decay properties in these levels may also exhibit the selectivity expected in the vibrator: the dominance of allowed (ΔN_{ph} = 1) to forbidden (ΔN_{ph} = 2) relative B(E2) values is more than an order of magnitude for both the two-phonon triplet and the candidates for the three-phonon quintuplet. However, recent measurements of the absolute $B(\text{E2}:0_2^+(1285) \rightarrow 2_1^+)$ value fail to show the collectivity required for a 2-phonon → 1-phonon transition. Either this strength is highly fragmented or the assignment of the 1285 and 1615 keV 0^+ states as two-phonon and intruder should be reversed, thereby substantially increasing the anharmonicity.

3.3. THE *SU*(3) SYMMETRY

Before discussing the empirical status of $SU(3)$, it is critical to stress that $SU(3)$ is a very particular case of the deformed symmetric rotor, and, although deformed nuclei abound, the vast majority differ substantially from $SU(3)$ and require large symmetry breaking to account for their observed properties. In fact, some of the properties most commonly associated with deformed nuclei are actually forbidden in $SU(3)$. To further discuss this we turn to Fig. 3.7 which summarizes important signatures of the $SU(3)$ limit and illustrates clearly why typical deformed nuclei cannot be described with $SU(3)$. The most obvious feature of Fig. 3.7 is the appearance of rotational band-like structures with energies within a band that vary as $J(J + 1)$. Different bands correspond to different intrinsic excitations, such as ones resembling the familiar β and γ vibrational modes with $K = 0$ and 2, respectively. (We shall use the terminology "β" and "γ" because they are so familiar but we caution the reader (and will discuss later) that the corresponding excitations in the IBM have rather specific properties quite different from these geometrical counterparts.)

The first, and most striking, criterion for $SU(3)$ stems from the fact that the γ and β vibrational modes belong to a different $SU(3)$ representation than the ground band. Hence, if the E2 operator is taken to be the $SU(3)$ generator of eq. (2.186), $\gamma \rightarrow g$ and $\beta \rightarrow g$ E2 transitions are strictly forbidden. While $\beta \rightarrow g$ transition strengths are often rather small in deformed nuclei, Fig. 1.12 in Chapter 1 shows that $\gamma \rightarrow g$ transitions are always significantly

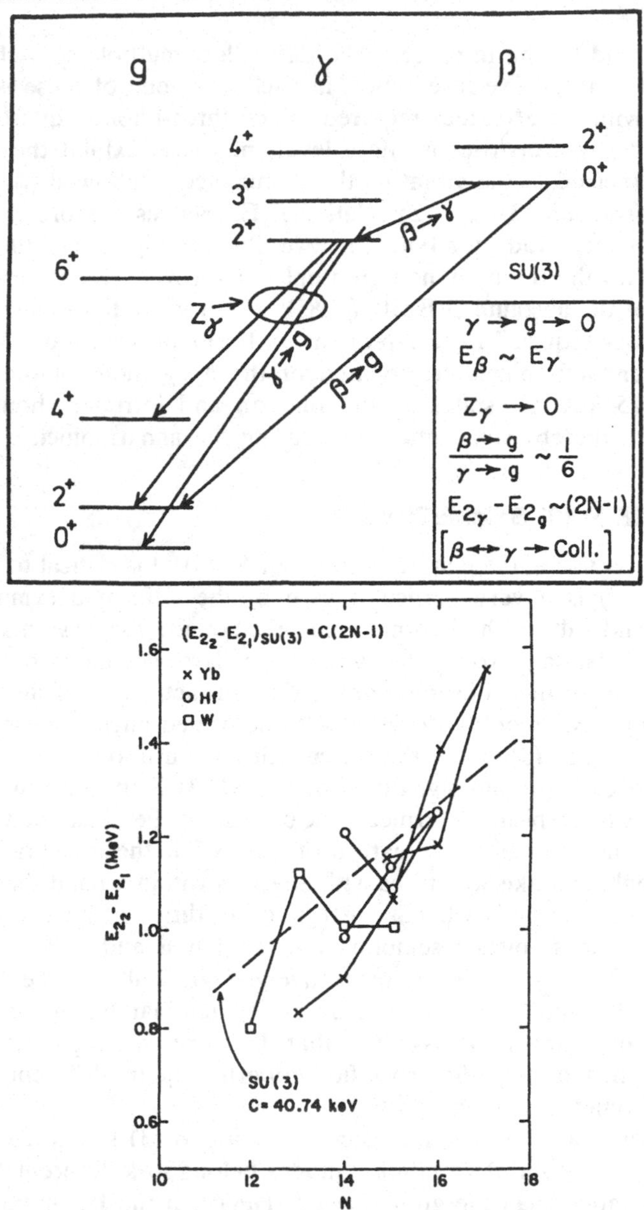

FIGURE 3.7. *Top: Principal signatures of the SU(3) limit of the* IBM. *Note that the last indicated, namely, collective* $\beta \rightarrow \gamma$ *transitions, is preserved in the presence of considerable SU(3) symmetry breaking and thus is not specifically a signature of the rigorous SU(3) limit. From Casten, von Brentano, and Haque* (1985). *Bottom: Energy levels of the* γ *band (relative to* $E(2_1^+)$) *for nuclei in the* $A = 170$ *region. The dashed line is the SU(3) prediction. From Casten and Warner* (1988).

stronger than single particle estimates for deformed nuclei. Hence, it is clear that an exact $SU(3)$ symmetry is never realized. Secondly, in $SU(3)$, levels of equal spin in the β and γ rotational bands must be degenerate. In actual deformed nuclei the most common situation is $E_\beta > E_\gamma$. A third criterion is related to the fact that, empirically, $\gamma \rightarrow g$ E2 branching ratios usually deviate from the Alaga rules. It has been traditional in geometrical models to describe such deviations in terms of band mixing whose strength is specified by a parameter Z_γ. In the pure $SU(3)$ limit there is no interaction, and no mixing, between either the β or γ bands and the ground state band. Thus, in $SU(3)$, $Z_\gamma \rightarrow 0$. This limit is never fully satisfied in real nuclei. A fourth criterion for identifying a nucleus with $SU(3)$ symmetry stems from a peculiarity of the E2 matrix elements connecting β and γ bands to the ground state band. $\beta \rightarrow g$ and $\gamma \rightarrow g$ intrinsic E2 matrix elements both vanish in the $SU(3)$ limit. However, if the $SU(3)$ condition on the E2 operator is relaxed, then for any χ value in the E2 operator, including the case where χ is arbitrarily close to $\chi_{SU(3)}$, the ratio $B(E2:2_\beta^+ \rightarrow 0_g^+)/B(E2:2_\gamma^+ \rightarrow 0_g^+)$ is finite and constant at a value $\approx 1/6$.

In general, most deformed nuclei do not display any of these particular $SU(3)$ features. [Occasionally, and presumably as a fortuitous consequence of evolving systematic trends, β and γ bands may lie close to each other and/or $\beta \rightarrow g/\gamma \rightarrow g$ $B(E2)$ ratios may approximate 1/6.] There is, however, one region which comes closest to displaying the four signatures cited above, namely, the rare earth nuclei near $N = 106$, especially Yb and Hf. The data are illustrated in Fig. 3.8 where it is seen that, near $N = 104, 106$, the systematics of each of these four observables passes near or through the $SU(3)$ predictions.

A further expectation for an $SU(3)$ region is a particular systematics of β and γ band energies. This arises because (if the coefficient of the $Q \cdot Q$ term in the Hamiltonian is constant), these energies scale as $(2N - 1)$. Thus, β and γ band energies should increase linearly with valence nucleon number and reach a maximum at midshell. Normally, this is not observed. However, in just this same region of Hf and Yb nuclei, such behavior is indeed observed as shown by the systematics of γ vibrational energies in the second half of the rare earth region shown at the bottom in Fig. 3.7. The empirical trend is consistent with $SU(3)$ although exaggerated.

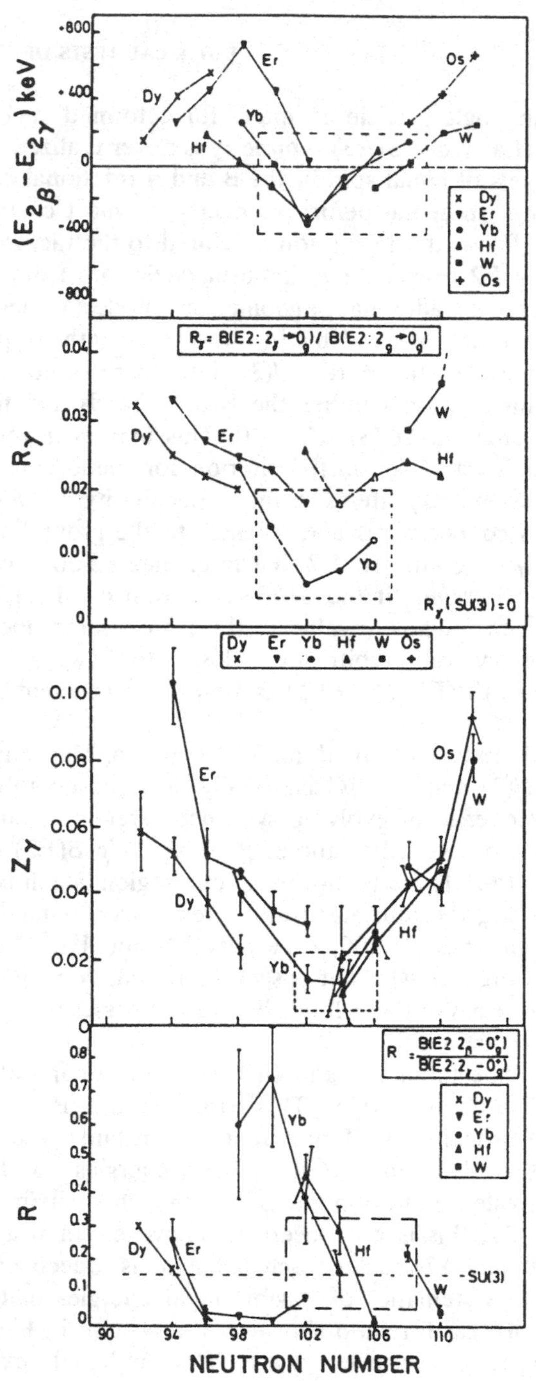

FIGURE 3.8. *Comparison of the data in the rare-earth region with the four signatures of the SU(3) limit summarized in Fig. 3.7(top). From Casten, von Brentano and Haque* (1985).

Moreover, as expected for $SU(3)$ near midshell, the γ band is particularly high lying.

Nevertheless, even in this region, an unambiguous interpretation is difficult since some of the $SU(3)$-like features could also occur if the quadrupole collectivity is low, as seems to be the case. Then, the β and γ bands would naturally be rather high lying, Z_γ would consequently be small, as would $\beta \rightarrow g$ and $\gamma \rightarrow g$ B(E2) values. The degeneracy of β and γ bands and the approximate 1/6 ratio of their B(E2) values to the ground state band are occasionally seen elsewhere and could also arise fortuitously. It is perhaps best to conclude that the status of the $SU(3)$ symmetry in this region is not yet settled.

3.4. NUMERICAL CALCULATIONS IN THE IBM

The distinctive structures of the three dynamical symmetries in the IBM provide three clear cut limits of the general Hamiltonian. Although evidence exists which suggests that some of the features of the pure symmetries are observed empirically in selected nuclei, in general, a realistic calculation will require a departure from the strict limits or indeed a transition between them. In this context the analytic limits emerging from the group theoretical treatment of the Hamiltonian can be viewed as "benchmarks" in constructing a more accurate description of the low lying collective structure of a particular nucleus, or series of nuclei. In general, then, most nuclei will require specific numerical calculations in which the IBM Hamiltonian is diagonalized after certain choices of parameter values are made. This approach can be illustrated schematically in the form of the symmetry triangle of Fig. 3.9. The three apexes represent the three exact symmetries, while the space enclosed by the three sides denotes a range of more general solutions which can be obtained numerically by diagonalizing the IBM-1 Hamiltonian (eq. (2.164) or (2.167)). A transition between two specific symmetries, without invoking any of the characteristics of the third, would correspond to a path along one of the three sides, but a more complex path between two limiting cases is clearly also possible. An important point is that, for a transition along the sides, the structure at any point will be determined solely by a single parameter, namely the ratio of the two coefficients which characterize the symmetries at the ends of that side. These parameters are also indicated in the figure and can be understood by reference to the Hamiltonians

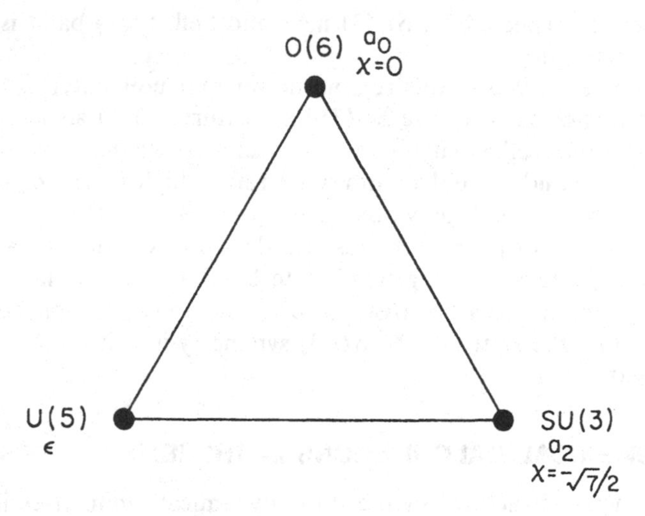

FIGURE 3.9. *Symmetry triangle of the* IBM *showing the symmetries, the transitions legs and* IBM *parameters relevant to each.*

giving each of the three limiting symmetries in Chapter 2 (eqs. (2.169), (2.172), and (2.176)). [The mention of χ in this context for the $SU(3) \leftrightarrow O(6)$ transition will be discussed below in the context of the Consistent Q Formalism.]

From this discussion, it is clear that one of the most appealing aspects of the IBM-1, aside from the basic symmetries themselves, is the ease with which transition regions can be calculated as a function of a single, physically intuitive, parameter. When it is recalled that such phase transitional regions have historically been considered the most challenging and complex testing grounds for nuclear models, because of the competition of rival degrees of freedom, it is clear that the group theoretical structure of the IBM-1 truncation offers a significant simplification. Whether this simplification is physically realistic, of course, can be assessed only by recourse to the data and this will be extensively discussed below.

For convenience, we repeat the multipole expansion of the Hamiltonian of eq. (2.167) here:

$$H = \epsilon'' \hat{n}_d + a_0 P^\dagger P + a_1 \mathbf{J}\cdot\mathbf{J} + a_2 Q\cdot Q + a_3 T_3\cdot T_3 + a_4 T_4\cdot T_4. \tag{3.3}$$

The effect of the various terms and their links with the limiting symmetries have already been discussed, and the latter aspect is contained in the definitions of the various Casimir operators. In the context of a general calculation, the first point to note is that the **J·J** term is always diagonal and simply gives a contribution proportional to $J(J + 1)$ to the energy of each level of spin J. Thus, not only has it no effect on state wave functions but it also has no effect on energy differences of states of the same spin. The term in $T_4 \cdot T_4$ does not appear explicitly in the definition of any of the Casimir operators and, indeed, has seldom been found necessary in actual applications. It will therefore be ignored in the discussion which follows. The terms in ϵ'' (or ϵ as it is often called), $P^{\dagger}P$, and $Q \cdot Q$ produce the characteristics of the $U(5)$, $O(6)$ and $SU(3)$ structures, respectively, while $T_3 \cdot T_3$, stemming from the $O(5)$ subgroup, is common to both the $U(5)$ and $O(6)$ chains.

The first step in an IBM-1 calculation is to determine the approximate position of the nucleus under study with regard to the three limiting symmetries. This can be done phenomenologically by reference to some of the characteristic energy and B(E2) ratios which distinguish the symmetries or by virtue of a theoretical understanding of the underlying microscopic shell model basis of the IBM and the geometrical approach. For example, nuclei where both neutrons and protons are near their respective closed shells (both particle-like or both hole-like) can be expected to show $U(5)$ or vibrational behavior, while Z and N values near midshell would suggest rotational structure, and hence $SU(3)$ as a starting point. The occurrence of axially asymmetric features can be expected in cases where the neutrons are particle-like and the protons hole-like, or vice-versa, and suggests the use of the $O(6)$ symmetry. Care must be taken in using these qualitative arguments, however, since the particle or hole-like character of the bosons can be influenced by significant subshell effects. Moreover, since, in practice, the $O(6)$ symmetry frequently occurs near the end of major shells, and since many of its predictions are very similar to those of $U(5)$, it is sometimes difficult initially to distinguish the two. It is particularly important to consider B(E2) values as well as energies in this case.

In treating nuclei deviating from the limiting symmetries we shall present two comparable but rather different practical approaches, one in terms of the multipole Hamiltonian of eq. (3.3) and the other in terms of the so-called Consistent Q Formalism (CQF): the latter is a special case of the IBM that sacrifices some generality but contains important simplifications, as well as a more direct physical interpretation, and fewer parameters. Moreover, it has been shown to provide sets of IBM predictions that agree remarkably well with the data.

We start with the multipole Hamiltonian approach and we first consider the broad class of deformed nuclei such as the rare earth region from $A \approx 150$–190 or the actinides. These nuclei have been studied extremely thoroughly experimentally, and it is a pre-requisite of any general model such as the IBM that it can effectively treat such structures.

We have seen that most deformed nuclei cannot be described by the $SU(3)$ limit. Clearly, it will be necessary to introduce some (considerable, it turns out) $SU(3)$ symmetry breaking. In most deformed nuclei the β band is above the γ band. Reference to Fig. 3.1 suggests that this can be produced by introducing a perturbation to $SU(3)$ in the direction of $O(6)$. An example of a calculation for a well-deformed nucleus is shown in Fig. 3.10 (right). On the left, the $SU(3)$ "starting point" is shown. The levels were obtained with the Hamiltonian $H = a_2 Q \cdot Q + a_1 J \cdot J$. The eigenvalues of $SU(3)$ were given in terms of those of the Casimir operators C_{2O3} and C_{2SU3} in eq. (2.173). For convenience, we give the eigenvalue equation for the Hamiltonian just cited:

$$E_{SU(3)} = \left(a_1 - \frac{3a_2}{8}\right) J(J+1) + \left(\frac{a_2}{2}\right)[\lambda^2 + \mu^2 + \lambda\mu + 3(\lambda + \mu)]. \tag{3.4}$$

Inserting values of J and (λ, μ) gives the useful relations

$$a_2 = -\frac{E(2_2^+) - E(2_1^+)}{3(2N-1)}, \qquad a_1 = \frac{E(2_1^+)}{6} + \frac{3}{8}a_2. \tag{3.5}$$

The resulting $SU(3)$ spectrum fits the 2_1^+ and 2_2^+ states by definition, has an exact $J(J+1)$ rotational band structure, and degenerate β and γ bands.

The next step is to break this degeneracy, without losing the rotational energy spacing. The terms in $\epsilon \hat{n}_d$ or $T_3 \cdot T_3$ will tend to decrease the ratio $E(4_1^+)/E(2_1^+)$. The reason is that $\epsilon \hat{n}_d$ alone produces a vibrator spectrum with $E(4_1^+)/E(2_1^+) = 2$ while $T_3 \cdot T_3$ gives

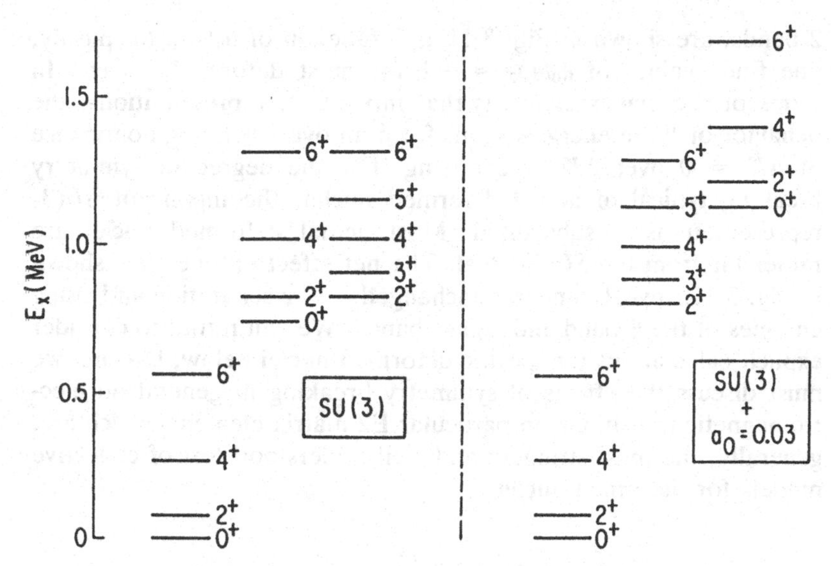

FIGURE 3.10. *Example of the effect of adding a term in $a_0P^\dagger P$ to an SU(3) Hamiltonian with a_1 = 0.0105 MeV, a_2 = −0.008 MeV and N = 16. a_0 is in* MeV *From Warner and Casten* (1982).

the $O(5)$ spectrum with its $\tau(\tau + 3)$ dependence (see eq. (2.178)) and, hence, $E(4_1^+)/E(2_1^+) = 2.5$. Thus, such terms must be kept small for well deformed nuclei. The $P^\dagger P$ term, however, has very little effect on $E(4_1^+)/E(2_1^+)$ (see Fig. 3.10) but does raise the β band, since it represents a symmetry breaking in the direction of $O(6)$, whose higher representations each begin with a 0^+ state.

The effects of this term can be understood rather easily if we consider the simple Hamiltonian

$$H = a_2Q\cdot Q + a_0P^\dagger P + a_1\mathbf{J}\cdot\mathbf{J}. \tag{3.6}$$

As noted earlier, the $\mathbf{J}\cdot\mathbf{J}$ term is diagonal and can be dropped for our present considerations. Thus, H can be rewritten

$$H = a_2[Q\cdot Q + (a_0/a_2)P^\dagger P] \tag{3.7}$$

from which it is clear that a_2 is now only a scale factor on the energies and the structure is fully specified by the ratio a_0/a_2. In light of the discussion of the symmetry triangle, this is not surprising: a_2 and a_0 are the coefficients giving the $SU(3)$ and $O(6)$ limits, respectively, and their ratio thus determines intermediate structures. The energies of the β and γ bands, and the next $K = 0$ and

2 bands, are shown in Fig. 3.11 as a function of $a_0/4a_2$. (Typically, one finds values of $a_0/4a_2 \approx -1$ for most deformed nuclei.) In terms of the matrix elements that mix $SU(3)$ representations, the behavior of these energies stems from an overwhelming dominance of $\Delta K = 0$ over $\Delta K = 2$ mixing. For the degree of symmetry breaking typical of actual deformed nuclei, the mixing of $SU(3)$ representations is substantial. Most actual deformed nuclei are rather far from the $SU(3)$ ideal. The net effect on energies, shown in Fig. 3.11, reveals a nearly unchanged $\gamma - g$ separation and rising energies of the β band and higher bands. We will return to consider explicit calculations for specific deformed nuclei below, but first we must discuss the effects of symmetry breaking in general on electromagnetic transitions, in particular E2 matrix elements which are, generally, the most stringent and well understood test of collective models for deformed nuclei.

3.5. E2 TRANSITIONS

The most general form of the E2 operator is given by eq. (2.179). Here, and in most of the literature, this is written

$$T(\mathrm{E2}) = e_B\{(s^\dagger\tilde{d} + d^\dagger s) + \chi(d^\dagger\tilde{d})^{(2)}\} \tag{3.8}$$

where e_B is the boson effective charge which determines the absolute $B(\mathrm{E2})$ scale. To understand the behavior of the E2 matrix elements for deformed nuclei, consider again the Hamiltonian of eq. (3.6), where the perturbation to the pure $SU(3)$ limit is supplied by the $P^\dagger P$ term, which can describe nuclei where the β band lies above the γ band in energy, as is in fact the case for the majority of the well deformed rare earth nuclei.

The characteristic empirical feature, and indeed signature, of deformed nuclei, illustrated in Fig. 1.12, is the remarkable stability of many of the $B(\mathrm{E2})$ systematics for the ground and γ bands across the entire region. Figure 1.12 also reveals the large fluctuations in the $\beta \rightarrow g$ $B(\mathrm{E2})$ strengths as well as their significantly reduced magnitude, relative to $\gamma \rightarrow g$ strengths. The most obvious conclusion from these systematics, in the context of the IBM, is that the finite $\gamma \rightarrow g$ and $\beta \rightarrow g$ $B(\mathrm{E2})$ values both require departures from the strict $SU(3)$ limit as they involve transitions, between representations, which cannot occur if $a_0 = 0$ and $T(\mathrm{E2})$ utilizes a Q operator with $\chi = -\sqrt{7}/2$.

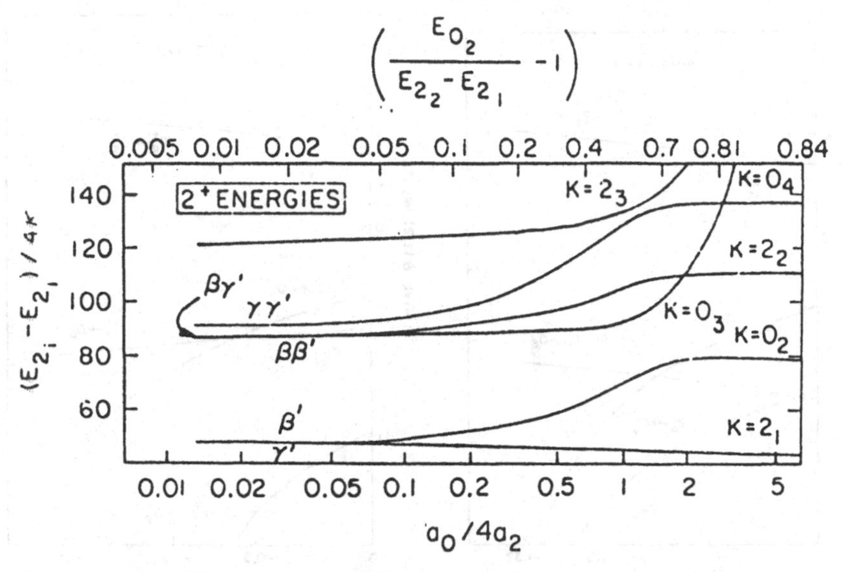

FIGURE 3.11. *The effects of SU(3) symmetry breaking in the* IBM. *The* 2^+ *energies of various rotational bands are shown relative to* $E(2_1^+)$. *The scale on the abscissa is such that 0 corresponds to SU(3), while the O(6) limit is approached when the abscissa tends toward large values. From Casten and Warner* (1988).

The B(E2) strengths which result from the $SU(3)$ Hamiltonian, and from a broken symmetry calculation with a typical ratio $a_0/4a_2 = -0.94$ are shown in Fig. 3.12 as a function of the parameter χ in T(E2). The *intra*-representation transitions are very little affected by χ, while the *inter*-representation transitions, which must tend to zero in the $SU(3)$ limit as $\chi \to -\sqrt{7}/2$, grow rapidly as χ deviates from this limit. For a given χ, the dominant effect of the $a_0 P^\dagger P$ symmetry breaking term is to decrease the magnitude of the $\beta \to g$ transitions, relative to other transition rates. All these results can be easily seen in the ratios plotted in Fig. 3.13 which also shows an additional striking feature, namely, the rigorous constancy of the ratio $B(\mathrm{E2}:2_\beta \to 0_g)/B(\mathrm{E2}:2_\gamma \to 0_g)$ for an $SU(3)$ Hamiltonian as a function of χ in the E2 operator.

This constancy is an example of a more general rule and can be understood simply in terms of the contributions to the E2 matrix element from the two parts of the E2 operator. The vanishing of the *inter*-representation transitions, for $\chi = -\sqrt{7}/2$, implies that,

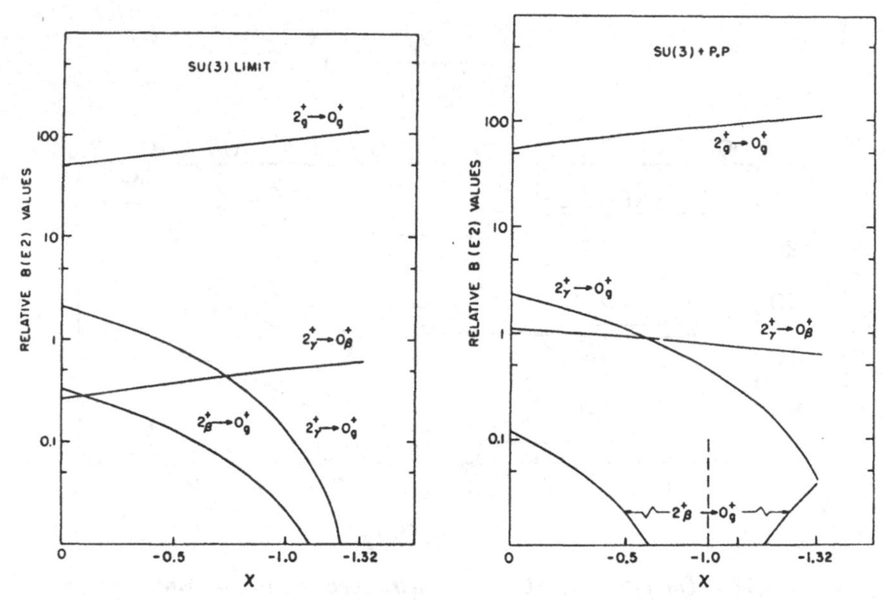

FIGURE 3.12. *Relative B*(E2) *values involving the* β, γ, *and ground bands plotted as a function of the constant* χ. (*Left*) *for the SU*(3) *Hamiltonian with N* = 16; (*Right*) *similar except with a* $P^{\dagger}P$ *perturbation to SU*(3). *The broken SU*(3) *predictions depend only on the ratio* a_0/a_2 *and on* χ. *Here, a typical value* $a_0/a_2 = -0.94$ *is used. From Warner and Casten* (1982).

of necessity, the contributions from the two terms in the E2 operator of eq. (3.8), that is, the $\Delta n_d = 0$ (or $s^{\dagger}\tilde{d} + d^{\dagger}s$) and $\Delta n_d = \pm 1$ (or $\chi d^{\dagger}\tilde{d}$) matrix elements, interfere destructively and cancel exactly. Thus, their magnitudes must be in the ratio $\sqrt{7}/2$ and their signs the same. But this in turn implies the more general result that any inter-representation transition for an *SU*(3) Hamiltonian is given by $(\chi - \sqrt{7}/2)\,\langle\psi^i|d^{\dagger}\tilde{d}|\psi^f\rangle$. Hence, we get the desired result that the *ratio* of any two *inter*-representation transitions will be a constant, independent of χ, for the *SU*(3) Hamiltonian. As we have seen, for the specific example of $\beta \to g/\gamma \to g$ branching ratios, this ratio takes the value of ≈ 1/6 which is in qualitative agreement with many empirical ratios (see Fig. 1.12). For *intra*-representation transitions the contribution from the first term in *T*(E2) ($\Delta n_d = \pm 1$) dominates and hence χ has very little effect on their magnitude.

The empirical data for the $\gamma \to g/g \to g$ *B*(E2) ratios (see Fig.

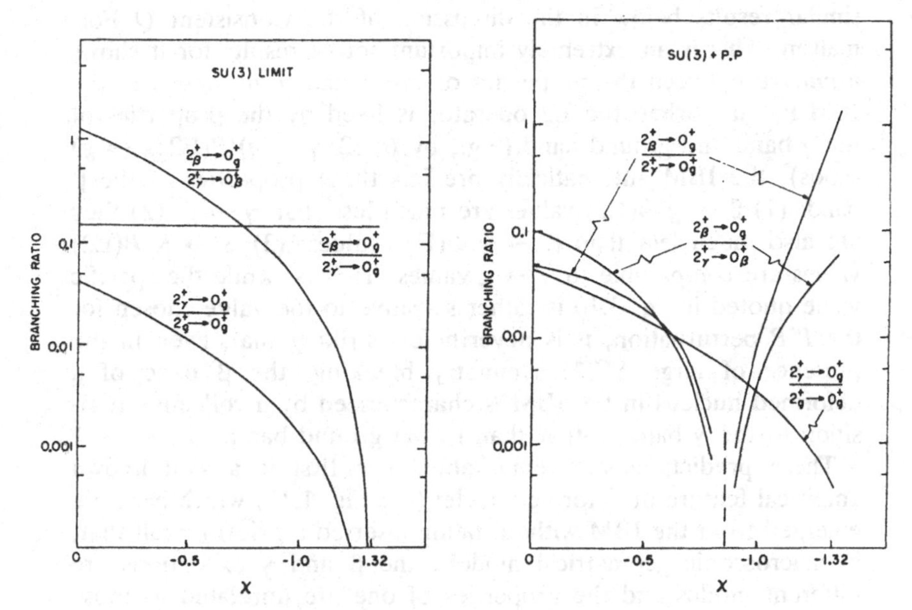

FIGURE 3.13. *B*(E2) *ratios based on Fig. 3.12.*

1.12) can now be combined with the results of Fig. 3.13 to extract a range of χ values capable of reproducing the B(E2) data in the rare earth region. Note that this step is possible because of the very small effect of the $P^\dagger P$ perturbation on both the $\gamma \rightarrow g$ and $g \rightarrow g$ B(E2) strengths. The result is

$$-0.54 < \chi < -0.22 \tag{3.9}$$

which reiterates the fact that the $SU(3)$ form of the E2 operator cannot reproduce the observed data. Moreover, one notes that, for most deformed nuclei, χ falls in a rather narrow range. Indeed, even with symmetry breaking in the Hamiltonian, the predicted magnitude of $\gamma \rightarrow g$ strengths would be two orders of magnitude too low if $\chi = -\sqrt{7}/2$ were used in T(E2). Thus the $SU(3)$ symmetry must always be broken in the E2 operator. In addition, the results of Fig. 3.13 and eq. (3.9) give rise to important predictions concerning the decay of the β band, namely

$$\frac{B(\mathrm{E2}:\beta \rightarrow g)}{B(\mathrm{E2}:\gamma \rightarrow g)} \simeq \frac{B(\mathrm{E2}:\beta \rightarrow g)}{B(\mathrm{E2}:\gamma \rightarrow \beta)} \approx 0.04. \tag{3.10}$$

(The specific values all refer to $2^+ \rightarrow 0^+$ transitions.) We shall see similar results below in the discussion of the Consistent Q Formalism. This is an extremely important set of results for it shows a *linkage* between the properties of the β and γ vibrations in the IBM in that, when the E2 operator is fixed by the properties of the γ band and ground band (e.g., by $B(\mathrm{E2}:\gamma \rightarrow g)/B(\mathrm{E2}:g \rightarrow g)$ ratios), the IBM automatically predicts three properties of the β band: (1) $\beta \rightarrow g$ B(E2) values are much less than $\gamma \rightarrow g$; (2) they are also much less than $\beta \rightarrow \gamma$ B(E2) values; (3) $\beta \rightarrow \gamma$ B(E2) values are comparable to $\gamma \rightarrow g$ values. That is, while the specific value quoted in eq. 3.10 is rather sensitive to the value chosen for the $P^{\dagger}P$ perturbation, it is nevertheless striking that, even in the presence of large $SU(3)$ symmetry breaking, the β band of a deformed nucleus in the IBM is characterized by a collective transition to the γ band, rather than to the ground band.

These predictions are remarkable. The first is a well known empirical feature of deformed nuclei (see Fig. 1.12) which has here emerged from the IBM without being inserted a priori (recall that, in macroscopic geometrical models, the β and γ excitations are different modes and the properties of one are unrelated to those of the other). The second and third results, though completely at variance with traditional understanding, with geometrical models, and with what had been considered the empirical situation, have now been experimentally verified. We shall return to this issue of β-γ linkage below after developing the Consistent Q Formalism (CQF).

To close this section, we now once again use the data of Fig. 1.12 on absolute B(E2) strengths for $2_g^+ \rightarrow 0_g^+$ transitions, in conjunction with the deduced range of χ values, to ascertain a corresponding range for the boson effective charge. The outcome is a phenomenologically defined form for T(E2) in well-deformed rare-earth nuclei:

$$T(\mathrm{E2}) = 0.145(15)\{(s^{\dagger}\tilde{d} + d^{\dagger}s) - 0.38(16)(d^{\dagger}\tilde{d})^{(2)}\}. \quad (3.11)$$

The situation in the actinides is much less clear. The ratio of $B(\mathrm{E2}:\gamma \rightarrow g)$ to $B(\mathrm{E2}:g \rightarrow g)$ values seems to be less than for the rare earth region and widely different E2 operators have been tried. In any case, this equation can be used as an initial guide to calculations and as a reasonable estimate. For detailed fits to a given nucleus it may be best to fine tune the coefficients e_B and χ.

3.6. THE CONSISTENT Q FORMALISM (CQF)

At this point it is appropriate to introduce an alternate approach to IBM calculations that offers substantial advantages and simplifications for many types of nuclei, most clearly for deformed and transitional cases, namely the CQF already referred to a couple of times.

In the previous section it was demonstrated that the empirical data on E2 transitions in the rare earth region mandate a rather well defined structure for the E2 operator which is not consistent with the $SU(3)$ form of Q assumed in the Hamiltonian. Therefore, different forms of the quadrupole operator were used in H and $T(\mathrm{E2})$ in order to describe the data. That is, χ was fixed at $-\sqrt{7}/2$ in H, where $SU(3)$ was broken by other terms, while χ was varied as a parameter in $T(\mathrm{E2})$. It is clearly of interest to ask whether a comparable description can be obtained within a framework where *consistent* forms of the quadrupole operator (i.e., the *same* χ) are used in both. This approach is referred to as the Consistent Q Formalism (CQF) and, in its simplest form, utilizes a Hamiltonian

$$H = a_2 Q \cdot Q + a_1 J \cdot J \tag{3.12}$$

and

$$T(\mathrm{E2}) = e_B Q \tag{3.13}$$

where, in both cases, Q is taken as

$$Q = (s^\dagger \tilde{d} + d^\dagger s)^{(2)} + \chi (d^\dagger \tilde{d})^{(2)} \tag{3.14}$$

and where χ is variable.

A reduction of $|\chi|$ in the CQF approach and the use of $P^\dagger P$ in the multipole Hamiltonian are both likely to generate the same type of perturbation to the $SU(3)$ spectrum, since both enhance the relative importance of terms in H which change the d-boson number by ± 2, i.e., they produce very similar symmetry breaking terms. More specifically, a reduction in the absolute magnitude of χ produces the same effect as the $P^\dagger P$ term in pushing the β band above the γ band in energy. Moreover, as with the $P^\dagger P$ term, the higher excitations are also pushed higher in energy relative to the γ band. These energies are shown as a function of χ in Fig. 3.14 and can be compared to those shown against $a_0/4a_2$ in Fig. 3.11 (up to $a_0/$

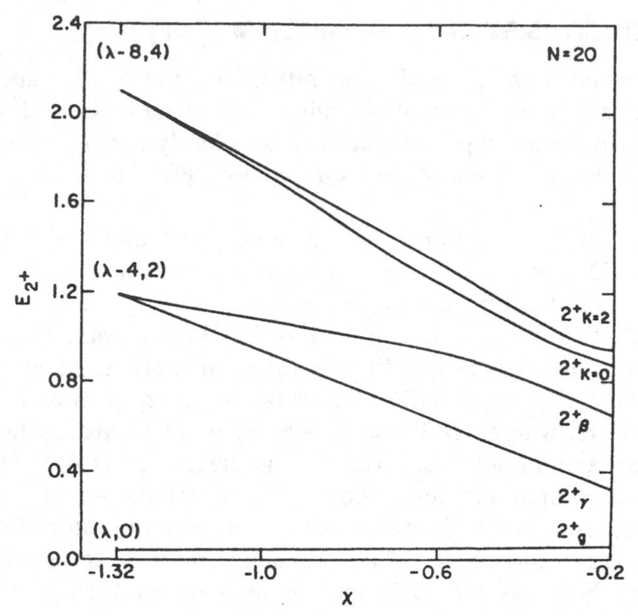

FIGURE 3.14. *Energies of various* 2^+ *states in the consistent-Q formalism for the transition from SU*(3) *(left-hand side) towards O*(6) ($\chi = 0$). *For* $N = 10$, *with* $a_2 = -0.02$ MeV. *From Casten and Warner* (1988).

$4a_2 \approx -1$). Although the behavior of the absolute excitation energies is different, their values relative to the γ band are quite similar.

Although the CQF is most used for transitional and deformed nuclei, it has an intimate and interesting relation to the $O(6)$ limit which we must briefly comment on first. We then turn to CQF predictions for deformed and transitional nuclei.

3.6.1. O(6) Nuclei in the CQF

Just as the dominance of the $a_0 P^\dagger P$ term in the original form of the multipole Hamiltonian results in an $O(6)$ spectrum, here too, in the CQF, when $\chi \rightarrow 0$, the quadrupole operator becomes a generator of $O(6)$ so that a spectrum with $O(6)$ symmetry must be produced. However, inspection of the original $O(6)$ Hamiltonian of eq. (2.176) shows that the CQF version, eq. (3.12), involves one less term (we neglect the $N(N + 4)$ term in eq. (2.176) since it is irrelevant for *excitation* energies). Hence, the eigenvalues for $O(6)$

can depend only on two parameters instead of the three appearing in eq. (2.178) or (3.1). The eigenvalue expression (eq. (3.1)) for $O(6)$ in the CQF becomes

$$E_{O(6)} = A[\sigma(\sigma + 4) + \tau(\tau + 3)] + CJ(J + 1) \qquad (3.15)$$

showing that the constrained form of $O(6)$ which emerges in the CQF embodies an equality of the $O(6)$ and $O(5)$ contributions with $A = B$ in eq. (3.1): the CQF requires that $A/B = 1$. This is a manifestation of an inherent relation, in the CQF, between the scales of degeneracy breaking occurring in the $O(6)$ and $O(5)$ stages of the group chain decomposition of eq. (2.92). It is therefore remarkable that of all possible A/B ratios, empirical fits to $O(6)$-like nuclei in *both* the $A = 130$ (Xe, Ba), and the Pt regions indeed give $A/B \approx l$. This is shown in Table 3.2. Thus, once again, the CQF, though simpler, seems very well adapted to the data.

3.6.2. Deformed Nuclei in the CQF

From our discussion it is apparent that the transition from $O(6)$ to $SU(3)$ can be accomplished in the CQF simply by varying χ between the values (0 and $-\sqrt{7}/2$) appropriate to each limit. As will be seen below, this treatment, which again is simpler than the earlier formalism and involves one less parameter, works at least as well.

It is interesting to understand the physical effect of varying χ. The basic idea is that $SU(3)$ is a symmetric rotor, while $O(6)$ is an asymmetric, γ-soft rotor with a mean or rms γ value of 30°. Thus, changes in χ should correspond somehow to the introduction of (dynamic) axial asymmetry. On the other hand, studies of the classical limit of the IBM-1 show that all γ dependence in the equivalent geometrical potential, $V(\beta, \gamma)$, enters through terms of the form $\cos 3\gamma$. Thus, $V(\beta, \gamma)$ can only have a minimum at $\gamma = 0°$ or 60° (prolate, oblate limits) but never for intermediate axially asymmetric values. Thus, the question arises as to how the IBM goes from the axially symmetric $SU(3)$ limit to the γ-soft ($\gamma_{rms} =$ 30°) $O(6)$ case. To this end it is useful to try to correlate IBM calculations in the CQF, where the only parameter is χ, to those in the Davydov model, which depend on γ. This is done in Fig. 3.15. For a given observable we see that the predictions of the CQF and Davydov models pass through exactly the same range of values. Moreover, the qualitative shapes are even similar. Therefore, for

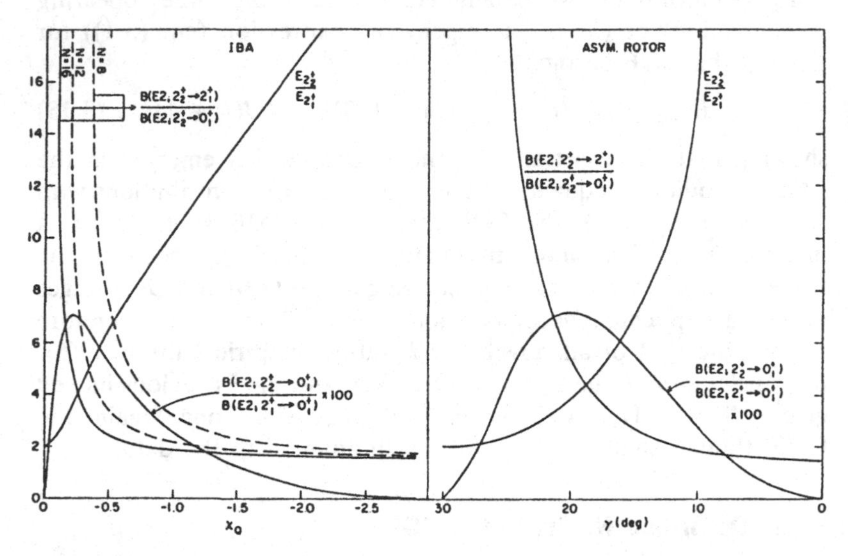

FIGURE 3.15. *Comparison of three observables in the IBM (as a function of* $\chi_Q \equiv \sqrt{5}\chi$*) and in the Davydov model (against* γ*). From Casten, Aprahamian, and Warner* (1984).

each of these observables it is possible to choose a γ value in the Davydov model that gives the identical prediction as the IBM for a particular χ and, thus, to establish a γ-χ correspondence or relation. When this procedure is done for general observables, similar γ-χ relationships result. As illustrated in Fig. 3.16, an effective γ value can be defined for each χ to within $\pm$ a few degrees. We also note in Fig. 3.16 that $\gamma_{\text{eff}} = 30°$ for $O(6)$ and that γ_{eff} drops smoothly as χ approaches its $SU(3)$ value of $-\sqrt{7}/2$. This establishes a simple, and physically intuitive, relation between an IBM CQF calculation and the geometrical concept of axial asymmetry. It is interesting that γ does not go to zero in $SU(3)$ but rather maintains a value near $\gamma = 10°$ (for realistic boson numbers). This has a simple interpretation. The $O(6) \rightarrow SU(3)$ transition corresponds to a potential changing from one that is completely γ flat to potentials that have a minimum at $\gamma = 0°$ and a finite slope as a function of γ (see Fig. 3.17). The slope becomes steeper (but remains finite) as χ approaches the $SU(3)$ value. (It only becomes infinitely steep if $N \rightarrow \infty$.) Thus, for finite N, γ_{eff} never vanishes, even in $SU(3)$. The γ values and the asymmetry that characterize

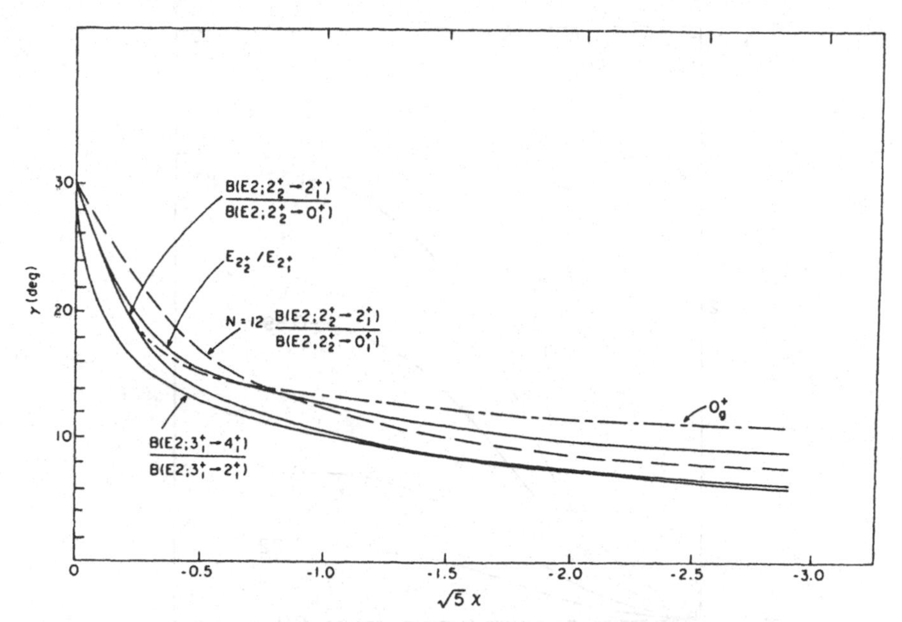

FIGURE 3.16. *$\gamma \leftrightarrow \chi$ relationship in the $SU(3) \leftrightarrow O(6)$ transition, for various observables in the* IBM. *For one observable the results for two different N values are shown. From Casten, Aprahamian and Warner* (1984). *The curve marked 0_g^+ was obtained analytically by Castanos, Frank and Van Isacker* (1984).

the IBM, even in $SU(3)$, but more so with deviations toward $O(6)$, are thus *dynamic*, resulting from zero point motion in a finite potential, $V(\beta, \gamma)$. [The above procedure could be subject to the criticism that it compares the Davydov model of rigid γ-asymmetry with one of substantial γ softness: this issue has been discussed in the literature by Castanos, Frank, and Van Isacker [1984]. It is outside the scope of the present chapter. Suffice it to say that the present discussion is validated, in particular the γ-χ relationship in Fig. 3.16.]

This figure correlates well with our knowledge of real deformed nuclei. We will show momentarily that, for typical deformed nuclei, $\chi \approx -0.5$. From Fig. 3.16, this corresponds to a γ value of $\gamma_{\text{eff}} \approx 11°$ which is exactly the magnitude often assigned to these nuclei in the past through geometrical model interpretations.

The principle advantage of the CQF approach lies in its considerable simplicity. Since the $J \cdot J$ term in eq. (3.12) is always diag-

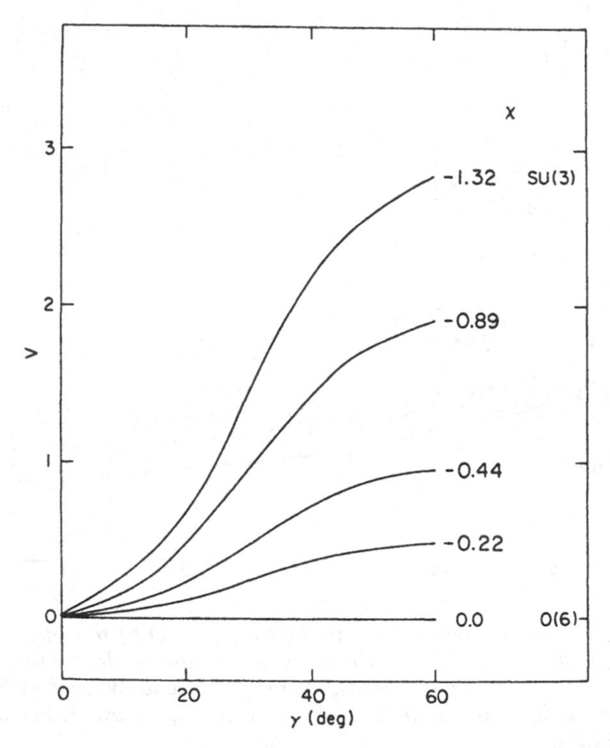

FIGURE 3.17. *The dependence of the classical potential corresponding to the* IBM *as a function of* γ *for several values of* χ, *ranging from the* $SU(3)$ *limit* ($\chi = -1.32$) *to* $O(6)$ ($\chi = 0$). *From Casten, Aprahamian and Warner* (1984).

onal, the wave functions of the CQF Hamiltonian are uniquely specified by χ (and the boson number N). The parameters a_2 and e_B both act only as scaling factors on the absolute energies and B(E2) values so that relative values of these observables are also uniquely specified only by χ and N. Thus in this framework, the energy and E2 properties are inexorably linked so that, if the χ value is determined from one, the other is determined (at least on a relative scale). In fact, ratios of energies or B(E2) values can be predicted in this framework as a function of χ and N and displayed in the form of contour plots as illustrated in Figs. 3.18 and 3.19. (The dots on these figures serve to emphasize the fact that the predictions are only valid for integral values of N.) These contour

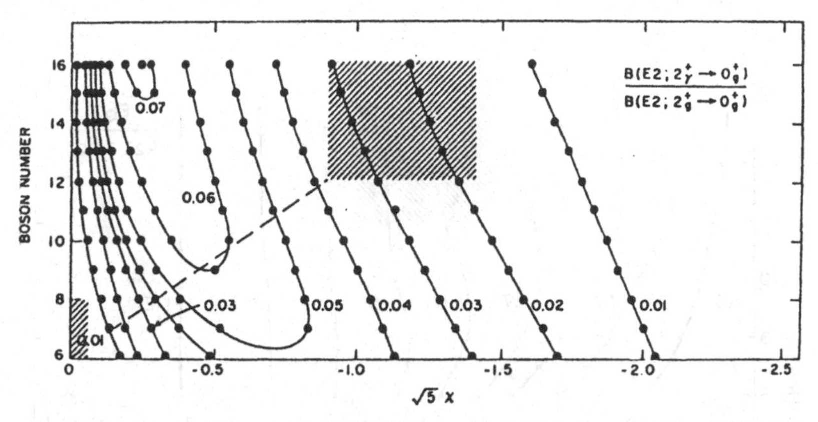

FIGURE 3.18. *Contour plot of the indicated* B(E2) *ratio in the* CQF *as a function of* N *and* χ. *From Warner and Casten* (1983).

plots are universal (in the context of eq. (3.12) and the CQF) and so can be used in a variety of ways. Some of these will be discussed in Section 3.9. Here we note that it is simple to use the ratio $R_\gamma = B(\mathrm{E2}:2_\gamma^+ \rightarrow 0_g^+)/B(\mathrm{E2}:2_g^+ \rightarrow 0_g^+)$ displayed in Fig. 3.18 as before, to determine the permissible range of χ values appropriate to the well deformed nuclei in the rare earth region by using the data of Fig. 1.12. The result is the hatched rectangle in the upper center of Fig. 3.18 which can then be transposed to other contour plots to provide predictions for other B(E2) ratios and for the relative energy spectrum. (The other hatched area in Fig. 3.18 (lower right) corresponds to $O(6)$-like nuclei.) The relatively narrow range of χ values appropriate to deformed nuclei reflects the relatively constant values of R_γ seen empirically in Fig. 1.12. Note that in the upper part of Fig. 3.19 the energy ratio plotted is effectively the ratio of the intrinsic excitation energies of the β and γ bands since the form of the denominator is only designed to remove the $J(J + 1)$ dependence. The contour plot involving $\beta \rightarrow \gamma$ transitions shows that these remain collective for large ranges of χ and N. Also, we see that the CQF gives $\beta \rightarrow g$ B(E2) values much less than $\gamma \rightarrow g$. We recall that the multipole Hamiltonian produced exactly these same results. Here, we note from the ranges of values seen in Figs. 3.18 and 3.19 that such predictions are an inherent feature of the IBM in the CQF: they cannot be avoided. If, for example, $\beta \rightarrow g$ B(E2) values were significantly larger than $\gamma \rightarrow g$ values, the CQF

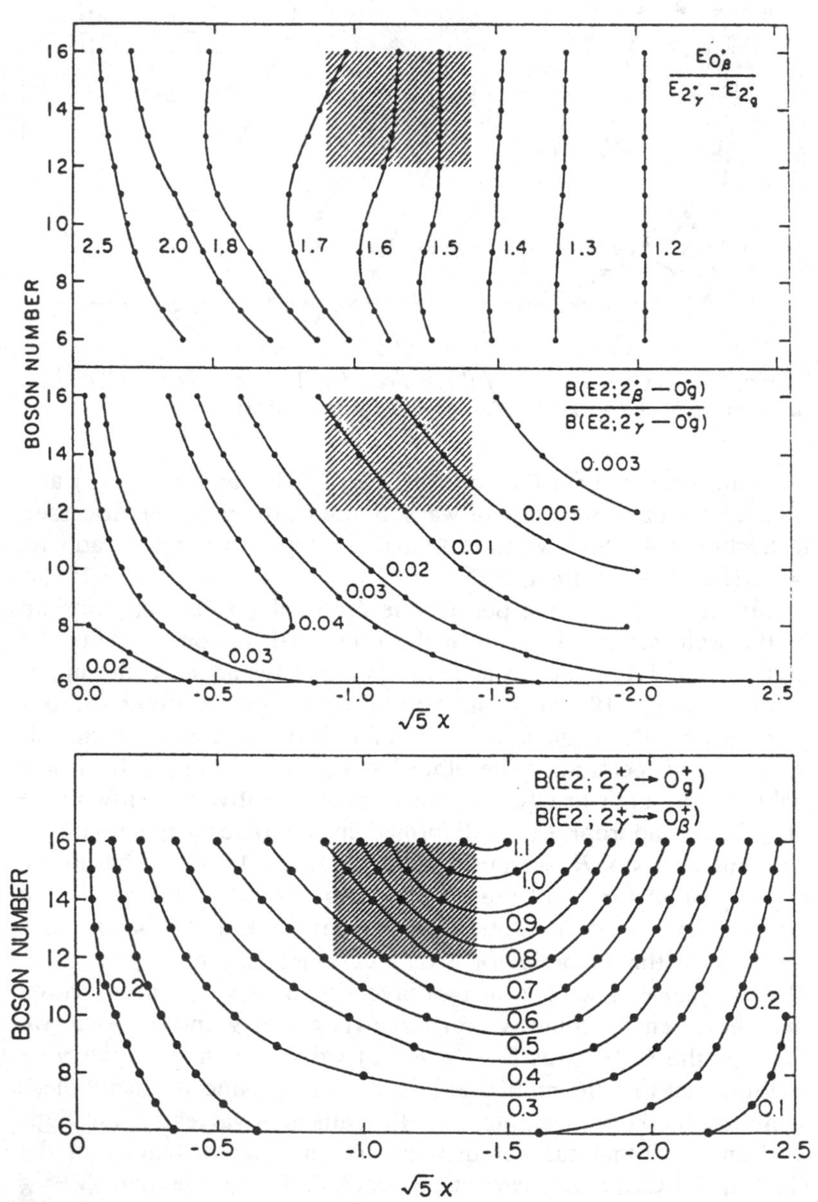

FIGURE 3.19. *Contour plot of the indicated ratios in the CQF as a function of N and* χ. *From Warner and Casten* (1983).

could not reproduce this. Thus, these qualitative predictions of the CQF are themselves a sensitive test of the IBM and the consistent Q formalism.

3.7. PRACTICAL CALCULATIONS

Having discussed many of the properties of broken $SU(3)$ calculations in general terms or in the context of the systematics of certain observables in deformed nuclei, it is useful at this point to consider, in some detail, an application of IBM calculations to a specific deformed nucleus. ^{168}Er is probably the best studied case, and has become a classic example for applications of collective models to deformed nuclei. This nucleus has been treated in both the original IBM formalism and in the CQF. The results are very similar in the two cases although, as we shall see, the CQF gives a better interpretation of E2 branching ratios and has, as is usual, one fewer parameter. For historical reasons, however, we will show a fit to the energy levels with the multipole Hamiltonian of eq. (3.3) since this was the first IBM calculation for ^{168}Er: indeed, it was the first extensive comparison for any deformed nucleus.

The predicted and measured levels for ^{168}Er are compared in Fig. 3.20. There are several essential points. First, it is evident that the overall agreement between experiment and theory seems excellent: the ground, γ band and β band energies are well reproduced as is the experimental sequence of higher lying bands. However, comparison of predicted and observed E2 branching ratios for the higher lying $K = 2$ bands suggests that the apparent agreement for these levels is partly fortuitous and that, even for the second excited $K = 0$ band, the IBM description only gives a semiquantitative interpretation. Also, there is one clear disagreement from the energy levels themselves. Since the experimental levels are complete for low spin states below about 1900 keV, due to the use of the (n, γ) ARC technique, it is clear that, empirically, there is no $K = 4$ band below ≈ 2 MeV while the calculations predict a $K = 4$, double γ vibration, at ≈ 1600–1700 keV. There is no easy way to repair this IBM prediction since, just as in geometrical models, where the $\gamma\gamma$ vibration is expected at $\approx 2E_\gamma$, it is difficult to introduce sufficient anharmonicity to obtain $E_{4^+} \approx 2.5E_\gamma$. It now appears likely that one must include a g boson or other extensions to the IBM to ameliorate the situation. Further treatment of the effects

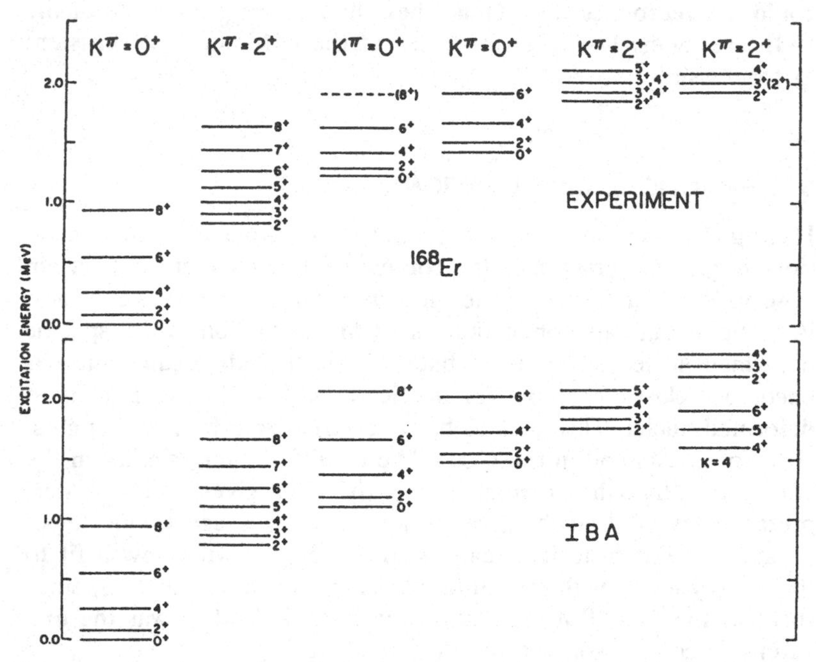

FIGURE 3.20. *Comparison of the experimental low-lying levels of* ^{168}Er *with the predictions of the* IBM *calculations of Warner, Casten and Davidson* (1980). *Note that, while all experimental levels of the low-lying* $K = 0, 2$ *bands are shown, the* IBM *also predicts an unobserved* $K = 4$ *band at approximately two times the energy of the* γ *band.*

of incorporating a g boson into the sd IBM-1 formalism will be presented in Chapter 6.

The most stringent tests of the IBM in deformed nuclei center on E2 transitions. We will first deal with the magnitudes of the reduced matrix elements connecting intrinsic excitations (g, β, γ bands) and then turn to relative B(E2) values or branching ratios that relate different transitions connecting the same two intrinsic excitations.

As we have just discussed in our treatment of the CQF, it is easy to choose a χ value that reproduces the intrinsic $\gamma \rightarrow g$ E2 matrix elements, that is, the empirical R_γ value. For a given N, one simply uses a contour plot of the type shown in Fig. 3.18. Indeed, this is often how χ values have been fixed. Then, one can inspect the predictions for the γ and β band excitation energies. Of course, χ can equally well be fixed from the energies and the E2 rates used as a test. In either case, good agreement is found. It is clear that, in fact, this must be so by comparing Fig. 3.18 and 3.19. The central cross hatched square in each, corresponding to typical χ values of ≈ -0.5 for deformed nuclei, gives $R_\gamma \approx 0.025$ and $E(2_\beta^+)/E(2_\gamma^+) \approx 1.6$, and both of these are in agreement with a large body of data: for R_γ this is evident from Fig. 1.12 and, for energies, it is verified by inspecting level schemes for typical deformed nuclei such as ^{168}Er.

This reiterates a point alluded to earlier. The properties of the β and γ bands are intimately linked in the IBM. Yet this is contrary both to common perception of the empirical situation and to the characteristics of most other models. The present results suggest that this linkage indeed appears in real nuclei. This linkage can be illustrated more explicitly. We use the Hamiltonian of eq. (3.12). As before, we ignore the $\mathbf{J}\cdot\mathbf{J}$ term. Then the parameter a_2 is just a scale factor and is of no importance for energy ratios. The structure is determined solely by χ which can be completely fixed by fitting the E2 properties of the γ band. Having done this, the energy ratio $R_{\beta\gamma} = E(0_\beta^+)/[E(2_\gamma^+) - E(2_g^+)]$ can be compared with experiment. The results are shown in Fig. 3.21 in which the experimental values of $R_{\beta\gamma}$ are compared with those calculated in the CQF. Agreement corresponds to points along the diagonal. Clearly most of the points lie close to this line, as indicated by the dashed lines which correspond to $\pm 20\%$ deviations from agreement.

Another central result of IBM calculations in deformed nuclei has been mentioned before and again arises automatically in the CQF, namely, the prediction that $\beta \rightarrow g$ $B(\mathrm{E2})$ values are much smaller than $\gamma \rightarrow g$ strengths. Of course, this is also a well known empirical feature. It appears in ^{168}Er and essentially all other deformed nuclei as Fig. 1.12 showed. Geometrical models cannot make a prediction here without guidance from a microscopic calculation since each vibration (β, γ octupole) mode is phenomenologically fit to the data.

A further key feature of both the earlier and the CQF calculations, referred to previously, is the characteristic prediction of strong transitions *between* the β and γ bands in deformed nuclei. Such transitions are forbidden in traditional harmonic geometric models, because they violate the phonon selection rule, but are

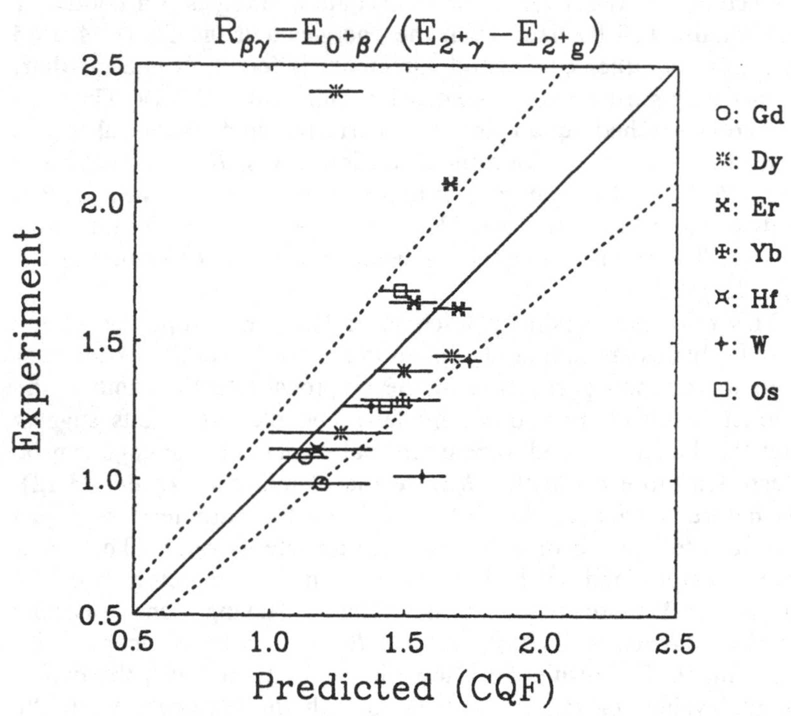

FIGURE 3.21. *Comparison of empirical values of* $R_{\beta\gamma}$ *with those calculated in the* CQF. *See Text. From Chou, Casten and von Brentano* (1992).

allowed and collective in the IBM because, in the parent $SU(3)$ scheme, the β and γ bands occur within the same (λ, μ) representation.

It was not realized for many years that collective $\beta \rightarrow \gamma$ transitions could characterize actual deformed nuclei. However, that situation was to some extent an artifact of the relation between $B(E2)$ values and transition rates $T(E2)$, namely $T(E2) \propto E_\gamma^5 B(E2)$. Since β and γ bands are much closer in energy than γ and g bands, β-γ transition rates are hindered by orders of magnitude and are hence very weak even if the $B(E2\!:\beta \rightarrow \gamma)$ values are collective. It is only in recent years, with the advent of highly sensitive instruments, in particular, the GAMS spectrometer at the Institut Laue-Langevin in Grenoble, France, that these transitions have been discovered and found to be collective indeed. The first such case to be studied in detail was ^{168}Er. Now many others are known. The prior prediction of this property by the IBM is one of the model's most remarkable successes.

These results also suggest a new way to look on the excitation we have been (somewhat glibly) calling the "β" band. More properly called, it is the lowest excited $K = 0^+$ band. It is clearly a collective mode, as evidenced by the (predicted and observed) enhanced $\beta \rightarrow \gamma$ $B(E2)$ values and by the smooth behavior of E_β/E_γ. Historically, however, that collectivity has often been called into question because of the weak $B(E2\!:\beta \rightarrow g)$ values and because of their highly erratic behavior (see Fig. 1.12). But the $\beta \rightarrow \gamma$ linkage in the IBM, and the data, suggest that it is misleading to focus on the $\beta \rightarrow g$ transitions since they are basically forbidden transitions, and forbidden transitions between collective states can, of course, fluctuate wildly. Simply put, the emphasis historically has been on the "noise" and not the "signal." The lowest $K = 0$ mode (we will continue to use the label "β") is collective but not collectively connected by E2 transition elements to the ground state.

Finally, still another important empirical characteristic of deformed nuclei is the fact that they exhibit interband E2 branching ratios that deviate from the Alaga rules. [These "rules" are simply the ratios of the squares of Clebsch–Gordan coefficients connecting initial and final states.] These deviations are well known to reflect 2-band mixing. The most studied case is that of γ-g, $\Delta K = 2$ mixing. In geometric models, such bandmixing has traditionally been parameterized in terms of a parameter Z_γ (or Z_β for $\beta \rightarrow g$ transitions). For any $B(E2)$ ratio connecting the same two intrinsic states (e.g., two $B(E2\!:\gamma \rightarrow g)$ values), Z_γ can be extracted from the deviation from the Alaga ratio. Generally, absolute $B(E2)$ values are not known and so branching ratios, with a common initial state [e.g., $B(E2\!:J_\gamma \rightarrow J_g)/B(E2\!:J_\gamma \rightarrow J'_g)$] are used. Often, a better way to extract Z_γ values is to exploit the technique of Mikhailov plots. This approach is discussed in standard texts and need not be elaborated here. The basic idea is that bandmixing (we speak here, for definiteness, of $\Delta K = 2$ γ-g bandmixing but the idea is easily extended to other cases) carries a natural spin dependence and that the full set of changes in $B(E2)$ values for all the transitions between two intrinsic excitations [e.g., for all $B(E2\!:J_\gamma \rightarrow J_g)$ values] can be written in terms of the same Z_γ value. The spin dependence is expressed by Clebsch–Gordan coefficients and universal spin functions. A simple analysis shows that one can write, for example, $B(E2\!:J_\gamma \rightarrow J_g) = B_0(E2)[1 + Z_\gamma f(J_\gamma, J_g)]^2$ where $B_0(E2)$

is the unperturbed value (proportional to the square of the appropriate Clebsch–Gordan coefficient) and the $f(J_\gamma, J_g)$ are functions of the initial and final spins. With further manipulation, it is standard to obtain:

$$\frac{\sqrt{B(\mathrm{E2}: J_\gamma \rightarrow J_g)}}{\sqrt{2}\langle J_\gamma 22 - 2|J_g 0\rangle} = M_1 - M_2[J_g(J_g + 1) - J_\gamma(J_\gamma + 1)] \quad (3.16)$$

where

$$M_1 = \langle \phi_\gamma | M(\mathrm{E2}) | \phi_g \rangle - 4M_2 \quad \text{and} \quad M_2 = (15/8\pi)^{1/2} Q_0 \epsilon_\gamma \quad (3.17)$$

where ϕ_i is an unperturbed wave function for the γ or ground band, Q_0 is the (identical) intrinsic quadrupole moment for the two bands, and ϵ_γ is the spin *in*dependent part of the mixing amplitude. Thus, M_1 is essentially the direct $\gamma \rightarrow g$ E2 matrix element while M_2 is proportional to the mixing amplitude. A plot of the left side of eq. (3.16) against the spin function on the right gives a straight line with intercept at $J_\gamma = J_g$ of M_1 and slope M_2. This is called a Mikhailov [1966] plot. Thus, if the *observed* $B(\mathrm{E2}: J_\gamma \rightarrow J_g)$ values do yield a straight line, *both* the intrinsic matrix element and the mixing can be extracted. If the data do not fall on a straight line in a Mikhailov plot it could be that more complex mixing (e.g., 3-band) is involved or that there are undetected M1 components in some of the $\Delta J = 0$ or 1 transitions.

For an example of this kind of analysis we present a Mikhailov plot for the $\gamma \rightarrow g$ transitions in ^{168}Er in Fig. 3.22. The data points indeed lie along a very good straight line sloping upward to the right. Before we comment on the IBM-1 calculations, also shown in the figure, it is interesting that, since points to the right of zero correspond to $J_g > J_\gamma$, these data disclose a sign of the mixing that favors such transitions and which *de*creases the strength of spin *de*creasing transitions. Since the magnitude of the effect increases with spin this implies substantial enhancement or hindrance of transitions from higher spin γ-band states depending on whether they increase or decrease the spin. This effect can easily reach order-of-magnitude levels (relative to the Alaga rules) even though the actual mixing amplitudes are less than 0.1. The mixing adds an *intra*-band matrix element coherently to the *inter*-band one and, thus, the effects of small mixing amplitudes are magnified. The data shown for ^{168}Er in Fig. 3.22 are typical of most deformed nuclei.

IBM calculations for ^{168}Er were done in the CQF with χ deter-

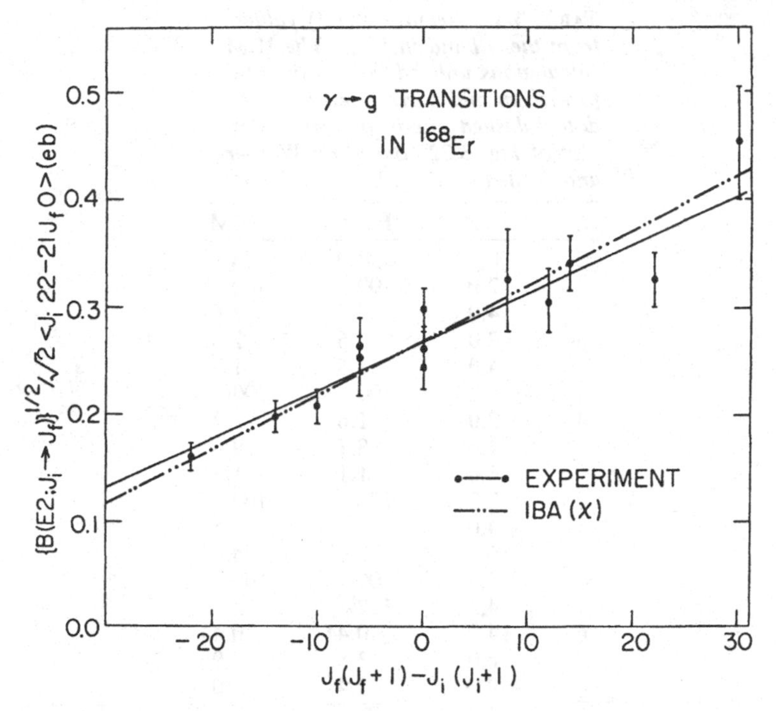

FIGURE 3.22. *Mikhailov plot for $\gamma \rightarrow g$ transitions in* ^{168}Er. *The solid line is a least squares fit to the data. The dot-dash line is the* CQF *predictions. From Warner and Casten* (1982).

mined from R_γ. The results for $\gamma \rightarrow g$ B(E2) values are shown in Fig. 3.22. There was no adjustment of parameters to fit the mixing and yet the agreement with the data is perfect. Table 3.3 presents a quantitative comparison of experimental and calculated B(E2) values for the decay of γ-band states. It also includes intraband transitions and shows that the model not only accounts for the relative $\gamma \rightarrow g$ B(E2) values but for the ratio of inter- and intra-band B(E2) values as well. (Though not shown in the figure or table, it is worth commenting that the calculations resulting from use of the multipole Hamiltonian approach are not as good even though there is an extra free parameter.) The meaning of Fig. 3.22 is simply that the IBM, in the CQF, can fully account for mixing between γ and g bands as well as provide a reasonable interpretation of the absolute matrix elements connecting different bands. Finally, it is worth noting that, while it is possible to parameterize geometrical models

TABLE 3.3. *Relative B(E2) values from the γ band in* ^{168}Er. *The* IBM *calculations utilized the consistent-Q formalism and correspond to the dotted-dashed line in the Mikhailov plot of Fig.* 3.22. *Based on Warner and Casten* (1982, 1983).

I_i	I_f, K_f	Expt.	IBM
2	0,0	54.0	54
	2,0	100	100
	4,0	6.8	7.6
3	2,0	2.6	2.6
	4,0	1.7	1.8
	2,2	100	100
4	2,0	1.6	1.7
	4,0	8.1	9.6
	6,0	1.1	1.5
	2,2	100	100
5	4,0	2.9	3.5
	6,0	3.6	4.4
	3,2	100	100
	4,2	122	95
6	4,0	0.44	0.44
	6,0	3.8	4.9
	8,0	1.4	1.0
	4,2	100	100
	5,2	60	57
7	6,0	0.7	1.9
	5,2	100	100
	6,2	59	36

to include bandmixing, such mixing is an inherent and *automatic* feature of the IBM in deformed nuclei resulting from the effects of finite boson number. Indeed, bandmixing in the IBM actually appears even in the exact $SU(3)$ limit (e.g., in ratios of $B(\text{E2}:\beta \rightarrow \gamma)$ strengths), but is more pronounced in broken $SU(3)$ calculations.

In realistic IBM calculations for deformed nuclei, there are two distinct mechanisms for deviations of $\gamma \rightarrow g$ transition strengths from the Alaga rules. One is similar to the geometrical model, namely $\Delta K = 2$ mixing of the $SU(3)$ γ and g intrinsic states when $SU(3)$ is broken. These matrix elements are, however, rather small. A more important source stems in an interesting, indirect, way from the allowed $\beta \rightarrow \gamma$ $SU(3)$ E2 transitions. When $SU(3)$ is broken, large $\Delta K = 0$ matrix elements mix substantial amplitudes of the $SU(3)$ β band into the $SU(3)$ g band, as discussed earlier. Therefore, the resultant $\gamma \rightarrow g$ E2 matrix elements will have important components of the type $\langle \beta_{SU(3)}|E2|\gamma_{SU(3)}\rangle$: but this allowed matrix element itself has characteristic deviations from the Alaga rules. Moreover, these deviations show a particularly strong N dependence, decreasing as $N \rightarrow \infty$. This fact has an interesting consequence that can be tested experimentally: because these $\beta_{SU(3)} \rightarrow \gamma_{SU(3)}$ E2 amplitudes are important contributions to $\gamma \rightarrow g$ B(E2) values, the effective Z_γ values that characterize the IBM predictions should also show a strong N dependence, with a parabolic variation throughout a shell, minimizing at mid-shell where N is largest. Most remarkably, this is in excellent agreement with the data, as illustrated in Fig. 3.23, for the rare earth region, where Z_γ values predicted by the IBM using the CQF are compared with the data.

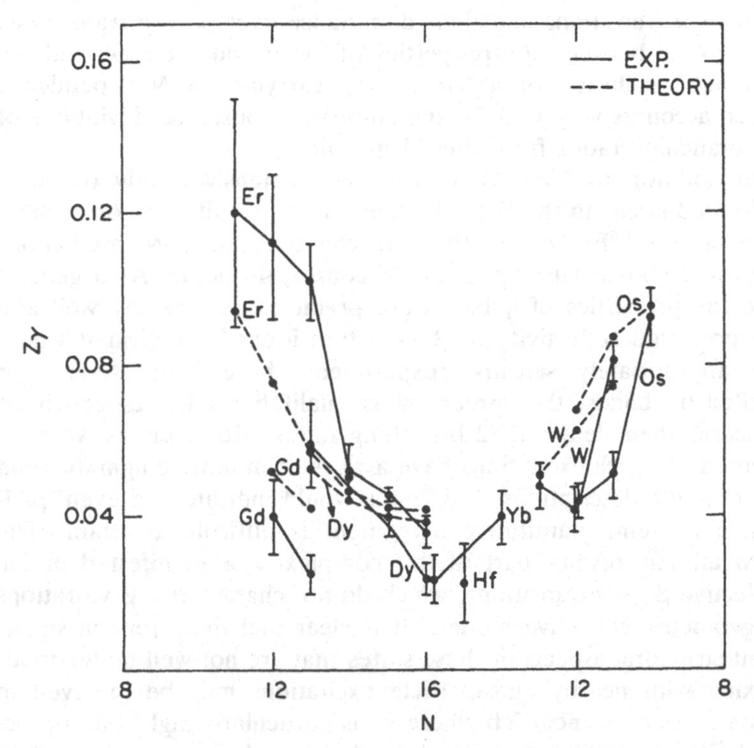

FIGURE 3.23. *Calculated and empirical values of Z_γ in the rare-earth nuclei. From Casten, Warner and Aprahamian* (1983).

This point is important for three reasons. It is typical of many "automatic" predictions of the IBM for deformed nuclei which cannot be avoided by reasonable parameter choices and which therefore are telling tests of the basic structure of the model. (Collective $\beta \rightarrow \gamma$ transitions are the most obvious other example.) Secondly, it is perhaps the most obvious example of the observable effects of finite N in the IBM: interestingly, it does not concern high spin states, nor small N values. Finally, it is an excellent example of a microscopic aspect of the phenomenological IBM-1 since, even with constant parameters, the model predicts a very specific valence nucleon number dependence of structure (that is verified empirically).

To summarize, there are several aspects of the IBM in deformed nuclei which agree with empirical observations and which are important because they are inherent features of the model, namely the dominance of $\gamma \rightarrow g$ over $\beta \rightarrow g$ $B(E2)$ values, the collectivity of $\beta \rightarrow \gamma$ transitions and their dominance over $\beta \rightarrow g$ transitions, the linkage between the properties of the β and γ bands, and the automatic inclusion of a bandmixing carrying an N dependence which accounts very well for the empirically observed deviations of E2 branching ratios from the Alaga rules.

In addition to ^{168}Er, there has been extensive study of many deformed nuclei in the IBM. In general, the results are comparably good as for ^{168}Er and confirm the characteristic IBM predictions discussed above. Discrepancies, of course, do occur. As a general rule the properties of γ bands are predicted remarkably well and the predicted collectivity of $\beta \rightarrow \gamma$ transitions is confirmed where the appropriately sensitive experiments have been done. For excited 0^+ bands, the agreement is qualitative only, especially as concerns their detailed E2 branching ratios. However, as we have pointed out such excitations have always been more enigmatic than their simple descriptions as β bands would indicate and even qualitative or semi-quantitative agreement is difficult to attain. The IBM already reveals part of this complexity as manifested in the collective $\beta \rightarrow \gamma$ transitions, which do not characterize β vibrations in geometric collective models. It is clear that there remain significant structural aspects of these states that are not well understood. Mixing with nearby quasiparticle excitations may be involved in some cases (e.g., near Yb where E_β is particularly high) but a practical IBM formalism that simply includes such degrees of freedom is not yet available.

3.8. M1 TRANSITIONS

Although the E2 properties of deformed nuclei are their best known, and most standard feature, important data and model tests also emerge from the study of other multipoles, in particular M1 transitions.

As pointed out in Chapter 2, in the lowest order description, the M1 operator is proportional to the total angular momentum, and hence does not give rise to transitions (although diagonal M1 matrix elements, that is, g factors, are finite). It is therefore necessary to consider higher order terms in a realistic calculations and, rather surprisingly, it is then in fact possible to extract some simple predictions for the behavior of E2/M1 mixing ratios in a variety of cases. Of course, these predictions must be tempered by the realization that, a priori, an explicit recognition of neutron and proton degrees of freedom would seem to be appropriate to the description of magnetic properties in general. Indeed, M1 transitions have recently taken on high importance in the context of the IBM-2 with the discovery of a new collective mode, the so-called isovector M1 excitation, which is characterized by collective M1 matrix elements connecting the ground state of deformed nuclei to 1^+ levels near 3 MeV. (See Chapter 4).

In the IBM-1, the expanded M1 operator becomes

$$T(\mathrm{M1}) = (g_B + A\hat{N})J + B(QJ)^{(1)} + C\hat{n}_d J \qquad (3.18)$$

where A and B are parameters (unrelated to the A and B appearing in the $O(6)$ eigenvalue expressions).

The first term remains diagonal and can be discarded in a discussion of transition properties. The quadrupole operator in the second term has the same structure as that defined in eq. (3.14) and can connect states which differ by $\Delta J = 0$ or ± 1, while the third term contains $\hat{n}_d$ and hence is proportional to the E0 operator and can only connect states with $\Delta J = 0$. In general the parameter χ in the quadrupole operator of the second term can be varied freely and a number of fits to M1 properties in collective nuclei have been attempted in this fashion with varying degrees of success. However, no well defined or systematic behavior of the fitted parameters in the M1 operator seems to emerge naturally from this procedure, nor does it give rise to any clearcut predictions concerning the behavior of M1 transitions in different regions. On the other hand, a number of simple predictions do result if the quadrupole operator is constrained to take the same form as that which is used to describe E2 transitions.

The matrix element of the M1 operator can be written as

$$\langle\phi' J_f \| T(\mathrm{M1}) \| \phi J_i\rangle = -Bf(J_i J_f)\langle\phi' J_f \| Q \| \phi J_i\rangle + C[J_i(J_i+1)(2J_i+1)]^{1/2} \times \langle\phi' J_f | \hat{n}_d | \phi J_i\rangle\, \delta_{J_i J_f} \quad (3.19)$$

where ϕ', ϕ denote additional quantum numbers. The spin dependence of the first term is contained in the factor

$$f(J_i J_f) = [(1/40)(J_i + J_f + 3)(J_f - J_i + 2) \cdot (J_i - J_f + 2)(J_i + J_f - 1)]^{1/2}. \quad (3.20)$$

In fact, for $J \rightarrow J \pm 1$ transitions, it is only this first term which contributes. With the assumption that the quadrupole operator has the same structure as that used in the E2 operator, the expression for the reduced mixing ratio then becomes

$$\Delta(\mathrm{E2/M1}) = \frac{\langle\phi' J_f \| T(\mathrm{E2}) \| \phi J_i\rangle}{\langle\phi' J_f \| T(\mathrm{M1}) \| \phi J_i\rangle} = \frac{-1}{Bf(J_i J_f)}. \quad (3.21)$$

The reduced mixing ratio is related to the quantity normally measured, $\delta(\mathrm{E2/M1})$, by

$$\delta(\mathrm{E2/M1}) = 0.835 E_\gamma \Delta(\mathrm{E2/M1}) \quad (3.22)$$

with E_γ in MeV. The second term in eq. (3.19) is proportional to the E0 operator and hence will be negligible in vibrational nuclei, or in deformed nuclei for the $\gamma \rightarrow \gamma$ and $\gamma \rightarrow g$ transitions. For these cases the E2/M1 mixing ratio will simply have the form of eq. (3.21).

This immediately leads to a prediction for the spin dependence of M1 $\gamma \rightarrow g$ transitions which in fact is the same as that which emerges from many other approaches. Moreover, the fact that the same spin dependence holds true for transitions within the γ band results in the additional prediction of a link between the reduced mixing ratios for $\gamma \rightarrow g$ transitions and $\gamma \rightarrow \gamma$ transitions, namely, that they should be identical for the same values of the initial and final state spins. In addition, the sign of the mixing ratios should be constant throughout the deformed region. In fact, all three predictions seem to be reasonably well borne out by the data.

For the case of $\beta \rightarrow g$ transitions it will be the second term in eq. (3.19) which dominates, since, while the E0 transitions are strong, the equivalent E2 transitions are weak, as discussed in the previous section. In fact, in this case, the spin dependence remains the same but, of course, the constant can change. Thus the sign of mixing ratios in $\beta \rightarrow g$ transitions may be different from those involving the γ band. The sparse data that exist do seem to support the idea that the sign of the mixing ratios for these transitions is constant across the rare earth region and opposite to that observed for $\gamma \rightarrow g$ transitions.

3.9. TRANSITION REGIONS

An important aspect of symmetry concepts is that pairs of symmetries act as benchmarks and, therefore, termini for nuclear transition regions. Sequences of nuclei in such a phase transitional region can be very simply calculated, generally by the variation of a single parameter which specifies their location along the appropriate leg of the symmetry triangle (Fig. 3.9). This parameter can usually be taken as the ratio of the coefficients in the Hamiltonian characteristic of the two symmetries occupying the vertices of the triangle at the termini of the transition leg. We will briefly discuss each of the three types of transitional regions, starting with $O(6) \rightarrow SU(3)$.

3.9.1. O(6) → SU(3) or Deformed Rotor

In an $O(6) \rightarrow$ rotor transition the $O(6)$ levels typical of a deformed γ-soft asymmetric rotor evolve into those of the deformed symmetric rotor as illustrated schematically in Fig. 3.24. The states of maximum spin in each τ multiplet become the ground state rotational band. The states of next lower spin become the γ band while the states of maximum spin for each τ value in the $\sigma = N - 2$ multiplet become the β vibrational band. Higher $O(6)$ sequences correlate with multi-phonon $SU(3)$ excitations such as the $K = 4^+$ double γ vibration.

Clearly, treating nuclei spanning structures between $SU(3)$ and $O(6)$ is just an extension of the approach discussed above for deformed nuclei, where terms in $a_0 P^\dagger P$, or χ values differing from $-\sqrt{7}/2$, were used to break $SU(3)$ and mix $SU(3)$ representations. Now, the deviations from their $SU(3)$ values are simply larger. We will first consider an approach utilizing the $P^\dagger P$ term, as applied to the Pt–Os region which is the best known $O(6) \rightarrow$ deformed tran-

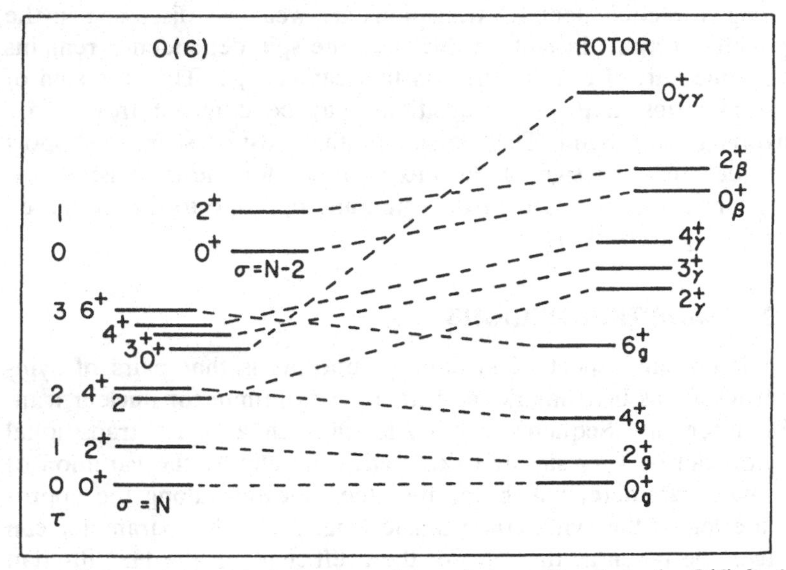

FIGURE 3.24. *Schematic indication of the relation of levels in the O(6) limit and in the deformed rotor. Note that the association of the excited* 0^+ *states is ambiguous and can be reversed. From Casten and Warner* (1988).

sition region. Then we will consider some general predictions, and their comparison to the data, in the context of the CQF.

The existing calculations for the Pt–Os transition region were done before the CQF was developed and hence utilized a Hamiltonian $H = a_2Q \cdot Q + a_1J \cdot J + a_3T_3 \cdot T_3 + a_0P^{\dagger}P$ with χ in the operator Q fixed at $-\sqrt{7}/2$. The $O(6) \rightarrow$ rotor transition was affected by varying the ratio a_2/a_0, which determines the essential structural change. The E2 transition operator included χ as a free parameter. Since the entire Pt–Os region is rather close to $O(6)$, though, χ was kept small throughout. Some results are shown in Figs. 3.25 and 3.26. The agreement is rather remarkable, especially considering the extreme simplicity of the calculations. One interesting feature is that these calculations reproduce the enigmatic decay of the excited 0^+ states, where, in Pt, some 0^+ levels decay predominantly to the 2_1^+ state but others decay to the 2_2^+ state while, in Os, all decay predominantly to the 2_2^+ states. These calculations also show the gradual emergence, with decreasing A, of branching ratios approaching the Alaga rules as opposed to those characteristic of the $O(5)$ symmetry observed in the Pt region. The

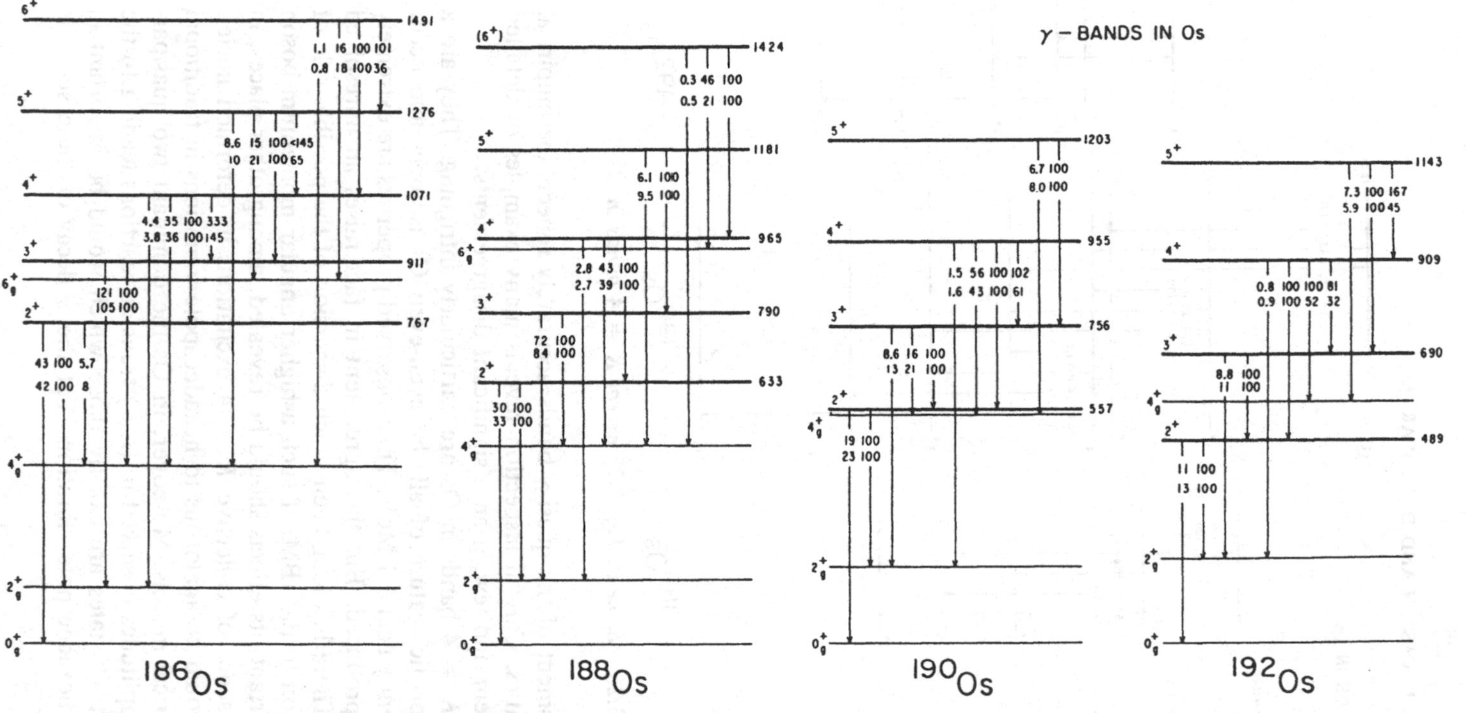

FIGURE 3.25. *Comparison of calculated and empirical relative B(E2) values for the γ bands, in an O(6) → SU(3) transition region, for the Os isotopes. The upper rows of numbers are the empirical relative B(E2) values, while the lower rows are the calculated ones. From Casten and Cizewski* (1978, 1984).

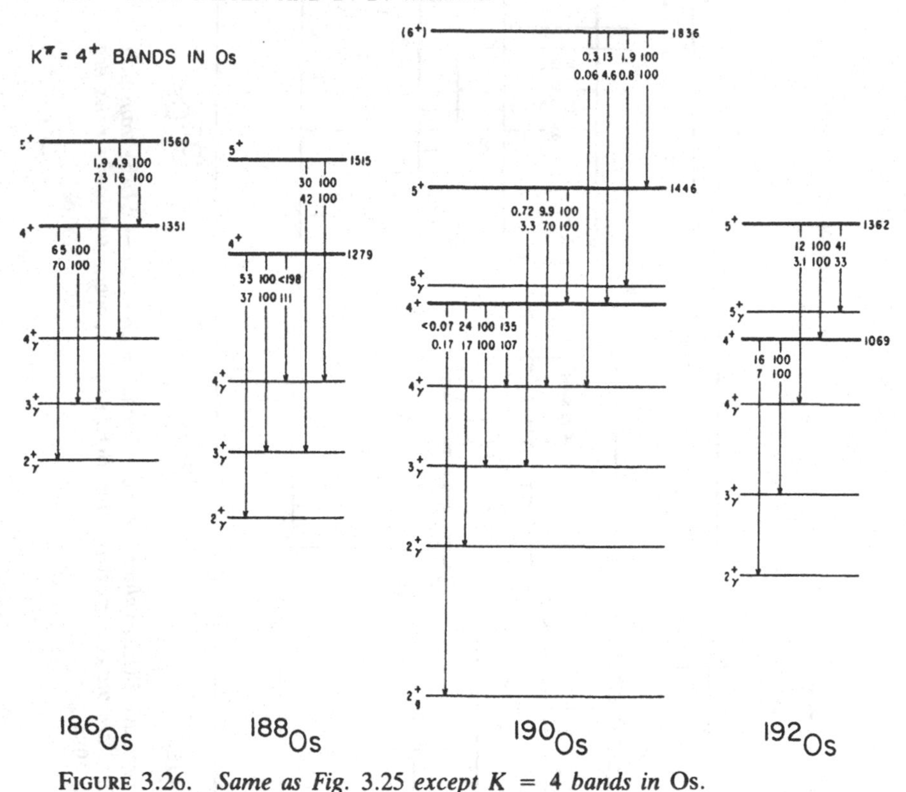

FIGURE 3.26. *Same as Fig.* 3.25 *except K* = 4 *bands in* Os.

development of the quasi-γ bands accurately reflects the empirical systematics. Careful inspection reveals many examples of detailed agreement and few, if any, significant disagreements.

The $K = 4$ bands in Os are particularly intriguing. They are a characteristic feature of all the even-even Os isotopes and occur consistently near 1.1 MeV. Their observed properties are extremely well reproduced. But this agreement in fact raises an unresolved point. There has long been much discussion of the possible role of a *g* boson in the IBM. If such a higher angular momentum boson is important, its effects should be revealed, among other places, in the presence of collective $K = 4$ excitations in deformed nuclei. They would be analogous to hexadecapole vibrations in traditional geometrical models. Moreover, in Os the dominant two quasiparticle amplitudes observed in (t, α) transfer reactions leading to the relevant 4^+ states are exactly those which would be important in such a hexadecapole vibration. Yet the γ decay of these states is reproduced in IBM calculations that do *not* include a g boson. Although the issue cannot be fully settled until more absolute transition rates are known, it seems that these states are admixtures of $\gamma\gamma$ or 2-phonon vibrational excitations (that is, in IBM language, $K = 4$ bands in the sd boson space) and hexadecapole vibrations, and that the collective E2 properties of these levels are dominated by their double γ-vibrational character while the transfer reaction properties are controlled by the two quasiparticle hexadecapole amplitudes. The measurement of the absolute $B(\mathrm{E2}:4^+K = 4 \rightarrow 2^+_\gamma\ K = 2)$ values would be the key experimental text of this idea, though it is very difficult to measure.

Another excellent example of an $O(6) \rightarrow$ deformed region is the $A = 130$ region, especially in the Xe and Ba isotopes: near $N = 74$ an $O(6)$ character predominates while recent heavy ion reaction data disclose a developing rotational character near $N = 68$ and 70. A detailed calculation of the phase transition in this region would be highly interesting.

The CQF provides further insight into an $O(6) \rightarrow SU(3)$ region. In an $O(6)$–$SU(3)$ transitional region there are obviously many routes which could be taken from the cross hatched area in the center of Fig. 3.18 to the bottom left-hand corner which represents thc $O(6)$ limit. However, some predictions are independent of the route. For example, it is clear that it is impossible to pass between the cross hatched regions, in both of which R_γ is small, without going through a maximum in R_γ. This prediction, though qualitative, is parameter independent and inescapable (within the CQF framework). Similar results characterize other $B(\mathrm{E2})$ ratios such as the quantity $B(\mathrm{E2}:3^+_\gamma \rightarrow 2^+_\gamma)/B(\mathrm{E2}:3^+_\gamma \rightarrow 2^+_g)$.

Finite values for such branching ratios therefore are an excellent signal of nuclear phase shape transitions. Even though it is an idealization, it is interesting to consider the simplest route between these cross hatched regions, namely that corresponding to the straight diagonal line indicated in Fig. 3.18. Calculations for such a route are affected with the CQF Hamiltonian of eq. (3.12) simply by varying χ. Some results are illustrated in Figs. 3.27 and 3.28 where the predicted behavior is compared with the experimental data for the relevant nuclei. The agreement is in general good although for specific nuclei in the transitional region it is no better than a factor of two. Figure 3.27 illustrates one of the branching ratios that vanishes in both the $O(6)$ limit and in the $N \rightarrow \infty$ limit of $SU(3)$, and which has a peak for intermediate structures, and finite N, as illus-

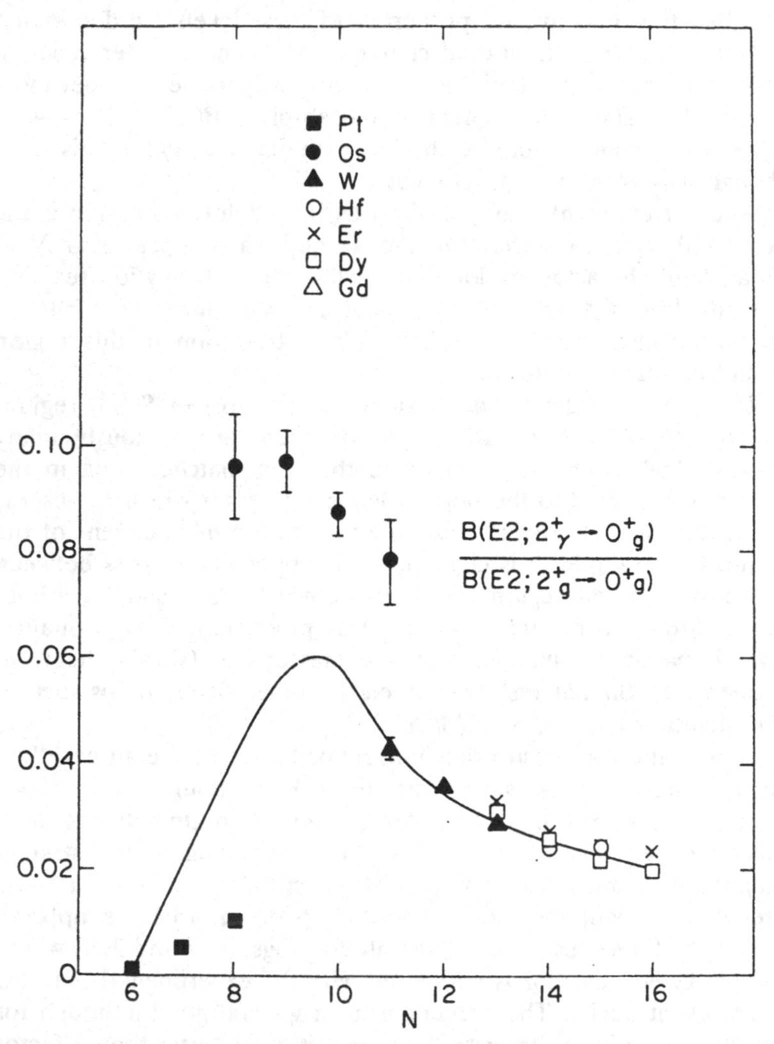

FIGURE 3.27. *Comparison of calculated* (CQF) *and empirical* $B(\mathrm{E2}: 2^+_\gamma \rightarrow 0^+_g)$ *values normalized to the* $2^+_1 \rightarrow 0^+_1$ *transition. The calculations utilized* χ *values corresponding to the dashed straight line trajectory in Fig.* 3.18. *From Warner and Casten* (1983).

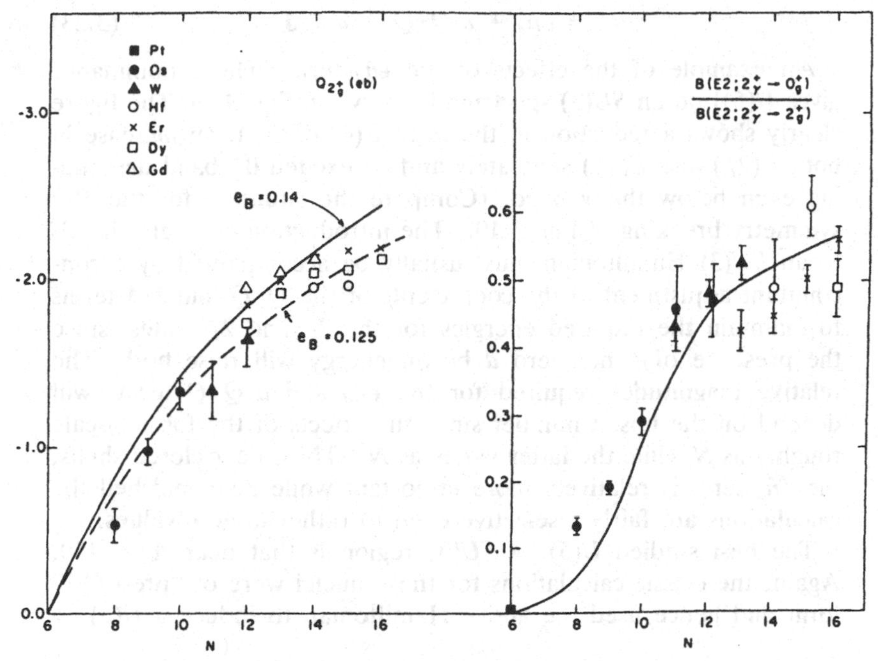

FIGURE 3.28. *Additional* IBM *predictions using the* CQF. *From Warner and Casten* (1983).

trated. The data very nicely reflect this same systematics although detailed quantitative agreement is not attained. On the left-hand side of Fig. 3.28, the predicted behavior of the quadrupole moment of the first 2^+ state is shown for two choices of effective charges (denoted α in place of e_B used in eq. (3.13)). The right-hand side shows the dependence on χ of a $\gamma \rightarrow g$ branching ratio. The interesting feature here is that, while the calculated curve clearly approaches the rotational value of 0.7 as N increases, it never attains it for realistic boson numbers. This behavior reflects both the automatic incorporation of band mixing effects within the IBM framework and the effects of finite boson number.

3.9.2. $U(5) \rightarrow SU(3)$

The vibrator → rotor transition region can be treated using the schematic multipole Hamiltonian (with $\chi = -\sqrt{7}/2$ in Q)

$$H = \epsilon\hat{n}_d + a_2 Q\cdot Q + a_1 \mathbf{J}\cdot\mathbf{J}. \qquad (3.23)$$

An example of the effects of the $\epsilon\hat{n}_d$ term (whose dominance gives $U(5)$) on an $SU(3)$ spectrum is shown in Fig. 3.29. The figure clearly shows a reduction in the ratio $E(4_1^+)/E(2_1^+)$, an increase in both $E(2_1^+)$ and $E(4_1^+)$ separately and an excited 0^+ band descending even below the γ band. (Compare the situation for the $P^\dagger P$ symmetry breaking of Fig. 3.10.) The introduction of a term in $\epsilon\hat{n}_d$ to an $SU(3)$ Hamiltonian must usually be accompanied by a concomitant adjustment of the coefficients of the $Q\cdot Q$ and $\mathbf{J}\cdot\mathbf{J}$ terms to maintain the required energies for the 2_1^+ and 2_2^+ states, since the presence of a non-zero d boson energy will raise both. The relative magnitudes required for the $\epsilon\hat{n}_d$ and $a_2 Q\cdot Q$ terms will depend on the boson number since the effects of the former scale roughly as N while the latter varies as N^2. Thus, near closed shells, the $\epsilon\hat{n}_d$ term is relatively more important while near midshell the calculations are fairly insensitive even to rather large ϵ values.

The best studied $U(5) \to SU(3)$ region is that near $A = 150$. Again, the classic calculations for these nuclei were of "pre-CQF" form and hence used the above Hamiltonian to induce a $U(5) \to$

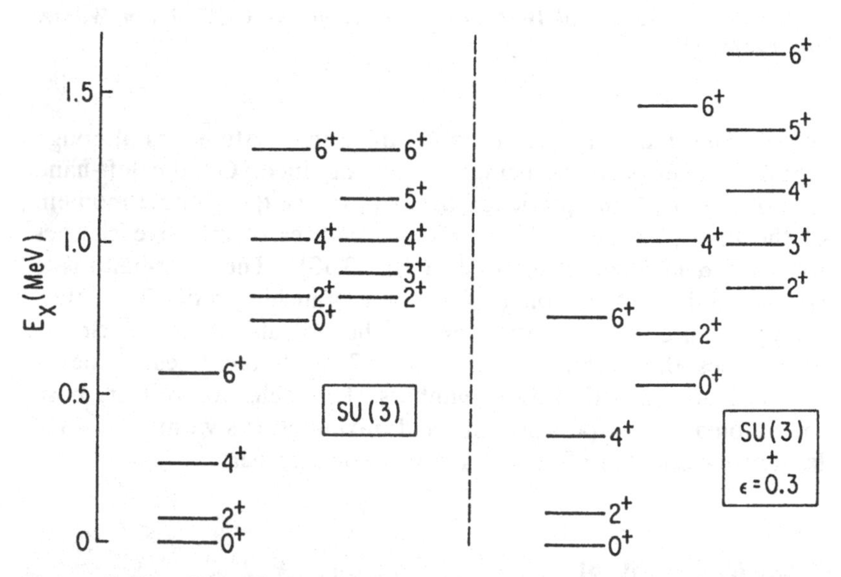

FIGURE 3.29. *Example of the effect of adding a term* $\epsilon\hat{n}_d$ *to an* $SU(3)$ *Hamiltonian with* $a_1 = 0.0087$ MeV, $a_2 = -0.013$ MeV, *and* $N = 10$. ϵ *is in* MeV. *From Warner and Casten* (1982).

$SU(3)$ phase transition. The parameter ϵ was constrained to decrease linearly with increasing boson number in going from nuclei close to the $N = 82$ closed shell towards the deformed nuclei with $N > 90$: that is, ϵ was written $\epsilon = \epsilon_0 - \theta N$. This decreases the ratio of ϵ/a_2. Good agreement with a wide variety of data was achieved. Examples of predicted energies are given in Fig. 3.30. As the phase transition develops (as ϵ decreases), the 2_2^+ state first decreases along with the compression of the entire scheme including, particularly, the first excited 0^+ state, which later evolves into the β band. Near the phase transitional point around $N = 90$, the $Q \cdot Q$ interaction, which varies as N^2, begins to dominate. One then sees that both the β and γ bands begin to increase in energy once deformation has set in. The 0^+ band, however, remains below the γ band because of the ϵ term.

Some E2 transition rates for this transition region are shown in Figs. 3.31 and 3.32 for transitions within the ground band and for those involving the interband transitions from the β and γ bands. The results are in good agreement with experiment and are typical of what is expected in a $U(5) \to SU(3)$ transition. The ground band $B(E2)$ values increase rapidly, changing from a proportionality to N in the $U(5)$ limit toward the N^2 dependence characteristic of $SU(3)$. At the same time the ground state transitions from the second and third 2^+ levels, which are forbidden in $U(5)$, become finite around $N = 90$ where the structure is intermediate between $U(5)$ and $SU(3)$. As expected from the earlier discussion, the predicted and experimental ground state transitions from the 2_3^+ γ bandhead state are stronger than those from the β band 2_2^+ level. Since the quadrupole operator in the Hamiltonian is, again, the $SU(3)$ form, both of these calculated $B(E2)$ values must vanish when the $Q \cdot Q$ term dominates since the Hamiltonian then yields the $SU(3)$ limit. This is seen in the figures. While there is no data in this regard for Sm it is well known (see Fig. 1.12) that the $\gamma \to g$ $B(E2)$ values do not vanish in the rare earth region. This simply reiterates the fact that the $SU(3)$ symmetry must be broken in calculation for real nuclei. Finally, the branching ratios shown in Fig. 3.32 increase from near 0 towards the rotational limit given by the appropriate Alaga rule as the deformed region sets in. This agrees reasonably well with the empirical trends. Some other branching ratios are not reproduced as well.

To treat a $U(5)$–$SU(3)$ transition in the CQF an $\epsilon\hat{n}_d$ term must be added to the Hamiltonian of eq. (3.12), giving the same Ham-

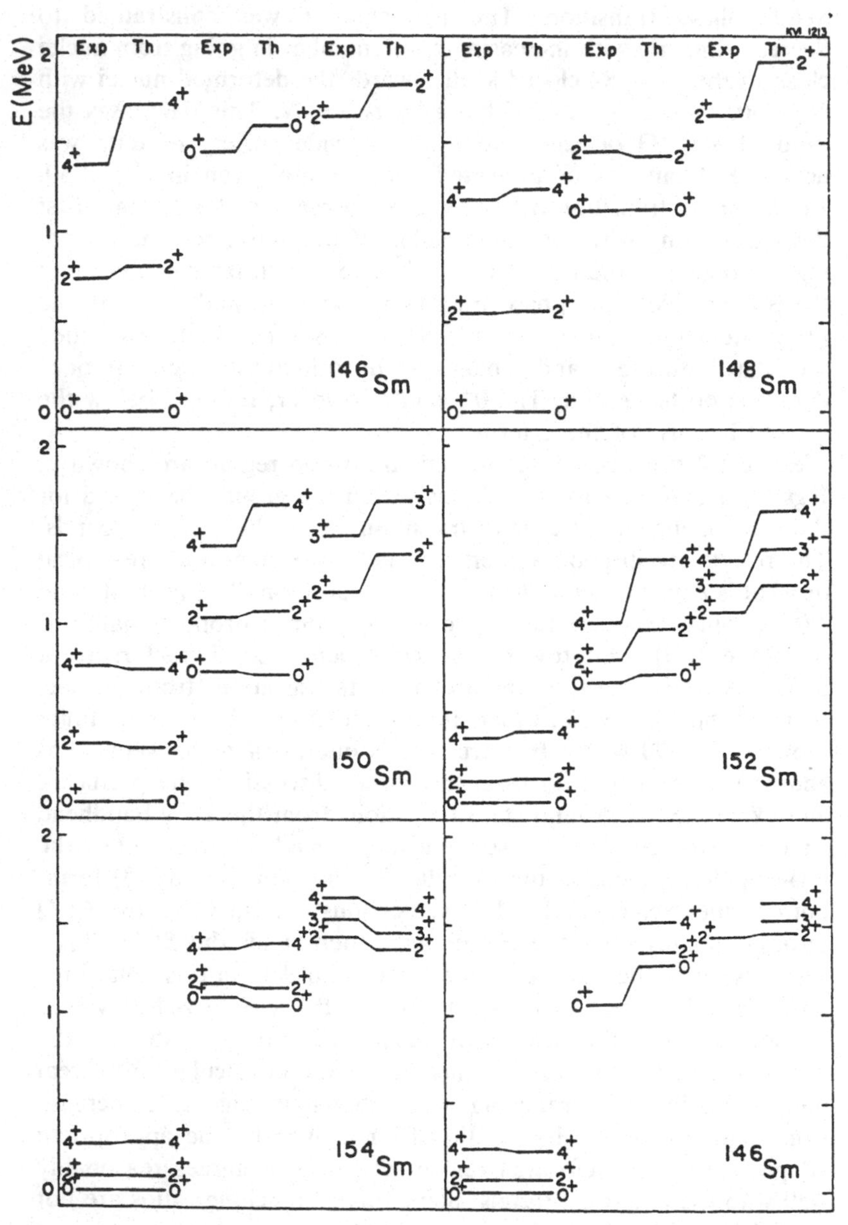

FIGURE 3.30. *Comparison of calculated and experimental energy levels in the* Sm *isotopes spanning a* $U(5) \rightarrow SU(3)$ *transition region. From Scholten, Iachello and Arima* (1978).

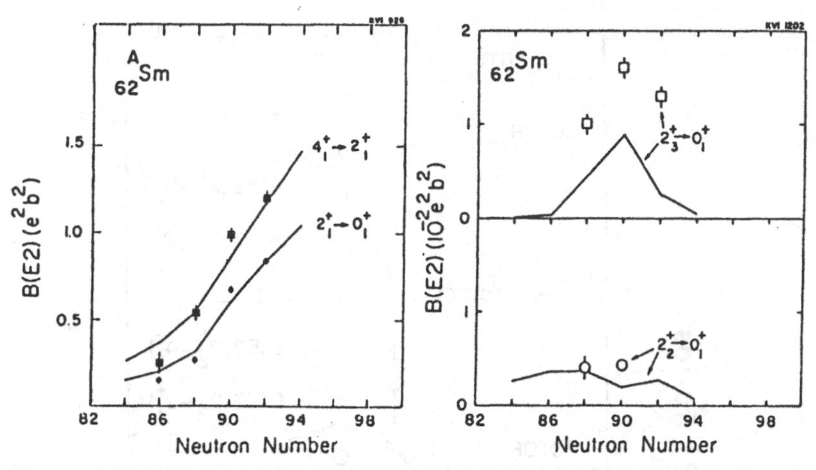

FIGURE 3.31. *Comparison of calculated and experimental* B(E2) *values in the* Sm *isotopes. From Scholten, Iachello and Arima* (1978).

iltonian as in eq. (3.23) but with the important difference that χ in the operator Q is now a free parameter and is set to the same value as in T(E2). This approach has been called the Extended Consistent Q Formalism or ECQF. This extension sacrifices some of the simplicity inherent in the original CQF Hamiltonian since, now, for a given boson number N, the wave functions will depend on both χ and ϵ/a_2 rather than on χ alone. Nevertheless, it is still possible to construct contour plots for energy and B(E2) ratios as functions of χ and ϵ/a_2, albeit now for a specific N value. Some examples are shown in Figs. 3.33 and 3.34. Clearly these figures are considerably more complex than their $SU(3) \rightarrow O(6)$ counterparts and of course two of them must be used in order to uniquely determine both χ and ϵ/a_2. There are still, however, a number of interesting signatures which can be discerned and which, together, emulate the known characteristics of a rotational-vibrational transition. One such signature is a lowering of the β bandhead below the γ bandhead and an accompanying increase in the strength of the $\beta \rightarrow g$ B(E2) transitions. This lowering of the β band immediately places the calculation outside of the scope of the normal CQF but reference to Fig. 3.33 shows that the addition of the $\epsilon\hat{n}_d$ term indeed generates a region in the $\chi - \epsilon/a_2$ plane where this occurs. Specifically, it involves the upper right-hand corner where the energy ratio $E_{0_\beta}/(E_{2_\gamma} - E_{2_g})$ is below unity. Then Fig. 3.34

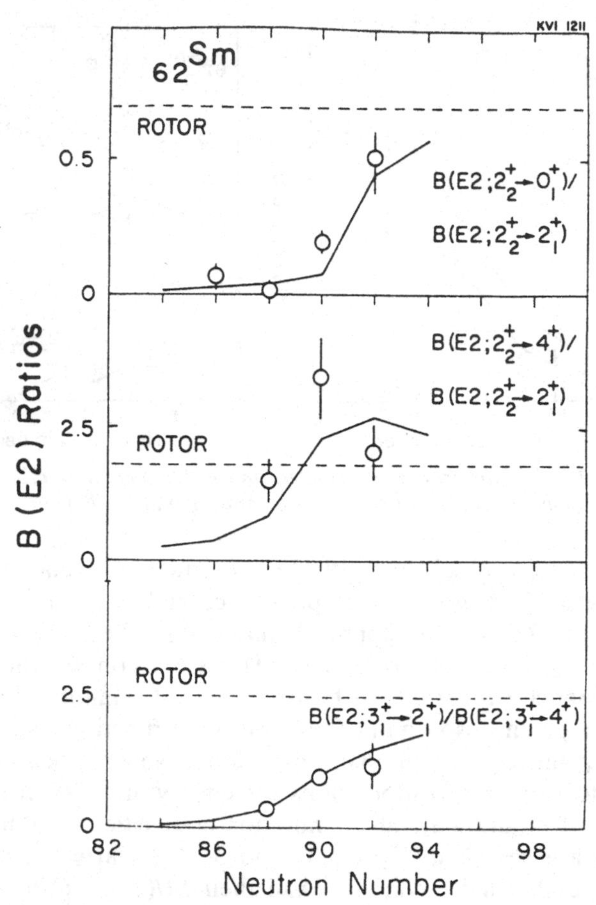

FIGURE 3.32. *Comparison of calculated and experimental* B(E2) *ratios in the* Sm *isotopes. The dashed lines labelled* ROTOR *give the Alaga values of these ratios. From Scholten, Iachello and Arima* (1978).

shows that, in the same quadrant, the $\beta \rightarrow g$ strength maximizes while the $\gamma \rightarrow g$ strength remains largely unaffected.

3.9.3. $U(5) \rightarrow O(6)$

There remains the question of the third leg of the symmetry triangle, namely a $U(5) \rightarrow O(6)$ phase transition. This type of tran-

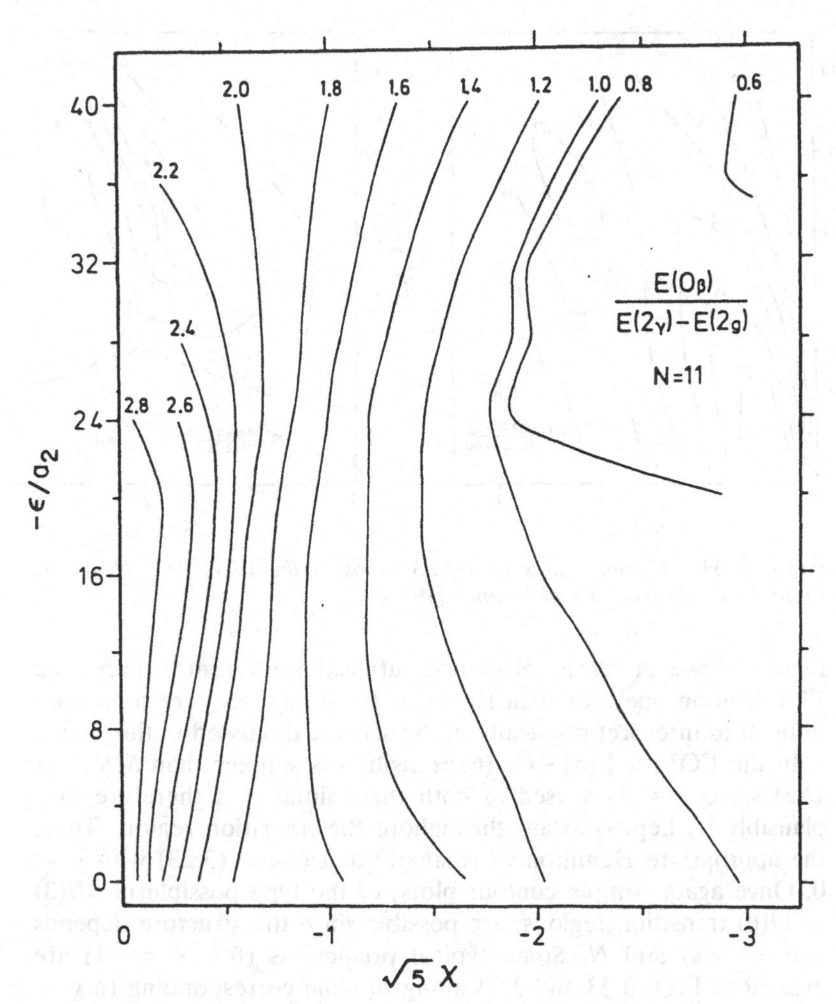

FIGURE 3.33. *Contour plot relating* β *and* γ *band energies in the* ECQF *for* $N = 11$. *From Lipas, Toivonen and Warner* (1985).

sition does not occur often and hence is not well established experimentally, at least in the sense of observing most stages of the evolution from one limit to the other. Nevertheless, the nuclei near ^{104}Ru may occupy an intermediate position along this transition leg since calculations for such a phase transition have obtained reasonably good agreement with empirical systematics in the Ru–Pd

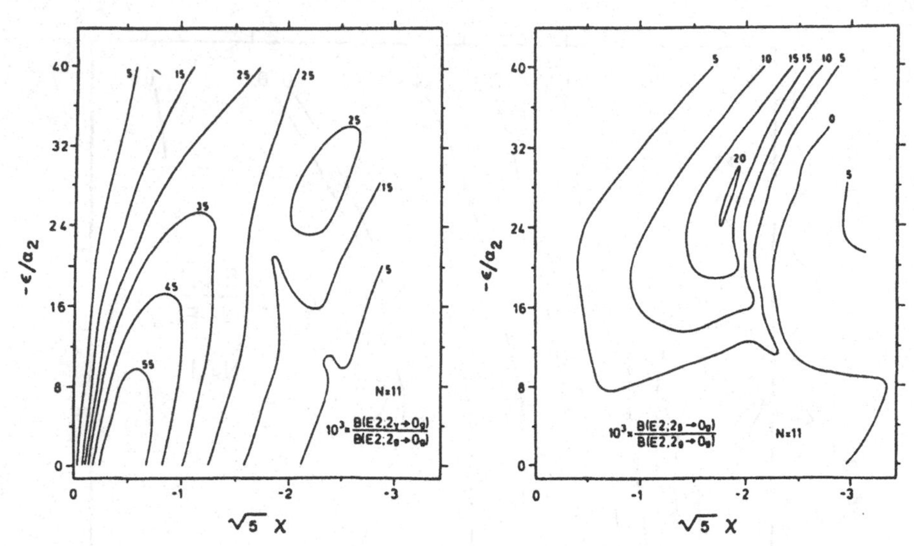

FIGURE 3.34. *Contour plots of B*(E2) *ratios in the* ECQF *for N* = 11. *From Lipas, Toivonen and Warner* (1985).

region. However, these calculations utilized a different form for the E2 transition operator than has been traditional and are somewhat difficult to interpret physically in the scheme discussed in this book.

In the CQF, a $U(5) \to O(6)$ transition is simpler than $SU(3) \to U(5)$ since $\chi = 0$ is used in both these limits and therefore may plausibly be kept constant throughout the transition region. Thus, the appropriate Hamiltonian is simply that of eq. (3.23) with $\chi = 0$. Once again, simple contour plots, of the type possible in $SU(3) \to O(6)$ transition regions, are possible since the structure depends only on ϵ/a_2 and N. Some typical predictions (for $N = 11$) are evident in Figs. 3.33 and 3.34 along the line corresponding to $\chi = 0$. The predicted energy and B(E2) ratios reflect the expected trends between $U(5)$ and $O(6)$. Note that as in the $SU(3) \to O(6)$ case, a number of branching ratios, such as those shown in Fig. 3.34, vanish in both limits and are finite only in the transition region.

To conclude and summarize this section, it is useful to recapitulate some basic guidelines for practical IBM-1 calculations. This is most easily done by reference to the symmetry triangle of Fig. 3.9. Calculations involving a large ϵ will tend towards a $U(5)$ structure, those with a large a_2 term will give deformed nuclei, in which the deviations from $SU(3)$ depend on the size of the symmetry breaking (e.g., on ϵ or on the deviations of χ from $\chi_{SU(3)}$). $O(6)$-like spectra are produced either by a large $P^{\dagger}P$ term or, in the CQF, by small $|\chi|$ values. As general guidance, it is also useful to recall that the N dependence of the $Q \cdot Q$ terms is $\approx N^2$ while the effects of $\epsilon \hat{n}_d$ go as N so that, even for constant ϵ/a_2, a monotonic tendency towards deformation will ensue as N increases. How far it goes depends on the actual values of ϵ/a_2. In an $SU(3)$–$O(6)$ region, it is useful to note that, while deviations from $SU(3)$ grow approximately linearly with $a_0P^{\dagger}P$, or with $|\chi - \chi_{SU(3)}|$, a good deformed structure nevertheless persists until these latter terms are quite large: for example, as the contour plots of Figs. 3.18 and 3.19 show, deformed character remains more or less intact for $N > 10$ even for $|\chi|$ values as small as 0.4.

3.10. SUMMARY

In this chapter we have tried to confront the IBM-1 with the data on medium and heavy mass nuclei, to show its successes and its failures, indicating where extensions to the model (some discussed in Chapter 6) may help alleviate the latter, and to outline numerical approaches to practical IBM-1 calculations.

Many topics had to be ignored to keep the discussion manageable. Hopefully, though, with the present background, the reader can now approach the literature to further pursue topics introduced here or to learn about others. Some other types of data (e.g., on high spin states or collective M1 transitions) are discussed at the appropriate places elsewhere in this book (e.g., Chapters 6 and 4, respectively). Finally, the whole realm of odd mass nuclei is the subject of a related model, the Interacting Boson-Fermion Model (IBFM) which is discussed in Chapter 5.

It might be best to conclude the present chapter by assessing where the IBM currently stands in the pantheon of nuclear models. The IBM, and other algebraic approaches (see Chapter 7 for a treatment of some of the most important of these) occupy an intermediate and complementary niche in nuclear structure, between the microscopic nuclear Shell Model (spherical for deformed) and the geometrical collective models. Algebraic models look to the Shell Model for a microscopic rationale and substrate and, ultimately, for their parameters, while providing often analytic, and usually simple, solutions for nuclei corresponding to different geometrical shapes. The IBM is a rather general model, incorporating, under one umbrella, the three symmetries and a wealth of intermediate structures including the phase transitional paths between symmetries. An appealing aspect is the model's simplicity: for example, calculations of phase transitional regions usually depend on only one or two parameters. In addition, the IBM, with its emphasis on the valence space and on the *finite* number of interacting nucleons, contains features absent from geometrical models or ones that are only incorporated in them from the outside when the data require it. These "finite N" effects confer an element of microscopic structure to the IBM, in that, even with constant parameters, its predictions will vary in specific, unavoidable, systematic ways with N and Z. The existence of these inherent features of the IBM which appear automatically and agree with the data is a key aspect and stringent test of the model: if the data were found to be the opposite, it would be difficult, if not impossible, to overturn these predictions. For example, as discussed in this chapter, the IBM contains no triaxial solutions: the $O(6)$ limit is γ flat and all axial asymmetry in the IBM-1 appears as γ softness (see Fig. 3.17). Experimentally, observed asymmetry at low spin is found to also arise from γ-softness. Related to this are the results (see Fig. 3.16) that the IBM suggests average γ values $\sim 10°$ for deformed nuclei, and the parabolic systematics for Z_γ, minimizing at mid-shell (see Fig. 3.23): both of these findings are well-known experimental features. Finally, the IBM (see Fig. 3.19) suggests a structure for the "β" band, radically different from earlier conceptions, which is nevertheless experimentally verified. The model predicts a linkage between the β and γ band properties which is evident, perhaps most vividly, in allowed, collective, $B(\mathrm{E2}:\beta \leftrightarrow \gamma)$ values, as well as in the relationship that $B(\mathrm{E2}:\beta \leftrightarrow g)$ values are much less than $B(\mathrm{E2}:\gamma \leftrightarrow g)$ values (compare Fig. 1.12), and in the predicted connection between β and γ vibrational energies (Fig. 3.21) which is observed, both in trend (Fig. 3.21) and magnitude (Fig. 1.11).

The algebraic approach, exemplified by the IBM, has thus provided new insights into the collective structure of nuclei and now forms a triad, along with the shell model and geometrical collective models, of approaches that have entered the standard lexicon of nuclear structure models.

ACKNOWLEDGEMENTS

We are grateful to all those who helped us in the writing of our article in Reviews of Modern Physics, upon which much of the present chapter is based.

This work was supported in part by contract DE-AC02-76CH00016 with the United States Department of Energy.

BIBLIOGRAPHY

A. Aprahamian, D. S. Brenner, R. F. Casten, R. L. Gill, and A. Piotrowski, *Phys. Rev. Lewtt.* **59**, 535 (1987). Discussion of ^{118}Cd in the context of the $U(5)$ limit.

A. Arima and F. Iachello, *Ann. Phys.* (N.Y.) **99**, 253 (1976), $U(5)$.

A. Arima and F. Iachello, *Ann. Phys.* (N.Y.) **111**, 201 (1978). $SU(3)$.

A. Arima and F. Iachello, *Ann. Phys.* (N.Y.) **123**, 468 (1979). $O(6)$.

D. M. Brink, A.F.R. de Toledo Piza, and A. K. Kerman, *Phys. Lett.* **19**, 413 (1965).

O. Castanos, A. Frank, and P. Van Isacker, *Phys. Rev. Lett.* **52**, 263 (1984).

R. F. Casten, *Phys. Lett. B* **152**, 145 (1985). The N_pN_n scheme.

R. F. Casten, A. Aprahamian, and D. D. Warner, *Phys. Rev. C* **29**, 356 (1984). Relation of IBM to axial asymmetry.

R. F. Casten, P. von Brentano, and A.M.I. Haque, *Phys. Rev. C* **31**, 1991 (1985). Discussion of the $SU(3)$ symmetry near $N = 104$.

R. F. Casten, P. von Brentano, K. Heyde, P. Van Isacker, and J. Jolie, *Nucl. Phys. A* **439**, 289 (1985). Introduction of γ-dependence to the $O(6)$ potential via cubic terms. (See Chapter 6.)

R. F. Casten and J. A. Cizewski, *Nucl. Phys. A* **309**, 477 (1978). The $O(6)$ $\rightarrow$ Rotor $[SU(3)]$ transition.

R. F. Casten and J. A. Cizewski, *Nucl. Phys. A* **425**, 653 (1984).

R. F. Casten and J. A. Cizewski, *Phys. Lett. B* **185**, 293 (1987).

R. F. Casten, D. D. Warner, and A. Aprahamian, *Phys. Rev. C* **28**, 894 (1983). Bandmixing and the IBM.

R. F. Casten and D. D. Warner, *Rev. Mod. Phys.* **60**, 389 (1988). Review article on the IBM. The present chapter is largely based on this reference.

W. T. Chou, R. F. Casten, and P. von Brentano, *Phys. Rev. C* **45**, R9 (1992). Discusses the linkage of the properties of β and γ bands.

J. A. Cizewski, R. F. Casten, G. J. Smith, M. L. Stelts, W. R. Kane, H. G. Börner, and W. F. Davidson, *Phys. Rev. Lett.* **40**, 167 (1978). Empirical manifestation of the $O(6)$ symmetry in ^{196}Pt.

A. S. Davydov and G. F. Filippov, *Nucl. Phys.* **8**, 237 (1958). The rigid triaxial rotor model.

J. N. Ginocchio and M. W. Kirson, *Nucl. Phys. A* **350**, 31 (1980). Basic article on the intrinsic state formalism.

F. Iachello and A. Arima, *The Interacting Boson Model*, Cambridge University Press, Cambridge, England (1987). Collection of basic ideas and formulas for the IBM.

W. Lieberz, A. Dewald, W. Frank, W. Krips, D. Lieberz, R. Wirowski, and P. von Brentano, *Phys. Lett. B* **240**, 38 (1990).

P. O. Lipas, P. Toivonen, and D. D. Warner, *Phys. Lett. B* **155**, 295 (1985). The extended CQF.

V. M. Mikhailov, Izv. Akad. Nauk. *SSST Ser. Fiz.* **30**, 1339 (1966). Graphical method of band mixing analysis.

O. Scholten, Ph.D. thesis, Kernfisisch Versneller Instituut, Groningen (1980). The code PHINT.

O. Scholten, F. Iachello, and A. Arima, *Ann. Phys.* (N.Y.) **115**, 325 (1978). $U(5) \rightarrow SU(3)$.

D. D. Warner, *Phys. Rev. Lett.* **47**, 1819 (1981). M1 transitions on the IBM.

D. D. Warner and R. F. Casten, *Phys. Rev. Lett.* **48**, 1385 (1982). The CQF approach to IBM calculations.

D. D. Warner and R. F. Casten, *Phys. Rev. C* **25**, 2019 (1982).

D. D. Warner and R. F. Casten, *Phys. Rev. C* **28**, 1798 (1983). More detailed article on the CQF.

D. D. Warner, R. F. Casten, and W. F. Davidson, *Phys. Rev. Lett.* **45**, 1761 (1980). The ^{168}ER calculations.

L. Wilets and M. Jean, *Phys. Rev. C* **102**, 788 (1956). The γ-soft rotor model.

N. V. Zamfir and R. F. Casten, *Phys. Lett. B* **260**, 265 (1991). Discusses empirical signatures of γ-softness or rigidity and the experimental evidence in favor of the former.

CHAPTER 4

Microscopic Basis and Introduction to IBM-2

TAKAHARU OTSUKA

We have seen, in previous chapters, the group theoretical view of the quadrupole collective motion by using the IBM, and the success of the phenomenological studies in terms of the IBM. One now should proceed to a question in what manner such a simple and useful model can be introduced from a more fundamental level. For this purpose, the microscopic foundation of the IBM is discussed in this chapter. In other words, we try to derive the IBM system from the multi-nucleon system. Although there could be different approaches to this goal, we focus on the standard one. When the IBM was proposed from the phenomenological viewpoint by Arima and Iachello, the microscopic picture of bosons of this model was not known. The microscopic theory has made a crucial contribution even to phenomenological studies by IBM, as will be discussed later.

4.1. SHELL STRUCTURE OF NUCLEUS

The nuclear force, which combines nucleons together into a nucleus, has two basic components, (i) a short-range attractive interaction between two nucleons, (ii) a repulsive core at even shorter distances. A schematic picture of the nuclear force is shown in Fig. 4.1 as a function of the distance r between two nucleons.

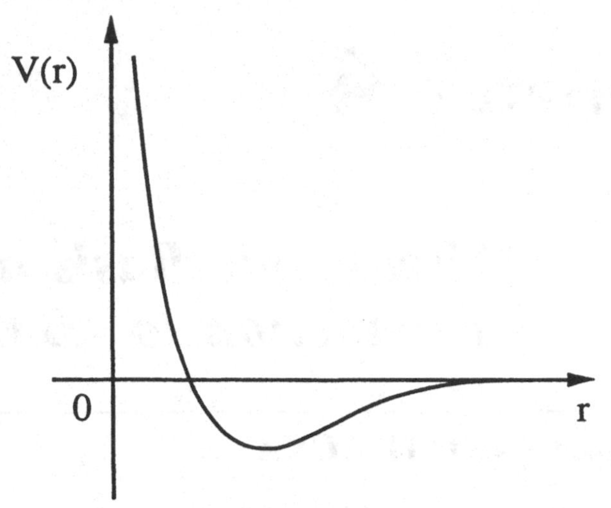

FIGURE 4.1. *Schematic picture of the nuclear force between two nucleons as a function of the distance* r.

The nuclear force should be of short range (a few fm) because of its origin in meson exchange effects. Property (i) produces the nucleus as a bound many-body system, while property (ii) keeps two nucleons away from each other. This prevents the nucleus from collapsing, and retains the nucleon density almost constant inside the nucleus, a property referred to as saturation (of density).

A nucleon inside the nucleus feels an attractive force from surrounding nucleons, and this effect is expressed as the mean potential for the nucleon. The range of the nuclear force is shorter than the size of the nucleus, and the density of nucleons is constant inside the nucleus as mentioned above. Therefore, the number of nucleons within the range of nuclear force from a specific nucleon is almost constant in the interior. Figure 4.2 gives a schematic picture of this; well inside the nucleus, each nucleon (marked by a cross in Fig. 4.2) feels the same potential induced by the surrounding nucleons (marked by dots in Fig. 4.2). Clearly the mean potential should be constant inside the nucleus, and is negative compared to a point infinitely far away.

A nucleon on the surface (marked by a triangle in Fig. 4.2) finds other nucleons only inside the nucleus as also shown in Fig. 4.2. Thus the surface nucleon feels a weaker potential. The potential should vanish far from the surface due to the short range character of the nuclear force. The mean potential therefore has the depen-

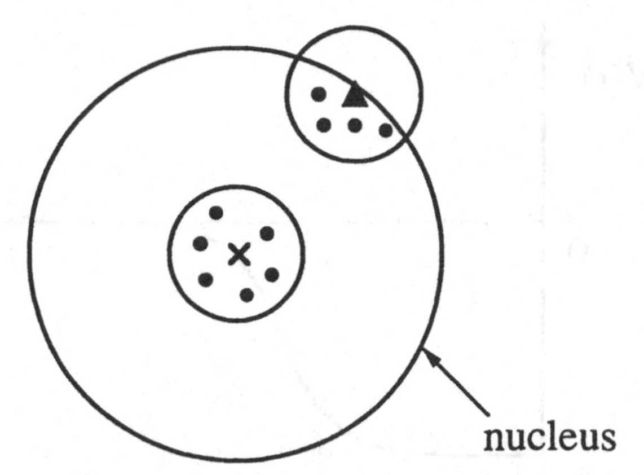

FIGURE 4.2. *Schematic picture of the nucleon-nucleon interaction in the nucleus. A nucleon marked by a cross interacts with nucleons (dots) within the range of the short-range nuclear force as shown by the small circle around the cross. A nucleon near the nuclear surface, marked by a triangle, interacts with a smaller number of nucleons (dots) in a small circle around this triangle.*

dence shown in Fig. 4.3 on the distance from the center of the nucleus.

The nucleons bound in the nucleus move through the mean potential in Fig. 4.3, forming single particle orbits. The energy levels of these orbits show a bunching pattern, resulting in the magic numbers and nuclear shell structure. Typical single particle energies and magic numbers were shown in Chapter 1 (see Figure 1.1). The orbits between two consecutive magic numbers form a (major) shell. For a given nucleus, protons and neutrons fill orbits starting from the bottom, rising through magic numbers. If the number of protons (neutrons) is equal to a proton (neutron) magic number, the protons (neutrons) constitute the closed shell. Otherwise, the highest shell partially filled by these nucleons is called the valence shell. The orbits in the valence shell are referred to as valence orbits, while protons (neutrons) in the valence shell are called valence protons (neutrons). The nucleons occupying shells below the valence shell constitute the inert core, which is nothing but a set of filled closed shells. Evidently, the valence shell is partly filled by nucleons, whereas the inert core is fully occupied. Excitation of a nucleon from the inert core to the valence shell requires a large amount of energy, and will be neglected in the following unless otherwise stated.

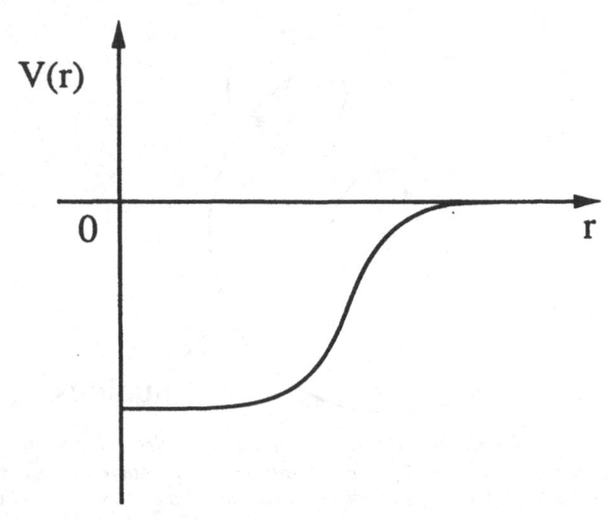

FIGURE 4.3. *Schematic picture of the mean potential of nucleons in the nucleus, where* r *indicates the distance from the center of the nucleus.*

4.2. NUCLEAR SURFACE AND NUCLEONS

Figure 4.2 suggests that the surface nucleons are less bound. Consequently, the less surface area, the more binding energy. The least surface shape for a given volume is known to be a sphere. Thus, by the above simple argument, it is inferred that a surface tension arises for the nucleus so as to keep the nuclear shape as close to a sphere as possible. On the other hand, the volume is conserved because of the saturation.

The inert core consists of orbits which are fully occupied by nucleons. If all magnetic substates, $m = -j, -j + 1, \ldots, j - 1, j$, of an orbital with the angular momentum j are completely occupied, this system is isotropic, i.e., spherical. Since the inert core is an assembly of such fully occupied orbitals, the inert core turns out to be spherical. Therefore the deformation should be due to the valence nucleons. Figure 4.4 shows by means of an intuitive picture that the valence nucleons produce the deformed shape, while the inert core remains spherical and carries angular momentum zero. In fact, the external rotation cannot change the state of the inert core. The valence nucleons, on the other hand, can be correlated spatially as shown in Fig. 4.4, resulting in "collective motion." The collective motions are of various types including rotation of the deformed ellipsoid, the oscillation of the surface from spherical

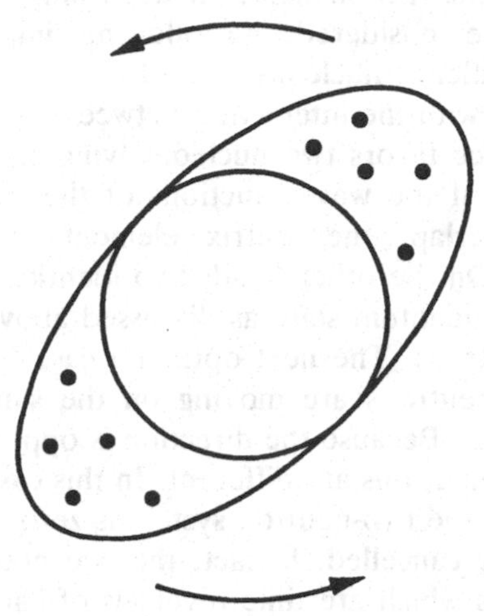

FIGURE 4.4. *Schematic picture of nuclear rotational motion. Nucleons (dots) outside the inner spherical core (small circle) constitute this motion.*

equilibrium, and intermediate situations. Figure 4.4 shows the case of rotation. These motions in general carry angular momentum, and produce a class of nuclear states called quadrupole collective states. The actual situation is slightly more complicated due to polarization of the inert core, but the basic picture remains.

If two nucleons move on the same orbit, the wave functions have the maximum overlap, gaining the largest binding energy from the short-range nuclear force. However, the Pauli exclusion principle does not allow two identical fermions in the same quantum state. This can be avoided by having a proton and a neutron which are evidently not identical. When two nucleons move together on the surface, density localization occurs and the nuclear shape is deformed from the sphere. Such density variations do not occur in the interior because of the saturation. Thus, one foresees that the proton-neutron correlation in the valence shell plays an important role in deforming the nucleus from a spherical shape.

4.3. EFFECTIVE NUCLEON-NUCLEON INTERACTION AND NUCLEAR DEFORMATION

The interaction between valence nucleons in the nucleus is different from that between free nucleons due to various medium effects, for instance, the core polarization. We do not go into details of the nucleon-nucleon interaction here. In describing the low-lying collective states to be considered, the following simple effective interaction between valence nucleons is used.

We shall begin with the interaction between neutrons. The short-range nuclear force favors two nucleons lying close to each other. This means that, if the wave functions of the two neutrons have large spacial overlap, the matrix element of this interaction becomes larger. On the other hand, two identical fermions cannot occupy the same quantum state as discussed above, and this is the case for two neutrons. The next optimum case for gaining energy is that the two neutrons are moving on the same orbital but in opposite directions. Because the direction is opposite, the quantum states of the two neutrons are different. In this case, the total angular momentum of the two-neutron system is zero, because the rotation is completely cancelled. In fact, the two neutrons take a pair of quantum states which are time reversals of each other. Because of this paired structure, this interaction is called the pairing interaction, as in the usual BCS theory. The pairing interaction scatters the two neutrons into other pairs of time reversal states. The pairing interaction does not favor any specific axis about which the two neutrons are moving around, and hence keeps the system spherical.

The pairing interaction is important also for the proton-proton channel. Thus, in the lowest-order approximation, the proton-proton and neutron-neutron interactions are dominated by the pairing interaction.

The proton-neutron interaction differs drastically in character. As stated briefly above, a proton moving around the nucleus tends to attract a neutron so that the two travel together. Such spatial correlations among valence nucleons cause density variations and then change the shape. The most basic deviation pattern from the sphere is the quadrupole deformation. In fact, the increase of the surface area becomes larger with higher multipoles. Once the proton distribution is varied towards a quadrupole shape, there should be some protons which produce this shape. Such deformed protons pull neutrons into a similar shape, deforming the neutron distribution into a quadrupole one. This kind of mechanism should be incorporated generally in terms of an effective proton-neutron interaction. In the present case, the proton quadrupole moment should be correlated to the neutron one. We thus use the quadrupole-quadrupole interaction which consists of the product of the proton quadrupole moment and the neutron quadrupole moment.

Through this interaction, the proton quadrupole moment enhances the neutron one and vice versa.

As empirical evidence of the decisive role of the proton-neutron interaction, one can plot the excitation energy of the first 2^+ states as a function of $N_p N_n$ where N_p and N_n denote respectively the numbers of valence protons and neutrons. Evidently, these 2^+ states are collective states, and reflect the deformation of the nucleus. Figure 4.5 demonstrates that many data fall onto one smooth line. Since the expectation value of the proton-neutron interaction can be scaled roughly as a function of $N_p N_n$, one can confirm from Fig. 4.5 the crucial role of the valence proton-neutron interaction for the collective states.

4.4. S AND D NUCLEON PAIRS—TRUNCATION OF THE SHELL MODEL

We have introduced in the preceding subsections an ansatz that the low-lying quadrupole collective states can be described in terms of valence nucleons. This ansatz leads us to a significant simplification for describing collective states. Even with this ansatz, however, the Hilbert space is still too large. As an example, we shall consider ^{154}Sm which is a typical rotational nucleus. Proton single particle orbits $0g_{7/2}$, $1d_{5/2}$, $1d_{3/2}$, $2s_{1/2}$, $0h_{11/2}$, and neutron single particle orbits $0h_{9/2}$, $1f_{7/2}$, $1f_{5/2}$, $2p_{3/2}$, $2p_{1/2}$, $0i_{13/2}$ are valence orbits. Having 12 protons and 10 neutrons in these valence orbits, one can generate various states. The dimension of the Hilbert space of such multi-nucleon system is enormous. For instance, the dimension for $J^\pi = 0^+$ is 41,654,193,516,917. This number becomes about ten times larger for the $J^\pi = 2^+$ subspace. Hilbert spaces with such large dimension are untractable, and the full treatment of such gigantic Hilbert spaces does not seem to make much physical sense.

We therefore truncate the Hilbert space into a smaller and tractable subspace with more physical significance. The truncated subspace is constructed by coherent pairs of valence nucleons (or holes) with angular momenta $J^\pi = 0^+$ (called the **S pair**) and $J^\pi = 2^+$ (called the **D pair**). These pairs are either of protons or of neutrons. The following discussion in this section is either for protons or for neutrons, but not for a mixture of them. In other words, we do not consider a pair consisting of a proton and a neutron.

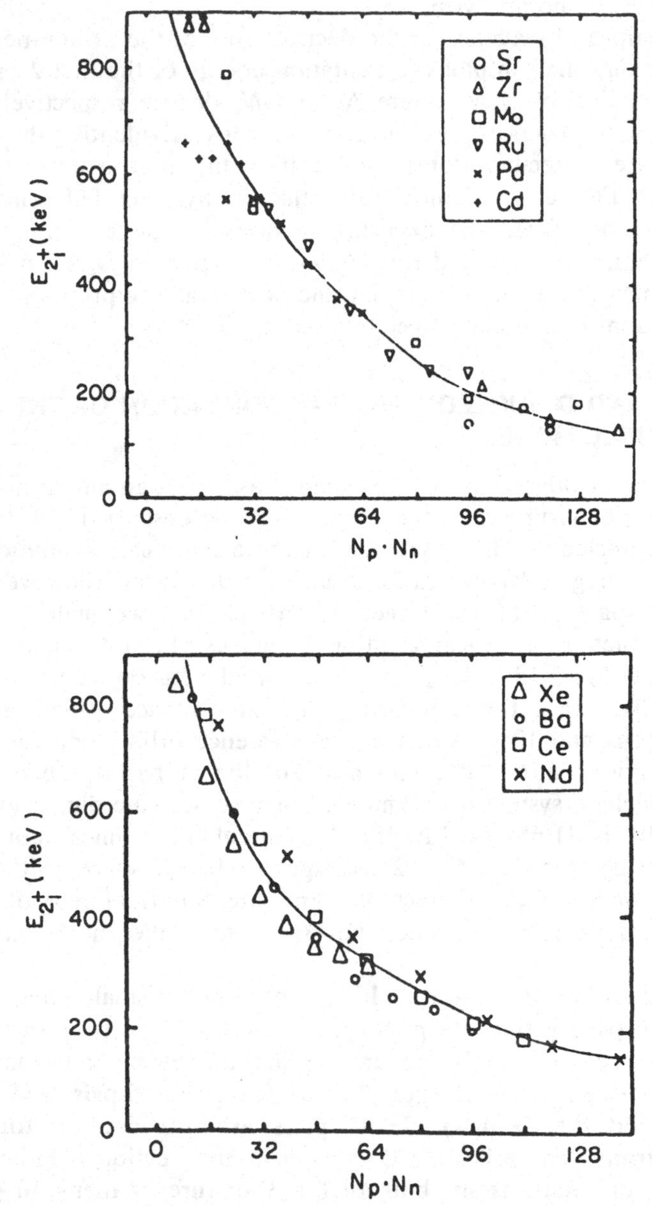

FIGURE 4.5. *Casten plots of energy levels as a function of the product of the number of valence protons and that of valence neutrons. From Casten (1985).*

As shown schematically in the left part of Fig. 4.6, two nucleons (i.e., two protons *or* two neutrons) are coupled to angular momentum zero in the S pair, whereas they are coupled to angular momentum two in the D pair. The creation operators of the S and D pairs are defined in general as

$$S^{\dagger} = \Sigma_j \alpha_j A^{\dagger}(jj; 0, 0) \tag{4.1}$$

and

$$D^{\dagger}_M = \Sigma_{jj'} \beta_{jj'} A^{\dagger}(jj'; 2, M) \tag{4.2}$$

where α_j and $\beta_{jj'}$ are normalized amplitudes, and the pair creation operator is defined as

$$A^{\dagger}(jj'; J, M) = \frac{1}{\sqrt{1 + \delta_{jj'}}} [a_j^{\dagger} a_{j'}^{\dagger}]^{(J)}_M \tag{4.3}$$

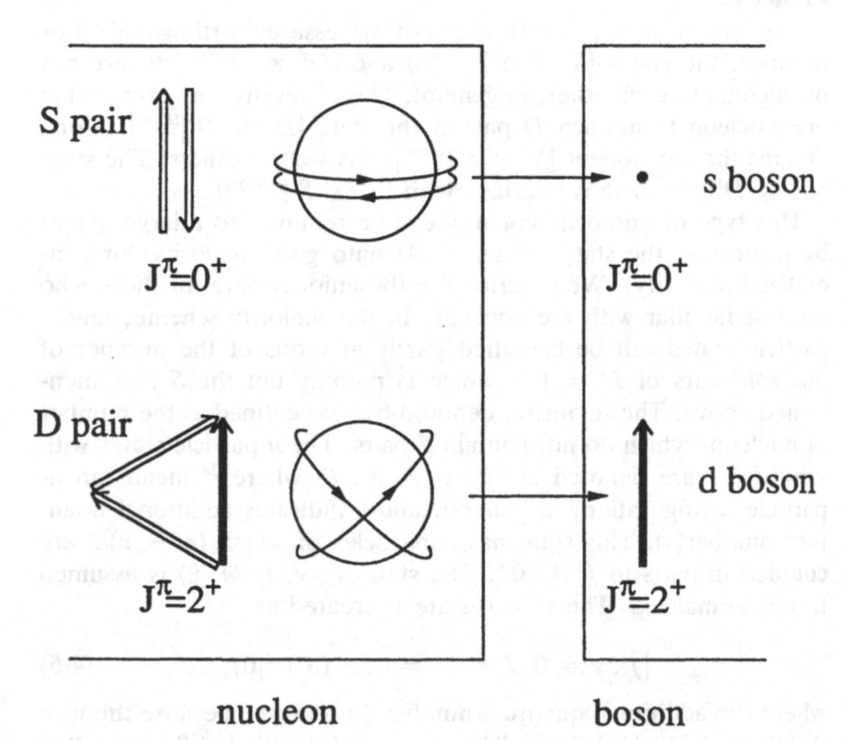

FIGURE 4.6. *Correspondence between nucleon pairs, S and D, and bosons, s and d.*

with $a_j^\dagger$ being the nucleon creation operator in the orbital j. The symbol $[\ \ ]_M^{(J)}$ indicates tensor coupling to angular momentum J and its z-component M. The letter j (or j') will be used to specify the nucleon orbit index as well as the angular momentum of this orbit.

The determination of the amplitudes α_j and $\beta_{jj'}$ is discussed in a subsequent section; they should have a coherent property so as to absorb as much binding energy as possible. Supposing that these amplitudes are given, the truncated subspace for describing the collective states with $2N$ valence nucleons is constructed by the state

$$(S^\dagger)^{N_S}[(D^\dagger)^{N_D}]_M^{(J)}|0\rangle \tag{4.4}$$

where $|0\rangle$ means the closed shell, N_S (N_D) counts the number of the S (D) pairs ($N_S + N_D = N$), J is the total angular momentum, and M is its z-component. The subspace is called the **SD subspace**.

The states in eq. (4.4) are not necessarily orthogonal. For instance, the states $[S^\dagger \times S^\dagger]^{(J=0)}|0\rangle$ and $[D^\dagger \times D^\dagger]^{(J=0)}|0\rangle$ are not orthogonal to each other, in general. This is because, by exchanging one nucleon from each D pair in the state $[D^\dagger \times D^\dagger]^{(J=0)}|0\rangle$, one obtains the component $[S^\dagger \times S^\dagger]^{(J=0)}|0\rangle$ as well as others. The state $[D^\dagger \times D^\dagger]^{(J=0)}|0\rangle$ thus overlaps with $[S^\dagger \times S^\dagger]^{(J=0)}|0\rangle$.

This type of non-orthogonality can be removed to a large extent by projecting the states in eq. (4.4) onto good seniority (or generalized seniority). We describe briefly seniority here for those who are not familiar with the concept. In the seniority scheme, many-particle states can be classified partly in terms of the number of nucleon pairs of $J^\pi = 0^+$, which is nothing but the S pair mentioned above. The seniority, denoted by v, is defined as the number of nucleons which do not contain S pairs. The n-particle states with seniority v are denoted as $|\bar{j}^n, v, J, M, \xi\rangle$ where $\bar{j}^n$ means an n-particle configuration, in general, and ξ indicates additional quantum number(s). This state has n particles of which $(n - v)/2$ are coupled in pairs to $J^\pi = 0^+$. The state $|\bar{j}^n, v, J, M, \xi\rangle$ is assumed to be normalized. The $v = 0$ state is created as

$$|\bar{j}^n, v = 0, J = M = 0\rangle \propto (S^\dagger)^{n/2}|0\rangle, \tag{4.5}$$

where the additional quantum number ζ is omitted because the $v = 0$ state is uniquely defined. The $v = 2$ states with $J \neq 0$ are written as

$$|\bar{j}^n, v = 2, J \neq 0, M, \zeta\rangle \propto (S^\dagger)^{(n-2)/2}|\bar{j}^2, v = 2, J \neq 0, M, \zeta\rangle. \quad (4.6)$$

The $v = 2$ states with $J = 0$ need additional care so that the state $|\bar{j}^n, v = 2, J = M = 0, \zeta\rangle$ should be orthogonal to $|\bar{j}^n, v = 0, J = M = 0\rangle$. In general, one ends up with the following definition:

$$|j^n, v, J, M, \xi\rangle \propto (S^\dagger)^{(n-v)/2}|j^v, v, J, M, \xi\rangle - \Sigma_{v'}\langle v, \zeta'|j^n, v', J, M, \xi'\rangle \times \langle j^n, v', J, M, \xi'|(S^\dagger)^{(n-v)/2}|j^v, v, J, M, \xi\rangle. \quad (4.7)$$

Note that the second term on the right-hand side simply indicates the Schmidt orthogonalization to lower seniority ($v' < v$) states. By definition, the state $|j^v, v, J, M, \zeta\rangle$ does not contain the S pair, and hence should satisfy the identity

$$S|j^v, v, J, M, \zeta\rangle = 0. \quad (4.8)$$

We shall now introduce the **SD states** as

$$|S^{N_S}, D^{N_D}; J, M, \xi\rangle \propto \frac{1}{\mathcal{N}_F}\mathcal{P}(S^\dagger)^{N_S}[(D^\dagger)^{N_D}]^{(J)}_M|0\rangle \quad (4.9)$$

where $\mathcal{N}_F$ denotes the normalization constant, $\mathcal{P}$ is the seniority projection operator onto $v = 2 \times N_D$, and ξ implies additional quantum number(s) as before. Note that the SD state on the left-hand side of eq. (4.9) is normalized. The SD states with no D pair will be shown as $|S^{N_S}; J = 0\rangle$ (i.e., D is dropped off). Although the seniority takes care of most of the non-orthogonality, the quantum number ξ is introduced so as to orthogonalize SD states within the same $v = 2N_D$, J and M. The quantum number ξ is not necessary for the states of lower values of N_d, where N_d, J and M determine the state uniquely; for example, $|S^{N_S}; J = 0\rangle$, $|S^{N_S}, D; J = 2, M\rangle$, etc. The quantum number ξ is needed, however, for larger values of N_D, as will be discussed in the next section. The quantum number M may be omitted hereafter, because it plays no essential role in most situations.

The first Ansatz for the microscopic foundation of the IBM is that **the low-lying quadrupole collective states are dominated by the SD states in eq. (4.9) with appropriately chosen amplitudes α_j and $\beta_{jj'}$**. The truncation from the full (valence) shell model space to the SD subspace is indicated schematically in the left part of Fig. 4.7.

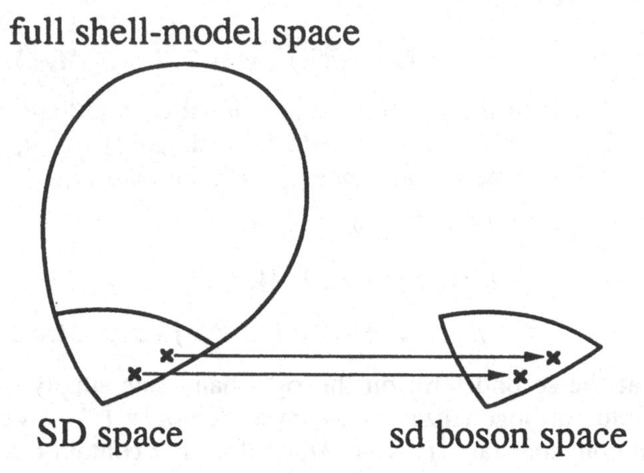

FIGURE 4.7. *Mapping from the SD subspace of the full shell-model space onto the sd boson space.*

This truncation has been proposed and developed by Arima, Otsuka, Iachello and Talmi (1977), and by Otsuka, Arima, Iachello and Talmi (1978).

Finally we note that, since we have protons and neutrons, there are **proton pairs, S_π and D_π, and neutron pairs, S_ν and D_ν**. Consequently, there are proton *SD* states and neutron *SD* states. The total space is spanned by the products of the proton *SD* and neutron *SD* states.

4.5. MAPPING TO SD BOSON STATES

In this section, we map the *SD* states onto boson states. The mapping scheme presented in Sections 4.4–4.8 is denoted as the OAI mapping (Otsuka, Arima and Iachello, 1978). Again, the discussion is either for protons or for neutrons, unless otherwise explicitly specified. We map the *S* and *D* pairs onto the $J^\pi = 0^+$ (s) and $J^\pi = 2^+$ (d) bosons, respectively, in the manner outlined below. Figure 4.7 shows this correspondence.

The *SD* states in eq. (4.9), which are correctly normalized, should be mapped onto *sd* boson states, which are also normalized. The general expression of the mapping is

$$|S^{N_S}, D^{N_D}, J, M, \xi\rangle = \frac{1}{\mathcal{N}_F}\mathcal{P}(S^\dagger)^{N_S}[(D^\dagger)^{N_D}]^{(J)}_M|0\rangle$$
$$\rightarrow |s^{N_s}, d^{N_d}, J, M, \xi) = \frac{1}{\mathcal{N}_B}(s^\dagger)^{N_s}[(d^\dagger)^{N_d}]^{(J)}_M|0) \tag{4.10a}$$

with

$$N_s = N_S \quad \text{and} \quad N_d = N_D \tag{4.10b}$$

where N_s and N_d imply, respectively, the numbers of the s and d bosons, and $\mathcal{N}_B$ denotes the boson normalization factor. In this Chapter, we use a convention that $|\rangle$ means a normalized fermion state, while $|)$ a normalized boson state. The state $|0)$ stands for the boson vacuum.

We shall consider this mapping more concretely, starting with the states where the additional quantum number ξ is not necessary. The multi-nucleon system is assumed to be comprised of N pairs. The closed shell is mapped onto the boson vacuum,

$$|0\rangle \rightarrow |0). \tag{4.11}$$

The pure S-pair state is mapped as

$$|S^N; J = 0\rangle \rightarrow |s^N; J = 0) \equiv \frac{1}{\sqrt{N!}}(s^\dagger)^N|0). \tag{4.12}$$

The state with one D pair is mapped as

$$|S^{N-1}D; J = 2\rangle \rightarrow |s^{N-1}d; J = 2) \equiv \frac{1}{\sqrt{(N-1)!}}(s^\dagger)^{N-1}d^\dagger|0). \tag{4.13}$$

The states with two D pairs are mapped as

$$|S^{N-2}D^2; J\rangle \rightarrow |s^{N-2}d^2; J) \equiv \frac{1}{\sqrt{2\cdot(N-2)!}}(s^\dagger)^{N-2}[d^\dagger d^\dagger]^{(J)}|0). \tag{4.14}$$

The total angular momentum of the SD states in eq. (4.14) can take the values 0, 2 or 4, whereas odd integers 1 and 3 are not allowed. This is because of the commutation relation

$$[D^\dagger_M, D^\dagger_{M'}] = 0. \tag{4.15}$$

In fact, because of this equation, one can deduce that

$$
\begin{aligned}
[D^\dagger D^\dagger]^{(J)}_{M+M'} &= \Sigma_{M,M'}(2M2M'|J\ M+M')D^\dagger_M D^\dagger_{M'} \\
&= \frac{1}{2}\Sigma_{M,M'}\{(2M2M'|J\ M+M')D^\dagger_M D^\dagger_{M'} \\
&\quad + (2M'2M|J\ M'+M)D^\dagger_{M'} D^\dagger_M\} \\
&= \frac{1}{2}\Sigma_{M,M'}\{(2M2M'|J\ M+M')D^\dagger_M D^\dagger_{M'} \\
&\quad + (-1)^{2+2-J}(2M2M'|J\ M+M')D^\dagger_M D^\dagger_{M'}\} \\
&= \frac{1}{2}(1 + (-1)^J)\Sigma_{M,M'}(2M2M'|J\ M+M')D^\dagger_M D^\dagger_{M'}
\end{aligned}
\tag{4.16}
$$

where $(2M2M'|J\ M+M')$ is the Clebsch–Gordan coefficient, and the relation $(2M2M'|J\ M+M') = (-1)^{2+2-J}\ (2M'2M|J\ M'+M)$ is used. Evidently, the state in eq. (4.16) vanishes for $J = 1$ and 3. This same result was obtained in a different way in Chapter 2.

The above selection rule on the total angular momentum remains after the seniority projection $\mathcal{P}$ in eq. (4.9). The commutation relation in eq. (4.15) imposes similar selection rules for the angular momenta of other multi-D-pair states. In the case of bosons, the boson commutation relation $[d^\dagger_M, d^\dagger_{M'}] = 0$ does exactly the same job as eq. (4.15). Consequently, for given N_d and J (and M), there are at most the same number of multi-D-pair states as multi-d-boson states. The reason for "at most" in the preceding sentence is that some multi-D-pair states may vanish due to the Pauli principle, whereas that never happens with bosons. The multi-d-boson states can be labelled in terms of $U(5)$ quantum numbers. Hence, in order to classify the D pair part of the SD states, one can make use of the $U(5)$ quantum numbers as the additional quantum number(s) ξ in eq. (4.9) with only slight modifications.

We shall discuss this $U(5)$-type labelling of the SD states, by way of a specific example. In the four d boson configuration, there are two 2^+ states, (i) $v_d = 2$, and (ii) $v_d = 4$. The former state is constructed as follows:

$$
|d^4;\ v_d = 2, J = 2) \propto [d^\dagger d^\dagger]^{(0)}\ [d^\dagger d^\dagger]^{(2)}|0). \tag{4.17}
$$

The latter can be created by orthogonalizing

$$
d^\dagger [d^\dagger [d^\dagger d^\dagger]^{(2)}]^{(0)}|0) \tag{4.18}
$$

to the state in eq. (4.17). This second state contains the three-d-boson triangle $[d^\dagger[d^\dagger d^\dagger]^{(2)}]^{(0)}$. The number of such d-boson triplets is given by the quantum number ν_Δ (See Section 2.6.2.2). For the *SD* states, we introduce "ν_d" and "ν_Δ" as the ξ's. The *SD* state which corresponds to eq. (4.17) is then written as

$$|S^{N-4}, D^4; \text{“}\nu_d\text{”} = 2, J = 2\rangle$$

$$= \frac{1}{\mathcal{N}_F} \mathcal{P}(S^\dagger)^{N-4}[D^\dagger D^\dagger]^{(0)}[D^\dagger D^\dagger]^{(2)}|0\rangle. \quad (4.19)$$

The "ν_Δ" quantum number can be introduced to the *SD* states, similarly as ν_Δ for bosons. The "ν_Δ" = 1 three-D-pair state is

$$|D^3; \text{“}\nu_d\text{”} = , \text{“}\nu_\Delta\text{”} = 1, J = 0 \propto \mathcal{P}[D^\dagger[D^\dagger D^\dagger]^{(2)}]^{(0)}|0\rangle. \quad (4.20)$$

The *SD* state $|S^{N-4}, D^4; \text{“}\nu_d\text{”} = 4, \text{“}\nu_\Delta\text{”} = 1, J = 2\rangle$ is then obtained from

$$\mathcal{P}(S^\dagger)^{N-4}D^\dagger[D^\dagger[D^\dagger D^\dagger]^{(2)}]^{(0)}|0\rangle, \quad (4.21)$$

with Schmidt orthogonalization to the state in eq. (4.19).

The process of constructing *SD* states thus resembles that of constructing *sd* boson states, except for the modification due to the fact that the actual amplitudes in the Schmidt orthogonalization can be different between the boson and fermion systems.

It is noted that some *SD* states may not exist due to the Pauli principle (i.e., the norm becomes zero after the seniority projection), while the *sd*-boson counterpart is present. This does not cause serious problems in practical calculations, because it does not happen in major components of states in the main part of the level scheme, to which most theoretical and experimental efforts are devoted.

4.6. PROTON BOSONS, NEUTRON BOSONS AND BOSON NUMBER

The proton pairs and neutron pairs were introduced in Section 4.4. As shown in Fig. 4.6, the proton pairs, S_π and D_π, are mapped, respectively, onto the $J^\pi = 0^+$ (s_π) and $J^\pi = 2^+$ (d_π) **proton bosons**. Similarly, the neutron pairs, S_ν and D_ν are mapped onto the $J^\pi = 0^+$ (s_ν) and $J^\pi = 2^+$ (d_ν) **neutron bosons.** We introduce, for later convenience, a convention that quantities related to the proton (neutron) bosons are indicated by the subscript $\pi(\nu)$. By the mapping, we are led to a new version of IBM, the **proton-neutron IBM** *or* **IBM-2**. The model without distinguishing proton bosons from neutron bosons, which has been the primary topic of Chapters 2 and 3, is called **IBM-1**. The relation between IBM-1 and IBM-2 will be discussed in Section 4.11.

The mapping discussed in Section 4.5 suggests a very important rule for determining the number of bosons. The proton boson number, denoted by N_π, should be equal to the number of pairs of valence protons. The neutron boson number, denoted by N_ν, should be equal to the number of pairs of valence neutrons. As we shall see in Section 4.11 and as discussed in Chapters 2 and 3, the boson number for the IBM-1 is given by $N_\pi + N_\nu$.

When the IBM was proposed by Arima and Iachello from the phenomenological viewpoint, the boson number was arbitrary. In fact, the boson number for ^{110}Cd was 6 in a very early paper, whereas it should be 7 according to the present rule. The boson number plays a decisive role even in phenomenology, as seen in Sections 4.9 and 4.10, and, earlier, in Chapter 3.

4.7. MAPPING OF NUCLEON OPERATORS ONTO BOSON OPERATORS

In the OAI mapping, the *sd* boson states can be considered as the boson (mapped) image of the *SD* states. The boson image of nucleon operators is discussed in this section. Similarly to previous sections, the following discussion is either for protons or for neutrons, unless otherwise explicitly specified.

The boson image of nucleon operators is determined so that the calculation for the *SD* states can be simulated by the *sd* boson system. The general principle is, therefore, that the boson image of a nucleon operator is determined by equating *SD*-state matrix elements of the nucleon operator with the corresponding *sd*-boson matrix elements, which are actually the boson images of the *SD* matrix elements. Thus, supposing that the *sd*-boson states φ and φ' are the boson images of *SD* states Ψ and Ψ', the boson image $\mathcal{O}^B$ of a nucleon operator $\mathcal{O}$ is obtained so as to satisfy

$$(\varphi|\mathcal{O}^B|\varphi') \cong \langle\Psi|\mathcal{O}|\Psi'\rangle. \tag{4.22}$$

The boson image $\mathcal{O}^B$ can include, in principle, many-body terms. We introduce the *k*-th order image; for an *i*-body nucleon operator, the *k*-th order boson image means an operator containing up to the

$k + i$ body terms. As an example, for a one-body operator of nucleons, the zeroth-order boson image consists also of constant and one-body terms. In practice, we shall restrict ourselves to the zeroth order image. This means that the zeroth-order boson image of a two-body nucleon-nucleon interaction is comprised of constant, single particle energy and two-body interactions. In fact, this is in accordance to the basic idea of the IBM, where one considers up to two-body interactions.

The process for determining $\mathcal{O}^B$ is illustrated in detail below. We shall consider spherical nuclei, where the states in eqs. (4.12)–(4.14) should be the dominant components in the lowest states, and are used as the *SD* states Ψ and Ψ' in eq. (4.22). The IBM Hamiltonian is written in general for one kind of bosons as

$$H^B = E_0^{(N)} + \epsilon \hat{N}_d + V, \qquad (4.23)$$

and

$$\begin{aligned} V = & \frac{1}{2} \Sigma_{L=0,2,4} c_L ([d^\dagger d^\dagger]^{(L)} \cdot [\tilde{d}\tilde{d}]^{(L)}) \\ & + \frac{1}{\sqrt{2}} y \{([d^\dagger d^\dagger]^{(2)} \cdot [s\tilde{d}]^{(2)}) + \text{h.c.}\} \\ & + \frac{1}{2} w \{[d^\dagger d^\dagger]^{(0)} [ss]^{(0)} + \text{h.c.}\}, \end{aligned} \qquad (4.24)$$

where $\hat{N}_d$ denotes the d-boson number operator, N stands for the total boson number, and u_0, u_1, u_2, ϵ, c_L, y and w are constants. Note that the total boson number is conserved. The quantity $E_0^{(N)}$ in eq. (4.23) is the energy of the state $|s^N; J = 0)$, and is fixed by the energy of the *SD* state $|S^N; J = 0\rangle$;

$$\begin{aligned} E_0^{(N)} &= (s^N; J = 0|H^B|s^N; J = 0) \\ &= \langle S^N; J = 0|H|S^N; J = 0\rangle \end{aligned} \qquad (4.25)$$

where H is the nucleon Hamiltonian. The quantity $E_0^{(N)}$ is a constant for a given nucleus. As long as one is interested in excitation energies and/or wave functions, one does not have to pay attention to $E_0^{(N)}$. We shall, in fact, ignore $E_0^{(N)}$ hereafter, while it should be regarded as the origin point from which to measure the energy. One should bear in mind, however, that $E_0^{(N)}$ has to be treated explicitly in describing binding energies.

The constant ϵ is determined by

$$\epsilon = \langle S^{N-1}D; J = 2|H|S^{N-1}D; J = 2\rangle - E_0^{(N)}, \quad (4.26)$$

because of

$$\begin{aligned}\langle S^{N-1}D; J = 2|H|S^{N-1}D; J = 2\rangle \\ = (s^{N-1}d; J = 2|H^B|s^{N-1}d; J = 2) = E_0^{(N)} + \epsilon.\end{aligned} \quad (4.27)$$

The constant c_L is determined similarly by

$$c_L = \langle S^{N-2}D^2; J = L|H|S^{N-2}D^2; J = L\rangle - 2\epsilon - E_0^{(N)}, \quad (4.28)$$

because of

$$\begin{aligned}\langle S^{N-2}D; J &= L|H|S^{N-2}D^2; J = L\rangle \\ &= (s^{N-2}d^2; J = L|H^B|s^{N-2}d^2; J = L) \\ &= E_0^{(N)} + 2\epsilon + c_L.\end{aligned} \quad (4.29)$$

The constants y and w are determined by

$$y = \langle S^{N-2}D^2; J = 2|H|S^{N-1}D^1; J = 2\rangle / \sqrt{N-1} \quad (4.30)$$

and

$$w = \langle S^{N-2}D^2; J = 0|H|S^N; J = 0\rangle / \sqrt{N(N-1)/2}, \quad (4.31)$$

because of

$$(s^{N-2}d^2; J = 2|H^B|s^{N-1}d; J = 2) = y\sqrt{N-1} \quad (4.32)$$

and

$$(s^{N-2}d^2; J = 0|H^B|s^N; J = 0) = w\sqrt{N(N-1)/2}. \quad (4.33)$$

The one-body operator is mapped similarly. For instance, the boson image of the quadrupole operator $Q = r^2Y^{(2)}(\theta, \varphi)$ should be of the following form:

$$Q \rightarrow Q^B = q_1(d^\dagger s + s^\dagger \tilde{d}) + q_2[d^\dagger \tilde{d}]^{(2)} \quad (4.34)$$

where q_1 and q_2 are coefficients. The value of q_1 is evaluated by the postulate

$$\begin{aligned}\langle S^{N-1}D^1; J &= 2\|Q\|S^N; J = 0\rangle \\ &= (s^{N-1}d^1; J = 2\|q_1 d^\dagger s\|s^N; J = 0) \\ &= q_1\sqrt{5} \times \sqrt{N},\end{aligned} \quad (4.35)$$

where $\langle \| \quad \| \rangle$ means the reduced matrix element, and $(d\|d^\dagger s\|s) = \sqrt{5}$ is used. Equation (4.35) gives rise to

$$q_1 = \langle S^{N-1}D^1; J = 2\|Q\|S^N; J = 0\rangle / \sqrt{5N}. \qquad (4.36)$$

Likewise, the value of q_2 is evaluated by

$$\begin{aligned}\langle S^{N-1}D^1; J = 2\|Q\|S^{N-1}D^1; J = 2\rangle \\ &= (s^{N-1}d^1; J = 2\|q_2[d^\dagger\tilde{d}]^{(2)}\|s^{N-1}d^1; J = 2) \qquad (4.37) \\ &= q_2\sqrt{5},\end{aligned}$$

yielding

$$q_2 = \langle S^{N-1}D^1; J = 2\|Q\|S^{N-1}D^1; J = 2\rangle / \sqrt{5}. \qquad (4.38)$$

Other operators can be mapped in the same way.

4.8. MAPPING IN A SINGLE-ORBIT SYSTEM

The previous section presented the general expression of the OAI mapped operators, which requires the value of SD matrix elements as input. The SD matrix elements contain effects of the Pauli principle, because they are multi-fermion matrix elements. The computation of SD matrix elements is not an easy task in general. On the other hand, if one is concerned with special cases, some of the many-body effects can be evaluated analytically. We shall discuss one such case in this section, namely that in which a valence shell consists of one orbit j, which will be referred to as a "**single-orbit system.**"

We first introduce three operators:

$$S_+ \equiv \sqrt{\Omega_j}\,A^\dagger(jj; 0, 0), \quad S_- \equiv (S_+)^\dagger, \quad S_0 \equiv (\Omega_j - \hat{n})/2, \qquad (4.39)$$

where

$$\Omega_j = j + \tfrac{1}{2}, \qquad (4.40)$$

and $\hat{n}$ is the nucleon number operator. The operators, S_+, S_-, and S_0 satisfy the commutation relations of the algebra of the group $SU(2)$ (i.e., the commutation relations of angular momentum);

$$[S_+, S_-] = 2S_0, \quad [S_+, S_0] = -S_+, \quad [S_-, S_0] = S_-. \qquad (4.41)$$

Because of this spin-like structure, the seniority in the single-orbit system is called the quasi-spin scheme. The quasi-spin brings about various simplifications in many-body problems. In fact, by using $[S, S^\dagger] = 1 - \hat{n}/\Omega_j$, one obtains

$$
\begin{aligned}
&\langle j^{v'}, v', J', M', \xi' | S^{(n-v')/2} (S^\dagger)^{(n-v)/2} | j^v, v, J, M, \xi \rangle \\
&= \langle j^{v'}, v', J', M', \xi' | S^{(n-v')/2-1} \\
&\quad \times \{(1 - \hat{n}/\Omega_j)(S^\dagger)^{(n-v)/2-1} + S^\dagger (1 - \hat{n}/\Omega_j)(S^\dagger)^{(n-v)/2-2} \\
&\quad + \cdots + (S^\dagger)^{(n-v)/2-1}(1 - \hat{n}/\Omega_j)\} | j^v, v, J, M, \xi \rangle \\
&= \langle j^{v'}, v', J', M', \xi' | S^{(n-v')/2-1} \\
&\quad \times \left\{ \frac{n-v}{2} - \frac{(n-2) + (n-4) + \cdots + v}{\Omega_j} \right\} \\
&\quad \times (S^\dagger)^{(n-v)/2-1} | j^v, v, J, M, \xi \rangle \qquad (4.42) \\
&= \frac{n-v}{2} \left\{ 1 - \frac{n-2+v}{2\Omega_j} \right\} \\
&\quad \times \langle j^{v'}, v', J', M', \xi' | S^{(n-v')/2-1} (S^\dagger)^{(n-v)/2-1} | j^v, v, J, M, \xi \rangle \\
&= \left(\frac{n-v}{2} \right)! \left[\left\{ 1 - \frac{n-2+v}{2\Omega_j} \right\} \right. \\
&\quad \times \left. \left\{ 1 - \frac{n-4+v}{2\Omega_j} \right\} \cdots \left\{ 1 - \frac{v}{2\Omega_j} \right\} \right] \\
&\quad \times \delta_{v,v'} \delta_{J,J'} \delta_{M,M'} \delta_{\xi,\xi'},
\end{aligned}
$$

where eq. (4.8) is used. This equation means that the orthogonalization in eq. (4.5) is not necessary in the quasi-spin scheme. The norm of eq. (4.42) is denoted for later convenience as

$$
\begin{aligned}
\mathcal{N}(n, v) &\equiv \langle j^v, v, J, M, \xi | S^{(n-v)/2} (S^\dagger)^{(n-v)/2} | j^v, v, J, M, \xi \rangle \\
&= \left(\frac{n-v}{2} \right)! \left[\left\{ 1 - \frac{n-2+v}{2\Omega_j} \right\} \right. \\
&\quad \cdot \left. \left\{ 1 - \frac{n-4+v}{2\Omega_j} \right\} \cdots \left\{ 1 - \frac{v}{2\Omega_j} \right\} \right].
\end{aligned} \qquad (4.43)
$$

The natural and usual way to construct orthogonal bases is, as already mentioned in Section 4.7, to start from the state of $v = 0$, and move up to v higher with Schmidt orthogonalization to lower seniority states. An alternative and elegant way is to modify the definition of the D pair as $D^\dagger \to \mathfrak{p}D^\dagger$, where $\mathfrak{p}$ is the projection operator onto the states of $v = n$ (i.e., no S pair; that is, states of highest seniority) but differs from $\mathcal{P}$ in eq. (4.9). The analytic expression for $\mathfrak{p}$ is given by OAI, but we shall not go into details about it here to avoid technical (and unnecessary) complexity. The applicability of the $\mathfrak{p}$ projection is restricted to the quasi-spin scheme.

Some SD matrix elements with many pairs can be written in terms of matrix elements with just one pair, by using the quasi-spin scheme for a single-orbit system. For instance, the matrix element in eq. (4.36) can be written as

$$\langle S^{N-1}D^1; J = 2\|Q\|S^N; J = 0\rangle = \sqrt{\frac{N(\Omega_j - N)}{\Omega_j - 1}}\langle D; J = 2\|Q\|S; J = 0\rangle, \tag{4.44}$$

while the one in eq. (4.38) is

$$\langle S^{N-1}D^1; J = 2\|Q\|S^{N-1}D^1; J = 2\rangle = \frac{\Omega_j - 2N}{\Omega_j - 2}\langle D; J = 2\|Q\|D; J = 2\rangle. \tag{4.45}$$

Substituting eqs. (4.44) and (4.45), respectively, into eqs. (4.36) and (4.38), one obtains

$$q_1(N) = \sqrt{\frac{\Omega_j - N}{\Omega_j - 1}}\frac{1}{\sqrt{5}}\langle D; J = 2\|Q\|S; J = 0\rangle \tag{4.46}$$

and

$$q_2(N) = \frac{\Omega_j - 2N}{\Omega_j - 2}\frac{1}{\sqrt{5}}\langle D; J = 2\|Q\|D; J = 2\rangle. \tag{4.47}$$

We shall derive eq. (4.44) to illustrate how such an expression arises. The norm of the state $(S^\dagger)^N|0\rangle$ is

$$\langle 0|S^N(S^\dagger)^N|0\rangle = \mathcal{N}(2N, 0). \tag{4.48}$$

Similarly,

$$\langle 0|DS^{N-1}(S^\dagger)^{N-1}D^\dagger|0\rangle = \mathcal{N}(2(N - 1), 2). \tag{4.49}$$

The quadrupole operator in a single-orbit system is given by

$$Q_M = -\frac{\langle j\|Q\|j\rangle}{\sqrt{5}}\,[a_j^\dagger \tilde{a}_j]^{(2)}_M \qquad (4.50)$$

where $\tilde{a}_{j,m} = (-1)^{j-m} a_{j,-m}$ and $\langle j\|Q\|j\rangle$ is the reduced matrix element. It is clear that

$$[Q_M, S^\dagger] = \langle D_M|Q_M|S\rangle\, D^\dagger_M, \qquad (4.51)$$

where $\langle D_M|Q_M|S\rangle$ does not depend on M. Omitting this M, one obtains

$$\begin{aligned}&\langle 0|DS^{N-1}Q(S^\dagger)^N|0\rangle\\ &\quad= N\,\langle 0|DS^{N-1}(S^\dagger)^{N-1}D^\dagger|0\rangle\,\langle 0|DQS^\dagger|0\rangle \qquad (4.52)\\ &\quad= N\mathcal{N}(2N, 2)\,\langle 0|DQS^\dagger|0\rangle.\end{aligned}$$

Finally,

$$\begin{aligned}\langle DS^{N-1}|Q|S^N\rangle &= \langle 0|DQS^\dagger|0\rangle\,\frac{N\mathcal{N}(2N, 2)}{\sqrt{\mathcal{N}(2N, 0)\,\mathcal{N}(2N, 2)}}\\ &= \langle D|Q|S\rangle\,\frac{\sqrt{N\{1 - N/\Omega_j\}}}{\sqrt{1 - 1/\Omega_j}}.\end{aligned} \qquad (4.53)$$

Equation (4.53) becomes identical to $\langle D|Q|S\rangle\sqrt{N}$ in the limit $\Omega_j \to \infty$, which means that the Pauli effect becomes negligible in an infinitely large system.

The N-dependent factor in eq. (4.47) can be derived similarly. The generalization to systems with more than one orbit is straightforward, if we take the degenerate-orbit assumption:

$$|\alpha_j| = \sqrt{\Omega_j}/\sqrt{\Omega} \qquad (4.54)$$

where

$$\Omega \equiv \Sigma_j \Omega_j, \qquad (4.55)$$

is satisfied for eq. (4.1). In such cases, we can use the commutation relation $[S, S^\dagger] = 1 - \hat{n}/\Omega$, which is the same as the one for the single-orbit system except for the replacement $\Omega_j \to \Omega$. The explicit formulas in the degenerate-orbit assumption are

$$q_1(N) = \sqrt{\frac{\Omega - N}{\Omega - 1}}\,\frac{1}{\sqrt{5}}\,\langle D; J = 2\|Q\|S; J = 0\rangle \qquad (4.56)$$

and

$$q_2(N) = \frac{\Omega - 2N}{\Omega - 2} \frac{1}{\sqrt{5}} \langle D; J = 2 \| Q \| D; J = 2 \rangle, \qquad (4.57)$$

which are indeed identical to eqs. (4.46) and (4.47) except for $\Omega_j \to \Omega$. If the degenerate-orbit assumption breaks down (i.e., $|\alpha_j| \neq \sqrt{\Omega_j/\Omega}$), the analytic treatment of many S pairs in SD matrix elements becomes impossible, and one has to evaluate numerically the SD matrix elements in eqs. (4.36) and (4.38).

In general cases with α_j differing from the values in eqs. (4.54), one loses some of the mathematical beauty and simplicity inherent in the quasi-spin (i.e., $SU(2)$) scheme of single-orbit or degenerate-multi-orbit systems. The concept of the seniority scheme, however, can still be used to classify the states according to the number of the S pairs. One can still introduce the SD states as the orthogonal basis of the truncated Hilbert space, while the Schmidt orthogonalization among the SD states has to be done numerically. For the sake of clarity, we note that, in the microscopic formulation of the IBM, the beauty and simplicity of the quasi-spin scheme help our understanding but are not essential ingredients.

So far, we have discussed the one-body operator. The treatment of the two-body interaction is certainly more complicated. We shall not show their general formulas here. On the other hand, we shall discuss below the mapping of the pairing interaction as the simplest example. The pairing interaction acts between two identical nucleons (i.e., two protons or two neutrons) which are in time-reversed states of each other. In other words, these two nucleons are moving on the same orbital but in opposite directions. These two nucleons are scattered by the pairing interaction into other pairs of time-reversed states with equal amplitudes. The effect of the pairing interaction does not depend on the initial states either. Because of these equalities for the initial and final states, the pairing interaction annihilates and creates nucleon pairs isotropically, and hence tends to keep the system spherical. The nucleon pair appearing in the pairing interaction is created by the $P^\dagger$ operator

$$P^\dagger = \frac{1}{\sqrt{\Omega}} \Sigma_{j,m>0} (-1)^{j-m} a^\dagger_{j,m} a^\dagger_{j,-m} \qquad (4.58)$$

where the phase factor $(-1)^{j-m}$ is needed to create two nucleons in a pair of time-reversed states, and the factor $1/\sqrt{\Omega}$ is introduced for the normalization $\langle 0 | PP^\dagger | 0 \rangle = 1$. Thus, all pairs of (j, m) and $(j, -m)$ are treated equally in eq. (4.58). Because of the Clebsch–

Gordan coefficient $(jmj\ -m|00) = (-1)^{j-m}/\sqrt{2\Omega_j}$, the $P^\dagger$ operator is equal to the $S^\dagger$ operator in eq. (4.1) except for the α_j. The mathematical expression of the pairing interaction discussed in Section 4.3 is given now by

$$\mathbf{V}_\mathbf{P} = -G\Omega P^\dagger P, \tag{4.59}$$

where G denotes the strength.

We shall consider the case of a single-orbit system, where $\alpha_j = 1$. The $P^\dagger$ operator then becomes identical to the $S^\dagger$ operator. Based on an argument similar to eq. (4.42) and the identity $P = S$ (and naturally $P^\dagger = S^\dagger$), it can be inferred that

$$\begin{aligned}
&\langle j^n, v', J', M', \xi' | \mathbf{V}_\mathbf{P} | j^n, v, J, M, \xi \rangle \\
&= -G\Omega \frac{1}{\sqrt{N(n,v')}} \frac{1}{\sqrt{N(n,v)}} \\
&\quad \times \langle j^{v'}, v', J', M', \xi' | S^{(n-v')/2} P^\dagger P (S^\dagger)^{(n-v)/2} | j^v, v, J, M, \xi \rangle \\
&= -G\Omega \frac{1}{\sqrt{N(n,v')}} \frac{1}{\sqrt{N(n,v)}} \\
&\quad \times \{\langle j^{v'}, v', J', M', \xi' | S^{(n-v')/2+1} (S^\dagger)^{(n-v)/2+1} | j^v, v, J, M, \xi \rangle \qquad (4.60) \\
&\quad + \langle j^{v'}, v', J', M', \xi' | S^{(n-v')/2} [S^\dagger, S] (S^\dagger)^{(n-v)/2} | j^v, v, J, M, \xi \rangle \} \\
&= -G\Omega \delta_{v,v'} \delta_{J,J'} \delta_{M,M'} \delta_{\xi,\xi'} \left[\frac{N(n+2,v)}{N(n,v)} + \frac{n-\Omega}{\Omega} \right], \\
&= -G \delta_{v,v'} \delta_{J,J'} \delta_{M,M'} \delta_{\xi,\xi'} \left[\frac{n-v}{2} \frac{2\Omega - n - v + 2}{2} \right],
\end{aligned}$$

where the norm $N(n, v)$ is defined in eq. (4.43), and Ω_j is replaced with Ω since $\Omega = \Omega_j$ for a single-orbit system.

We now use this expression for computing the SD matrix elements in eqs. (4.25), (4.26), (4.28), (4.30), and (4.31). Since the pairing interaction conserves the seniority, the number of D pairs should be conserved in the SD subspace. Therefore, the seniority changing matrix elements vanish, which implies that $y = w = 0$ in eqs. (4.30)–(4.31). The other quantities are given as follows:

$$E_0^{(N)} = -GN(\Omega - N + 1), \tag{4.61}$$

$$\epsilon = -G(N-1)(\Omega - N) - E_0^{(N)} = G\Omega, \tag{4.62}$$

$$c_L = -G(N-2)(\Omega - N - 1) - 2\epsilon - E_0^{(N)} = -2G. \tag{4.63}$$

This expression can be generalized for degenerate-orbit systems, with the identity in eq. (4.54).

One of the most significant features in eqs. (4.61)–(4.63) is that ϵ and c_L do not depend on the number of bosons, N. We thus obtain, from the pairing interaction, an N-independent single particle energy of the d boson. Another interesting point is that $c_L/\epsilon \to 0$ for $\Omega \to \infty$, which indicates that the boson image of the pairing interaction is characterized by the term ϵN_d with a constant ϵ.

4.9. SPECTRUM OF MAPPED BOSON SYSTEM

The boson image of the quadrupole operator and the pairing interaction was presented in the previous section. As discussed in Section 4.3, the principal part of the effective nucleon-nucleon interaction for quadrupole collective states consists of the proton-proton and neutron-neutron pairing interactions and the proton-neutron quadrupole interaction. In Section 4.6, the IBM-2 system was introduced as a system comprised of N_π proton bosons and N_ν neutron bosons. The proton-proton pairing interaction is mapped onto a proton boson Hamiltonian, following the prescription of the previous sections. In a degenerate-orbit case, eqs. (4.61)–(4.63) show that the mapping is

$$\mathbf{V}_{\mathbf{P}\pi} \to \mathbf{V}^{\mathbf{B}}_{\mathbf{P}\pi} = E_{0_\pi}^{(N_\pi)} + \epsilon_\pi \hat{N}_{d_\pi} + c_\pi \tfrac{1}{2} \hat{N}_{d_\pi}(\hat{N}_{d_\pi} - 1), \quad (4.64a)$$

$$\epsilon_\pi = G_\pi \Omega_\pi, \quad (4.64b)$$

$$c_\pi = -2G_\pi, \quad (4.64c)$$

where $\hat{N}_{d_\pi}$ denotes the proton d-boson number operator, $E_{0_\pi}^{(N_\pi)}$ is the c-number part given by $-G_\pi N_\pi(\Omega_\pi - N_\pi + 1)$, G_π denotes the pairing strength for protons, and Ω_π implies Ω in eq. (4.55) for protons. Likewise, the neutron-neutron pairing interaction is mapped onto a neutron boson Hamiltonian obtained by $\pi \to \nu$ for eq. (4.64);

$$\mathbf{V}_{\mathbf{P}\nu} \to \mathbf{V}^{\mathbf{B}}_{\mathbf{P}\nu} = E_{0_\nu}^{(N_\nu)} + \epsilon_\nu \hat{N}_{d_\nu} + c_\nu \tfrac{1}{2} \hat{N}_{d_\nu}(\hat{N}_{d_\nu} - 1). \quad (4.65)$$

The proton-neutron quadrupole interaction is written as a scalar product of the proton quadrupole operator and neutron quadrupole operator which are mapped according to eq. (4.34). The boson image is then given as

$$\mathbf{V}_{\pi\nu} = -f(Q_\pi \cdot Q_\nu) \rightarrow \mathbf{V}^{\mathbf{B}}_{\pi\nu} = -f(Q^B_\pi \cdot Q^B_\nu) \qquad (4.66)$$

where f is the strength, and Q^B_π and Q^B_ν denote the boson images in eq. (4.34) for protons and neutrons, respectively. The mapping is more analytic for single- or degenerate-orbit systems as shown in eqs. (4.46)–(4.47) and (4.56)–(4.57).

Assuming degenerate orbits, we can derive an IBM-2 Hamiltonian starting from a multi-nucleon system. Although this calculation is schematic to a certain extent due to the degenerate-orbit assumption, we can see systematic changes as a function of boson number. The nuclei to be considered are the Ba isotopes, where $Z = 56$. Since $Z = 50$ is a good closed shell, the Ba isotopes constitute a system of six valence protons, which is mapped onto a system of three proton bosons, $N_\pi = 3$. The neutron shell is assumed to be closed at the neutron numbers 50 and 82, and we vary the number of neutrons from 52 to 80, increasing the valence neutron number from 2 to 30. The middle of the major shell is at the neutron number 66, where the neutron boson number is $N_\nu = 8$. Between the neutron numbers 66 and 82, we take the hole representation. This is because the particles and holes are equivalently transformed into each other, while the mapping is simpler for smaller number of fermions. In the hole representation, a boson corresponds to a pair of holes. Otherwise, the mapping process is identical between the particle system and the hole system. Thus, in going from 50 to 82, the number of neutron bosons N_ν is increased from 0 to 8, and then is decreased from 8 to 0 in the hole region.

We shall begin with the boson image of the pairing interaction between neutrons. Since there are valence neutrons but no valence protons in the Sn isotopes (where $Z = 50$), we first look at these nuclei. No valence protons implies that the proton-neutron quadrupole interaction produces no effect. Experimentally, the excitation energy of the first 2^+ state is essentially constant $E_x \sim 1.2$ MeV in all the even-even Sn isotopes between neutron numbers 52 and 80. This property is reproduced well by eq. (4.65) with $\epsilon_\nu =$ 1.2 MeV in eq. (4.62). In this description, the first 2^+ state of the even-even Sn isotopes corresponds to the state in eq. (4.13). Note that eq. (4.65) holds in the hole region. Although ϵ can take a different value for protons, we shall assume $\epsilon_\pi = \epsilon_\nu = 1.2$ MeV for the sake of simplicity. Since $\Omega = 16$ in the present case, eqs. (4.62)–(4.63) indicate $c_L = -0.15$ MeV for both protons and neutrons.

The mapping of the quadrupole operator is considered next. For simplicity, we determine the value of $\beta_{jj'}$ by the commutation relation

$$D^{\dagger}_{M} \propto [Q_M, S^{\dagger}], \tag{4.67}$$

where the value of $\beta_{jj'}$ does not depend on M. The D pair thus defined absorbs all the strength of the quadrupole operator from the S pair. Using this $\beta_{jj'}$, we calculate the values of q_1 and q_2 in eqs. (4.56)–(4.57) for $Q = r^2 Y^{(2)}(\theta, \varphi)$.

The sign of q_1 is fixed as positive in the usual convention. If it is negative, one can always carry out the transformation $s \Rightarrow -s$ in the boson space, which does not change any physics. The sign of q_2 cannot be controlled this way.

In the hole region, the fermion quadrupole operator expressed in terms of holes changes its sign due to the particle-hole transformation. We then map the operator $-Q$ rather than Q where Q is still given by eq. (4.50) while the a_j and $a^{\dagger}_j$ denote hole annihilation and creation operators. Due to this sign change, the overall negative phase factor is multiplied to the boson image of the fermion operator Q. At this stage, we introduce the phase convention for the q_1 term as above; the sign of q_1 is set to be positive. On the other hand, the sign of q_2 is not affected by this convention, and is in fact altered due to the particle-hole transformation. As a consequence of this sign alternation, q_2 vanishes at mid-shell as seen from eq. (4.57), because q_2 must be invariant at mid-shell under this sign change.

Because of the above sign convention, eqs. (4.56) and (4.57) are modified. The sign of q_1 is always positive, while its value is given by the absolute magnitude of eq. (4.56). The boson number N in eq. (4.56) is given either by the number of particle pairs or hole pairs depending on whether the nucleus is before or after mid-shell, respectively. The parameter q_2 is determined by eq. (4.57) on the particle side, while q_2 for N "hole-bosons" is given by the negative of q_2 for N "particle-bosons."

We introduce the ratio,

$$\chi_\tau = q_{2_\tau}/q_{1_\tau}, \qquad \tau = \pi, \nu. \tag{4.68}$$

This quantity has the following N_τ dependence as obtained from eqs. (4.56) and (4.57),

$$\chi_\tau = \mathring{\chi}_\tau \frac{\Omega_\tau - 2N_\tau}{\Omega_\tau - 2}\sqrt{\frac{\Omega_\tau - 1}{\Omega_\tau - N_\tau}}, \qquad \tau = \pi, \nu, \tag{4.69}$$

where $\mathring{\chi}$ denotes χ for the one-boson system, i.e., two-particle or two-hole system. The numerical calculation indicates $\mathring{\chi} = -0.71$ on the particle side for the degenerate orbits between the magic numbers 50 and 82, assuming harmonic-oscillator-potential single particle wave functions for nucleons. On the hole side, $\mathring{\chi} = 0.71$ due to the particle-hole transformation as discussed just above.

The proton-neutron boson interaction is written as

$$\mathbf{V}^{\mathbf{B}}_{\pi\nu} = -\kappa(\mathbf{Q}_\pi \cdot \mathbf{Q}_\nu) \tag{4.70}$$

where

$$\begin{aligned}\kappa &\equiv -fq_{1_\pi}(N_\pi)q_{1_\nu}(N_\nu) \\ &= \kappa_0 \sqrt{\frac{\Omega_\pi - N_\pi}{\Omega_\pi - 1}}\sqrt{\frac{\Omega_\nu - N_\nu}{\Omega_\nu - 1}},\end{aligned} \tag{4.71}$$

$$\kappa_0 \equiv -fq_{1_\pi}(N_\pi = 1)q_{1_\nu}(N_\nu = 1), \tag{4.72}$$

and the $\mathbf{Q}$ operator is defined as

$$\mathbf{Q}_\tau \equiv d^\dagger_\tau s_\tau + s^\dagger_\tau \tilde{d}_\tau + \chi_\tau [d^\dagger_\tau \tilde{d}_\tau]^{(2)}, \qquad \tau = \pi, \nu. \tag{4.73}$$

The value of f in eq. (4.72) is adjusted to give $\kappa_0 = 0.3$ MeV, so that the final result reproduces basic trends of observed energy levels.

The total boson Hamiltonian is given by

$$\begin{aligned} H &= \mathbf{V}^{\mathbf{B}}_{\mathbf{P}_\pi} + \mathbf{V}^{\mathbf{B}}_{\mathbf{P}_\nu} + \mathbf{V}^{\mathbf{B}}_{\pi\nu} \\ &= E^{(N_\nu)}_{0_\nu} + \epsilon_\nu N_{d_\nu} + c_\nu \tfrac{1}{2} N_{d_\nu}(N_{d_\nu} - 1) \\ &\quad + E^{(N_\pi)}_{0_\pi} + \epsilon_\pi N_{d_\pi} + c_\pi \tfrac{1}{2} N_{d_\pi}(N_{d_\pi} - 1) - \kappa(\mathbf{Q}_\pi \cdot \mathbf{Q}_\nu). \end{aligned} \tag{4.74}$$

The value of κ with $N_\pi = 3$ is shown in Fig. 4.8 as functions of N_ν. The magnitude of the parameter κ decreases gradually towards the middle of the major shell due to Pauli blocking. The parameter χ_ν depends only on N_ν as seen in eq. (4.69), and its value is shown also in Fig. 4.8. The parameter χ_ν changes almost linearly as a function of N_ν, reversing its sign as discussed above.

The energy levels obtained by solving the Hamiltonian in eq.

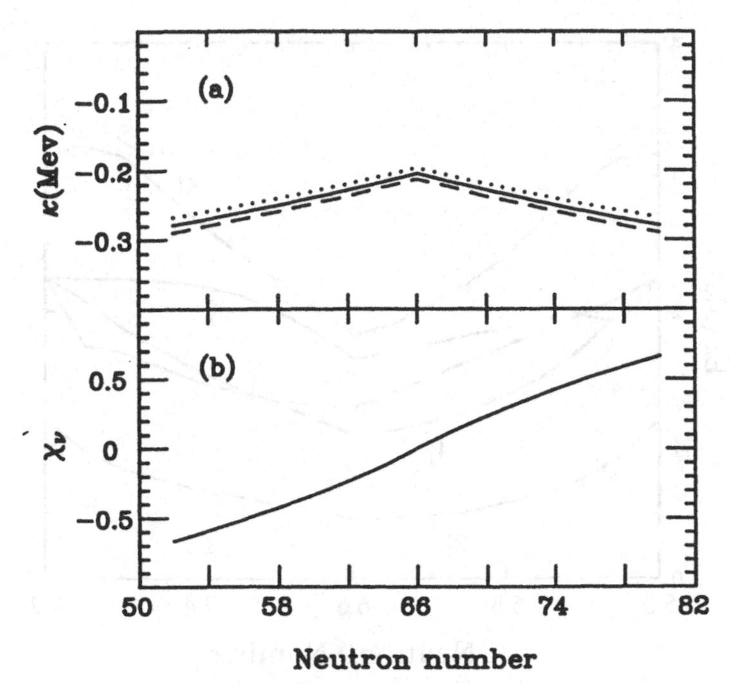

FIGURE 4.8. *Values of the parameters* (a) κ *and* (b) χ_ν *given by the* OAI *mapping in the degenerate-orbit approximation. In* (a), *dotted, solid and dashed lines correspond to* N_π = *2, 3 and 4, respectively.*

(4.74) are shown in Fig. 4.9. There are several important points to be noted. The 2_1^+ state decreases in energy from both sides towards the middle of the major shell. The same pattern is seen for the 4_1^+ level. This trend is consistent with experiment. The ratio of excitation energies, $R = E(4_1^+)/E(2_1^+)$ shows its maximum around mid-shell. All these patterns indicate developing rotational structure, i.e., stronger quadrupole deformation towards the middle of the major shell.

We shall consider the mechanism for the trends shown in Fig. 4.9. The structure of the eigenstates is determined basically by the two terms in eq. (4.74); (i) $\epsilon_\nu \hat{N}_{d_\nu} + \epsilon_\pi \hat{N}_{d_\pi}$ and (ii) $-\kappa(\mathbf{Q}_\pi \cdot \mathbf{Q}_\nu)$. The $c_\nu \frac{1}{2} \hat{N}_{d_\nu}(\hat{N}_{d_\nu} - 1) + c_\pi \frac{1}{2} \hat{N}_{d_\pi}(\hat{N}_{d_\pi} - 1)$ term is not important. Since positive ϵ_ν and ϵ_π prevent the nucleus from developing d-boson admixtures in low lying states, the term (i) tends to retain the nucleus in a spherical shape where the s bosons dominate the wave functions. Therefore, if the term (ii) has minor effects compared

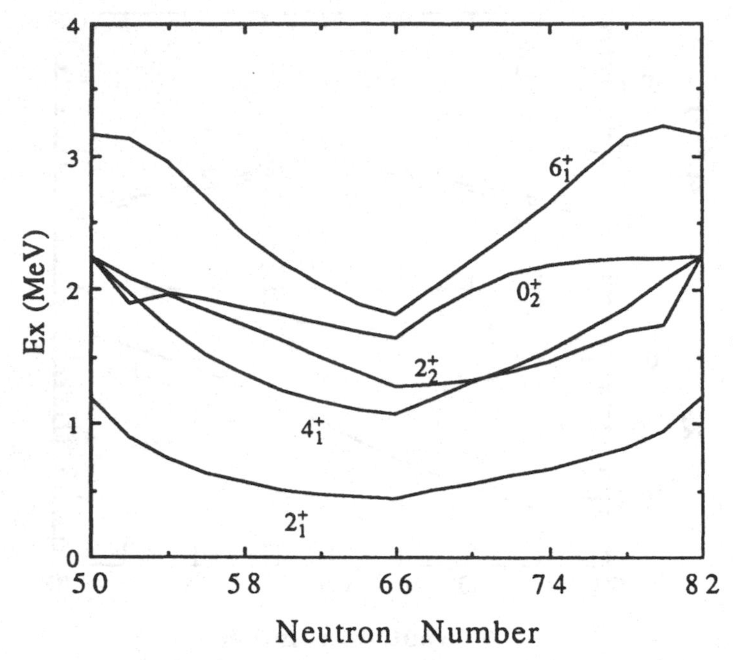

FIGURE 4.9. *Excitation energies of N_π = 3 IBM-2 systems as a function of the neutron number. The κ and χ_ν parameters in Fig. 4.8 are used, while ϵ is kept constant at* 1.2 MeV. *For the other parameters, see the text.*

to the term (i), the system becomes spherical, showing a $U(5)$-type level pattern. This is the case with a small number of bosons, as one can see below. As the neutron boson number increases, matrix elements of $\mathbf{Q}_\nu$ increase in general, primarily because $\mathbf{Q}_\nu$ is a neutron one-body operator. This drives the system so that the matrix elements of $\mathbf{Q}_\pi$ become larger and more binding energy can be gained from the $(\mathbf{Q}_\pi \cdot \mathbf{Q}_\nu)$ interaction. Note that the structure of the wave function is changed in this direction, even though the proton boson number is fixed. The situation with large matrix elements of $\mathbf{Q}_\pi$ and $\mathbf{Q}_\nu$ is referred to as a deformed shape or strong deformation. In order that the $\mathbf{Q}_\tau$ operator has a larger matrix element for a given boson number, the relevant states should contain more d bosons because of the s-d and d-d terms in eq. (4.73). The stronger deformation implies the higher expectation value of the term (i), which resists deformation. Terms (i) and (ii) thus have opposite effects, and hence the equilibrium solution is determined by the competition between them. Once the proton deformation is increased, for instance by a larger proton boson number, it favors, or reinforces, a stronger neutron deformation through the $(\mathbf{Q}_\pi \cdot \mathbf{Q}_\nu)$ interaction. In the equilibrium, these kinds of dynamical effects, which can be quite non-linear, are included self-consistently.

One can look at the points discussed above from a slightly different and more quantitative perspective. The ground state is assumed to be parametrized in terms of a parameter (or a set of parameters) η. The ground-state expectation value of terms (i) and (ii) is then expressed approximately by the functions $\mathcal{F}_1(N_\pi + N_\nu, \eta)$ and $\mathcal{F}_2(N_\pi N_\nu, \eta)$, respectively. These specific boson-number dependences are employed because the terms (i) and (ii) consist of a one-body operator and the proton-neutron interaction, respectively. For the former, $\epsilon_\pi \sim \epsilon_\nu$ is anticipated. The equilibrium solution is given by

$$\frac{\partial}{\partial \eta} \mathcal{F}_1(N_\pi + N_\nu, \eta) + \frac{\partial}{\partial \eta} \mathcal{F}_2(N_\pi N_\nu, \eta) = 0. \qquad (4.75)$$

In this equation, if $N_\pi N_\nu$ is sufficiently small, the solution is governed by $\mathcal{F}_1(N_\pi + N_\nu, \eta)$, producing spherical wave functions. On the other hand, if $N_\pi N_\nu$ is sufficiently large, $\mathcal{F}_2(N_\pi N_\nu, \eta)$ dominates the equilibrium, giving rise to the deformed solution. In the transitional situation, both terms contribute. For a fixed value of N_π, therefore, as N_ν increases, $\mathcal{F}_2(N_\pi N_\nu, \eta)$ plays a more crucial role and the equilibrium point is shifted towards stronger deformation. The deformed system shows the rotational spectrum, as seen in the $SU(3)$ symmetry. On the other hand, if N_π and/or N_ν are sufficiently small, the $(\mathbf{Q}_\pi \cdot \mathbf{Q}_\nu)$ interaction is not effective enough to overcome positive ϵ_ν and ϵ_π values. This is the basic reason why the system goes from spherical to deformed shapes in Fig. 4.9 as N_ν is varied. In changing N_π for a fixed value of N_ν, one finds a similar trend. The change between spherical and deformed structures is often referred to as the spherical-deformed phase transition.

It should be emphasized that the proton and neutron boson numbers are the most crucial parameters in describing the phase transition. The values of the coefficients in the Hamiltonian do not vary so much, and can be regarded as constant in lowest approximation. In fact, the boson number dependences of the parameters shown in Fig. 4.8 actually tend to slow down the development of the phase transition. For instance, κ in eq. (4.74) has the smallest magnitude in the middle of the major shell where the deformation is maximal. In other words, as stated, the phase transition is not due to the N_ν dependence of κ and χ_ν. Thus, as stated the boson number is the crucial quantity in describing the phase transition. We emphasize that the boson number has been given microscopically in terms of the number of valence nucleons. This is one of the most crucial consequences of the microscopic basis of IBM discussed so far, and without such study the boson number can remain arbitrary. As already pointed out, the counting rule for the boson number was not known when the IBM was proposed.

If one looks at the far right part of Fig. 4.9, the situation where $E(2_2^+) < E(4_1^+)$ should be noticed. This is one of the most characteristic patterns of the $O(6)$ dynamical symmetry. The high lying 0_2^+ state is also consistent with $O(6)$. In the left half of Fig. 4.9, the 2_2^+ state is well above the 4_1^+ state, and the level pattern shows a resemblance to the $SU(3)$ symmetry rather than to $O(6)$. Near the left edge, the level structure becomes similar to that of $U(5)$ with degenerate 0_2^+, 2_2^+ and 4_1^+ states. Thus, all three dynamical symmetry limits, or at least their "precursors," show up in Fig. 4.9.

It should be stressed that Fig. 4.9 is obtained from a "fixed nucleon" Hamiltonian. It is remarkable that many aspects, indeed, actually all essential aspects, of the low-lying quadrupole collective states appear as the neutron boson number changes in the full span of the major shell.

4.10. STANDARD IBM-2 HAMILTONIAN: PHENOMENOLOGY

The IBM-2 Hamiltonian was derived from a nucleon Hamiltonian in Section 4.9. Although the calculated energy levels show the basic pattern of the observed levels, there are discrepancies in quantitative comparisons to experiment. This is partly due to effects of states which are not formed by the S and D pairs. Such states are referred to as non-S-D states. Their effects are incorporated by renormalizing mainly ϵ_ν and ϵ_π. Although these are interesting discussions, this problem is beyond the scope of this textbook, and we shall change the values of ϵ_ν and ϵ_π only phenomenologically hereafter. For the sake of simplicity, the equality $\epsilon_\nu = \epsilon_\pi$ is assumed, knowing that only the mean value $(\epsilon_\pi + \epsilon_\nu)/2$ is important in most cases.

The standard phenomenological IBM-2 Hamiltonian is then written as, apart from terms that contribute only to binding energies,

$$H = \epsilon \hat{N}_d - \kappa(\mathbf{Q}_\pi \cdot \mathbf{Q}_\nu) \tag{4.76}$$

where $\hat{N}_d \equiv \hat{N}_{d_\nu} + \hat{N}_{d_\pi}$ and $\epsilon \equiv \epsilon_\nu = \epsilon_\pi$. Figure 4.10 presents a nice fit to Xe–Ba–Ce nuclei, where eqs. (4.69) and (4.71) are retained with $\mathring{\chi}_\tau$ fitted, and ϵ is adjusted phenomenologically. Note that $N_\pi = 2$ for Xe and $N_\pi = 4$ for Ce. The values of the ϵ parameter are shown in Fig. 4.11. By allowing ϵ to vary, one obtains an almost perfect fit to the experimental energies. It may be worth noting that more than 100 energy levels of 23 nuclei are described well in this calculation.

Usually, the parameters κ, χ_π and χ_ν, as well as ϵ, are fitted for each nucleus. As generalized from eq. (4.69), it is still assumed that χ_π depends only on the proton number, and χ_ν only on the neutron number. The terms like $c_\nu \frac{1}{2} \hat{N}_{d_\nu}(\hat{N}_{d_\nu} - 1)$ and $c_\pi \frac{1}{2} \hat{N}_{d_\pi}(\hat{N}_{d_\pi} - 1)$ are less important, and are considered only as a fine tuning of the fit.

Fig. 4.12 presents a result of such a fit to Xe–Ba–Ce nuclei performed by Puddu, Scholten and Otsuka (1980). Note that energy levels of neutron number less than 72 in Fig. 4.12(b) have been observed after the calculation. The calculation for Fig. 4.12 was carried out with different fitting policy from that for Fig. 4.10, while the agreement for energy levels is as good as that seen in Fig. 4.10. Fig. 4.13 shows the values of κ and χ_ν parameters used in the fit of Fig. 4.12, indicating systematic variations of the parameters κ and χ_ν as functions of N_ν which deviate from Fig. 4.8. The breaking of the degenerate-orbit approximation can be one of the primary reasons for this deviation. Precise microscopic calculations based on realistic single-particle energies and nucleon-nucleon interaction should clarify this question, but are still to come. It therefore remains an open problem which fit between Fig. 4.10 or 4.12 is more appropriate.

The IBM-2 E2 transition operator is defined as

$$T(\mathrm{E2}) = e_\pi^B \mathbf{Q}_\pi + e_\nu^B \mathbf{Q}_\nu \tag{4.77}$$

where e_π^B and e_ν^B are parameters called boson effective charges, while $\mathbf{Q}_\pi$ and $\mathbf{Q}_\nu$ are defined in eq. (4.73). In the degenerate-orbit approximation, the boson charges are given by

$$e_\tau^B = \mathring{e}_\tau^B \sqrt{\frac{\Omega_\tau - N_\tau}{\Omega_\tau - 1}}, \qquad \tau = \pi, \nu, \tag{4.78}$$

with

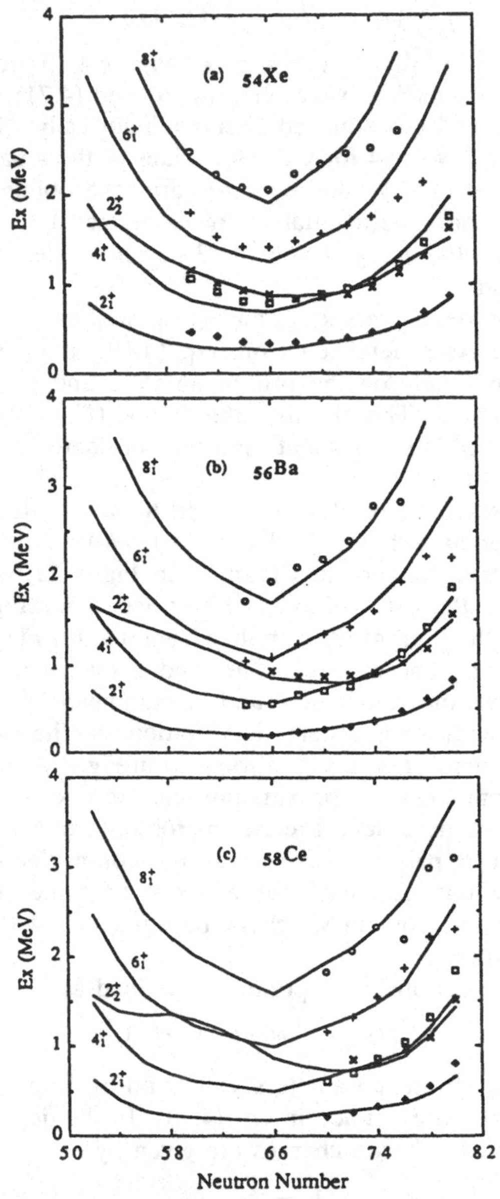

FIGURE 4.10. *Excitation energies of* (a) Xe, (b) Ba *and* (c) Ce *isotopes. Lines are calculations obtained by using* κ *and* χ_ν *in Fig. 4.8 and* ϵ *in 4.11 (see below). For the other parameters, see the text. Points are from experiment for the* 2_1^+ *(diamond),* 4_1^+ *(square),* 6_1^+ *(plus),* 8_1^+ *(circle) and* 2_2^+ *(cross) levels. The calculated results are from X. W. Pan (personal communication).*

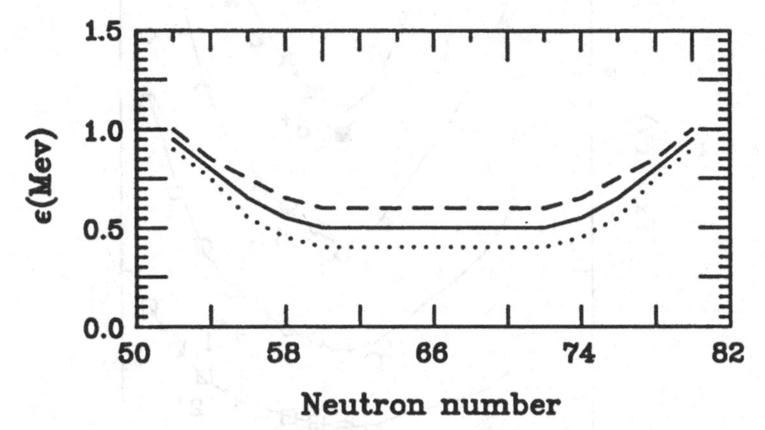

FIGURE 4.11. *Values of the parameter* ϵ *used for fitting to* Xe *(dashed),* Ba *(solid) and* Ce *(dotted) isotopes.*

$$\overset{\circ}{e}{}^{B}_{\tau} = e_{\tau} \langle D \| r^2 Y^{(2)}(\theta, \varphi) \| S \rangle / \sqrt{5} \tag{4.79}$$

where $\langle d \| d^{\dagger} s \| s \rangle = \sqrt{5}$ is used, and the e_{τ}'s are nucleon effective charges in the shell model. The boson effective charge includes matrix elements of r^2, because of which the dimension of e^B_{τ} is area (i.e., fm^2). Note that the N_{τ} dependence in eq. (4.78) is essentially equal to that in eq. (4.71). The $B(\mathrm{E2}; 0^+_1 \to 2^+_1)$ values calculated for the states in Fig. 4.10 are shown in Fig. 4.14 in comparison with experimental data. One finds growing E2 collectivity towards the middle of the shell. Since the boson effective charge decreases in this direction, the E2 enhancement is solely due to the increase of the boson number and the evolution of the collectivity in the wave function. In a purely phenomenological treatment of transitions, the e_{τ}'s can be adjusted for each nucleus.

4.11. F-SPIN

The proton boson and neutron boson are introduced in IBM-2. We shall discuss a symmetry associated with this proton-neutron degree of freedom (A. Arima, T. Otsuka, F. Iachello and I. Talmi, 1977; T. Otsuka, A. Arima and F. Iachello, 1978). This symmetry is called *F*-spin. While *F* spin has been discussed briefly in Section 2.4 from the group theoretical viewpoint, we discuss *F* spin in this section in more detail and from a somewhat more elementary level without using the Young tableaux.

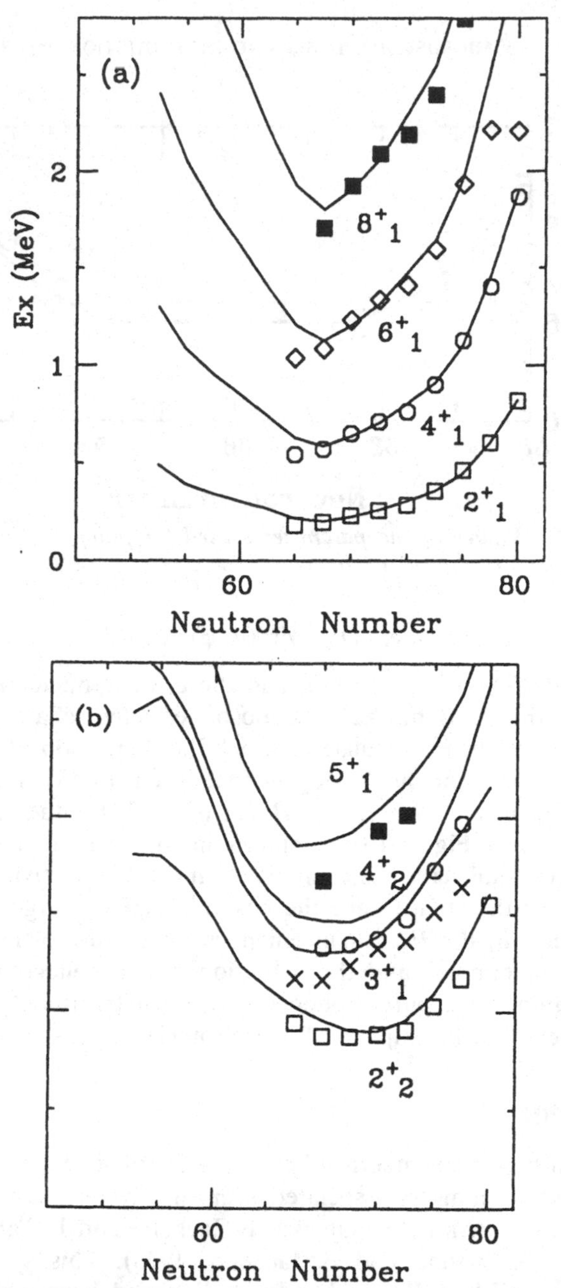

FIGURE 4.12. *Excitation energies of* Ba *isotopes. Lines are calculations obtained with the parameters in Fig. 4.13 (See below). Points are from experiment for* (a) 2_1^+ *(open square),* 4_1^+ *(open circle),* 6_1^+ *(open diamond) and* 8_1^+ *(closed square), and* (b) 2_2^+ *(open square),* 3_1^+ *(cross),* 4_2^+ *(open circle) and* 5_1^+ *(closed square) levels. See Puddu, Scholten and Otsuka (1980) for details.*

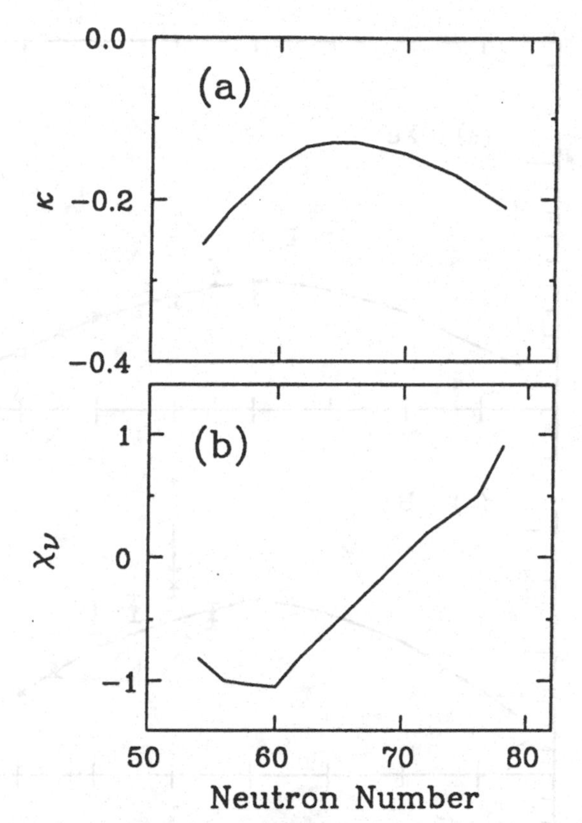

FIGURE 4.13. *Values of the parameters* (a) κ *and* (b) χ_ν *used for the fitting calculation shown in Fig. 4.12. See Puddu et al.* (1980) *for details.*

F-spin has similarities and differences compared to isospin. In the isospin scheme, each nucleon carries isospin $T = \frac{1}{2}$, and protons and neutrons are treated as the $T_z = \frac{1}{2}$ and $T_z = -\frac{1}{2}$ states, respectively. F-spin $F = \frac{1}{2}$ is assigned to each boson in IBM-2, although each boson corresponds to a pair of either neutrons or protons which carries the isospin $T = 1$. The proton boson is the $F_z = \frac{1}{2}$ state, while the neutron boson the $F_z = -\frac{1}{2}$ state. To recapitulate:

	F	F_z
proton bosons s_π and d_π	$\frac{1}{2}$	$\frac{1}{2}$
neutron bosons s_ν and d_ν	$\frac{1}{2}$	$-\frac{1}{2}$

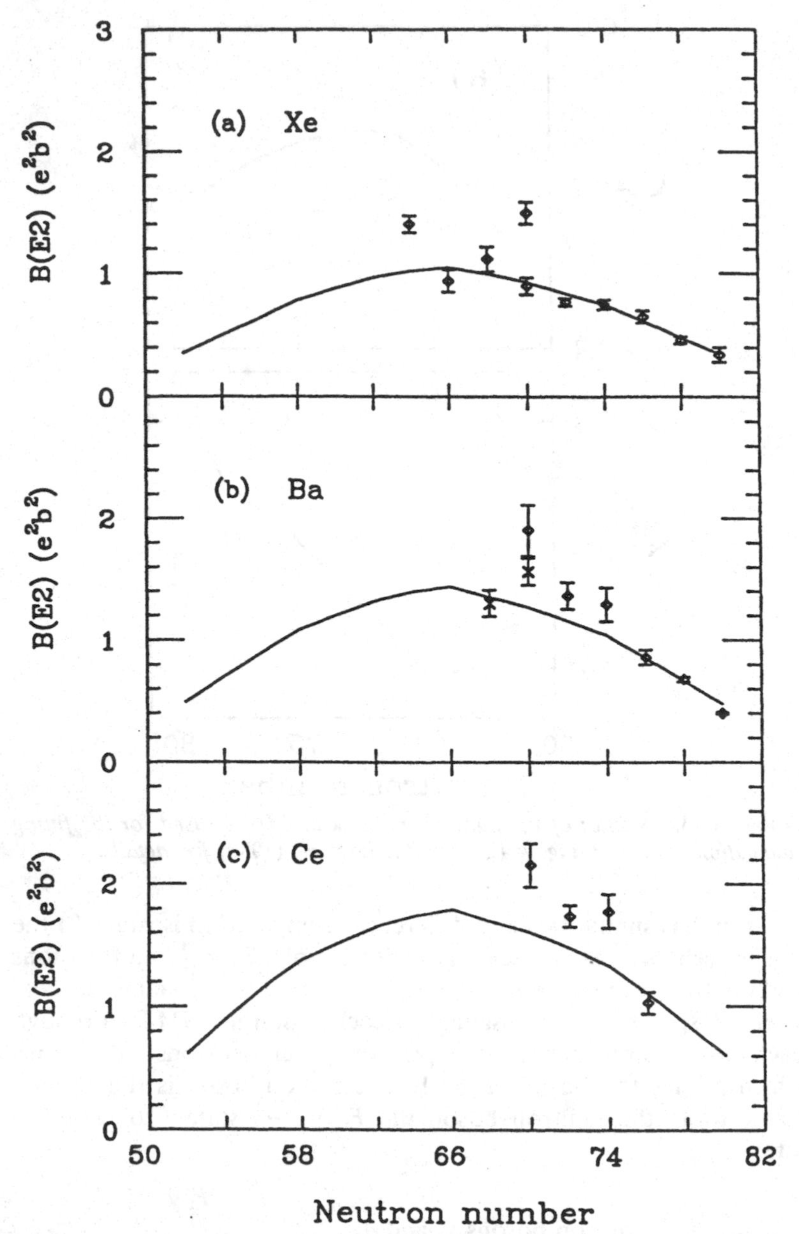

FIGURE 4.14. *B(E2; $0_1^+ \rightarrow 2_1^+$) values of* (a) Xe, (b) Ba *and* (c) Ce *isotopes. Points are experimental values compiled by Raman et al.* (1987).

F-spin is formulated as a scheme mathematically equivalent to the angular momentum algebra with the unit spin $\frac{1}{2}$, in the same way as the isospin is formulated. One can introduce the F-spin raising and lowering operators as well as its z-component;

$$F_+ \equiv s_\pi^\dagger \tilde{s}_\nu + (d_\pi^\dagger \cdot \tilde{d}_\nu), \tag{4.80}$$

$$F_- \equiv s_\nu^\dagger \tilde{s}_\pi + (d_\nu^\dagger \cdot \tilde{d}_\pi), \tag{4.81}$$

$$F_Z \equiv (N_\pi - N_\nu)/2. \tag{4.82}$$

These three operators satisfy the algebra of $SU(2)$, that is the algebra of the angular momentum (See Chapter 2, Section 2.2.2),

$$[F_\pm, F_Z] = \pm F_Z, \qquad [F_+, F_-] = 2F_Z. \tag{4.83}$$

It is clear that the system with N_π proton bosons and N_ν neutron bosons has a good quantum number $F_Z = (N_\pi - N_\nu)/2$. The magnitude of the F-spin is given as usual in terms of

$$(F \cdot F) = F_+F_- + F_Z^2 - F_Z. \tag{4.84}$$

This operator has eigenvalues $F(F + 1)$ similarly to the angular momentum, where F denotes a half integer or an integer depending on the boson number. The IBM-2 states can then be classified in terms of this quantum number. Since the F-spin obeys the same vector addition rules as the angular momentum, the total F-spin of the system with N_π proton bosons and N_ν neutron bosons varies from $F = |N_\pi - N_\nu|/2$ to $F = (N_\pi + N_\nu)/2$.

We shall demonstrate how IBM-2 states can be classified in terms of F-spin. For this purpose, we consider, as an example, two-boson systems. Table 4.1 lists the F-spin classification for two bosons.

It is clear that the $F = 1$ multiplet states $|s_\pi^2; 0^+)$, $|s_\pi s_\nu; 0^+)$ and $|s_\nu^2; 0^+)$ are connected by the F_+ and F_- operators. For instance, one can show $F_+|s_\pi s_\nu; 0^+) = \sqrt{2}|s_\pi^2; 0^+)$. Similar relations hold for other multiplets. The $F = 1$ states in Table 4.1 are symmetric with respect to the interchange between proton and neutron bosons. On the other hand, the $F = 0$ states are antisymmetric. We note that $|d_\pi d_\nu; L^+) = -|d_\nu d_\pi; L^+)$ for $L = 1$ and 3.

Denoting the total boson number $N \equiv (N_\pi + N_\nu)$, the possible maximum total F-spin, which is referred to as F_{max}, is given by $F_{max} = N/2 = (N_\pi + N_\nu)/2$ for systems with N_π proton bosons and N_ν neutron bosons. The total F spin of the states with $F = F_{max}$ is formed by adding the F spin of each boson successively, in such a

TABLE 4.1. *F-spin classification of states with two bosons.*

	F	F_Z
$\vert s_\pi^2; 0^+)$	1	1
$\vert s_\pi s_\nu; 0^+)$	1	0
$\vert s_\nu^2; 0^+)$	1	−1
$\vert d_\pi s_\pi; 2^+)$	1	1
$\frac{1}{\sqrt{2}}\{\vert d_\pi s_\nu; 2^+) + \vert d_\nu s_\pi; 2^+)\}$	1	0
$\vert d_\nu s_\nu; 2^+)$	1	−1
$\frac{1}{\sqrt{2}}\{\vert d_\pi s_\nu; 2^+) - \vert d_\nu s_\pi; 2^+)\}$	0	0
$\vert d_\pi^2; L^+)$ $(L = 0, 2, 4)$	1	1
$\vert d_\pi d_\nu; L^+)$ $(L = 0, 2, 4)$	1	0
$\vert d_\nu^2; L^+)$ $(L = 0, 2, 4)$	1	−1
$\vert d_\pi d_\nu; L^+)$ $(L = 1, 3)$	0	0

way that the F spin is maximally stretched in each step of this vector addition process. Any pair of bosons in the $F = F_{max}$ states therefore carries $F = 1$, otherwise $F = F_{max}$ cannot be formed. All $F = 1$ boson pairs are symmetric with respect to protons and neutrons as discussed above. The $F = F_{max}$ states are hence symmetric for exchanging any pair of proton and neutron bosons, and are referred to as totally symmetric states.

The totally symmetric states can be obtained by N_ν consecutive actions of F_- on the states with N proton bosons only,

$$|F = N/2, F_Z = (N_\pi - N_\nu)/2 \propto (F_-)^{N_\nu}|F = N/2, F_Z = N/2). \quad (4.85)$$

Since the state on the right-hand side consists of proton bosons only, this state is nothing but an IBM-1 state. Thus, the totally symmetric IBM-2 states and the corresponding IBM-1 states can be transformed into each other by F-spin raising or lowering operators, and are in the same F-spin multiplet of $F = F_{max}(= N/2 = (N_\pi + N_\nu)/2)$.

The states with $F < F_{max}$ contain $F = 0$ pairs of bosons which are antisymmetric as shown in Table 4.1. These states hence should be partially antisymmetric, and are called mixed-symmetry states. These states are not in IBM-1, and highlight the fact that this is a key difference between IBM-1 and IBM-2.

The IBM-2 Hamiltonian always commutes with the F_Z operator, since N_π and N_ν are conserved. If the Hamiltonian commutes also with $F_\pm$;

$$[F_\pm, H] = 0, \tag{4.86}$$

one can deduce, from the analogy to the relation between the usual angular momentum and the three-dimensional rotation, that the Hamiltonian is invariant with respect to "rotation" in F spin space. In this case, as easily understood again from this analogy, the Hamiltonian is an F-spin scalar, and conserves F spin. All eigenstates of the Hamiltonian then have good F values. If this F-scalar Hamiltonian favors coherent motion of proton and neutron bosons as expected in practice, the lowest eigenstates are totally symmetric (i.e., $F = F_{\max}$). Mixed-symmetry states are excited states, and the lowest lying mixed-symmetry states have $F = F_{\max} - 1$, because lower F values imply less proton-neutron symmetry.

In realistic situations, (for instance, with the Hamiltonian of eq. (4.76)), the commutation relation in eq. (4.86) does not hold. However, there still remain mechanisms which lower totally symmetric states and/or raise mixed-symmetry states. The most important mechanism for this effect originates in a term in the $(\mathbf{Q}_\pi \cdot \mathbf{Q}_\nu)$ interaction,

$$-\kappa\{(d_\pi^\dagger s_\pi \cdot s_\nu^\dagger \tilde{d}_\nu) + (d_\nu^\dagger s_\nu \cdot s_\pi^\dagger \tilde{d}_\pi)\}. \tag{4.87}$$

In order to see this mechanism more precisely, we shall consider the matrix elements of this term with respect to the following two states in Table 4.1:

$$\varphi_s = \frac{1}{\sqrt{2}}\{|d_\pi s_\nu; 2^+) + |d_\nu s_\pi; 2^+)\} \tag{4.88}$$

$$\varphi_a = \frac{1}{\sqrt{2}}\{|d_\pi s_\nu; 2^+) - |d_\nu s_\pi; 2^+)\}. \tag{4.89}$$

Straightforward calculation shows

$$(\varphi_s| -\kappa\{(d_\pi^\dagger s_\pi \cdot s_\nu^\dagger \tilde{d}_\nu) + (d_\nu^\dagger s_\nu \cdot s_\pi^\dagger \tilde{d}_\pi)\}|\varphi_s) = -\kappa, \tag{4.90}$$

and

$$(\varphi_a| -\kappa\{(d_\pi^\dagger s_\pi \cdot s_\nu^\dagger \tilde{d}_\nu) + (d_\nu^\dagger s_\nu \cdot s_\pi^\dagger \tilde{d}_\pi)\}|\varphi_a) = \kappa. \tag{4.91}$$

Since κ is positive, the $F = 1$ state φ_s comes down, whereas the $F = 0$ state φ_a goes up. We thus separate the totally symmetric state from the mixed-symmetry state, and obtain a symmetry energy with respect to the F spin. More explicit symmetry energy terms, called the Majorana force, will be discussed in the next section. We comment that these symmetry-energy terms favor states with larger F values, whereas the symmetry energy for the isospin favors shell-model states with smaller T values. The nucleon wave functions must be antisymmetric and, therefore, by making the isospin part antisymmetric (i.e., favoring lower T values), the other part can be symmetric and increases the binding energy. In the case of bosons, the boson wave function is symmetric, and hence, by making the F-spin part symmetric (i.e., favoring higher F values), the other part becomes also symmetric. This differentiates favourable values of F and T.

We further note that, if $\chi_\pi = \chi_\nu$ holds in eq. (4.76), another $F = 1$ state $|d_\nu d_\pi; 2^+)$ is coupled to the state φ_s which is $F = 1$ as well. By this mixing, an $F = 1$ level is pushed down, whereas no such partner exists for the state φ_a. There are similar mechanisms in general. Most of the lowest states are thus dominated by totally symmetric components with respect to the F spin, although the standard IBM-2 Hamiltonian in eq. (4.76) is not an F-scalar.

We shall consider F-spin properties of the IBM-2 Hamiltonian. The single-particle energy term of the IBM-2 Hamiltonian consists of a boson creation operator and a boson annihilation operator. Each boson creation or annihilation operator carries $F = \frac{1}{2}$. By the F-spin algebra, the single particle energy can be decomposed into $F = 0$ (F-scalar) and $F = 1$ (F-vector) terms. The F-scalar operator does not change F spin. The F-vector carries $F = 1$, and thereby can change the F spin up to $\Delta F = 1$. The general single d-boson energy term is

$$H_1 = \epsilon_\pi \hat{N}_{d_\pi} + \epsilon_\nu \hat{N}_{d_\nu}. \tag{4.92}$$

which can be rewritten as

$$H_1 = \tfrac{1}{2}(\epsilon_\pi + \epsilon_\nu)(\hat{N}_{d_\pi} + \hat{N}_{d_\nu}) + \tfrac{1}{2}(\epsilon_\pi - \epsilon_\nu)(\hat{N}_{d_\pi} - \hat{N}_{d_\nu}). \tag{4.93}$$

From the commutation relation

$$[F_\pm, \hat{N}_{d_\pi} + \hat{N}_{d_\nu}] = 0, \tag{4.94}$$

one sees that $\hat{N}_{d_\pi} + \hat{N}_{d_\nu}$ is F-spin invariant, and is an F scalar operator. If one operates with $(\hat{N}_{d_\pi} - \hat{N}_{d_\nu})$ on the state in eq. (4.89), one obtains only the state in eq. (4.88). The former is an $F = 0$ state, while the latter an $F = 1$ state. This operator therefore should change the F spin, and must be F vector. We thus assign the F-scalar and F-vector components to the first and second terms, respectively, in eq. (4.93).

The boson-boson interaction is comprised of two creation operators and two annihilation operators. It is hence decomposed into F-scalar, F-vector and F-tensor terms. The F-tensor term carries $F = 2$, and can change the F spin up to $\Delta F = 2$. In general, the IBM-2 Hamiltonian in eq. (4.76) contains all of these terms. However, because totally symmetric components are dominant in the lowest states and therefore F-spin mixing terms should play minor roles for these states, the energy levels can be reproduced to a good approximation by an appropriately chosen F-scalar Hamiltonian as far as these lowest states are concerned. If the Hamiltonian is an F-scalar, then the structure does not depend on F_Z, and it can be described by the IBM-1. Thus, the IBM-1 works for the lowest states, while it is the IBM-2 that can be derived microscopically.

For an F scalar IBM-2 Hamiltonian with constant parameters, the energy levels depend on the F-spin magnitude F, but not on $F_Z = (N_\pi - N_\nu)/2$. In such a case, F-spin multiplets can arise in real nuclei. The F-spin multiplet implies a group of energy levels that show up identically in the spectra of several nuclei and can be assigned the same F value. The F spin multiplets can take various F values in principle, although primarily those of $F = F_{max}$ have been reported so far. In this case, the F-spin multiplet appears as the lowest states in nuclei of the same value of $N = N_\pi + N_\nu$. The F-spin multiplet is analogous to the isospin multiplet, for instance, $T = 1$ states in ^{14}O–^{14}N–^{14}C as shown in Fig. 4.15. The nucleon-nucleon interaction excluding the electromagnetic force is isoscalar and conserves the isospin. The F-spin analog to this is the F-scalar Hamiltonian, which seems to appear as an effective Hamiltonian for the lowest states, as discussed above. An F spin multiplet seen experimentally, as reported by von Brentano et al. in 1985, is shown in Fig. 4.16 for $N = 13$ from ^{158}Dy.

At higher excitation energies, the states dominated by $F = F_{max} - 1$ components appear. We shall turn to this subject in the next section.

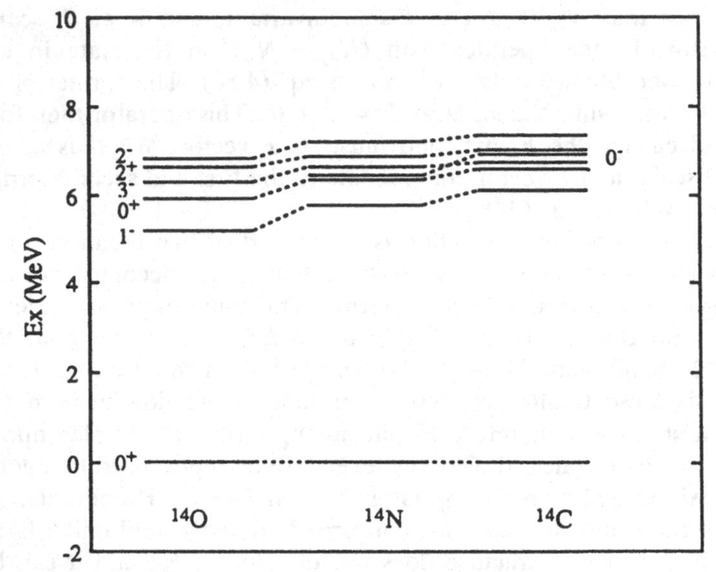

FIGURE 4.15. *Isospin multiplet of* A = 14 *isobars with* ^{14}O (T_Z = 1), ^{14}N (T_Z = 0) *and* ^{14}C (T_Z = −1). *Observed excitation energies of* T = 1 *levels are shown.*

4.12. SCISSORS MODE AND OTHER MIXED-SYMMETRY STATES

The most distinct mixed-symmetry states are 1^+ states, because no 1^+ state is allowed in IBM-1. The lowest 1^+ state in usual IBM-2 calculations corresponds to the scissors mode which has been observed by Richter and his collaborators in 1984. The scissors mode is a contra-oscillation in relative angle between proton and neutron deformed ellipsoids as shown schematically in Fig. 4.17. The 1^+ state of the scissors mode is a one-phonon excitation of this contra-oscillation from the ground state. As a rotational band is built on the ground band, a rotational band is anticipated on the scissors 1^+ state, although it has not been observed so far.

We here introduce an operator

$$J_v^B = J_\pi^B - J_\nu^B \tag{4.95}$$

where J_π^B and J_ν^B denote, respectively, the proton and neutron boson angular momentum operators defined as

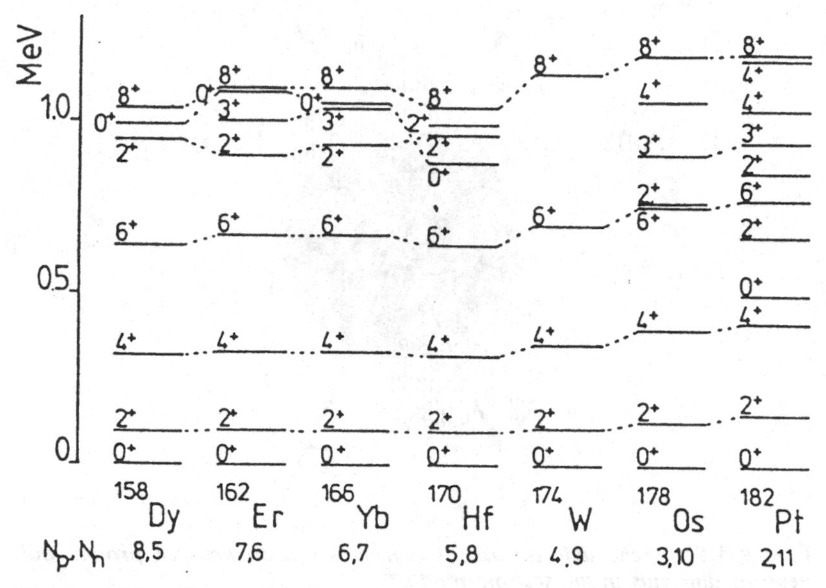

FIGURE 4.16. *Observed excitation energies of the F-spin multiplet of N = 13 equal-boson-number nuclei. Taken from Lipas, von Brentano and Gelberg (1990).*

$$J_\nu^B = \sqrt{10}\,[d_\tau^\dagger \tilde{d}_\tau]^{(1)}, \qquad \tau = \pi \text{ or } \nu. \tag{4.96}$$

Since the angular momentum operator causes infinitesimal rotation, the J_ν^B operator rotates the proton bosons and the neutron bosons in opposite angles. This is nothing but the contra-rotation mentioned above. Note that the operator in eq. (4.96) has the property of a vector and carries angular momentum unity. The total angular momentum operator is given by

$$J_T^B = J_\pi^B + J_\nu^B. \tag{4.97}$$

We shall consider the scissors mode 1^+ state in terms of the states in Table 4.1. Although the system of $N_\pi = N_\nu = 1$ is too simple to study the scissors mode in a realistic situation, the basic features can be seen. The general 0^+ ground state of $N_\pi = N_\nu = 1$ is given by

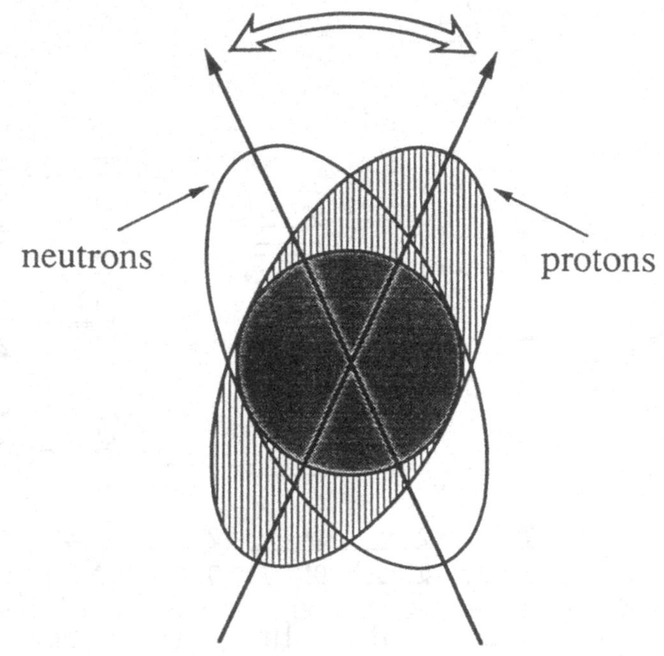

FIGURE 4.17. *Schematic picture of contra-oscillation between proton and neutron ellipsoids in the scissors mode.*

$$\varphi_g = x|s_\pi s_\nu; 0^+) + y|d_\pi d_\nu; 0^+) \quad (4.98)$$

where x and y are amplitudes ($x^2 + y^2 = 1$). This state is symmetric with respect to the interchange between the proton boson and the neutron boson, and has $F = 1$. We then act with the J_ν^B operator in eq. (4.95) on this state. Since the J_ν^B operator is a d-boson one-body operator, it simply annihilates the $|s_\pi s_\nu; 0^+)$ component. The J_ν^B operator carries the angular momentum one as pointed out above, and the state in eq. (4.98) has the angular momentum zero. Hence the action of the J_ν^B operator on the state in eq. (4.98) should produce a state of the angular momentum one. In fact, one can show that

$$J_\nu^B\{x|s_\pi s_\nu; 0^+) + y|d_\pi d_\nu; 0^+)\} \propto y|d_\pi d_\nu; 1^+) \quad (4.99)$$

where $|d_\pi d_\nu; 1^+)$ is the 1^+ state in Table 4.1. We now see that the proton-neutron contra-rotation (i.e., scissors excitation) on the ground state creates this 1^+ state. The left-hand side of eq. (4.99) alters its sign if the proton and neutron bosons are interchanged, which is consistent with the antisymmetric structure of the state $|d_\pi d_\nu; 1^+\rangle$. The J_T^B operator shows that

$$J_T^B\{x|s_\pi s_\nu; 0^+\rangle + y|d_\pi d_\nu; 0^+\rangle\} = 0, \tag{4.100}$$

because this operator is the total angular momentum which is zero for this state. Comparing eqs. (4.99) and (4.100), one can see the importance of the anti-phase combination in eq. (4.95).

The physical process which does the job of the operator J_ν^B is the magnetic dipole (M1) transition. The magnetic dipole operator for nucleons has the form

$$M = g_\pi^s S_\pi + g_\pi^\ell L_\pi + g_\nu^s S_\nu + g_\nu^\ell L_\nu, \tag{4.101}$$

where the g's are nucleon g-factors, and s and ℓ denote, respectively, spin and orbital angular momenta; M is expressed in nuclear magneton units ($e\hbar/2Mc$). The proton part of the right-hand side of eq. (4.101) is mapped onto the proton boson magnetic dipole operator, $g_\pi^B J_\pi^B$ with g_π^B being the proton boson effective g-factor. This boson g-factor is fixed by the OAI mapping as

$$\begin{aligned}\langle S_\pi^{N_\pi-1} D_\pi^1; J_\pi = 2\|g_\pi^s S_\pi + g_\pi^\ell L_\pi\|S_\pi^{N_\pi-1} D_\pi^1; J_\pi = 2\rangle \\ = (s_\pi^{N_\pi-1} d_\pi^1; J_\pi = 2\|g_\pi^B J_\pi^B\|s_\pi^{N_\pi-1} d_\pi^1; J_\pi = 2).\end{aligned} \tag{4.102}$$

In collective pairs of nucleons like S and D, the total intrinsic spin of two nucleons tends to be $S \sim 0$ (i.e., spin saturation). In fact, by coupling two nucleons to $S = 0$, the antisymmetry of the wave function is satisfied by the spin part of the wave function, and one can maximize the symmetry in the orbital part, resulting in maximum energy gain from the spatial overlap. In the lowest approximation, therefore, the spin matrix elements on the left hand side of eq. (4.102) can be neglected. In other words, for collective states, one can use an approximation such as $J_\pi = L_\pi$. Combining this with the identities

$$\begin{aligned}(s_\pi^{N_\pi-1} d_\pi^1; J = 2\|J_\pi^B\|s_\pi^{N_\pi-1} d_\pi^1; J = 2) \\ = (d_\pi\|J_\pi^B\|d_\pi) = \sqrt{30},\end{aligned} \tag{4.103a}$$

and

$$\begin{aligned}\langle S_\pi^{N_\pi-1} D_\pi^1; J = 2\|J_\pi\|S_\pi^{N_\pi-1} D_\pi^1; J = 2\rangle \\ = \langle D_\pi\|J_\pi\|D_\pi\rangle = \sqrt{30},\end{aligned} \tag{4.103b}$$

one obtains

$$g_\pi^B = g_\pi^\ell. \tag{4.104}$$

The above argument can be applied to neutrons. By dropping the spin terms, the boson magnetic dipole operator gives rise to

$$M^B = g_\pi^B J_\pi^B + g_\nu^B J_\nu^B \approx g_\pi^\ell J_\pi^B + g_\nu^\ell J_\nu^B. \tag{4.105}$$

It is known that $g_\pi^\ell \sim 1$ and $g_\nu^\ell \sim 0$, which can be understood naturally in terms of the current causing the magnetic moment. One therefore finally obtains a simple relation

$$M^B \approx J_\pi^B. \tag{4.106}$$

We now rewrite eq. (4.106) as

$$\begin{aligned} M^B &\approx \tfrac{1}{2}\,(J_\pi^B + J_\nu^B) + \tfrac{1}{2}\,(J_\pi^B - J_\nu^B) \\ &= J_T^B + J_\nu^B. \end{aligned} \tag{4.107}$$

The first term on the right-hand side is the total angular momentum, which is diagonal and causes no transition. The second term has been discussed in connection to the scissors mode. One thus sees that the magnetic dipole operator does indeed excite the scissors mode.

The mixed-symmetry 2^+ state is also expected to be observed. We shall consider this state again in terms of the states in Table 4.1. In spite of the simplicity of $N_\pi = N_\nu = 1$, the basic features can be studied. In this case, the mixed-symmetry 2^+ state is the state

$$\varphi_a = \frac{1}{\sqrt{2}}\,\{|d_\pi s_\nu;\, 2^+) - |d_\nu s_\pi;\, 2^+)\} \tag{4.108}$$

which appeared in eq. (4.89). The mixed-symmetry 2^+ state is interpreted as an out-of-phase oscillation of the proton surface and the neutron surface, whereas the usual surface oscillation is in phase between protons and neutrons. The state in eq. (4.108) can be excited by the E2 operator in eq. (4.77) from the state $|s_\pi s_\nu;\, 0^+)$ which is included in Table 4.1 and is the ground state in the limit of large ϵ's in eqs. (4.64)–(4.65). The excitation matrix element is given by

$$(\varphi_a|e_\pi^B \mathbf{Q}_\pi + e_\nu^B \mathbf{Q}_\nu|\varphi_g) \propto (e_\pi^B - e_\nu^B) \tag{4.109}$$

where $|\varphi_g) = |s_\pi s_\nu;\, 0^+)$. Note that this is proportional to the difference of proton and neutron boson effective charges. The values of e_π^B and e_ν^B are evaluated by the OAI mapping procedure as

$$e_\tau^B = e_\tau \langle S_\tau^{N_\tau - 1} D_\tau^1; J = 2 \| Q_\tau \| S_\tau^{N_\tau}; J = 0 \rangle / (\sqrt{5} \times N_\tau), \qquad (4.110)$$

$$\tau = \pi \text{ or } \nu,$$

where the e's are effective charges in the shell model, Q was introduced in eq. (4.67) as the second quantized form of the one-body operator $r^2 Y^{(2)}(\theta, \varphi)$. Since nucleon effective charges e_π and e_ν are not very different from each other due to core polarization, the difference $e_\pi^B - e_\nu^B$ in eq. (4.109) is not large. Therefore, the state in eq. (4.108) is only weakly excited by E2 transition. On the other hand, the M1 transition connects the state in eq. (4.88) to this state, and seems to be a reasonable probe for the mixed-symmetry 2^+ state.

As far as mixed-symmetry states are concerned, if one surveys various combinations of a proton boson and a neutron boson, there should be at least one such pair that belongs to the $F = 0$ states in Table 4.1; $(1/\sqrt{2}) \{|d_\pi s_\nu; 2^+\rangle - |d_\nu s_\pi; 2^+\rangle\}$ or $|d_\pi d_\nu; L^+\rangle$ ($L = 1, 3$). The structure of the mixed-symmetry states are therefore characterized by a mixture of these $F = 0$ two-boson states. The energies of the states of $F = 0$ in Table 4.1 can be shifted by the Majorana interaction defined as

$$V_M = \xi_1 ([d_\pi^\dagger d_\nu^\dagger]^{(1)} \cdot (\tilde{d}_\nu \tilde{d}_\pi]^{(1)})$$
$$+ \xi_2 \tfrac{1}{2} \{(d_\pi^\dagger s_\nu^\dagger - d_\nu^\dagger s_\pi^\dagger) \cdot (\tilde{d}_\pi s_\nu - \tilde{d}_\nu s_\pi)\} \qquad (4.111)$$
$$+ \xi_3 ([d_\pi^\dagger d_\nu^\dagger]^{(3)} \cdot [\tilde{d}_\nu \tilde{d}_\pi]^{(3)})$$

where the ξ's are parameters. By using this interaction, one can adjust phenomenologically the excitation energy of mixed-symmetry states. The origin of the Majorana interaction is still under investigation.

4.13. GINOCCHIO MODEL

We have discussed the microscopic basis of the IBM in terms of the S and D pairs. Although small admixtures of other pairs can be handled in perturbation or a similar method, the wave functions of collective states should be dominated by the S and D pairs. It is therefore of much interest to look for a fermion system which is precisely closed within a subspace constructed by the S and D pairs only. A trivial example is the quasi-spin group shown in eq. (4.41), where the S pair generates a closed algebra. Ginocchio found a non-trivial example, and this is known as the Ginocchio model. In this model, the pseudo-spin, $\tilde{\mathbf{s}}$, and pseudo-orbital angular momentum, $\tilde{\boldsymbol{\ell}}$, are introduced. The total single-particle angular momentum $\mathbf{j}$ is given by the vector coupling $\mathbf{j} = \tilde{\mathbf{s}} + \tilde{\boldsymbol{\ell}}$. The pseudo-spin, $\tilde{\mathbf{s}}$, is assumed to take a value of $\frac{3}{2}$, instead of $\frac{1}{2}$ for the usual spin. The pseudo-orbital angular momentum, $\tilde{\boldsymbol{\ell}}$, takes various values. In creating nucleon pairs, the total pseudo-spin of two nucleons can be coupled to the magnitude $\tilde{\mathcal{S}} = 0 \sim 3$. The wavefunction of two nucleons of this pair should be antisymmetric, because we are considering a pair of two identical nucleons (i.e., the two nucleons are either both neutrons or both protons). In order to fulfill this condition, Ginocchio assumed that the wavefunction is given by the product of the pseudo spin part and the pseudo-orbital angular momentum part, and that the pseudo-spin part is antisymmetric while the pseudo-orbital angular momentum part symmetric. This assumption was made in analogy to the fact that the spatial overlap of the two nucleons is larger with the symmetric wavefunction for the (real) orbital angular momentum than with the antisymmetric wavefunction. Such a symmetric wavefunction is obtained by coupling the orbital angular momenta of the two nucleons to an even integer. Although the pseudo-orbital angular momentum does not refer to the wavefunction in the coordinate space and therefore one cannot directly apply this argument to the pseudo-orbital angular momentum, we simply use the same argument.

In order that the pseudo-spin part of the wavefunction is antisymmetric, the magnitude of the total pseudo-spin of the two nucleons should be $\tilde{\mathcal{S}} = 0$ or 2. This can be understood in terms of the structure of the spin singlet ($S = 0$) wavefunction of real spin. We now introduce another analogy to real nuclei. If the nucleon-nucleon interaction acting on the two nucleons of the pair is of short range, as expected usually, the most favored total (real) orbital angular momentum is zero. Thus, extending this argument to the present pseudo-system, we assume that the total pseudo-orbital angular momenta of the two nucleons is $\tilde{\mathcal{L}} = 0$. Thus, we end up with the system of $\tilde{\mathcal{S}} = 0$ or 2 and $\tilde{\mathcal{L}} = 0$. One can then create only pairs of total angular momentum $J = 0$ and 2, which can be related, respectively, to the S and D pairs of the IBM. These pairs constitute a closed algebra, and the states constructed by these pairs consist of eigenstates for a set of Hamiltonians comprised of properly defined monopole and quadrupole pairing interactions. This is the outline of the Ginocchio model. It is remarkable that there exists a fermion scheme which is closed exactly with "*S*" and "*D*" pairs only. On the other hand, it is unlikely that there is an origin in realistic multi-nucleon systems for this specific truncation in terms of the pseudo-spin and pseudo-orbital angular momenta. Owing to its exact analytic solutions, however, the Ginocchio model still provides us with an excellent testing ground for many-body theories. Finally, it should be mentioned that the Ginocchio model has the same group theoretical limit as the $O(6)$ limit of the IBM, although one is a fermionic model and the other is a bosonic one. This might indicate certain profound reason for the presence of $O(6)$ structure. The concept of pseudo-spin is further discussed in Chapter 7.

4.14. INTERACTING BOSON MODELS 3 AND 4

The Interacting Boson Model is normally applied to nuclei with masses larger than about 80. In these nuclei, valence protons and valence neutrons are mainly in different orbits, and therefore the isospin does not give us any information. On the other hand, in lighter nuclei, protons and neutrons may occupy the same orbits, and the isospin plays a decisive role. To accommodate the isospin into the IBM, IBM-3 and IBM-4 were introduced, respectively, by Elliott and White (1980), and Elliott and Evans (1981). In these versions of IBM, bosons corresponding to pairs consisting of a proton and a neutron are included. In IBM-3, the bosons are assumed to carry implicitly the isospin $T = 1$, while the bosons carrying implicitly the isospin $T = 0$ are included in IBM-4 in addition to the "$T = 1$" bosons mentioned above. The IBM-3 and -4 clarify logical relationship between the IBM and the shell model with the isospin, rather than present new quantitative descriptions of light nuclei.

4.15. CONCLUDING REMARKS

In summary, we have seen, in the first half of this Chapter, the basic concepts on which the microscopic picture of the IBM can be built. In the second half of this Chapter, the proton-neutron IBM, i.e. IBM-2, was discussed with applications to real nuclei and formulation of the proton-neutron symmetry in the boson space, i.e. F-spin.

At this stage, I would like to mention that the purpose of the microscopic study is twofold. The first is to prove that the IBM can be derived in principle from underlying multi-nucleon systems. The second is to compute parameters in the IBM Hamiltonian and operators from the nucleonic degrees of freedom. These two are related to each other, but are also different. The former is a more basic issue and does not depend on details of the nucleon-nucleon interaction, whereas the latter is sensitive to the interaction. For the first purpose, we have learned that the IBM can be formulated starting from collective pairs (S and D) of nucleons. This study leads us further to the introduction of the IBM-2. For the second purpose, we have seen that the degenerate-orbit approximation already provides us with a reasonable description of variations of structure of collective states as functions of proton and neutron numbers. Thus, these two purposes have been fulfilled to a certain extent. It is, however, true that neither of them has been completed yet. Actually, considerable efforts are being made for them, and the microscopic studies of the IBM remain as an active field of nuclear physics.

I here would like to compare the study of the microscopic foundation for the IBM with that for the shell model. In order to grant the shell model a microscopic basis, one usually starts from the fundamental nucleon-nucleon interactions (say those obtained from meson theories). After a number of approximations and technical treatments, one can end up with a model similar to the shell model used empirically. Even such elaborate work can produce only partial (microscopic) justification of the shell model. For instance, core polarization effects in effective two-body interactions have been evaluated extensively only for the cases with two particles (or holes) outside the closed core, and we just assume that the same renormalization process persists in cases with many valence nucleons or holes. Thus, the complete derivation of the shell model is still to come. On the other hand, thc validity of the shell model has been well established for empirical usage. In fact, it is widely recognized that the shell model works well as an empirical framework, and the shell model is used even to test fundamental symmetries. Moreover, because of partial justification from meson theories and empirical successes in many applications, it is believed by a large number of physicists that the derivation of the shell model as a formalism should be possible, in principle, while the two-body matrix elements cannot be calculated presently to a great accuracy from a more profound basis.

An analogous situation seems to hold for the IBM. We have learned that the bosons in the IBM can be introduced from nucleon pairs. This picture provides us with some insight of the basic relation to the underlying nuclear structure and a guideline for the values of the parameters in the boson Hamiltonian. This guideline is not perfect, but can produce basic patterns of collective level schemes as mentioned above. Clearly, the present achievement of the microscopic study of the IBM has not reached the level of the sophistication of the shell model, and more work is needed. However, it is not very productive to refrain, because of only partial success, from the recognition of the IBM as a useful scheme until the complete derivation is presented eventually.

Among remaining major problems in the microscopic study of IBM, the most important ones are (i) *best* determination of the intrinsic structure of S and D pairs, (ii) fermion–boson mapping suitable for deformed nuclei, (iii) improved renormalization of various effects. Although (some of) these problems have been discussed by Scholten, Zirnbauer and Brink, von Egmond and Alaart, Barrett, Pittel, Duval, Druce and their collaborators, Otsuka and Yoshinaga, etc. (see a more advanced textbook for their references), we are still not very close to the goal. For (i), it should be mentioned that the structure of S and D pairs can be determined by variational methods so that one obtains the optimum description. Although there has been recent progress, such discussions are not included, because one must utilize many-body theories, which are beyond the scope of this introductory textbook. The microscopic study of the IBM should be extended to the Interacting Boson–Fermion Model. This model will be discussed in Chapter 5, and it is not appropriate to discuss its microscopic basis before the model itself is shown.

In the early stage of the microscopic foundation of the IBM, Talmi utilized the double-commutator approach in the generalized seniority scheme. This approach is instructive for understanding the effect of the interaction between like nucleons, i.e., the seniority conserving force. Because Talmi's generalized seniority is designed for seniority-conserving nuclei but the proton-neutron interaction breaks the seniority, Talmi's generalized seniority has not been mentioned in this Chapter, despite its relevance to IBM. There are many works concerning boson expansions, Belyaev–Zelevinsky, Marumori, etc. One can refer to the book by Bonatsos (1988).

The problems about the proton-neutron symmetries in low-lying collective states have been highlighted by *F*-spin in the IBM-2. The discovery of the scissors mode by Richter and his collaborators (1984, 1991) is a good example. There will be many new developments in this direction, in particular in the relation to magnetic properties.

ACKNOWLEDGEMENTS

I would like to thank Adrian Gelberg for a number of useful discussions and for his careful reading of this chapter. I am also very grateful to Akito Arima and Francesco Iachello for numerous discussions and collaborations over many years.

BIBLIOGRAPHY

A. Arima and F. Iachello, The interacting boson model, in *Advances in Nuclear Physics* (Plenum, New York, 1984), Vol. 13, p. 139. First comprehensive review on the IBM up to 1983. A compact review of the microscopic basis of the IBM and an introduction to IBM-2, but very little about *F*-spin.

A. Arima, T. Otsuka, F. Iachello and I. Talmi, "Shell model description of interacting bosons," *Phys. Lett.* **66B**, 205 (1977). The prototype of the totally symmetric states in IBM-2 was presented based on the discussion by the generalized seniority.

D. Bohle *et al.*, "New magnetic dipole excitation mode studied in the heavy deformed nucleus ^{156}Gd by inelastic electron scattering," *Phys. Lett.* **137B**, 27(1984); A Richter, "Electron scattering and elementary excitations," *Nucl. Phys.* **A522**, 139c (1991). The scissors mode was discovered.

A. Bohr and B. R. Mottelson, *Nuclear Structure* (Benjamin, New York, 1969, 1975), Vols. 1 and 2. An excellent and practical introduction to the shell structure of nuclei (Chapter 3, Vol. 1), and the most extensive description of nuclear deformations (Chapters 4–6, Vol. 2).

D. Bonatsos, *Interacting Boson Models of Nuclear Structure* (Oxford, Clarendon Press, Oxford, 1988). A review on IBM. Enormous reference list (11 pages) of papers related to the microscopic basis of the IBM.

R. Casten, *Nuclear Structure from a Simple Perspective* (Oxford, New York, 1990). A compact and easy-access introduction to nuclear collective motion, nuclear deformation, Interacting Boson Model and other models.

R. F. Casten, *Phys. Rev. Lett.* **54**, 1991 (1985). The first paper on the Casten plot.

J. P. Elliott and J. A. Evans, "An intrinsic spin for interacting bosons," *Phys. Lett.* **101B**, 216 (1981). The original paper for IBM-4.

J. P. Elliott and A. P. White, "An isospin invariant form of the interacting boson model," *Phys. Lett.* **97B**, 169 (1980). The original paper for IBM-3.

A. de Shalit and I. Talmi, *Nuclear Shell Theory* (Academic Press, New York, 1963). An introduction to the angular momentum algebra and seniority.

A. de Shalit and H. Feshbach, *Theoretical Nuclear Physics* (John Wiley, New York, 1974), Vol. 1. A general and (comparatively) modern introduction to nuclear stucture and nuclear many-body problems.

F. Iachello and A. Arima, *The Interacting Boson Model* (Cambridge, 1987). An extensive description of mathematical aspects of IBM.

F. Iachello and I. Talmi, "Shell-model foundations of the interacting boson model," *Rev. Mod. Phys.*, **59**, 339 (1987). First specialized review of the microscopic basis of IBM. A nice description of basic concepts. Less extensive than this Chapter.

P. O. Lipas, P. von Brentano and A. Gelberg, "Proton-neutron symmetry in boson models of nuclear structure, *Rep. Prog. Phys.* **53**, 1355 (1990). First review specialized in IBM-2. Some emphasis on magnetic properties and *F*-spin multiplets.

T. Otsuka, A. Arima, F. Iachello and I. Talmi, "Shell model description of interacting bosons," *Phys. Lett.* **76B**, 139 (1978). The IBM-2 description covering a full major shell was presented.

T. Otsuka, A. Arima and F. Iachello, "Nuclear shell model and interacting bosons," *Nucl. Phys.* **A309**, 1 (1978). The original paper where the OAI mapping was presented in detail. *F*-spin operators were shown.

G. Puddu, O. Scholten and T. Otsuka, *Nucl. Phys.* **A348**, 109 (1980). The first paper on systematic IBM-2 phenomenology.

S. Raman *et al.*, *At. Data and Nucl. Data Tables* **36**, 1 (1987). A compilation of experimental $B(\mathrm{E2}: 0_1^+ \rightarrow 2_1^+)$ values.

N. Yoshinaga, "Present status of the microscopic foundations of the interacting boson model," *Nucl. Phys.* **A522**, 99 (1991). A review talk discussing microscopic studies of the IBM up to 1990.

CHAPTER 5

The IBFM and Bose–Fermi Symmetries

DAVID D. WARNER

5.1. CORE-PARTICLE COUPLING

In the preceding chapters, the discussion focussed on the application of algebraic models in general, and the IBM in particular, to the collective modes in even-even nuclei. The possible coupling of these modes to single-particle excitations was not considered because, in even-even nuclei, such excitations occur at higher energies owing to the need to overcome the attractive pairing interaction. However, in order to extend the techniques to the description of odd-even or odd-odd nuclei, it is necessary to account for the coupling of the nucleonic and collective modes of motion. The simplest extension of a collective model to treat a nucleus with odd mass number ($A + 1$) would describe the states of such a nucleus as stemming from the excitations of a *core,* represented by one of the neighbouring even-even nuclei with mass A or $A + 2$, coupled to the motion of a single valence nucleon. Under the usual adiabatic approximation, we assume that the collective Hamiltonian H_c, describing the slow surface oscillations of the nuclear shape, can be separated from the much faster motion of the valence particle. Moreover, this particle motion is described in terms of a potential which is directly related to the nuclear shape and hence to the collective variables of the core. In this approximation, the change in shape of the nucleus is small during the time a particle takes to complete several cycles of the orbit and the shape variables in the single-particle potential can be taken as fixed, representing average values of the "mean field" generated by all the other nucleons in the core. This assumption should be valid for the majority of problems encountered in nuclear structure, with the exception of phenomena involving very high spin states. Thus we attempt to separate the overall Hamiltonian into two parts, viz:

$$H = H_c + H_p \tag{5.1}$$

with H_c describing the collective, coherent motion of the nucleus as a whole while H_p describes the single-particle motion in some average potential. The effects of the residual pairing interaction can be included by treating the particles as quasiparticles, within the usual pairing formalism. However, in physically reasonable systems, a general Hamiltonian will not separate purely in this way, so that an additional term H_{int}, is needed which couples the core and single-particle degrees of freedom via a core-particle interaction. Thus

$$H = H_c + H_p + H_{\text{int}}. \tag{5.2}$$

H_{int} essentially takes into account the effect on the odd particle of that part of the core field which is not already incorporated in H_p.

The basic approach outlined above is referred to as the *Unified Model* and detailed discussions of it can be found in the original papers of Bohr (1952) and Bohr and Mottelson (1953) and in many books. Here, however, we will proceed to look at some specific examples which throw some light on its application.

Let us consider first the situation for nuclei near closed shells where the spectrum of the even-even core is vibrational, consisting of a set of equally spaced levels corresponding to the excitation of one, two, three, etc., quadrupole phonons, as shown on the left of Fig. 5.1. In this case, the appropriate single-particle potential would be the spherical shell model and, if the single-particle state has spin j, then coupling this to the core spectrum will produce a ground state of spin j, a multiplet of states at the one-phonon energy with spins lying between $|j - 2|$ and $|j + 2|$, and so on for the multiphonon states. The resulting odd-A spectrum is shown on the right of Fig. 5.1 for the case $j = \frac{1}{2}$. Thus the basis states of the problem can be written as

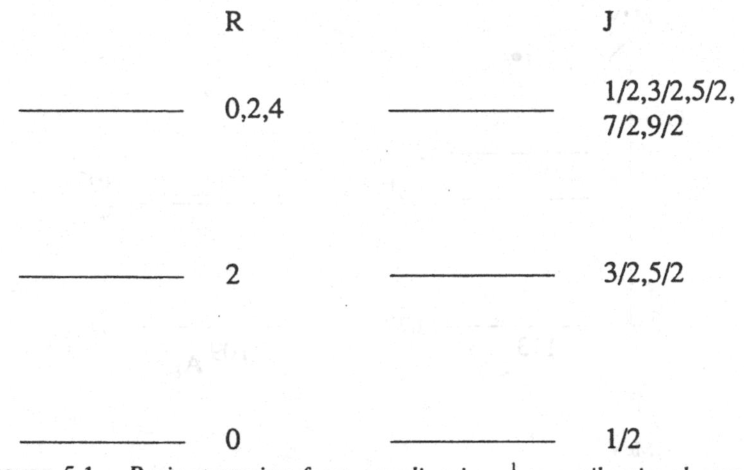

FIGURE 5.1. *Basis stemming from coupling $j = \frac{1}{2}$ to a vibrational core.*

$$|j, NR; JM\rangle = \sum_{m_j M_R} \langle jm_j RM_R|JM\rangle |jm_j\rangle |NRM_R\rangle \tag{5.3}$$

where the N-phonon state of the core with angular momentum R and projection M_R is coupled to the single particle state $|jm_j\rangle$ to give the total $|JM\rangle$.

In the limit in which H_{int} can be neglected, and H_c is the Hamiltonian of the quadrupole harmonic oscillator, then the total Hamiltonian (5.2) will be diagonal in the basis (5.3). This is the weak-coupling limit and the extent to which it pertains in actual nuclei can be judged from the empirical splitting of the members of the phonon multiplets. Such a splitting will be produced by a term in H_{int} which, in the simple example considered here, mixes states of the core. Of course, in the more general case where more than one single-particle orbital is considered, H_{int} may also mix the single-particle states. If the splitting induced by H_{int} is small compared to the separation of the core states, then the coupling can be thought of as weak; if not then the strong-coupling limit is approached and the basis states (5.3) are strongly mixed.

Some examples of odd-mass nuclei which might be expected to show vibrational characteristics are shown in Fig. 5.2. In each case, the ground state single-particle orbit has $j = \frac{1}{2}$, implying a couplet with $J = \frac{3}{2}, \frac{5}{2}$ at the energy of the first 2^+ (one phonon) state of the even-even core. Note that, in each case, the core nucleus is that with one nucleon more or less than the odd-A nucleus, accord-

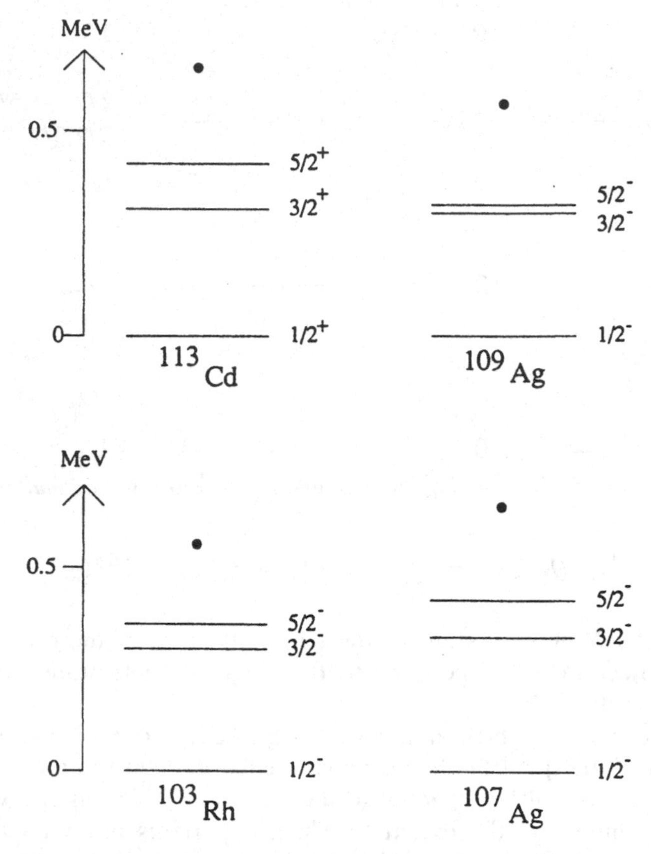

FIGURE 5.2. *Examples of vibrational odd-*A *nuclei with ground state spin* $\frac{1}{2}$. *The black circles show the energies of the first* 2^+ *states in the corresponding core nuclei.*

ing to whether the single-particle state is hole-like or particle-like. Thus, if the last nucleon in the $A + 1$ odd-mass nucleus is above mid-shell, the core has $A + 2$; if it is below, the core has A. While the basic vibrational structure is evident, so also is the splitting of the photon multiplets, so that the coupling term cannot be ignored. Moreover, the simplest form of the vibrational model would also yield zero quadrupole moments but, experimentally, this is not the case.

If we turn now to nuclei lying between major shell closures, such as those in the rare-earth region, the empirical situation is very different. Both even-even and odd-even nuclei exhibit the well-known rotational band structures associated with a stable quadrupole deformation of the nuclear shape. An example of the ground state bands of ^{167}Er and its adjacent even core is shown in Fig. 5.3. In this case, the appropriate single-particle potential is the deformed shell model, or Nilsson model, in which the single-particle angular momentum j, is no longer a conserved quantity. Nevertheless, in the example chosen here, the state of interest stems from the $i_{13/2}$ shell-model orbital, which is a so-called "unique parity" state. That is, it is an $N = 6$ positive-parity orbital situated

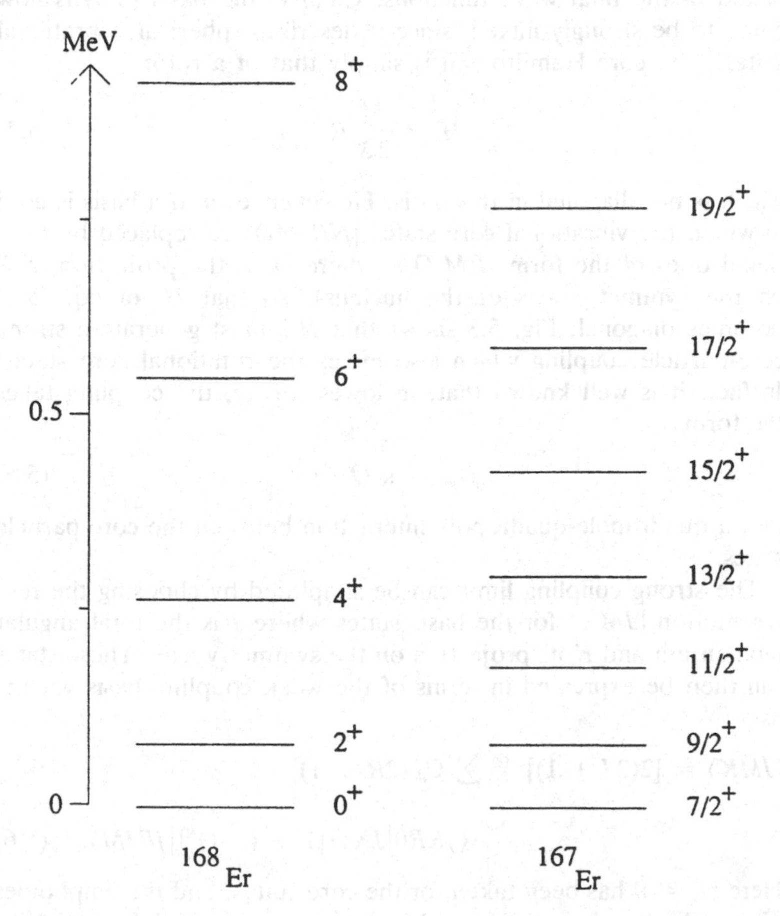

FIGURE 5.3. *Example of a rotational odd-A nucleus and the corresponding even-even core.*

in the middle of the $N = 5$ negative-parity shell, as a result of the l^2 and $l.s$ terms in the shell-model potential. As such, it cannot mix via the quadrupole interaction with the neighbouring single-particle states and so effectively maintains a pure $j = \frac{13}{2}$ character.

With the above observations in mind, even a cursory inspection of Fig. 5.3. reveals that the weak coupling picture no longer applies. Not only is there no trace of multiplets associated with the underlying core states, but even the ground state does not have the spin of the single-particle state. In fact, of course, this is an example of the strong-coupling limit, in which the core states are strongly mixed in the final wave functions. Clearly, the basis (5.3) is now going to be strongly mixed, since it describes spherical, vibrational states. The core Hamiltonian is simply that of a rotor,

$$H_c = \frac{\hbar^2}{2\mathfrak{J}} R^2 \tag{5.4}$$

which is not diagonal in this basis. However, even if a basis is used in which the vibrational core states $|NR\ M_R\rangle$ are replaced by rotational ones of the form $|RM\ \Omega_c\rangle$ (where Ω_c is the projection of R on the symmetry axis of the nucleus), so that H_c of eq. (5.4) becomes diagonal, Fig. 5.3 shows that H_{int} must generate a strong core-particle coupling which also mixes the rotational core states. In fact, it is well known that, in lowest order, the coupling takes the form

$$H_{\text{int}} = \kappa\ Q_c \cdot q_p \tag{5.5}$$

i.e., a quadrupole-quadrupole interaction between the core-particle states.

The strong coupling limit can be simplified by choosing the representation $|JMK\rangle$ for the basis states where J is the total angular momentum and K its projection on the symmetry axis. These states can then be expressed in terms of the weak coupling basis vectors as

$$|JMK\rangle = [2(2J + 1)]^{-1/2} \sum_{jR} C_{jl}\,(2R + 1)^{1/2} \cdot\langle jKR0|JK\rangle\cdot[1 + (-1)^R]\,|jRJM\rangle. \tag{5.6}$$

Here $\Omega_c = 0$ has been taken for the core states, and the amplitudes C_{jl} are the usual ones involved in the expansion of deformed Nilsson orbitals in a spherical basis. Equation (5.6) shows immediately the way in which the core states are admixed in the strong-coupling limit. Evaluation of the matrix elements of H_c between the states (5.6) then gives the standard rotational energy eigenvalues of the Bohr–Mottelson model, complete with decoupling and band-mixing terms, as discussed in Chapter 1. Of course, this result can also be obtained by the more usual route of replacing R in eq. (5.4) by $(J - j)$.

The examples discussed above focus on two limits of the core-particle coupling Hamiltonian which correspond to the extremes of weak coupling and strong coupling. While in many cases, the empirical situation is close to one of these extremes, in many others the nuclei of interest are intermediate in structure, often referred to as "transitional." Moreover, we have learned in Chapters 2 and 3 that the IBM has revealed a third limit of collective structure, the $O(6)$ symmetry, corresponding to a potential which is independent of the γ shape variable. Until relatively recently, odd-mass nuclei in regions of $O(6)$ structure have also been regarded as transitional, since they correspond to neither the weak nor strong-coupling limits. Obviously, one attractive goal in the treatment of odd-A nuclei would be to describe the full gamut of structure within a single comprehensive framework. The algebraic approach embodied in the IBM makes considerable strides towards this end for even-even nuclei, and we shall now see how this framework can be extended to encompass the coupling of an odd nucleon. The model which results is known as the Interacting Boson Fermion Model, or IBFM.

5.2. THE IBFM HAMILTONIAN

By analogy with the general discussion of the preceding section, we can start by writing the IBFM Hamiltonian as the sum of three terms:

$$H = H_{\mathrm{B}} + H_{\mathrm{F}} + H_{\mathrm{BF}} \tag{5.7}$$

where H_{B} is the core Hamiltonian describing the s-d boson space, the form and properties of which have been dealt with extensively in Chapters 2 and 3. H_{F} is the single-particle Hamiltonian and, at this stage, we must introduce the odd-nucleon creation (annihilation) operators $a_j^{\dagger}(\tilde{a}_j)$ into the problem, in addition to the s- and d-boson operators discussed earlier. To be correct, the operator $a_j^{\dagger}$ is not a nucleon creation operator in the shell-model sense, but is rather a generalized seniority raising operator. However the distinction is a technical one, mainly of importance in considering a microscopic theory of the IBFM, and can be neglected here. Thus, for a single odd nucleon which can occupy different shell model orbits j, H_F is given by

$$\begin{aligned} H_F &= \sum_j \epsilon_j (a_j^\dagger \cdot \tilde{a}_j) \\ &= \sum_j \epsilon_j \hat{n}_j \end{aligned} \tag{5.8}$$

where $\hat{n}_j$ is the fermion number operator for the orbit j.

The final part of the IBFM Hamiltonian is H_{BF} which represents the core-particle interaction. The most general form of H_{BF} is too complex for a phenomenological treatment. On the basis of microscopic considerations it has been found that a simpler form can account for the majority of observed properties. This form is written in terms of three components referred to as the monopole, quadrupole and exchange terms, respectively,

$$H_{BF} = H_{BF}^M + H_{BF}^Q + H_{BF}^E \tag{5.9}$$

which have the following form:

$$H_{BF}^M = \sum_j A_j [(d^\dagger \tilde{d})^{(0)}] (a_j^\dagger \tilde{a}_j)^{(0)}]_0^{(0)}, \tag{5.10}$$

$$H_{BF}^Q = \sum_{jj'} \Gamma_{jj'} [Q_B^{(2)} (a_j^\dagger \tilde{a}_{j'})^{(2)}]_0^{(0)}, \tag{5.11}$$

$$H_{BF}^E = \sum_{jj'j''} \Lambda_{jj'}^{j''} : [(d^\dagger \tilde{a}_j)^{(j'')} (\tilde{d} a_{j'}^\dagger)^{(j'')}]_0^{(0)} : \tag{5.12}$$

with

$$Q_B^{(2)} = (s^\dagger \tilde{d} + d^\dagger s)^{(2)} + \chi (d^\dagger \tilde{d})^{(2)} \tag{5.13}$$

The notation used here for tensor products is the same as that in earlier chapters, while the : denotes normal order.

In most phenomenological applications to date, the strength of the monopole interaction has been taken as a constant $A_j = A_0$ and is sufficiently small that it hardly affects the structure of the predicted spectrum. In fact, it can be seen from eq. (5.10) that it is proportional to $\hat{n}_d \hat{n}_j$ and will thus have the same effect as changing the d-boson energy. The second term in eq. (5.11) represents the "standard" quadrupole-quadrupole interaction between core and particle while the exchange term of eq. (5.12) stems from the explicit recognition that the bosons themselves are built up from fermions which can occupy the same orbits as the odd fermion. Thus the interaction (5.12) proceeds via an exchange between the odd fermion and a fermion in the d-boson pair.

Clearly, one could also construct exchange interactions involving the s-boson. For a single j-orbit, the effects of these are already incorporated in the other terms and they have not been used in numerical treatments. However, it will become apparent later in the context of symmetries that, for multi-j applications, they can be important. The difference between the exchange and direct types of boson–fermion interaction is displayed in Fig. 5.4. The bosons (fermion pairs) are denoted by double lines and the fermions by single lines, while the interaction is shown as a dashed line.

The physical origins of the various interactions can dictate a certain j-dependence of the coefficients which appear in eqs.

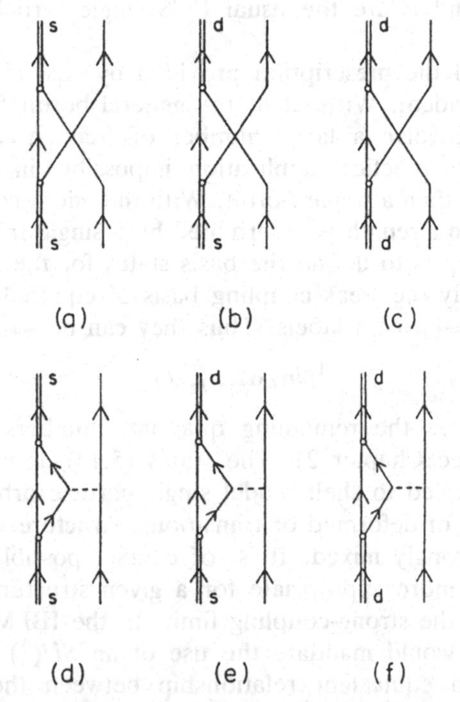

FIGURE 5.4. *Schematic representation of the boson–fermion interaction (taken from Iachello and Van Isacker* (1991)*. The exchange interaction is depicted by* (a), (b) *and* (c) *while* (d), (e) *and* (f) *show the direct interaction.*

(5.10)–(5.12) Use of the generalized seniority scheme in particular leads to the following expressions:

$$A_j = \sqrt{5(2j + 1)}A_0, \tag{5.14}$$

$$\Gamma_{jj'} + \sqrt{5}\,\Gamma_0(u_ju_{j'} - v_jv_{j'})Q_{jj'}, \tag{5.15}$$

$$\Lambda^{j''}_{jj'} = -2\sqrt{\frac{5}{2j'' + 1}}\,\beta_{jj''}\,\beta_{j''j'}\Lambda_0 \tag{5.16}$$

with

$$\beta_{jj'} = (u_jv_{j'} + v_ju_{j'})\,Q_{jj'}. \tag{5.17}$$

The $Q_{jj'}$ above are the single-particle matrix elements of the quadrupole operator

$$Q_{jj'} = \langle j||Y_2||j'\rangle \tag{5.18}$$

while the u_j and v_j are the usual BCS single-particle occupation amplitudes.

The value of the prescription provided by eqs. (5.14)–(5.18) is immediately evident. Without it, the general boson–fermion interaction would involve a large number of free parameters which would make its practical application impossible in any situation involving more than a single j-orbit. With the microscopic relations, each interaction strength is determined by a single free parameter.

The final step is to define the basis states for the problem, and these are simply the weak-coupling basis of eq. (5.3), couched in terms of boson–fermion labels. Thus they can be written as

$$|Nn_d\,\alpha L;\, j;\, J\rangle \tag{5.19}$$

where α denotes the remaining quantum numbers of the $U(5)$ boson basis (see Chapter 2). The states (5.19) describe spherical core states coupled to shell model single-particle orbits. Hence in the description of deformed or transitional structure we can expect them to be strongly mixed. It is, of course, possible to define a basis which is more appropriate for a given structure, just as we did earlier for the strong-coupling limit. In the IBFM, the strong-coupling limit would mandate the use of an $SU(3)$ basis for the bosons and an equivalent relationship between the strong and weak-coupling bases can then be obtained.

By now it should be evident, at least initially, that the IBFM is simply a core-particle coupling model with the advantage of a rather versatile and extensive description of the collective core states by virtue of the IBM Hamiltonian. However, it was shown in the earlier chapters that it is the existence and role of symmetries in the IBM framework which lend it its most unique and characteristic features. Such a comment is no less true for the IBFM. The three fundamental symmetries which emerge from the algebraic treatment of the boson problem re-appear in the odd-A formalism although, as will be seen, their existence and properties now depend critically on the single-particle space available to the odd fermion. Thus we could discuss the IBFM as a numerical Hamiltonian for the treatment of a broad range of odd-A structure, in terms of the role and effect of different terms in the Hamiltonian, or we can start from the analytic symmetry limits which define benchmark structures in the odd-A problem. Since the focus of this volume is on the algebraic approach to nuclear structure, and since it is this aspect which renders the IBFM unique, we will concentrate here on the symmetry properties of the Hamiltonian.

5.3. THE ROLE OF SYMMETRY

We have seen how the IBFM can be thought of as a general core-particle coupling problem. In Chapters 2 and 3 we saw how the algebraic approach revealed three basic symmetries in the IBM description of the even-even (core) nucleus and how these three symmetries control and characterize the nuclear structure properties which emerge from the formalism. One can then ask whether, in the IBFM case, the presence of a particular symmetry in H_B leads to any simple predictions for the structure of the odd-A nucleus. There are two points to remember here. The first is that, for a given H_B, there are a plethora of possible odd-A structures, dependent on the specific orbits available to the odd particle. The second point stems from our discussion of the weak-coupling and strong-coupling limits of the odd-A Hamiltonian. Only in the former case do the odd-A states conserve the character of the underlying core states. Thus in the $U(5)$ limit, we could expect the coupling of spherical single-particle states to result in energies and transition probabilities which still reflect the symmetry structure of the core. In all other cases, the core states will be mixed and, with no symmetry constraint on the form of the core-particle interaction or the energies and nature of the single-particle orbits, there is no obvious prospect of any analytic solution to the odd-A problem. Thus it becomes clear that to explore further the possible role of symmetry in the IBFM, we need to study the algebraic properties of a system built from both boson and fermion operators. Such a problem has also been a focus of attention in the field of elementary particle physics over recent years, where theoreticians have postulated the existence of a larger class of symmetry transformations in nature, involving products of boson and fermion operators. These are known as supersymmetry transformations and, at first glance, have little relevance in the nuclear domain, since they correspond to the exchange of a boson for a fermion, or vice-versa, which would violate the total particle number. Nevertheless it will be shown in the final section of this chapter that supersymmetric schemes can be invoked for nuclei. At this stage, however, and for the majority of the chapter, we will concentrate on bose-fermi symmetries, where the boson and fermion operators each appear as bilinear products which conserve the boson and fermion numbers separately.

The algebraic structure of bilinear products of fermion operators, where j can assume more than one value, is much richer than that for boson operators and hence the study of possible fermion algebras, and their coupling to the boson degrees of freedom, becomes much more complex than the problem of Chapter 2. No attempt will be made here to go into the details of the possible algebras, but the interested reader is referred to the recent book by Iachello and Van Isacker (1991) which deals with the topic in far more depth. Rather we will simply use some of the concepts introduced earlier of groups, dynamical symmetries, and their associated Casimir operators to show how bose–fermi symmetries arise and to discuss the constraints they imply on the overall form of H_{BF}. Moreover, we will concentrate on a particular class of such symmetries, which has proved to be the most generally applicable to date.

5.3.1. Coupling of Boson and Fermion Groups

The basic group structures associated with the IBFM are $U^{B}(6)$ and $U^{F}(m)$ where $m = \Sigma_i (2j_i + 1)$. $U^{B}(6)$ is the usual boson group describing the collective excitations in the core Hamiltonian; its properties have been discussed extensively in Chapter 2. $U^{F}(m)$ is the corresponding group of unitary transformations describing the single-particle space, the dimension m being the total number of quantum states available. The starting point for the boson–fermion problem is then the direct product group

$$U^{B}(6) \times U^{F}(m) \tag{5.20}$$

which contains the two sets of generators, boson and fermion. The next step is to ask how the boson and fermion degrees of freedom can be coupled, and whether a chain of subgroups can be constructed which will lead to one or more dynamical symmetries, as is the case for $U^{B}(6)$ alone.

The problem of coupling two group structures has already arisen in Chapter 2 in the context of IBM-2. There, the constituents were the proton and neutron degrees of freedom. Nevertheless, the basic approach is the same. Indeed, it will become apparent later that, for the class of bose–fermi symmetries that will be explored in detail in this chapter, the algebraic structure of the problem is totally analogous to that obtained in the neutron-proton basis.

Recall that the basic approach to defining dynamical symmetries centers on the search for subgroups, each of which contains a subset of the generators of its predecessor and each of which will have linear and/or quadratic Casimir operators associated with it which commute with all the generators of the subgroup. In addition, there will be an irreducible representation (irrep) associated with each group labelled by one or more characteristic quantum numbers. The basic physical requirement in this process is that the chain ends with the subgroup whose generators are those of the total angular momentum, so that rotational invariance is assured. In the bose–fermi case, this is the $\text{Spin}^{BF}(3)$ group, with Casimir operator proportional to J^2 where $J = L + j$, the vector sum of the boson and fermion angular momenta. The superscript BF will be used throughout to denote the direct sum group whose generators are formed by adding the boson and fermion generators.

We already know that the decomposition of the group $U^{B}(6)$ can proceed by one of three possible routes. It is thus now necessary to consider the possible decompositions of the fermion group. In general, the answer depends on the constituents of $U^{F}(m)$; i.e., which j orbits are included. Moreover, it is necessary to identify in the fermion subgroup chain a subgroup which is identical or isomorphic to (i.e., has the same algebraic structure as) one of the boson subgroups, so that the two spaces can be coupled. As mentioned above, this coupling is represented by introducing the sum group, and the point at which the boson and fermion chains are coupled algebraically reflects, in physical terms, the strength and form of the boson–fermion coupling H_{BF} in eq. (5.7).

To take an extreme example, if the coupling were only done at the lowest possible level, one would have boson and fermion group chains which proceed independently and end with the step

$$O^{B}(3) \times \mathrm{Spin}^{F}(3) \supset \mathrm{Spin}^{BF}(3) \tag{5.21}$$

which represents the coupling of boson and fermion angular momenta (note that Spin(3) is isomorphic to $O(3)$). However, in such a chain, there would be no contributions to H_{BF}, since only separate boson and fermion Casimir operators would appear other than the final (diagonal) J^2 term. This would therefore represent the ultimate (and trivial) weak-coupling limit which, as we have seen, is not physically applicable.

The existence of symmetries, and their type, depends critically on the single-particle orbits available which determine the content of $U^{F}(m)$. As mentioned earlier, here we will concentrate on the properties of a particular class of such symmetries which gives rise to a rich structure encompassing all three of the IBM limits.

5.3.2. The Pseudo-Spin Decomposition

It was pointed out many years ago that it is possible to transform a single-particle basis described in terms of angular momenta j into a scheme involving pseudo-orbital and pseudo-spin variables $\tilde{l}$ and $\tilde{s}$ describing integer and half-integer quantities, respectively. (From now on, we shall use the tilde superscript to denote angular momentum variables in the pseudo scheme). The new variables are referred to as "pseudo" because they do not, in general, correspond to the real orbital or spin angular momenta. As an example, a single-particle basis involving the $s_{1/2}$ and $d_{3/2}$ orbits could be described as having $\tilde{l} = 1$, $\tilde{s} = \frac{1}{2}$. The correct values of j are conserved in the pseudo-scheme, and the two schemes, pseudo and real, are related by a simple transformation. For some examples there is more than one pseudo-scheme possible. The set of orbits with $j = \frac{1}{2}, \frac{3}{2}, \frac{5}{2}, \frac{7}{2}$ could be described as either

$$\tilde{l} = 1, 3; \qquad \tilde{s} = \tfrac{1}{2} \tag{5.22}$$

or

$$\tilde{l} = 2; \qquad \tilde{s} = \tfrac{3}{2}. \tag{5.23}$$

The change to pseudo-variables is just a mathematical transformation. It does not per se change the physics. However, the transformation can prove crucial in constructing fermion algebras which lead to symmetries. In the case where $\tilde{s} = \frac{1}{2}$, the transformation is equivalent to the reduction

$$U^F(m) \supset U^F\left(\frac{m}{2}\right) \times SU^F(2). \tag{5.24}$$

A symmetry of great significance occurs whenever the reduction (5.24) above leads to a set of $\tilde{l}$-values corresponding to an irrep (λ_F, μ_F) of $SU(3)$. This implies a single particle space with $j = \frac{1}{2}, \frac{3}{2}, \frac{5}{2}, \ldots, (n + \frac{1}{2})$, i.e., a "complete set" of j-values from $j = \frac{1}{2}$ to some maximum. The group $U^F(m/2)$ then always contains $SU^F(3)$ as a subgroup and can thus be coupled at this level to the boson group $SU^B(3)$:

$$\begin{aligned} U^B(6) \times U^F(m) &\supset U^B(6) \times U^F\left(\frac{m}{2}\right) \times SU^F(2) \\ &\supset SU^B(3) \times SU^F(3) \times SU^F(2) \\ &\supset SU^{BF}(3) \times SU^F(2) \\ &\supset O^{BF}(3) \times SU^F(2) \\ &\supset \mathrm{Spin}^{BF}(3). \end{aligned} \tag{5.25}$$

This is a general result for the boson–fermion basis. However this property also allows the fermion space itself to be treated in terms of $SU(3)$ quantum numbers and leads to the pseudo-$SU(3)$ fermion model, which is treated in Chapter 7. Note that the pseudo-spin, represented by the group $SU^F(2)$ in eq. (5.25) cannot be coupled until the final stage (the groups $O(3)$ and $SU(2)$ are isomorphic). Physically, this means that the pseudo-spin is "decoupled" from the problem, so that the Hamiltonian is independent of $\tilde{s}$, with the exception of the $J(J + 1)$ splitting introduced by the J^2 Casimir operator of $\mathrm{Spin}^{BF}(3)$. This "pseudo-spin symmetry" is a crucial underlying assumption of such schemes. It requires a concomitant degeneracy in the single-particle space between states with $j = \tilde{l} \pm \frac{1}{2}$, or "pseudo-spin-orbit partners." It is the fact that the shell model, by chance or by design, provides us with pairs of orbits which, in the transformed basis, come close to realizing this degeneracy that allows us to employ these symmetry concepts.

The situation for the major shells with $N = 4$ and 5 is illustrated in Fig. 5.5. The set of normal parity states in each shell N (which excludes the state of largest j for each N) is seen to form the complete set of states belonging to the pseudo-oscillator shell with $\tilde{N} = N - 1$. Moreover, the spin-orbit splitting is dramatically reduced in the pseudo scheme. Clearly, if our transformation results in a scheme in which $\tilde{l} \cdot \tilde{s} \rightarrow 0$, then we have a single-particle Hamiltonian which is independent of pseudo-spin. This is the microscopic basis of pseudo-spin symmetry. The concept will be discussed again in far more detail in Chapter 7.

As yet, we have only produced $SU(3)$ symmetry in the boson–fermion framework. To induce the appearance of additional symmetries, it is necessary to further restrict the fermion space. Specifically if we choose $j = \frac{1}{2}, \frac{3}{2}$ and $\frac{5}{2}$, then the reduction (5.24) becomes

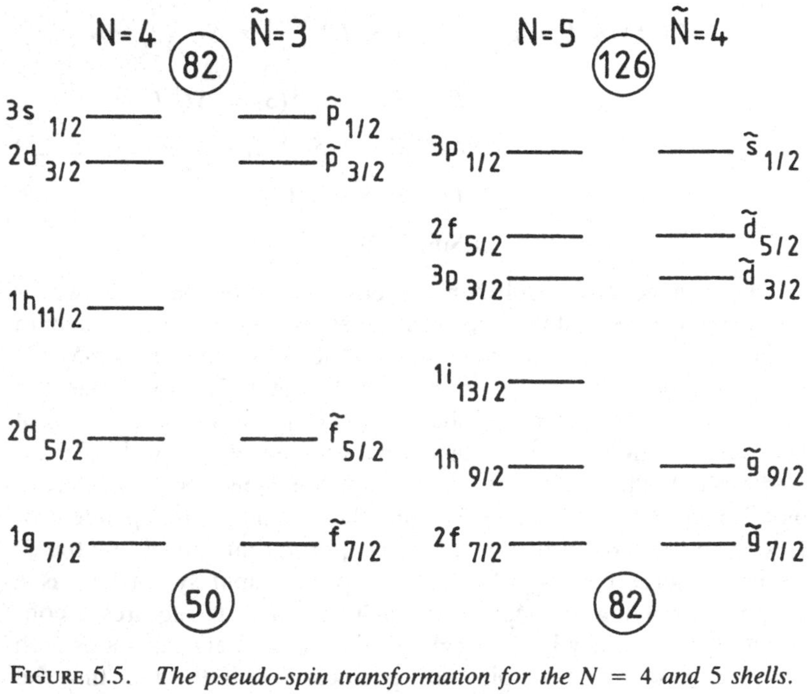

FIGURE 5.5. *The pseudo-spin transformation for the $N = 4$ and 5 shells.*

$$U^F(12) \supset U^F(6) \times SU^F(2). \tag{5.26}$$

The group $U(6)$ will now appear for the fermion pseudo-orbital variables as well as for the bosons. Moreover, the subgroup structure for $U^F(6)$ is identical to that for $U^B(6)$ with the result that all three limiting symmetries can be realized. In fact, we are now treating the fermion space as if it arises from the coupling of $\bar{l} = 0, 2$ (*or s, d*) states to an inert pseudo-spin of $\frac{1}{2}$ (see Fig. 5.6). Thus our boson–fermion coupling problem has become, algebraically, almost identical to the neutron-proton problem of Chapters 2 and 3. We will therefore choose this example to explore in more detail the predicted properties and structure of bose–fermi symmetries. A review of the experimental evidence for a more extensive range of symmetry schemes has been given by Vervier (1987).

5.4. SYMMETRIES ASSOCIATED WITH $U^B(6) \times U^F(12)$

The starting point for this class of symmetries can be written as

$$U^B(6) \times U^F(12) \supset U^B(6) \times U^F(6) \times SU^F(2). \tag{5.27}$$

As mentioned earlier, both the boson and fermion $U(6)$ structures can give rise to one of the three dynamical symmetry chains, $U(5)$,

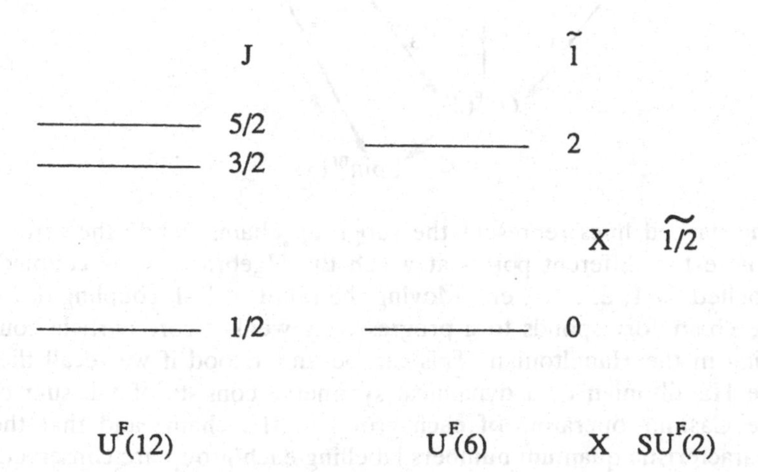

FIGURE 5.6. *Schematic representation of the reduction* $U^F(12) \supset U^F(6) \times SU^F(2)$.

$SU(3)$ or $O(6)$. For each symmetry, the boson and fermion groups can be coupled at any point in the two chains where the algebras coincide (or are isomorphic). In practice, in the present example, this means that there are several ways of achieving this coupling, each of which leads to a different physical description of the nuclear states.

This aspect is best illustrated by a specific example. Consider the $SU(3)$ case. The dynamical symmetry chain can be constructed as follows:

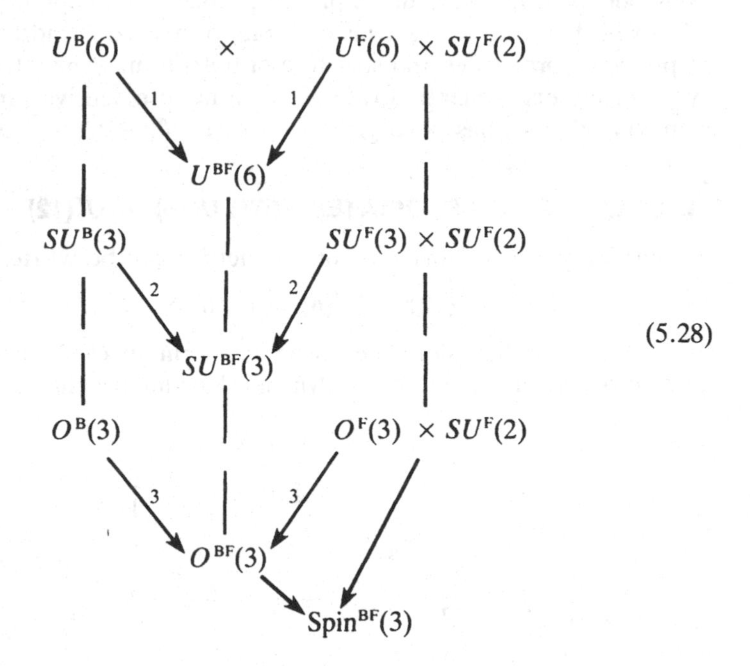

(5.28)

The dashed lines represent the subgroup chains, while the arrows denote the different points at which the algebras can be coupled, labelled by 1, 2, . . . , etc. Moving the point of B–F coupling down the chain corresponds to a progressively weaker core-particle coupling in the Hamiltonian. This can be understood if we recall that the Hamiltonian of a dynamical symmetry consists of the sum of the Casimir operators of each group in the chain, and that the characteristic quantum numbers labelling each group are conserved. Then, for example, coupling route 3, at the level of $O^{BF}(3)$, would correspond to an interaction which generates no mixing between core states, since the presence of $O^{B}(3)$ in the Hamiltonian would mean that the angular momentum of the cores states must remain a good quantum number. Coupling at level 2 relaxes this constraint but maintains the $SU(3)$ quantum numbers of the core, and hence implies no mixing between configurations built on the ground band and higher (e.g., β,γ) intrinsic excitations. A subset of states in this scheme would have a core basis corresponding to that of the Nilsson model, where the core-particle wave functions involve only the ground state band. Coupling route 1 at the $U(6)$ level allows, a priori, mixing between all core states in the final odd-A wave function.

At this point, it is worthwhile noting that exactly the same coupling options exist in the neutron-proton algebra of IBM-2. In fact, algebraically, the problems are identical, apart from the decoupled pseudo-spin group in IBFM. In IBM-2, it was found that experiment in general mandates the strongest level of coupling, level 1, and it will be seen presently that the same is true in the bose–fermi case. We will therefore proceed to concentrate here on the predicted properties of the following three bose–fermi dynamical symmetries which, for completeness, are written out in full.

$$\begin{aligned} U^{B}(6) \times U^{F}(12) &\supset U^{B}(6) \times U^{F}(6) \times SU^{F}(2) \\ &\supset U^{BF}(6) \times SU^{F}(2) \supset \\ \text{(I) } U^{BF}(5) \times SU^{F}(2) &\supset O^{BF}(5) \times SU^{F}(2) \\ &\supset O^{BF}(3) \times SU^{F}(2) \supset \text{Spin}^{BF}(3); \end{aligned} \tag{5.29}$$

$$\text{(II) } SU^{BF}(3) \times SU^{F}(2) \supset O^{BF}(3) \times SU^{F}(2) \supset \text{Spin}^{BF}(3); \tag{5.30}$$

$$\begin{aligned} \text{(III) } O^{BF}(6) \times SU^{F}(2) &\supset O^{BF}(5) \times SU^{F}(2) \\ &\supset O^{BF}(3) \times SU^{F}(2) \supset \text{Spin}^{BF}(3). \end{aligned} \tag{5.31}$$

We now need to find the generators and Casimir operators of the subgroups appearing in (5.29)–(5.31). Since the subgroup structure emanating from $U^{BF}(6)$ is the same as that of IBM-1, the only new features which appear are the groups $SU^{F}(2)$, and $\text{Spin}^{BF}(3)$, and $U^{BF}(6)$ itself. While this latter group has been dealt with in Chapter 2 in the context of IBM-2, here it incorporates fermion generators and therefore warrants some further brief discussion. A fuller treatment can be found in Van Isacker, Frank and Sun (1984).

The generators of $U^{F}(6)$ and $SU^{F}(2)$ are constructed via the transformation from the *jj* to the pseudo-*ls* coupling scheme. They are defined by

$$A^{(\lambda)}_{\mu}(\tilde{l},\tilde{l}') = -\sqrt{2}\sum_{jj'}\sqrt{(2\lambda+1)(2j+1)(2j'+1)} \times \begin{Bmatrix} \tilde{l} & \frac{1}{2} & j \\ \tilde{l}' & \frac{1}{2} & j' \\ \lambda & 0 & \lambda \end{Bmatrix} (a^{\dagger}_{j}\tilde{a}_{j'})^{(\lambda)}_{\mu}$$
$$= -\sum_{jj'}\sqrt{(2j+1)(2j'+1)}(-1)^{\lambda+j'+1/2} \times \begin{Bmatrix} \tilde{l} & j & \frac{1}{2} \\ j' & \tilde{l}' & \lambda \end{Bmatrix} (a^{\dagger}_{j}\tilde{a}_{j'})^{(\lambda)}_{\mu}, \quad (5.32)$$

where the generators of the orbital group $U^{\mathrm{F}}(6)$ have been obtained by coupling the pseudo-spin to zero. The generators of $SU^{\mathrm{F}}(2)$ are formed in a similar fashion by coupling the pseudo-orbital part to zero.

$$S_{\mu} = \sum_{\tilde{l}}\sum_{jj'}\sqrt{(2j+1)(2j'+1)}\,(-1)^{j+3/2} \times \begin{Bmatrix} \frac{1}{2} & j & \tilde{l} \\ j' & \frac{1}{2} & 1 \end{Bmatrix} (a^{\dagger}_{j}\tilde{a}_{j'})^{(1)}_{\mu}. \quad (5.33)$$

The generators of eqs. (5.32) and (5.33) can be shown to close under commutation, the commutation rules for the $A^{(\lambda)}_{\mu}(\tilde{l},\tilde{l}')$ being exactly the same as those for the boson operators $(b_{l}b_{l'})^{(\lambda)}_{\mu}$ discussed in Chapter 2. The generators of the sum groups are just the sum of the generators of the constituents. Hence for $U^{\mathrm{BF}}(6)$ we get

$$G^{(\lambda)}_{\mu}(\tilde{l},\tilde{l}') = B^{(\lambda)}_{\mu}(\tilde{l},\tilde{l}') + A^{(\lambda)}_{\mu}(\tilde{l},\tilde{l}') \quad (5.34)$$

where $B^{(\lambda)}_{\mu}(l,l') = (b^{\dagger}_{l}\tilde{b}_{l'})^{(\lambda)}_{\mu}$ and $b_0 \equiv s$, $b_2 \equiv d$. Similarly, the generators of $\mathrm{Spin}^{\mathrm{BF}}(3)$ are constructed from those of $O^{\mathrm{BF}}(3)$ and $SU^{\mathrm{F}}(2)$

$$J_{\mu} = -\frac{1}{\sqrt{2}}S_{\mu} + \sqrt{10}\,G^{(1)}_{\mu}(2,2) \quad (5.35)$$

where J is the total angular momentum operator.

The form of the generators (5.32) means that the Casimir operators describing the three symmetry chains can now be inferred directly from the results of IBM-1. One simply replaces $(s^\dagger\tilde{d})^{(2)}$ by $G^{(2)}(0, 2)$, $(d^\dagger\tilde{d})^{(2)}$ by $G^{(2)}(2, 2)$, etc. As an example, let us consider the quadratic Casimir operator of $SU(3)$. Its form, from Chapter 2, is

$$C_{2SU3} = \tfrac{4}{3}\, Q\cdot Q + \tfrac{1}{2}\, L\cdot L. \tag{5.36}$$

This is still the form applicable to $SU^{\rm BF}(3)$ but now the boson operators Q and L are replaced by

$$Q_{\rm BF} = G^{(2)}(0, 2) + G^{(2)}(2, 0) - \frac{\sqrt{7}}{2}\, G^{(2)}(2, 2) \tag{5.37}$$

and

$$L_{\rm BF} = \sqrt{10}\, G^{(1)}(2, 2).$$

This is just equivalent to writing

$$Q_{\rm BF} = Q_{\rm B} + Q_{\rm F}; \qquad L_{\rm BF} = L_{\rm B} + \bar{l}_{\rm F}. \tag{5.38}$$

Note also that, in keeping with the analogy between the boson and fermion spaces implied in Fig. 5.6, boson operators of the form $(d^\dagger\tilde{d})^{(\lambda)}$ produce bose–fermi operators involving coupling between the $j=\frac{3}{2}$ and $\frac{5}{2}$ single particle states (with $\bar{l} = 2$), but not the $j = \frac{1}{2}$. Similarly the counterpart of $(s^\dagger\tilde{d})^{(2)}$ involves fermion operators linking $j = \frac{1}{2}$ with $j = \frac{3}{2}$ and $\frac{5}{2}$, i.e. $\bar{l} = 0$ to $\bar{l} = 2$. The detailed structure in this latter case is

$$G^{(2)}(0, 2) = (s^\dagger\tilde{d})^{(2)} - \sqrt{4/5}(a^\dagger_{1/2}\tilde{a}_{3/2})^{(2)} - \sqrt{6/5}(a^\dagger_{1/2}\tilde{a}_{5/2})^{(2)}. \tag{5.39}$$

The explicit forms for the other generators of $U^{\rm BF}(6)$ can be found in Van Isacker, Frank and Sun (1984), along with those for the Casimir operators, which can, for the most part, be inferred directly from the results for IBM-1. The only new Casimir operators which appear are the quadratic Casimir operators of $U^{\rm BF}(5)$ and $U^{\rm BF}(6)$, given by

$$C_{2U6} = \sum_{\bar{l},\bar{l}'} \sum_{\lambda} G^{(\lambda)}\cdot(\bar{l}, \bar{l}')\cdot G^{\lambda}(\bar{l}', \bar{l}), \tag{5.40}$$

$$C_{2U5} = \sum_{\lambda} G^{(\lambda)}(2, 2)\cdot G^{(\lambda)}(2, 2). \tag{5.41}$$

5.4.1. Energy Spectra

We can now turn to the predicted energy spectra in the three limits. The Hamiltonian in each case is written as a sum of the relevant Casimir operators, each multiplied by an arbitrary constant, and the energy eigenvalue expression then stems from the eigenvalues of each Casimir operator. These are given in Table 5.1 along with the quantum numbers which label the irreps of each subgroup. The eigenvalues shown correspond to the definition of the Casimir operators given in Chapter 2. Definitions differing to within a multiplicative constant will be found in the literature.

The structure of the solutions thus follows those in IBM-1 with the important difference that additional quantum numbers have appeared for many of the subgroups and their physical significance has changed. Both features, of course, result from the fact that our symmetry groups now describe the coupled system of bosons and fermions. In IBM-1, the only irreps which enter the problem are those which are fully symmetric in the boson space. For coupled systems, such as IBM-2 or IBFM, non-symmetric irreps can occur. The values of the labels describing $U^{B}(6)$ and its subgroups have been dealt with in Chapter 2. The corresponding values for $U^{F}(6)$ are given in Table 5.2 and can be understood intuitively by realizing (again see Fig. 5.6) that the $\tilde{l} = 0$ and 2 states available to the fermions are analogous to the s and d states available to a system with one boson. The labels for the sum groups, and their range of possible values, then stem from standard coupling rules for the various algebras. An example is given for the unitary groups using

TABLE 5.1. *Casimir operators, quantum numbers and eigenvalues for the three* B–F *symmetry limits.*

Casimir operator	Quantum numbers	Eigenvalue	Symmetry chain
C_{2U6}	N_1, N_2	$N_1(N_1+5)+N_2(N_2+3)$	I, II, III
C_{1U5}	n_1, n_2	n_1+n_2	I
C_{2U5}	n_1, n_2	$n(n_1+4)+n_2(n_2+2)$	I
C_{2SU3}	λ, μ	$2/3\,(\lambda^2+\mu^2+\lambda\mu+3\lambda+3\mu)$	II
C_{2O6}	σ_1, σ_2	$2\sigma(\sigma_1+4)+2\sigma_2(\sigma_2+2)$	III
C_{2O5}	τ_1, τ_2	$2\tau_i(\tau_1+3)+2\tau_2(\tau_2+1)$	I, III
C_{2O3}	$\bar{L}$	$2\bar{L}(\bar{L}+1)$	I, II, III
$C_{2\mathrm{Spin}3}$	J	$J(J+1)$	I, II, III

TABLE 5.2. *Fermion quantum numbers for subgroups of $U^F(6)$.*

Subgroup	Quantum numbers	Values
$U(6)$	N_1, N_2	1, 0
$U(5)$	n_1, n_2	1, 0 or 0, 0
$SU(3)$	λ, μ	2,0
$O(6)$	σ_1, σ_2	1, 0
$O(5)$	τ_1, τ_2	1, 0 or 0, 0
$O(3)$	$\tilde{l}$	0 or 2

Young tableaux in Fig. 5.7. For $U(6)$, there is only one boson representation and one for the fermions. There are then only two possible irreps of $U^{BF}(6)$ as shown in the figure. This result can be contrasted with that obtained in IBM-2 where many more possibilities can, in general, arise from the coupling of $\{N_\nu, 0\}$ to $\{N_\pi, 0\}$. The result here is simply that which would be obtained in IBM-2 if $N_\nu = N$ and $N_\pi = 1$ (or vice-versa). For $U(5)$, Fig. 5.7 shows the coupling of the boson representation with $N = n_d$ to the two possible orbital states in $U^F(5)$. Clearly, in this case, there are many more boson irreps with $N = n_d - 1, n_d - 2, \ldots, 0$ which will give rise to an accompanying chain of values for $\{n_1, n_2\}_{BF}$. In $SU(3)$, the convention for the Young tableaux is slightly different and the boson system is represented by $2N$ boxes, arranged among three rows, while the fermion system is denoted as two boxes with $\{\lambda, \mu\}_F = \{2, 0\}$. Figure 5.7 shows the case for the ground state boson representation, which gives rise to three possible structures in the coupled system. Again, the coupling can be performed with the higher boson states, the next having $\{\lambda, \mu\}_B = \{2N - 4, 2\}$, to give the full set of $SU^{BF}(3)$ quantum numbers.

The coupling rules for the orthogonal groups differ somewhat, but the basic principle remains the same. It is worth mentioning specifically the $O^{BF}(3)$ group which appears at the end of each symmetry chain. This represents the coupling of $O^B(3)$ and $O^F(3)$ and hence the coupling of the total boson angular momentum with the pseudo-orbital angular momentum of the fermions. The new quantum number which emerges, $\tilde{L}$, is referred to as the total pseudo-orbital angular momentum. Its appearance in all three symmetry limits is a direct consequence of our ab initio assumption of pseudo-spin symmetry and it gives rise to a number of characteristic predictions which will be discussed presently.

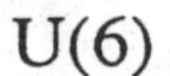

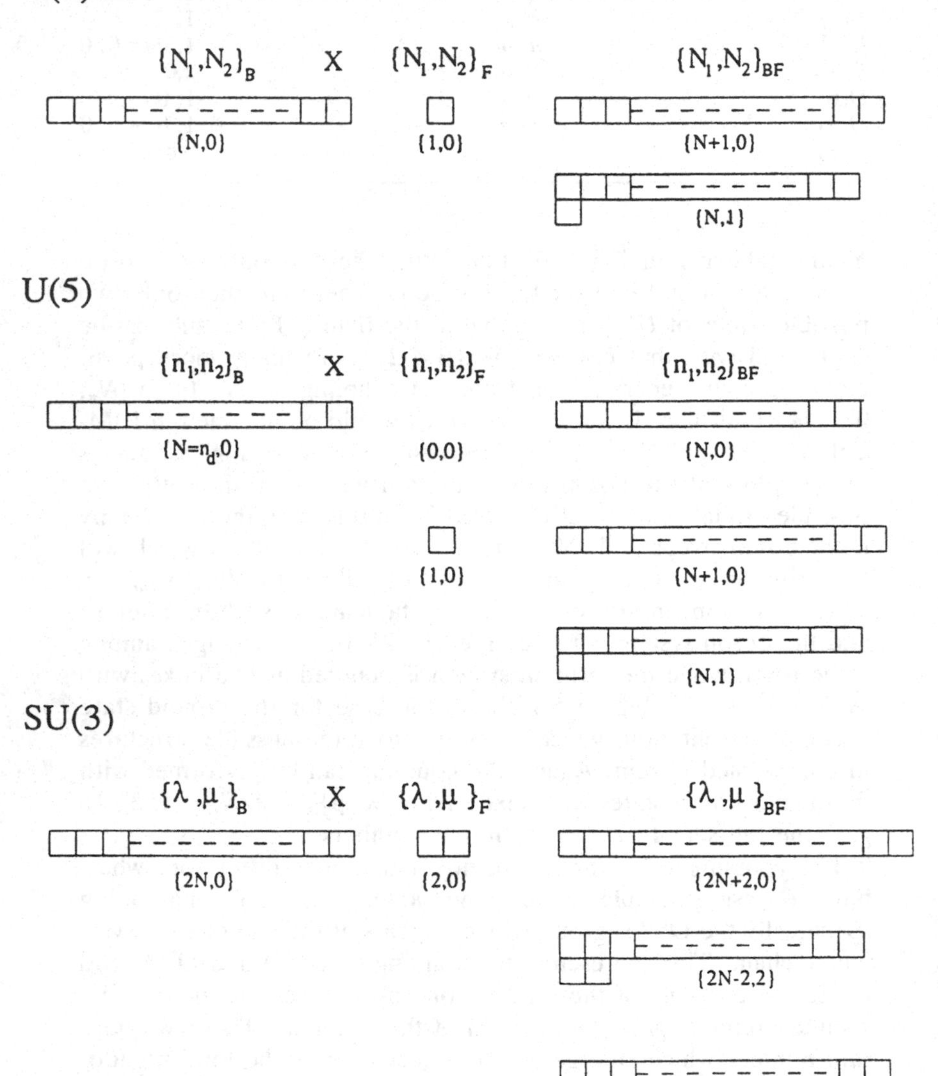

FIGURE 5.7. *Young tableaux for boson–fermion coupling.*

The final step is to define the reduction rules; that is, to determine which representations of, for example, $SU^{BF}(3)$ are contained in each of the two representations of $U^{BF}(6)$, and so on down the three chains. These results, along with the content of the other orthogonal groups not discussed above, are given in detail in Iachello and Van Isacker (1991) or Van Isacker, Frank and Sun (1984) and will now become apparent in the discussion of the predicted energy spectra in the three symmetry limits.

The spectra predicted for the three limits are shown in Figs. 5.8–5.10. The total structure contained within each is, in the gen-

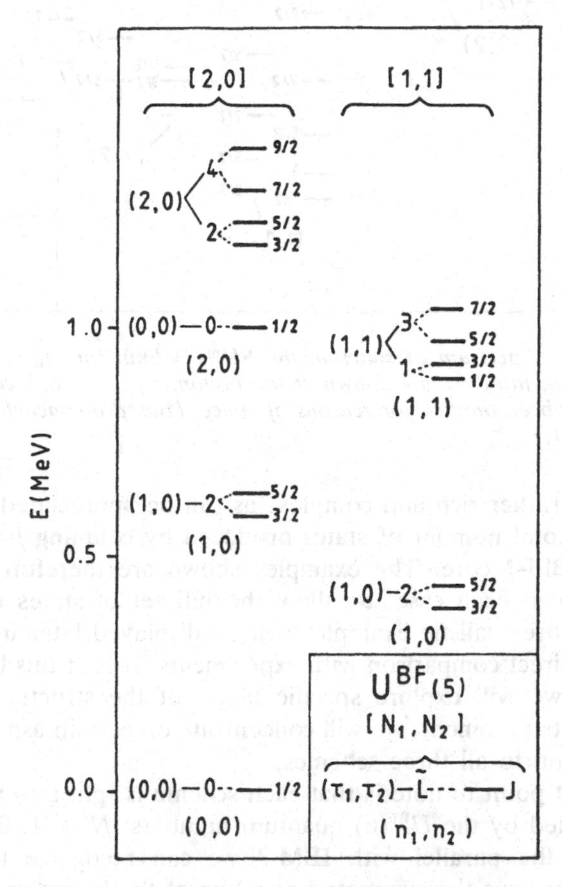

FIGURE 5.8. *Spectrum of states in the $U^{BF}(5)$ limit for $N_B = 1$, $N_F = 1$. The quantum numbers are shown at the bottom right. Based on Iachello and Van Isacker* (1991).

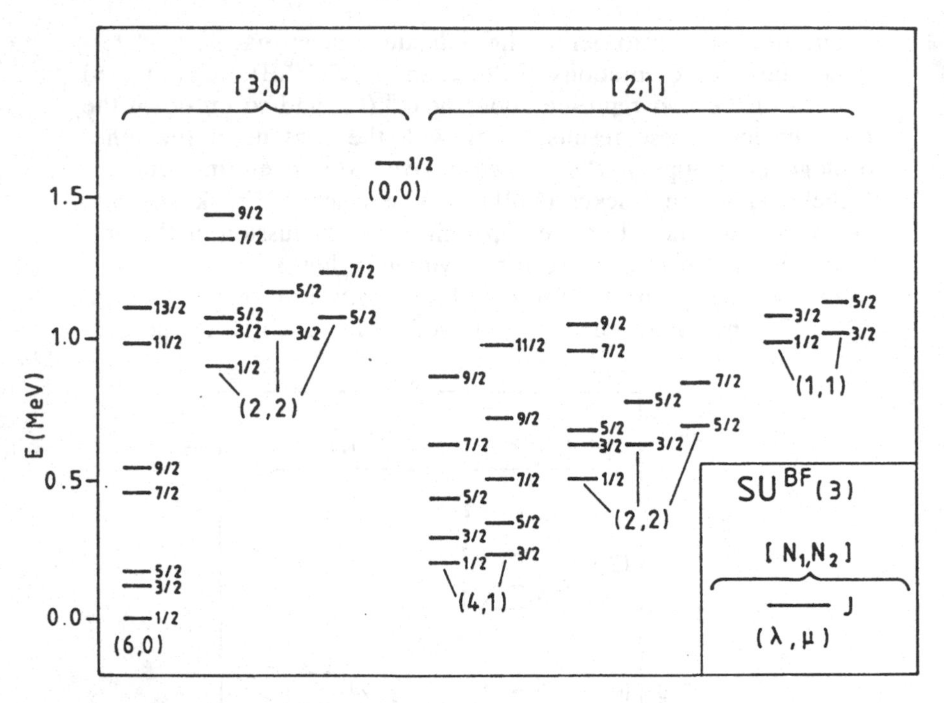

FIGURE 5.9. *Spectrum of states in the* $SU^{BF}(3)$ *limit for* $N_B = 1$, $N_F = 1$. *The quantum numbers are shown at the bottom right. In this case, the* $\tilde{L}$-*values have been omitted for reasons of space. (Based on Iachello and Van Isacker* (1991).)

eral case, rather rich and complex, as can be appreciated by imagining the total number of states produced by coupling $j = \frac{1}{2}$, $\frac{3}{2}$ and $\frac{5}{2}$ to an IBM-1 core. The examples shown are therefore only for one boson in each case, to allow the full set of states to be displayed. More realistic examples will be displayed later in the context of a direct comparison with experiments. It is at this later stage also that we will explore specific facets of the structure of each limit. For the moment, we will concentrate on certain aspects which are common to all three schemes.

The first point to note is that each scheme is split into two parts, distinguished by the $U^{BF}(6)$ quantum numbers $[N + 1, 0]$ and $[N, 1]$. From the parallel with IBM-2 we can recognize that these groups represent the symmetric coupling of the boson and fermion orbital motion, and the non-symmetric coupling. However, there is now a crucial physical difference between the IBFM case and IBM-

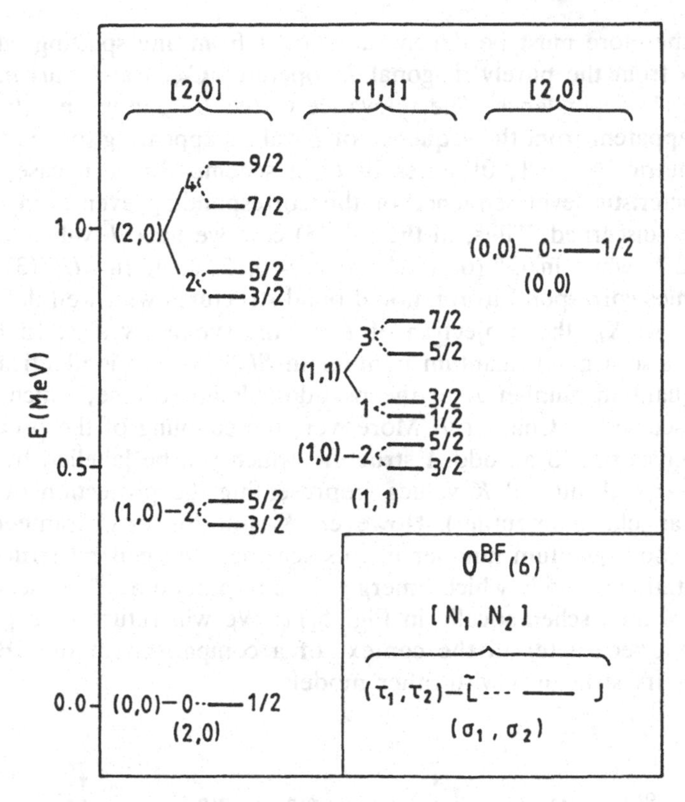

FIGURE 5.10. *Spectrum of states in the* $O^{BF}(6)$ *limit for* $N_B = 1$, $N_F = 1$. *The quantum numbers are shown at the bottom right. (Based on Iachello and Van Isacker* (1991).)

2. In the latter, we expect, and find empirically, that the contribution of C_{2U6} must be used to generate the "symmetry energy" which places the non-symmetric *n-p* modes at a significantly higher energy than the fully-symmetric states. We saw in Chapters 2 and 4 that this energy typically amounts to some 3 MeV. In the odd-*A* case, there are no grounds for an equivalent assumption. On the contrary, we will see that the non-symmetric states frequently constitute the ground and lowest states of the system.

The other global characteristic which is immediately evident is the appearance of closely spaced couplets of levels with $J = \tilde{L} \pm \frac{1}{2}$. This is the result of assuming a pseudo-spin independent Hamiltonian. The two states in a couplet differ only in the way in which the pseudo-spin is coupled to the rest of the angular momentum and therefore must be degenerate, apart from any splitting which arises from the purely diagonal J^2 operator $C_{2\text{Spin}3}$ and thus has a $J(J + 1)$ dependence. The relevance of the $\tilde{L}$ quantum number is also apparent from the sequence of $\tilde{L}$-values appearing for the fully symmetric $[N + 1, 0]$ states in each scheme. In each case, the characteristic level sequence of the corresponding even-even core can be discerned. Thus, in the $U^{\text{BF}}(5)$ case we have $\tilde{L}$-values of 0, 2, 0-2-4, while in $O^{\text{BF}}(6)$ we have $\tilde{L} = 0, 2, 2, 4$. In $SU^{\text{BF}}(3)$ the $\tilde{L}$-values correspond to rotational band structures with well defined values of K_L, the projection of $\tilde{L}$ on the symmetry axis. In fact, K_L is also a good quantum number in $SU^{\text{BF}}(3)$ and is identical to the quantum number $\tilde{\Lambda}$ of the pseudo-Nilsson scheme, which will be discussed in Chapter 7. Moreover, the coupling of the pseudo-spin gives rise to an odd-A structure which can be labelled by the familiar half-integral K values (representing the projection of the total angular momentum). However, K is no longer guaranteed to be a good quantum number in this scheme. The general structure of rotational bands which emerges in any pseudo-$SU^{\text{BF}}(3)$ scheme is illustrated schematically in Fig. 5.11. We will return to explore these aspects later in the context of a comparison of the IBFM symmetry structures with other models.

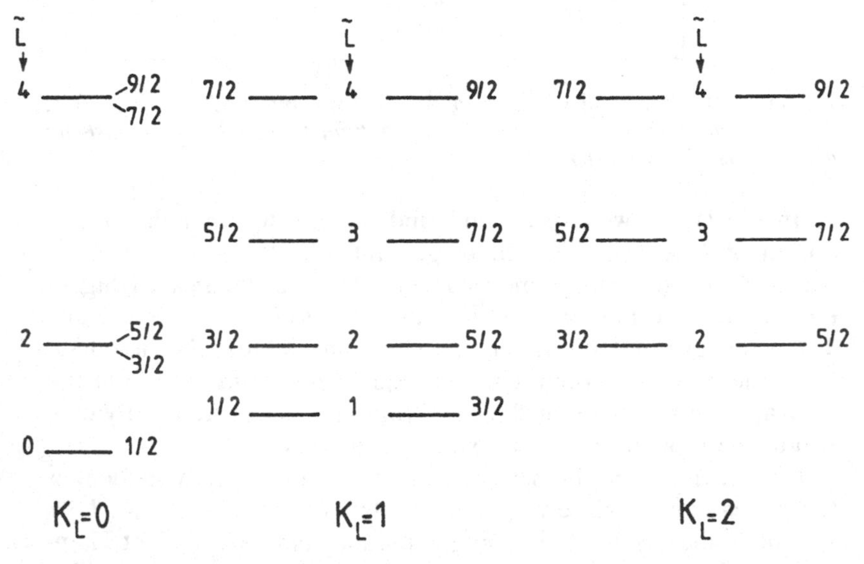

FIGURE 5.11. *Rotational band structures in a pseudo $SU^{\text{BF}}(3)$ scheme for $K_L = 0$, 1 and 2.*

It is, of course, evident from Figs. 5.8–5.11 that some sequences of $\tilde{L}$-values occur which are not characteristic of even-even nuclei. The $K_L = 1$ band of Fig. 5.11 and the $\tilde{L} = 1$ and 3 states of Fig. 5.10 are obvious examples. In the IBM-1 description of even-even nuclei, such states are absent because of the fully symmetric nature of the group representations. In the odd-even nucleus, however, the orbital motions of the core and particle may couple symmetrically or non-symmetrically, the latter giving rise to the "abnormal" $\tilde{L}$-values. Thus, comparison of Figs. 5.9 and 5.11 shows that the band structure with $[N_1, N_2] = [2, 1]$ and $(\lambda, \mu)_{BF} = (4N, 1)$ in Fig. 5.9 has $K_L = 1$ and is the counterpart of the "scissors" mode in the IBM-2 framework.

5.4.2. Electromagnetic Transition Rates

The symmetry assumed in the Hamiltonian can be extended to encompass the transition operators by writing the latter in terms of the generators of the group(s) under consideration. It is then possible to derive simple analytic expressions for the corresponding matrix elements between states which carry the quantum numbers of the group representations. This procedure entails a severe constraint in the overall description and, for the $U^B(6) \times U^F(12)$ case under consideration here, has only been applied to E2 moments. For magnetic dipole moments, the utility of a framework which makes no distinction between neutron and proton degrees of freedom seems questionable. In addition, the fermion part of the M1 operator cannot, to a good approximation, be written in terms of generators of the subgroups. As a result, the evaluation of M1 matrix elements is rather complex and will be omitted.

For E2 transitions, the relevant operator would be written

$$T(E2) = e_{BF}[Q_B + Q_F] \tag{5.42}$$

where the boson and fermion quadrupole operators each represent generators of the relevant group. Note also that in deriving analytic expressions and selection rules, the boson and fermion effective charges are assumed to be equal. Thus the transition operators required are just the appropriate generators of the sum groups, i.e.

$$O(6) : T(\mathrm{E2}) = e_{\mathrm{BF}}[G^{(2)}(0, 2) + G^{(2)}(2, 0)], \qquad (5.43)$$

$$SU(3) : T(\mathrm{E2}) = e_{\mathrm{BF}}[G^{(2)}(0, 2) + G^{(2)}(2, 0) - (\sqrt{7}/2)\, G^{(2)}(2, 2)] \qquad (5.44)$$

where $G^{(2)}(\tilde{l}, \tilde{l}')$ are given in (5.34). Thus the fermion operators of (5.43) involve couplings of the type $(j, j') = (\frac{1}{2}, \frac{5}{2})$ or $(\frac{1}{2}, \frac{3}{2})$ but *not* $(\frac{3}{2}, \frac{3}{2})$ or $(\frac{5}{2}, \frac{5}{2})$ or $(\frac{3}{2}, \frac{5}{2})$. For $U(5)$, the generators would be $(d^{\dagger}\tilde{d})^{(2)}$ and hence $G^{(2)}(2, 2)$. However, as in the even-even case, this would give rise to predictions of zero transition strengths and large quadrupole moments, at variance with the expectations of a vibrational nucleus. It has been the practice, therefore, to use the operator of eq. (5.43) to derive analytic predictions, although a more realistic approach would undoubtedly involve all the components of the quadrupole operator. In the $SU(3)$ case, while eq. (5.44) gives the exact generator, it is also possible, just as in IBM-1, to vary the relative strength of the two terms. This in turn suggests the relevance of a "consistent Q" approach in odd-A nuclei which has, in fact, been explored.

The E2 selection rules associated with each of the symmetry groups are summarized in Table 5.3 in terms of the quantum numbers appropriate to each group. Note that, in some cases, the forbiddenness of a particular transition stems from an exact cancellation of the boson and fermion contributions; the σ selection rule in $O^{\mathrm{BF}}(6)$ is one such example and, as we shall see, can be easily broken by relaxing the equality of boson and fermion effective changes. In other cases, the selection rule stems from both the boson and fermion E2 matrix elements being zero.

The fact that the transition operator is independent of the pseudo-spin gives rise to a further simplification. The dependence of E2 matrix elements on the total angular momentum J can be

TABLE 5.3. *E2 selection rules for the bose–fermi groups.*

$U(6)$	$\Delta N_1 = \Delta N_2 = 0$
$U(5)$	$\Delta (n_1 + n_2) = \pm 1$
$SU(3)$	$\Delta\lambda = \Delta\mu = 0$
$O(6)$	$\Delta\sigma_1 = \Delta\sigma_2 = 0$
$O(5)$	$\Delta(\tau_1 + \tau_2) = \pm 1$
$O(3)$	$\Delta\tilde{L} = 0, 1, 2$

described in terms of a simple geometrical factor $F\ (\bar{L}_iJ_i;\ \bar{L}_fJ_f)$. Thus

$$\langle N\alpha\bar{L}_iJ_i||T(\mathrm{E2}||N\alpha'\bar{L}_fJ_f\rangle \propto F(\bar{L}_iJ_i;\ \bar{L}_fJ_f) \tag{5.45}$$

where α, α' denote other quantum numbers and

$$F(\bar{L}_iJ_i;\ \bar{L}_fJ_f) = (-1)^{\bar{L}_i+1/2+J_f}\sqrt{(2J_i+1)(2J_f+1)} \cdot \begin{Bmatrix} \bar{L}_i & J_i & \frac{1}{2} \\ J_f & \bar{L}_f & 2 \end{Bmatrix}. \tag{5.46}$$

This feature gives rise to an extension of the Alaga rules used in rotational nuclei. In the $SU^{\mathrm{BF}}(3)$ case, the following general expression for transitions within an irrep can be obtained in the large-N limit

$$B(\mathrm{E2};\ J_i\bar{L}_iK_{L_i} \rightarrowtail J_f\bar{L}_fK_{L_f}) = \left(\frac{5}{16\pi}\right) e\ Q_0^2(2J_f+1)(2\bar{L}_i+1) \cdot\langle\bar{L}_iK_{L_i}2K_{Lf} - K_{L_i}|\bar{L}_fK_{Lf}\rangle^2 \begin{Bmatrix} \bar{L}_i & J_i & \frac{1}{2} \\ J_f & \bar{L}_f & 2 \end{Bmatrix}^2. \tag{5.47}$$

The Alaga rule Clebsch–Gordan has now been replaced by one describing pseudo-orbital variables.

5.4.3. Single-Particle Transfer

Two situations can be met in the single-nucleon transfer problem. In the first, the transfer takes place between an even-even and an odd-even nucleus with the same number of bosons, i.e. ($N_{\mathrm{B}} = N$, $N_{\mathrm{F}} = 0$) $\leftrightarrow$ ($N_{\mathrm{B}} = N$, $N_{\mathrm{F}} = 1$). The lowest order operator in this case is simply

$$T_j^\dagger = \xi_j a_j^\dagger \tag{5.48}$$

or its hermitian conjugate. Clearly, the transfer takes place between the $J = 0^+$ ground state of the even-even nucleus and a state with $J = j$ in the odd-even nucleus. Thus in the $U^{\mathrm{B}}(6) \times U^{\mathrm{F}}(12)$ scheme discussed earlier, three constants have to be determined, corresponding to $j = \frac{1}{2}$, $\frac{3}{2}$ and $\frac{5}{2}$. Note, however, that the ratio of transfer matrix elements (or "structure factors" as they are sometimes called) for two states of the same spin is uniquely determined.

The second class of transfer reaction is that in which the boson number changes by one ($N_B = N$, $N_F = 1$) $\leftrightarrow$ ($N_B = N + 1$, $N_F = 0$). The lowest order operator takes a more complex from

$$T_j^\dagger = \theta_j(s^\dagger\tilde{a}_j)^{(j)} + \sum_{j'} \theta_{jj'}(d^\dagger\tilde{a}_{j'})^{(j)}. \tag{5.49}$$

In this case, the number of parameters rapidly becomes prohibitive when more than one orbit is available. However, certain simplifying assumptions can be made to render the operator more practical. Staying with our example of the $U^B(6) \times U^F(12)$ scheme, it has been proposed by Bijker (1984) that, since the single-particle orbits therein are characterised by good values of the pseudo-orbital angular momentum, $\tilde{l} = 0$ and 2, it may be reasonable to assume that the operator for $j = \frac{1}{2}$ transfer transforms as a spherical tensor of rank 0, and that for $j = \frac{3}{2}$ and $\frac{5}{2}$ as a spherical tensor of rank 2. This yields the expressions

$$\begin{aligned}
T_{1/2}^\dagger &= \theta_{1/2}(s^\dagger\tilde{a}_{1/2})^{(1/2)} \\
&\quad + \theta'_{1/2}\{-\sqrt{2}\,(d^\dagger\tilde{a}_{3/2})^{(1/2)} + \sqrt{3}\,(d^\dagger\tilde{a}_{5/2})^{(1/2)}\}, \\
T_{3/2}^\dagger &= \theta_{3/2}\,\{(s^\dagger\tilde{a}_{3/2})^{(3/2)} - (d^\dagger\tilde{a}_{1/2})^{(3/2)}\} \\
&\quad + \theta'_{3/2}\,\{\sqrt{7/2}\,(d^\dagger\tilde{a}_{3/2})^{(3/2)} + \sqrt{3/2}\,(d^\dagger\tilde{a}_{5/2})^{(3/2)}\}, \\
T_{5/2}^\dagger &= \theta_{5/2}\,\{(s^\dagger\tilde{a}_{5/2})^{(5/2)} - (d^\dagger\tilde{a}_{1/2})^{(5/2)}\} \\
&\quad + \theta'_{5/2}\,\{-(d^\dagger\tilde{a}_{3/2})^{(5/2)} + 2\,(d^\dagger\tilde{a}_{5/2})^{(5/2)}\}.
\end{aligned} \tag{5.50}$$

With the further simplifying assumption that $\theta_{1/2} = \theta'_{1/2}$, the number of free parameters in the final form of the operator is reduced to five.

The pseudo-spin symmetry also gives rise to a rather simple general selection rule for the transfer matrix elements, namely

$$\Delta\tilde{L} = \tilde{l} \tag{5.51}$$

where $\Delta\tilde{L}$ is the change in total pseudo-orbital angular momentum and $\tilde{l}$ is the pseudo-orbital angular momentum of the transferred nucleon. Thus, for example, for $\tilde{l} = 0$ and 2 and transfers involving a 0^+ ground state, it is clear that states in the odd-even nucleus with odd values of $\tilde{L}$ cannot be populated. This rule gives rise to rather distinctive patterns of transfer strength and we will return to this feature later when we compare with data.

The treatment of the transfer operator given above is algebraically motivated and thus gives rise to a number of free parameters but maintains the symmetry properties of the system. It is possible, within the general IBFM framework, to derive expressions for the various coefficients of eqs. (5.48) and (5.49) in terms of the occupation amplitudes of the spherical single particle orbits. However, treatments to date have been largely based on generalized seniority schemes and hence their applicability in well-deformed nuclei remains open to question.

5.5. PHYSICAL REQUIREMENTS FOR BOSE–FERMI SYMMETRY

The discussion of the preceding sections showed how, in appropriate circumstances, dynamical symmetry schemes can be constructed for odd-A nuclei. While one class of symmetry was chosen as an example to pursue in detail, there are many other symmetry schemes which have already been studied, and undoubtedly many more possible, albeit of steadily increasing complexity. However, up to this point, the constraints cited in searching for possible symmetries have been purely algebraic, namely, the ability to couple the boson and fermion group structures at some stage. Moreover, the schemes themselves then represent essentially algebraic constructs. The next question to be addressed is to what extent the physical structure contained within each is applicable to the collective structure of real odd-mass nuclei.

To consider where, and if, a particular symmetry might be valid, it is most instructive to think in terms of the effective Hamiltonian which each represents. In group theoretical language, this corresponds to a sum over the Casimir operators C^i appropriate to the specific group chain, i.e.:

$$H_{\text{sym}} = \sum_i \alpha_i C^i. \tag{5.52}$$

However, to obtain a physical insight into the structure implied by the symmetry, it is more useful to regroup the various interactions represented by the right-hand side (r.h.s.) of (5.52) into the more familiar form of the general core-particle IBFM Hamiltonian of eq. (5.7), viz.

$$H_{\text{sym}} = H_{\text{F}} + H_{\text{B}} + H_{\text{BF}}. \tag{5.53}$$

Clearly, since the form of the Casimir operators in (5.52) are uniquely specified for any group chain in terms of the basic annihilation and creation operators for bosons and fermions, it is always possible to rewrite H_{sym} in the form of eq. (5.53). The resultant structure and strengths of the three terms on the r.h.s. will then be inextricably linked via the specific Casimir operators involved, and the associated α_i. The specific relationship between the two forms has been given in detail elsewhere for various symmetry schemes. Here it is sufficient to point out that writing the symmetry Hamiltonian in the form of (5.53) immediately yields three general characteristics which constrain the regions in which one can expect to find an empirical realization of a particular scheme.

5.5.1. Core Structure

The core structure implied by H_B in (5.53) is invariably that of one of the three limiting boson symmetries, i.e., $SU(5)$, $SU(3)$ or $O(6)$. Thus a first criterion for a boson-fermion symmetry to be applicable is that the neighbouring even-even nuclei should display the corresponding type of symmetry structure. However, it is only necessary that H_B yield an exact description of a neighbouring even nucleus if the question of supersymmetry is being explored (see final section). A boson–fermion symmetry scheme represents only a description of an odd-A nucleus, and it is certainly possible that the core structure assumed in an odd-A Hamiltonian is different in detail from that of the neighbouring even-even nuclei, particularly in a situation involving strong core-particle coupling. Nevertheless, clearly the type of structure (vibrational, rotational, γ-unstable) implicit in the scheme under consideration should be in evidence.

5.5.2. Single-Particle Energies

The particular Casimir operators appropriate to a given boson-fermion group chain mandate a very specific single-particle structure in the equivalent core-particle Hamiltonian, both in terms of the j-values involved, and the relative energies of the single-particle orbits. Indeed, this constraint is undoubtedly the most severe in terms of the limitations it imposes on the likely regions of applicability of the various boson–fermion symmetry schemes. Firstly, most existing schemes are restricted to a few single-particle orbits, less than would be necessary to include an entire major shell, for instance. Thus it is necessary to find nuclei and/or subsets of states within nuclei where only these orbits are likely to play a dominant role. Clearly this problem is most serious in the presence of a strong deformed field, which will mix single-particle states throughout a major shell and can also bring in components from neighbouring shells.

The second restriction implied by a symmetry involves the relative energies of the single-particle states. In the most general form of the IBFM Hamiltonian, the pairing interaction is accounted for by taking spherical quasiparticle energies at the input to the calculation, and also by modifying the various boson–fermion interactions by factors dependent on the spherical occupation amplitudes of the orbits involved (as discussed in Section 5.2). Thus the conditions required to generate a particular dynamical symmetry are best thought of in terms of a set of quasiparticle energies and occupation amplitudes in the spherical limit.

In general, the fermionic part of the Hamiltonian arising from a symmetry scheme can be deduced by considering the various groups occurring in the chain, and the fermionic quantum numbers appropriate to each group, as given in Table 5.2. The assumed energy of each single-particle orbit is then obtained from the symmetry eigenvalue expression, in the same way as for the states in the nucleus.

An example will serve to illustrate the points discussed above. Consider an $O(6)$ symmetry for a boson–fermion structure of the type $U^{B}(6) \times U^{F}(12)$. Then, assuming coupling of the boson and fermion degrees of freedom at the level of $U(6)$, the group chain decomposition has been given in eq. (5.31).

The Hamiltonian is thus formed from the Casimir operators as

$$H_{\text{sym}} = AC_{2U6} + BC_{2O6} + CC_{2O5} + DC_{2O3} + EC_{2\text{Spin}3}. \tag{5.54}$$

Retaining only those terms which contribute to relative energies, the excitation energy spectrum is given in terms of the quantum numbers of each subgroup as

$$\begin{aligned} E_x = A[N_1(N_1+5) + N_2(N_2+3)] + 2B[\sigma_1+4) + \sigma_2(\sigma_2+2)] \\ + 2C[\tau_1(\tau_1+3) + \tau_2(\tau_2+1)] \\ + 2D[L(L+1)] + E[J(J+1)]. \end{aligned} \tag{5.55}$$

The values of these quantum numbers for the fermionic space have been given in Table 5.3, and are illustrated again in Fig. 5.12.

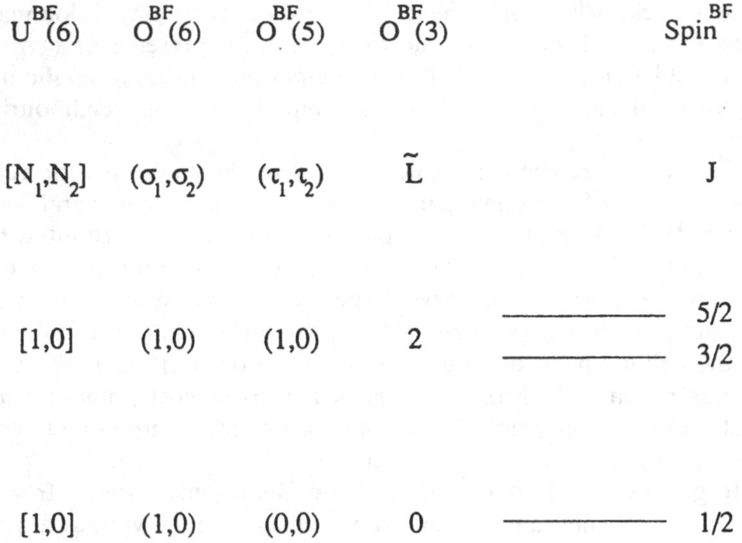

$U^{BF}(6)$	$O^{BF}(6)$	$O^{BF}(5)$	$O^{BF}(3)$		Spin^{BF}
$[N_1,N_2]$	(σ_1,σ_2)	(τ_1,τ_2)	$\tilde{L}$		J
				———	5/2
[1,0]	(1,0)	(1,0)	2	———	3/2
[1,0]	(1,0)	(0,0)	0	———	1/2

FIGURE 5.12. *Fermion quantum numbers in the $O^{BF}(6)$ limit.*

Substitution of these into eq. (5.55) leads to the following single-quasiparticle energies:

$$\begin{aligned} \epsilon_{1/2} &= 6A + 10B + (3/4)E, \\ \epsilon_{3/2} &= 6A + 10B + 8C + 12D + (15/4)E, \\ \epsilon_{5/2} &= 6A + 10B + 8C + 12D + (35/4)E. \end{aligned} \tag{5.56}$$

Thus in this case, the fermionic energy spectrum is simply that which would result from coupling a pseudo-spin $\frac{1}{2}$ to a 1 boson system obeying eq. (5.55). There is, however, one additional indirect contribution to H_F which arises from the $U^{BF}(6)$ Casimir operator. This operator, aside from its direct contribution to the single-particle energies evident in eq. (5.56), also gives rise to so-called monopole terms in the boson–fermion interaction, which are proportional to

$$2A[(s^\dagger s)^{(0)}\,(a^\dagger_{1/2}\tilde{a}_{1/2})^{(0)}]^{(0)}, \qquad 2A[(d^\dagger\tilde{d})^{(0)}(a^\dagger_{3/2}\tilde{a}_{3/2})^{(0)}]^{(0)}$$

and

$$2A[(d^\dagger\tilde{d})^0(a^\dagger_{5/2}\tilde{a}_{5/2})^0]^{(0)}. \tag{5.57}$$

The origin of these contributions is evident from the explicit form of C_{2U6} given in eq. (5.40). The three terms of eq. (5.57) are not independent since the bosonic part is simply the number operator for s (or d) bosons. Thus, recalling that

$$\hat{N} = \hat{n}_s + \hat{n}_d = (s^\dagger s)^{(0)} + \sqrt{5}(d^\dagger \tilde{d})^{(0)} \tag{5.58}$$

then if we choose to retain the d-boson monopole term in H_{BF}, the s term can be removed by virtue of (5.58), but this procedure produces an additional contribution to the $j = \frac{1}{2}$ single-particle energy, namely

$$\Delta\epsilon_{1/2} = 2AN. \tag{5.59}$$

In fact, this contribution is crucial in obtaining the correct empirical spacing of single-particle states in the Pt region.

The implicit single-particle energies for other symmetry chains can be deduced in a similar manner to the example outlined above, if the fermionic quantum numbers of the various subgroups are known. The example shows that, in general, the fermionic Hamiltonian is closely linked to the bosonic one, so that the single-particle energies are linked to the energies of the first few core states. For instance, the energies given by eq. (5.56) and illustrated in Fig. 5.12 correspond to a situation where the separation of the $j = \frac{1}{2}$ and $\frac{3}{2}$, $\frac{5}{2}$ states is simply the 2^+ energy in the core, although the indirect contribution from the $U^{BF}(6)$ group relaxes this constraint. This feature highlights a major reason for the likely scarcity of good examples of bose–fermi symmetry. The formalism requires a specific relationship between the energies of the low-lying states in the even-even core, and the underlying single-quasiparticle energies in the odd-A nucleus. This can be contrasted with a general IBFM Hamiltonian, or indeed any core-particle Hamiltonian, where the two sets of quantities can be chosen independently.

5.5.3. Boson–Fermion Interaction

The final part of the general Hamiltonian which is constrained by the assumption of a symmetry is H_{BF}, which represents the core-particle interaction. In the general IBFM Hamiltonian, we have seen that H_{BF} is normally written in terms of three components, monopole, quadrupole and exchange, which have the form given in eqs. (5.10)–(5.12).

In general, each Casimir operator in a given symmetry group chain will make contributions to one or more of these terms. As an example, the presence or absence of such contributions in the case of the three symmetries stemming from the decomposition of $U^{B}(6) \times U^{F}(12)$ are indicated in Table 5.4. However, there is an additional interaction arising from the Casimir operator of $U^{BF}(6)$ which is *not* present in the general IBFM Hamiltonian defined earlier. This takes a form proportional to

$$\sum_{j} (2j + 1)^{1/2}[(s^{\dagger}\tilde{d})^{(2)}(a_{j}^{\dagger}\tilde{a}_{1/2})^{(2)}]^{(0)} + \text{h.c.} \tag{5.60}$$

and, as such, is not symmetric in the boson part and, in fact, represents an *s-d* boson exchange force.

Examination of Table 5.4 shows that a given symmetry chain will give rise to a specific relationship between the various terms in the equivalent IBFM Hamiltonian, which in turn implies constraints on the relative occupation amplitudes of the single-particle orbits, and on the overall multiplying constants in the interaction.

Thus it is apparent that the algebraically derived symmetry Hamiltonian, can, in principle, be "dissected" to reveal the underlying physical assumptions inherent in it, in terms of core structure, single-particle energies and occupation amplitudes and interaction strengths, and hence judged with regard to its applicability in a particular region of the nuclear chart. The constraints discussed in this section are considerable, and it is therefore likely that good examples of an exact symmetry will be rare.

5.6. RELATIONSHIP WITH OTHER MODELS

One of the less satisfying aspects of the study of boson–fermion symmetries is that the purely algebraic derivation of the predicted properties does not carry with it an immediate physical interpre-

TABLE 5.4.

H_{IBFM}	C_{2U6}	C_{2U5}	C_{2SU3}	C_{2O6}	C_{2O5}	C_{2O3}	C_{2Spin3}
H_{BF}^{M}	X	X					
H_{BF}^{Q}	X		X	X			
H_{BF}^{E}	X	X	X	X	X	X	X

tation. This situation can be contrasted with geometrical approaches to the problem, which tend to assume a specific limiting shape for the core nucleus, and hence the average potential felt by the single particle, so that the resulting characteristics of the odd-A spectrum can be understood on a more intuitive basis. It is thus useful to consider the relationship between the two approaches in certain limits and to use the comparison to gain a more physical insight into the symmetry structures. Such a comparison is also important in establishing the principal advantages or disadvantages of each framework, and in highlighting where the most important differences are likely to appear.

We have already seen that the three symmetries which may be relevant to any boson–fermion group structure are $U(5)$, $SU(3)$, and $O(6)$, which generate core-particle Hamiltonians involving spherical, rotational, or γ-unstable core structures. The three analogous approaches would thus be the weak or intermediate coupling, Nilsson and particle γ-unstable models. The last of these is not generally familiar and does not therefore lend itself to the role of providing a straightforward physical interpretation for the algebraic scheme. In the case of $U(5)$, we remarked in Section 5.3 that, in the presence of spherical symmetry, one might expect to maintain the symmetry structure of the core in the odd-A nucleus. This limit should therefore represent rather well the weak coupling of spherical single-particle orbits and a vibrational core. Indeed, since the quantum number labelling the irreps of $U^{B}(5)$ is n_d, the d-boson number, the core states should remain rather pure in a $U(5)$ bose–fermi description. However, bose–fermi coupling at the level of $U(6)$ introduces the Casimir operator of eq. (5.40) into the Hamiltonian which does contain $\Delta n_d = \pm 1$ terms. Thus, just as the even-even $U(5)$ description offers the recipe to introduce a considerable degree of anharmonicity into the basic description of a spherical vibrator, so the odd-even $U(5)$ scheme can go beyond the strict conditions of weak coupling. It was pointed out in the introductory discussion of core-particle coupling that the data suggests that the core-state mixing term in the Hamiltonian cannot be ignored. A comparison of $U(5)$ bose–fermi symmetry with experiment will be undertaken in the next section.

One advantage of $U(5)$ then, is that the basis states are not strongly mixed, and that the low-lying states of the predicted scheme can be associated with one, or a few, core-particle components of the type of eq. (5.3). The case of $SU(3)$ symmetry is very different. Here, from our earlier discussion, we can expect the strong-coupling limit in which the spherical basis states are strongly mixed to form the description of a rotational (deformed) nucleus. One obvious route to establishing a more physical understanding of the structure of the symmetry scheme in this case is via a direct (if possible) comparison with the Nilsson model.

5.6.1. The *SU*(3) Symmetry and the Nilsson Model

It has proved possible to establish a semi-quantitative link between the Nilsson model and the $SU(3)$ limit of the $U^{B}(6) \times U^{F}(12)$ group structure. This link leads to a greatly enhanced understanding of the symmetry itself and of the role of the various Casimir operators, particularly that of $U^{BF}(6)$. The salient features of the method used will therefore be outlined here, since they define a general approach which can be applied to link any of the possible boson–fermion $SU(3)$ symmetries to the appropriate region of the Nilsson scheme. The reader is referred to the orginal paper of Warner and Bruce (1984) for more detail.

The $SU(3)$ limit of $U^{B}(6) \times U^{F}(12)$ requires a rotational core structure coupled to single-particle orbits with $j = \frac{1}{2}$, $\frac{3}{2}$, and $\frac{5}{2}$. The aim is now to connect the resultant band structure with that of the Nilsson model orbits stemming from the same three shell-model states. The $p_{1/2}$, $p_{3/2}$ and $f_{5/2}$ neutron orbits in the $N = 82$–126 shell represent an appropriate choice, since they are well separated from the remaining negative-parity orbits. This then points to the *W* nuclei around $A = 184$ as the likely region of applicability. In nuclei of lower mass in the well deformed rare earth region, the Fermi surface will be progressively further from the single-particle orbits of interest, while for higher masses the core structure begins to depart from axial symmetry.

The structure of the $SU(3)$ level scheme is illustrated in Fig. 5.13, the corresponding Hamiltonian being constructed from the Casimir operators of the groups $U^{BF}(6)$, $SU^{BF}(3)$, $O^{BF}(3)$ and Spin(3) (see eq. (5.30)). The states thus group into the various (λ, μ) representations of $SU^{BF}(3)$ and within each representation, the equivalent of one or more odd-A rotational bands can be assigned and labelled by an appropriate value of the K quantum number which, in a geometrical framework, denotes the projection of the total angular momentum along the symmetry axis of the deformed nucleus.

In order to establish any sort of quantitative link between the

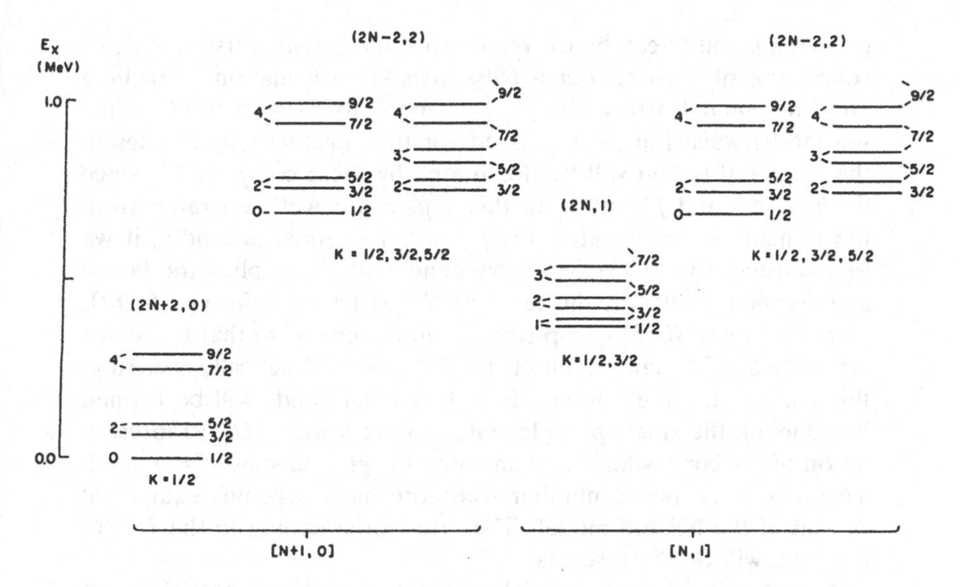

FIGURE 5.13. *The SU(3) scheme of* $U^B(6) \times U^F(12)$ *for N bosons. Levels are labelled to the left by* $\bar{L}$ *and to the right by J. The quantum numbers* $[N_1, N_2]$ *of* $U^{BF}(6)$ *and* (λ, μ) *of* $SU^{BF}(3)$ *are also shown. The approximate K-values of the bands contained within each irrep are indicated.*

symmetry structure and that predicted by the Nilsson model, it is first necessary to examine the bases used in the two approaches. Both methods represent core-particle coupling Hamiltonians applicable to regions of strong deformation, but if the core and particle states included in each differ substantially, it will not be possible to couch one description in terms of the other. In fact, at first glance, it seems that this is indeed the situation. The Nilsson model includes a far more complete single particle space, involving all relevant major shells, while the symmetry basis is limited to three specific orbits. Conversely, the symmetry framework includes a far more complete collective description of the core, incorporating many intrinsic excitations, while the Nilsson model only considers the ground-state rotational band. [Generally the Nilsson model is not portrayed explicitly in terms of core-particle coupling, but rather in terms of intrinsic single-particle states in a deformed potential. However, the introduction of the rotational motion in this framework in fact corresponds to coupling the single-particle states to the ground state rotational band of an axially symmetric rotor, as shown in eq. (5.6).] Nevertheless, these apparent sources of incompatibility can be overcome to some extent. Firstly, it is, of course, possible to restrict a Nilsson model calculation to include only shell-model orbits with $j = \frac{1}{2}, \frac{3}{2}$ and $\frac{5}{2}$. More realistically, Nilsson-model wave functions derived for the negative parity states in the $A = 180$ region will be dominated by these components, since the $p_{1/2}$, $p_{3/2}$ and $f_{5/2}$ orbits in this region are well separated from the remaining $N = 5$ states, the $f_{7/2}$ and $h_{9/2}$ orbits. Secondly, if we first consider the $SU(3)$ symmetry generated by coupling the boson and fermion groups at the level of $SU(3)$ (route 2 in eq. (5.28)), then the boson $SU(3)$ group remains in the chain, so that the states carry the $SU(3)$ quantum numbers of the core. Practically speaking, this implies that the lowest odd-A rotational bands will be formed by coupling the single-particle states to the lowest $SU(3)$ representation of the core, which contains only the ground state $K = 0$ band. Thus, for large boson number, the core basis becomes equivalent to that of the Nilsson model. The effect of switching to the $U^{BF}(6)$ coupling will be discussed later.

A summary of the essential ingredients and approach in each framework is given in Fig. 5.14. The relative energies of the single-particle states shown for the $U^B(6) \times U^F(12)$ system are to be considered as quasiparticle energies in the spherical limit and result from the considerations outlined in Section 5.5.2. The starting basis then comprises these states coupled to the $SU(3)$ core states and, for the lower lying bands, these latter will be restricted to the ground state rotational band, for the reasons mentioned above. The interaction applied involves quadrupole-quadrupole and exchange terms. Note that in this framework, the pairing occupation amplitudes refer to spherical orbits, and it is the exchange term which must allow for the effects of deformation on these quantities.

In the Nilsson treatment, the spherical single-particle states are first mixed via a quadrupole interaction to generate the Nilsson orbits shown on the right. The solid lines denote orbits stemming from the $p_{1/2}$, $p_{1/3}$ and $f_{5/2}$ shell model states while the dashed lines indicate orbits originating from the $f_{7/2}$ and $h_{9/2}$ states, which are outside the $U^B(6) \times U^F(12)$ basis. The core states are then coupled via the introduction of the rotational motion, and the Coriolis and pairing interactions are also included. In this case, the occupation amplitudes are computed for the orbit of the *deformed* single-particle potential.

The link between the two bases can be established by comparing the single-particle structure of the wave functions in each case, as

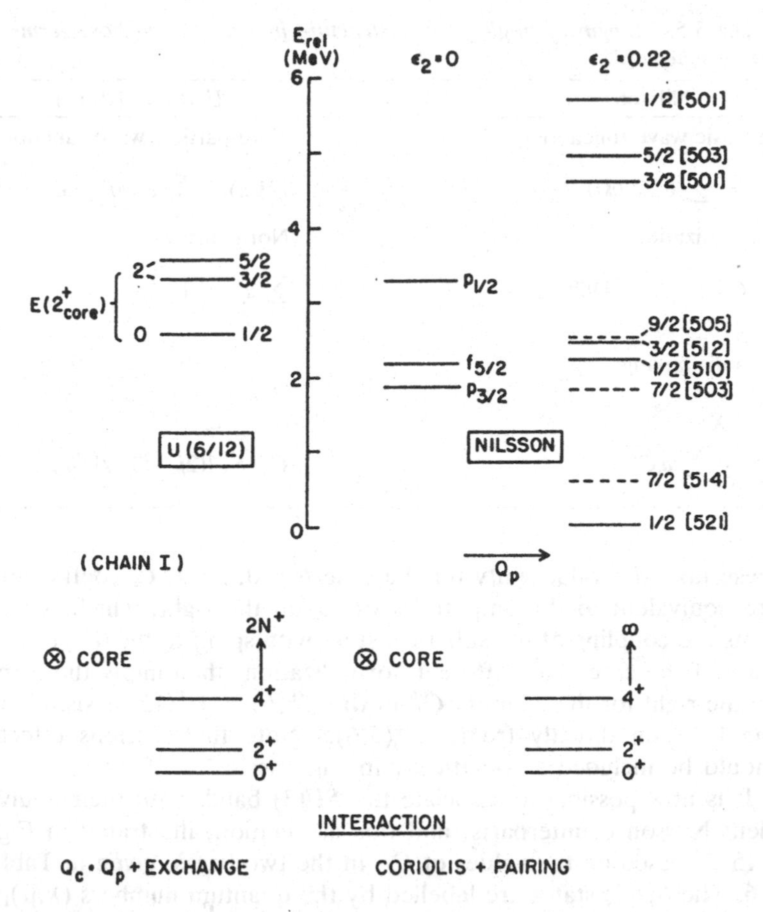

FIGURE 5.14. *Comparison of the single particle energies assumed in the* $U^B(6) \times U^F(12)$ *(left) and Nilsson (right) schemes. In the Nilsson scheme, both spherical and deformed states are shown. The dashed levels denote orbits stemming from the* $f_{7/2}$ *and* $h_{9/2}$ *shell model states.*

indicated in Table 5.5. In the Nilsson scheme, the single-particle structure for each orbit is contained in the intrinsic wave function, χ_Ω, in terms of the spherical amplitudes C_{jl}. The Coriolis interaction produces a mixing of the pure orbits, represented by the amplitudes a_i, so that the final structure can be represented by the quantity C_{jl}^{eff} on the left. In terms of a core-particle basis, as used in the IBFM scheme, the appropriate coupling coefficients are simply proportional to the Clebsch–Gordan coefficients $\langle j\Omega R0 | JK\rangle$. Since

TABLE 5.5. *Origin of single-particle structure in Nilsson and bose–fermi SU(3) schemes.*

Nilsson	$U^B(6) \times U^F(12)$
Intrinsic wave functions:	Core-particle wave function:
$\chi_\Omega = \sum_{jl} C_{jl} \vert Nlj\Omega\rangle$	$\vert JK\alpha\rangle = \sum_{jR} a_{jR} \vert jR; J\alpha\rangle$
Normalization:	Normalization:
$\sum_\Omega C_{jl}^2 = (2j + 1)/2$	$\sum_{J,K} a_{jR}^2 = 1$
Coriolis mixing:	
$\sum_i a_i \chi_\Omega^i$	
$C_{jl}^{\mathrm{eff}} = \sum_i a_i C_{jl}^i$	$C_{jl}^{\mathrm{eff}} = \{(2j + 1)/2\}^{1/2} a_{j0}$

these take the value unity for the case $R = 0$, $j = J$, C_{jl} coefficients are equivalent to the amplitudes of a_{j0} on the right, which represents the coupling of the spherical state with spin j to the 0^+ ground state of the core. The different normalizations then imply the form on the right for the quantity C_{jl}^{eff} in the $U^B(6) \times U^F(12)$ basis. (This can be seen directly from eq. (5.6).) Note that Coriolis effects should be included automatically in this case.

It is now possible to associate the $SU(3)$ bands with their equivalent Nilsson counterparts, and this association, illustrated in Fig. 5.15, is based on the values of C_{jl}^{eff} in the two bases, given in Table 5.6. The $SU(3)$ states are labelled by the quantum numbers $(\lambda,\mu)_{BF}$ and the total pseudo-orbital angular momentum $\tilde{L}$ and form band structures of the type discussed in Section 5.4.1 and illustated in Fig. 5.11. For the Nilsson model bands, the wave functions correspond to Coriolis mixing amplitudes which were deduced many years ago for ^{185}W by Casten et al. (1972). Note that the very specific structure of the symmetry scheme, which is characterized by zero values of C_{jl}^{eff} for the states with odd values of the pseudo-orbital momentum $\tilde{L}$, emerges for the $\frac{1}{2}$[510] and $\frac{3}{2}$[512] bands in the Nilsson basis specifically as a result of their mutual Coriolis interaction. For the three higher-lying Nilsson bands no Coriolis mixing calculations are available, since these bands have not been identified empirically. Nevertheless, from the unperturbed wave functions, the association of bands is again clear. The C_{jl}^{eff} values

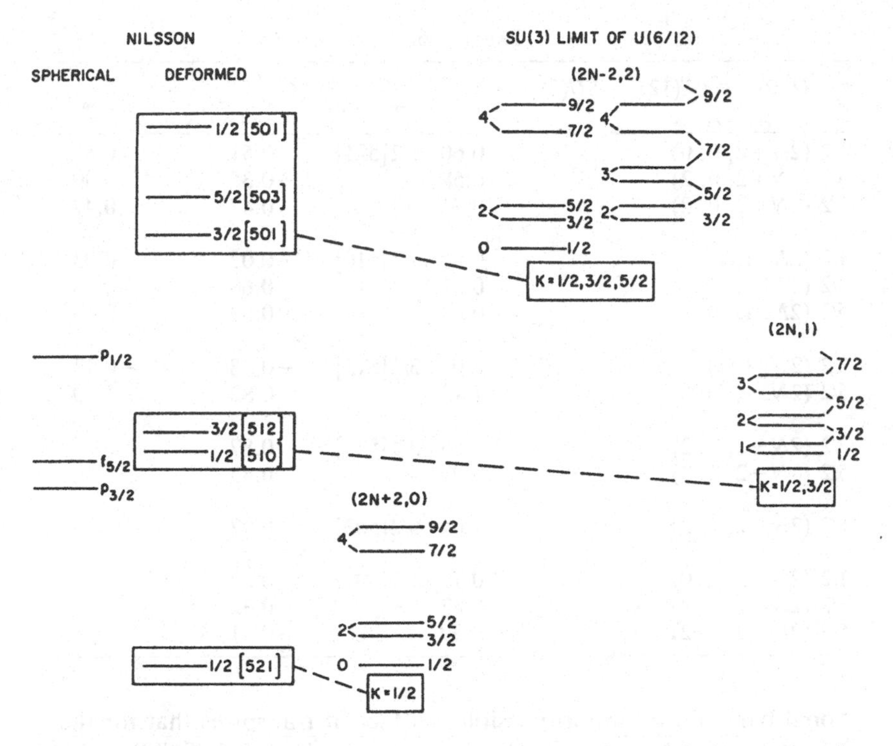

FIGURE 5.15. *Correspondence between Nilsson orbits and the rotational bands contained within the irreps of $SU^{BF}(3)$. Quantum numbers of the latter are as in Fig.* 5.13.

describing the $\frac{1}{2}[501]$ and $(2N - 2, 2)$ $K = \frac{1}{2}$ bands are almost identical in the two models, while the situation for the $\frac{3}{2}[501]$ and $\frac{5}{2}[503]$ bands mirrors that in the $\frac{1}{2}[510]$ and $\frac{3}{2}[512]$ case. For the two $\frac{5}{2}$ states, the symmetry structure incorporates a Coriolis interaction which shifts all the single-particle strength into one state. In fact, in the Nilsson scheme, the separation of, and matrix element between, the $\frac{3}{2}[501]$ and $\frac{5}{2}[503]$ orbits are essentially the same as for the $\frac{1}{2}[510]$ and $\frac{3}{2}[512]$ orbits.

The preceding discussion was based on the assumption of boson–fermion coupling at the level of $SU(3)$ in the $U^{B}(6) \times U^{F}(12)$ scheme, so that the core states involved in the lowest-lying odd-A bands were restricted to those of the ground state band of the core. However, it has been pointed out in earlier sections that experiment seems to favour coupling at the $U(6)$ level, so that admixtures of higher-lying core $SU(3)$ representations (i.e., vibra-

TABLE 5.6.

J	$U^{B}(6) \times U^{F}(12) - SU(3)$ $(\lambda, \mu, \tilde{L})$	C_{jl}^{eff}	$K[Nn_z\Lambda]$	C_{jl} unperturbed	C_{jl}^{eff}
1/2	$(2N+2, 0, 0)$	0.60	1/2[521]	−0.51	0.51
3/2	$(2N+2, 0, 2)$	−0.58		−0.35	−0.30
5/2	$(2N+2, 0, 2)$	0.71		0.47	0.47
1/2	$(2N, 1, 1)$	0.0	1/2[510]	−0.02	0.03
3/2	$(2N, 1, 2)$	−0.91		0.68	−0.78
5/2	$(2N, 1, 3)$	0.0		0.59	0.15
3/2	$(2N, 1, 1)$	0.0	3/2[512]	−0.38	−0.13
5/2	$(2N, 1, 2)$	1.12		0.82	1.00
3/2	$(2N-2, 2, 2)$	0.85	3/2[501]	0.87	
5/2	$(2N-2, 2, 3)$	0.0		0.43	
5/2	$(2N-2, 2, 2)$	1.04	5/2[503]	0.97	
1/2	$(2N-2, 2, 0)$	0.79	1/2[501]	0.82	
3/2	$(2N-2, 2, 2)$	0.32		0.32	
5/2	$(2N-2, 2, 2)$	−0.39		−0.41	

tional bands) are a priori possible. In fact, it transpires that for the lowest three Nilsson bands, the wave functions of their $SU(3)$ counterparts remain unchanged in the presence of $U^{BF}(6)$ coupling, so that the results derived above remain valid. For the higher three bands, however, the alternative scheme indeed results in a certain degree of fragmentation of the single-particle strength across a number of states, and it is encouraging to find that the empirical presence of such fragmentation has been pointed out in earlier studies of the W, Os nuclei.

The major deficiency of the $SU(3)$ limit of the $U^{B}(6) \times U^{F}(12)$ scheme is that, in a well deformed system, a larger spherical single-particle space is needed to achieve a complete description. It was pointed out in Section 5.3.2 that the pseudo-spin transformation always provides an $SU^{BF}(3)$ symmetry of the form of eq. (5.25) for a complete set of j-values from $j = \frac{1}{2}$ to some maximum j. Thus, for example, in the $N = 5$ neutron shell, the complete set of normal parity j-values, namely, $j = \frac{1}{2}, \frac{3}{2}, \frac{5}{2}, \frac{7}{2}$ and $\frac{9}{2}$, could be used. If one considers coupling of the single-particle space to only the ground state representation of $SU^{B}(3)$, with $(\lambda, \mu)_B = (2N, 0)$, one obtains a one-to-one correspondence with the Nilsson scheme for the same major shell. An example is shown in Fig. 5.16 for the $N = 5$ shell. The λ-values for the various orbits will depend on the boson number but will, in general, be large.

5.6.2. The Role of the Casimir Operator of $U^{BF}(6)$

The aim of establishing a link between the Nilsson model and the structure of an $SU(3)$ bose–fermi symmetry was to provide a physical insight into the algebraic scheme. One excellent example of how an improved understanding can emerge from this connection centers on the role which the operator C_{2U6} plays in the odd-A Hamiltonian. Recall from Section 5.4 that the group $U^{BF}(6)$ enters simply by virtue of the fact that it represents the first point at which the boson and fermion groups can be coupled in the algebra. Nevertheless, its presence is found to be essential in reproducing the majority of experimental examples. Why is this? What effect does C_{2U6} have on the predicted spectra? The link which we have established with the Nilsson scheme allows us to obtain a qualitative interpretation, which can then be put on a firm quantitative basis.

The situation is illustrated in Fig. 5.17. The upper part shows that the lowest Nilsson orbits from the $p_{1/2}$, $p_{3/2}$ and $f_{5/2}$ shell model states, and the lower part shows the corresponding bandheads in the $SU(3)$ scheme discussed earlier. The coefficients α and β represent the constants multiplying the Casimir operators of $U^{BF}(6)$ and $SU^{BF}(3)$, respectively, in the Hamiltonian. As the strength of the $U^{BF}(6)$ term is increased, relative to that of $SU^{BF}(3)$, the $K = \frac{1}{2}$ and $\frac{3}{2}$ bands come lower in energy, and eventually constitute the ground state structure as shown on the lower right. The corresponding situation in the Nilsson scheme is indicated schematically in the upper part of the figure, where the Fermi surface moves from below, or in the vicinity of, the $\frac{1}{2}[521]$ orbit, to a position between the $\frac{1}{2}[510]$ and $\frac{3}{2}[512]$ orbits. Thus the implication is that the strength of the $U^{BF}(6)$ term is related to the position of the Fermi surface in the corresponding Nilsson scheme. Without such a term, the ordering of the three bands cannot be changed, although their separation can be controlled by the magnitude of β, since the Casimir operator of $SU^{BF}(3)$ translates into a quadrupole-quadrupole interaction in the IBFM Hamiltonian. The $U^{BF}(6)$ Casimir, on the other hand, contributes *exchange* terms to the Hamiltonian, not only of the *d-d* boson type normally included, but also of the *s-d* form. It is, in fact, the competition between the $Q \cdot Q$ and exchange

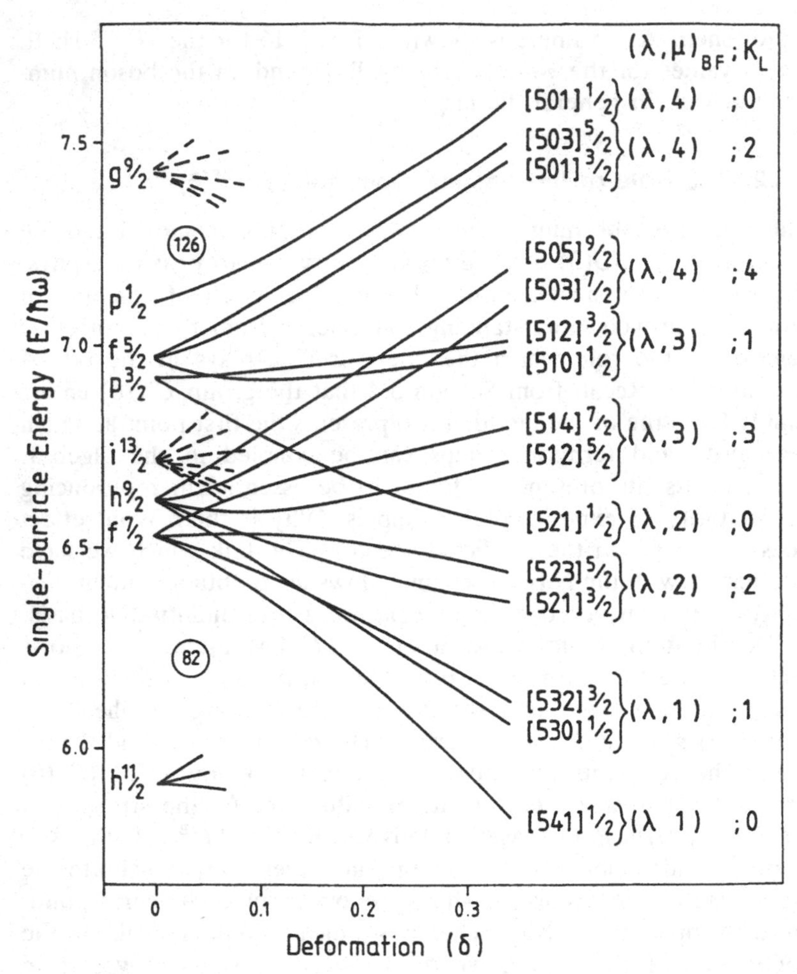

FIGURE 5.16. *Nilsson* $[Nn_z\Lambda]$ *and the SU(3)* (λ, μ)*-*K_L *classification of deformed orbits originating from the* 82–126 *shell.* (*Based on Bijker and Kota,* 1988).

terms in the general IBFM Hamiltonian which controls the predicted ordering of rotational bands in a deformed nucleus, and thus the position of the effective Fermi level in the deformed potential.

The role of, and necessity for, the $U^{BF}(6)$ decomposition is thus clear. It permits the position of the Fermi surface in a non-spherical potential to be varied, thus allowing the empirical situation to be attained in many cases.

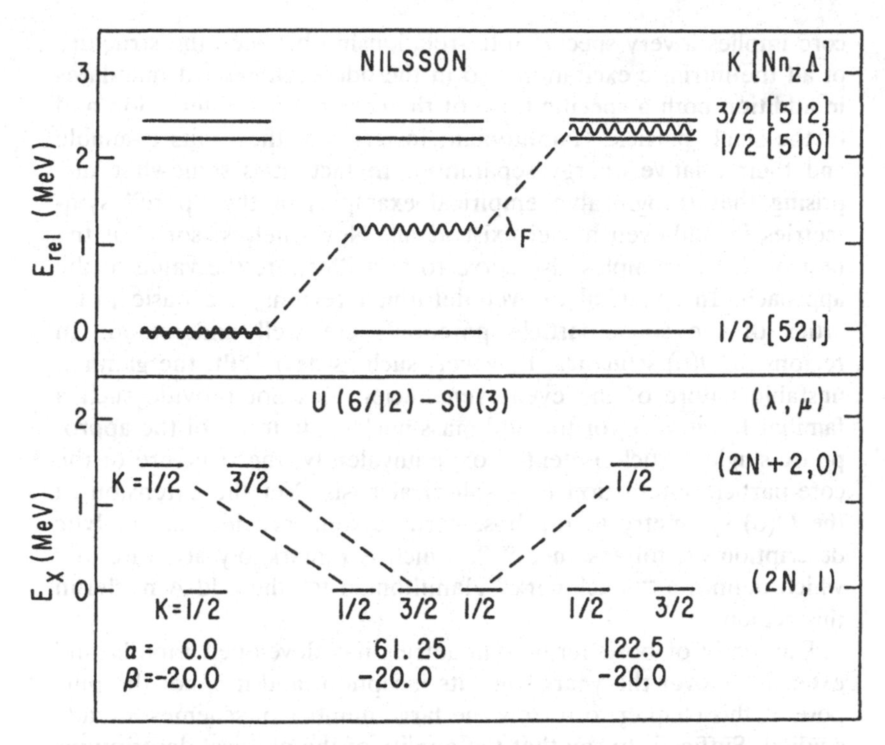

FIGURE 5.17. *Bottom: Low-lying bandheads in the SU(3) scheme as a function of α and β. (See text.) Top: Corresponding Nilsson orbits and a schematic indication of the effective position of the Fermi level. (Taken from Scholten and Warner, 1984.)*

5.7. EXPERIMENTAL EXAMPLES

We can now proceed to review some of the experimental evidence for the boson–fermion symmetries. It should be clear from the discussion of Section 5.5 that the task of finding good examples of symmetries in odd-*A* nuclei will not be an easy one. The extension of the dynamical symmetry concept to odd-even systems can, in the first instance, simply be regarded as ensuring consistency between the symmetry structure in the core and the core-particle interaction and hence mean field for the particle. The symmetry thus provides a convenient starting point in an IBFM calculation. However, as we have seen, the constraints implied by symmetries are far more severe than, for instance, simply the assumption of axial symmetry. Just as in the *SU*(3) case for even-even nuclei, the symmetry in the core implies a very specific inter-relationship between the structure of all the intrinsic excitations, so in the odd-even case, it mandates in addition both a specific form of the core-particle interaction and of the single-particle Hamiltonian, in terms of the orbits available and their relative energy separation. In fact, it is somewhat surprising that recognizable empirical examples of the "pure" symmetries in odd-even nuclei exist at all! Nevertheless, some of the best of these examples also serve to best illustrate the value of the approach. In spherical or well-deformed regions, the basic properties of the single-particle potentials are well understood. In regions of $O(6)$ structure, however, such as near ^{196}Pt, the gamma-unstable nature of the even-even nuclei does not provide such a familiar framework for the odd-mass nucleus, in terms of the appropriate single-particle potential or, equivalently, the structure of the core-particle interaction in a spherical basis. Yet the extension of the $O(6)$ symmetry to the bose–fermi system provides an analytic description of, for instance, ^{195}Pt which is remarkably accurate and which defines a "benchmark" Hamiltonian for the odd-A nuclei in this region.

The study of bose–fermi symmetries has developed rapidly and extensively over the years since its inception, and it is not the purpose of this chapter to review the large number of schemes already studied. Suffice it to say that the quality of the physical descriptions offered by many of these schemes indicates that their basis, though conceptually algebraic, must be founded on some degree of reality. Here, the major emphasis will continue to be on the $U^{B}(6) \times U^{F}(12)$ structure, and the remainder of the discussion will be ordered in terms of the three basic core structures and hence symmetry chains.

5.7.1 *U*(5) Symmetry

We have already seen that this limit closely resembles the standard vibrational picture which stems from coupling spherical single-particle orbits to a vibrational core. It can therefore be expected *a priori* to be the most likely one for which the symmetry criteria will be met empirically, since the coupling is weak, the core structure is largely maintained and the resultant clustering of odd-A levels around the energies of the first few core states is a well known empirical feature in odd-A nuclei near closed shells. The simplest example is shown in Fig. 5.18 which shows the lowest lev-

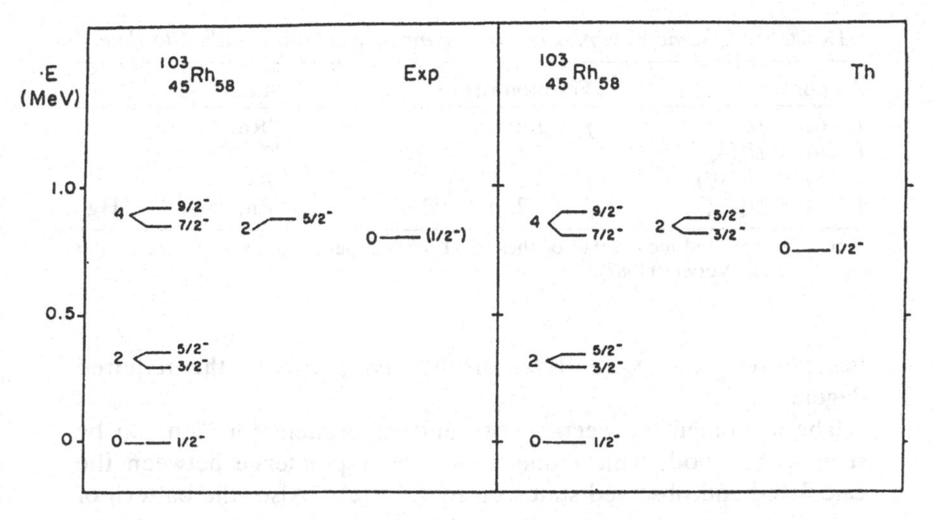

FIGURE 5.18. *A comparison of* ^{103}Rh *with the predictions of the* $U(5)$ *symmetry from* $U^B(6) \times U^F(2)$. (*Taken from Iachello and Van Isacker*, 1991.)

els of ^{103}Rh and the corresponding result using a $U(5)$ symmetry originating from a $U^B(6) \times U^F(2)$ group structure. Note that the odd proton in this case is in the $2p_{1/2}$ orbit and is therefore described by $U^F(2)$. Use of this simple scheme implies neglect of coupling to the higher lying $2p_{3/2}$ and $1f_{5/2}$ orbits and thus limits its applicability to the lower-lying negative-parity states.

Some of the other $U(5)$ schemes which have been compared with experiment are listed in Table 5.7. Not all of these examples are as apparently trivial as that of ^{103}Rh. Figure 5.19 shows an example originating from the $U^B(6) \times U^F(12)$ structure that we have discussed in more detail in the preceding sections. The group chain giving rise to the theoretical structure is thus that of eq. (5.29), containing $U^{BF}(5)$. Recalling the discussion of Section 5.5.2, both a vibrational core structure and a specific set of single-quasiparticle energies is required for a $U^{BF}(5)$ symmetry. The latter constraint was illustrated in Fig. 5.12 which shows that near degenerate $j = \frac{3}{2}$ and $\frac{5}{2}$ orbits are required, separated from the $j = \frac{1}{2}$ state by (approximately) the energy of the first 2^+ state in the core. However, the separation required also depends on the $U^{BF}(6)$ term. There are several regions in the periodic table where these conditions appear to be fulfilled, as indicated in Table 5.7. In all cases the orbitals involved are the $p_{1/2}$ and $p_{3/2}$, $f_{5/2}$ and the Fermi surface

TABLE 5.7. *Some examples of U(5) symmetry schemes studied to date.*

Algebra	Fermion orbits	Nuclei[a]
$U^{B}(6) \times U^{F}(2)$	$j = 1/2$	^{103}Rh, ^{103,109}Ag
$U^{B}(6) \times U^{F}(4)$	$j = 3/2$	^{63,65}Cu
$U^{B}(6) \times U^{F}(10)$	$j = 3/2, 5/2$	^{75}As
$U^{B}(6) \times U^{F}(12)$	$j = 1/2, 3/2, 5/2$	^{63}Zn, ^{75}As, $^{193-199}$Hg

[a]For references and more detail of these and other experimental studies, the reader is referred to Vervier (1987).

lies between the $p_{3/2}$ and $f_{5/2}$ orbits, giving rise to the required degeneracy.

The agreement between theory and experiment for ^{63}Zn can be seen to be good, with a one-to-one correspondence between the calculated and observed states up to 1.5 MeV. Also, the pattern of pseudo-spin couplets is apparent. Note, however, that the pattern of low-lying levels does not follow the simple weak coupling expectations. Instead, the $U^{BF}(6)$ term in the Hamiltonian has been used to bring the $[N, 1]$ non-symmetric representation below the symmetric one. Thus, here, the achievement of the correct ground state structure mandates bose–fermi coupling at the level of $U(6)$.

As far as single nucleon transfer reactions are concerned, only a few comparisons with experiment have been attempted so far, but the symmetry schemes predict that the spectroscopic strength in the odd-A nucleus is concentrated in one or two states for each j-value, as would be expected in a spherical nucleus. Some extensive comparisons with B(E2) data have been made and one example of such results for a $U^{B}(6) \times U^{F}(4)$ structure ($j = \frac{3}{2}$ for the odd nucleon) is given in Table 5.8.

5.7.2. *SU*(3) Symmetry

We have already introduced an empirical example of this symmetry, in the context of our comparison of the symmetry scheme with the Nilsson model. Reference to Section 5.6.1 indicates that the nuclei of interest for the $SU^{BF}(3)$ chain stemming from $U^{B}(6) \times U^{F}(12)$ (see eq. (5.30)) must be the odd-neutron W isotopes in the region of $A = 183$–187. In fact, we saw that a correspondence between the single-particle structure in the two models could be obtained for ^{185}W. In contrast, the specific Coriolis interaction necessary in the

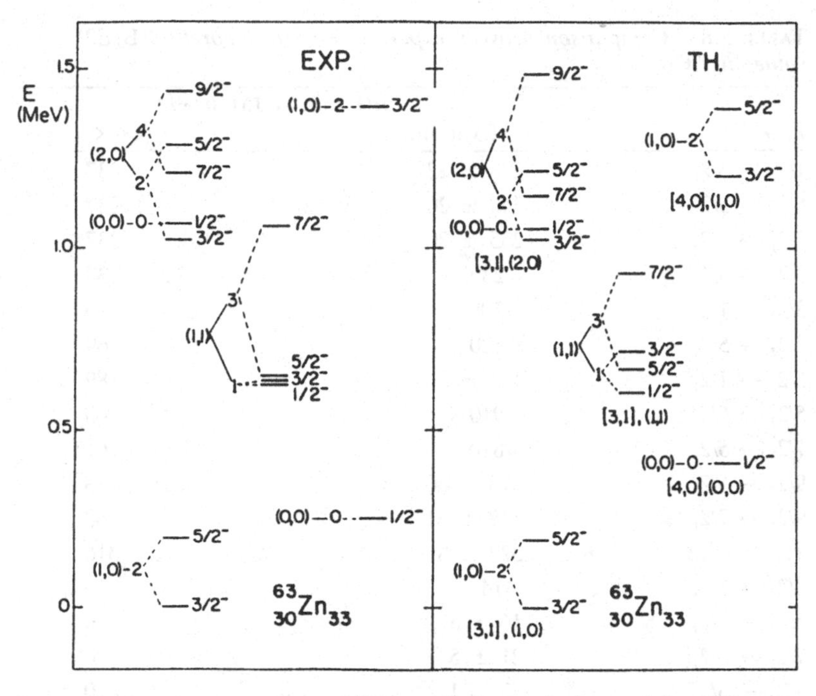

FIGURE 5.19. *A comparison of* ^{63}Zn *with the predictions of the U*(5) *symmetry from* $U^{B}(6) \times U^{F}(12)$. (*Taken from Bijker, 1984*).

Nilsson framework to generate the structure of $SU^{BF}(3)$ is not found in ^{183}W and ^{187}W.

The predictions of the $U^{B}(6) \times U^{F}(12)$ scheme for energies can now be simply compared with the known empirical structure in ^{185}W. The assignment of the lowest Nilsson bands is well established in this nucleus, and the comparison with the corresponding $SU^{BF}(3)$ predictions is shown in Fig. 5.20. There are several important points to note. On the positive side, the same Coriolis mixing which gives rise to the distinctive "fingerprint pattern" of single-particle structure also results in a near degeneracy of states in the lowest lying $K = \frac{1}{2}$ and $\frac{3}{2}$ bands, which appears naturally in the algebraic description. Both of these features imply that the intrinsic matrix elements $\langle j \rangle$ in the Nilsson framework are well reproduced in the alternative scheme. However, it is also evident in Fig. 5.20 that certain observed bands (denoted by dashed lines) cannot be reproduced from the $U^{B}(6) \times U^{F}(12)$ basis because they stem from the $f_{7/2}$ and $h_{9/2}$ shell model states. Moreover, only previously assigned

TABLE 5.8. *Comparison between experimental and theoretical* B(E2) *values in* ^{63}Cu[a]

$J_i \to J_f$	$B(\mathrm{E2}; J_i \to J_f)(10^{-4}e^2b^2)$ Experiment	$U^{B}(6) \times U^{F}(4)$
$1/2_1 \to 3/2_1$	228 ± 12	237
$5/2_1 \to 3/2_1$	240 ± 20	237
$7/2_1 \to 3/2_1$	255 ± 20	237
$3/2_3 \to 3/2_1$	<239	237
$3/2_2 \to 1/2_1$	<208	110
$3/2_2 \to 5/2_1$	<520	169
$5/2_2 \to 1/2_1$	135 ± 36	189
$5/2_2 \to 5/2_1$	<910	48
$7/2_2 \to 5/2_1$	<810	161
$9/2_1 \to 5/2_1$	273 ± 60	248
$9/2_1 \to 7/2_1$	148 ± 81	68
$11/2_1 \to 7/2_1$	322 ± 83	316
$3/2_2 \to 3/2_1$	<14	0
$5/2_2 \to 3/2_1$	16 ± 5	0
$7/2_2 \to 3/2_1$	21 ± 5	0
$7/2_3 \to 3/2_1$	3 ± 1	0
$5/2_3 \to 3/2_1$	<47	0
$1/2_2 \to 3/2_1$	<9	0

[a]Taken from Bijker (1984).

states are shown in the figure. There are additional states observed which have not proved amenable to a Nilsson-model description, and which cannot all be accounted for by the symmetry scheme either. Again, the deficiency probably lies in the limited single-particle space assumed in the bose–fermi case, and the limited core space in the Nilsson picture.

A comparison of predicted and empirical (d, t) cross sections is also shown at the top of Fig. 5.20. In the IBFM description of ^{185}W, since the neutron number is past midshell, the number of bosons is equal to that in ^{186}W. Thus the lowest-order operator describing the (d, t) reaction can be taken simply as $\xi_j a_j^\dagger$ where ξ_j are constants. While the results look promising, it must be emphasized that such a simple approach to the single-particle transfer problem is not likely to succeed as well in the general case in deformed nuclei.

The basic problem can be easily understood in the following way.

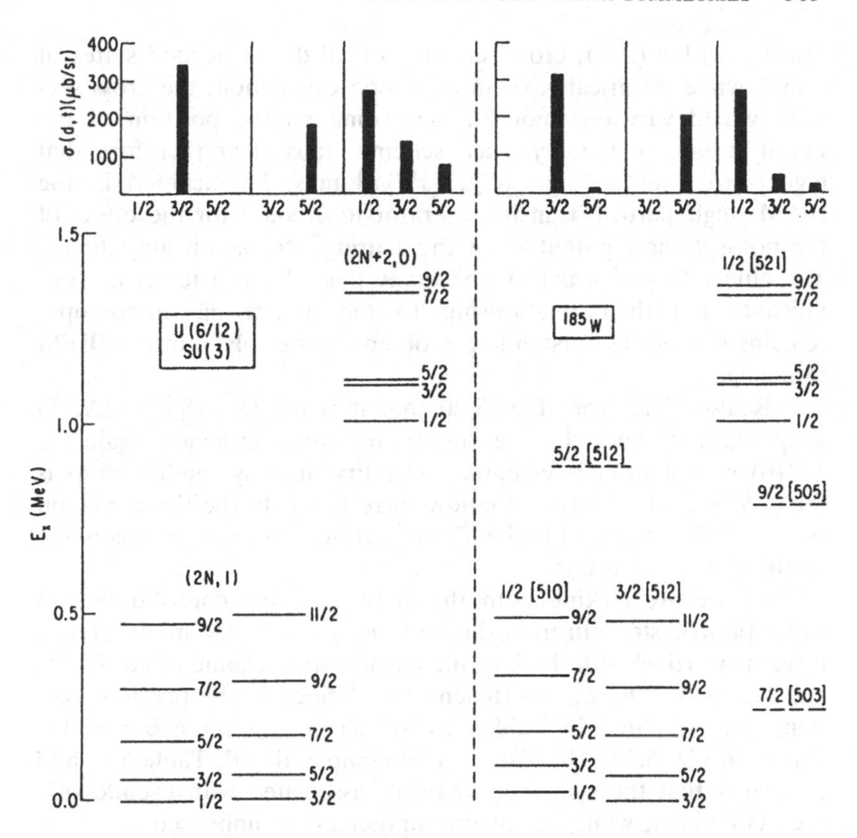

FIGURE 5.20. *Predicted and empirical energies and σ(d, t) (θ = 90°) for assigned Nilsson bands in* ^{185}*W. Dashed levels denote bandheads whose origin is outside the* $U^B(6) \times U^F(12)$ *basis. (Taken from Warner and Bruce, 1984.)*

In the Nilsson approach, the structure factor relevant to (d, p) or (d, t) cross sections would be given by the sum $\Sigma_i a_i u_i(v_i) C^i_{jl}$ where the a_i are Coriolis amplitudes as before, and the occupation amplitudes $u_i(v_i)$ are those derived from the *deformed* single particle scheme. In the IBFM approach, there is only one constant, ξ_j, for each *spherical* state j, which can be regarded as being related to the spherical occupation amplitudes. If we now consider, for example the case of a unique-parity state of spin j, then the values of C^{eff}_{jl} for all the bands stemming from it would simply be unity in either basis, in the absence of Coriolis mixing. In the IBFM calculation, however, this would then lead to the prediction of identical (d, p) [or (d, t)] cross sections for all the associated states of spin j, while empirically, or in a Nilsson calculation, the cross sections would vary enormously, depending on the position of the Fermi surface in the *deformed* scheme. It is clear therefore that higher-order terms [e.g., [e.g., $(s^\dagger \tilde{d})a_j^\dagger$] must be included in the IBFM single-particle transfer operator to account for the effect of the non-spherical potential on the pairing occupation amplitudes. The elucidation of which of the many possible such terms are significant, and their relationship to the underlying microscopy, remains one of the outstanding problems to be solved in the IBFM formalism.

It is also clear from Fig. 5.20 that it is the $(\lambda, \mu)_{BF} = (2N, 1)$ irrep which is required to be the ground state and hence, again the $U^{BF}(6)$ contribution is essential. The first fully-symmetric irrep is the $(2N + 2, 0)$ and the situation corresponds to the three Nilsson orbits of Fig. 5.17, with the Fermi surface mid-way between the $\frac{1}{2}[510]$ and $\frac{3}{2}[512]$ orbits.

Since we are lacking both the states and the contributions to single-particle strength from the $j = \frac{7}{2}$ and $\frac{9}{2}$ $N = 5$ shell-model states, it seems worthwhile to look at the pseudo-spin scheme of eq. (5.25) and Fig. 5.16. The C_{jl}^{eff} coefficients, as defined in the previous section, are compared in Table 5.9 for the lowest three bands displayed in Fig. 5.20. This can also be compared with Table 5.6, and it is clear that the symmetry features associated with pseudo-spin are maintained, while the overall agreement is improved.

The problem which arises in the $U^B(6) \times U^F(30)$ scheme is the lack of a $U^{BF}(6)$ term in the group chain of eq. (5.25). As we have seen, without such a term, one cannot change the sequence of rotational bands to correspond to the changing Fermi surface in the deformed potential. It is possible to overcome this obstacle by adding quadrupole and exchange terms to the symmetry Hamiltonian. Moreover, the problem becomes particularly simplified if a strong-coupling, large-N limit is taken.

A general expression for $B(E2)$ values in an $SU(3)$ scheme was given in eq. (5.47). This yields identical results to the normal "good K" basis for $K_L = 0$, $K = \frac{1}{2}$ but for $K_L \neq 0$ the results differ. Most importantly, because two or more K-bands occur within a single representation of $SU^{BF}(3)$, the E2 transitions between such bands emerge on an equal footing with intraband transitions in the symmetry scheme; they are all "intrarepresentation" transitions.

TABLE 5.9. *Effective spherical single particle amplitudes for* ^{185}W *in Nilsson and* $U^B(6) \times U^F(30) - SU(3)$ *schemes.*[a]

J	$U^B(6) \times U^F(30) - SU(3)$ $(\lambda, \mu)K_L$	L	C_{jl}^{eff}	Nilsson $K[Nn_z\Lambda]$	C_{jl}^{eff}
1/2	$(2N, 2)0$	0	0.53	1/2[521]	0.51
3/2		2	0.21		0.30
5/2		2	0.25		0.47
7/2		4	0.55		0.38
9/2		4	0.61		0.45
1/2	$(2N-2, 3)1$	1	0.00	1/2[510]	0.03
3/2		2	0.70		0.78
5/2		3	0.00		0.15
7/2		4	0.53		0.47
9/2		5	0.00		0.03
3/2	$(2N-2, 3)1$	1	0.00	3/2[512]	0.13
5/2		2	0.86		1.00
7/2		3	0.00		0.12
9/2		4	0.60		0.43

[a]Based on Bijker and Kota (1988).

This situation is best illustrated by an example. The two sets of predictions are compared with available data for the low lying $K = \frac{1}{2}$ and $\frac{3}{2}$ bands in ^{187}Os in Table 5.10. These bands are based on the same deformed orbits that have appeared in our earlier example, namely, the $\frac{1}{2}[510]$ and $\frac{3}{2}[512]$ Nilsson states and hence are described by a single structure with $K_L = 1$ in the symmetry framework. In the normal "good K" scheme, and in the absence of mixing, transitions between the $K = \frac{1}{2}$ and $\frac{3}{2}$ bands are predicted to be of neligible strength, because they are single particle in nature, rather than collective. In the scheme in which the quantum numbers $\tilde{L}$ (and K_L) are good, such transitions are of comparable strength to in-band transitions. The forbidden transitions will instead be those for which $\Delta\tilde{L} > 2$. An example would be the $(j = \frac{7}{2}\ \tilde{L} = 4) \rightarrow (j = \frac{3}{2}, \tilde{L} = 1)$ transition. Unfortunately there is as yet little detailed B(E2) data available to test such predictions.

The symmetry scheme of eq. (5.25) can be applied to the complete set of normal parity orbits of any of the major shells. Taken to the limit of large boson number N, its predictions coincide with those of the pseudo-Nilsson scheme which will be discussed in more detail in Chapter 7. Very recent studies seem to suggest that the

TABLE 5.10. *Relative B(E2) values in ^{187}Os*[a]

J_iK_i	J_fK_f	Relative B(E2; $J_i \rightarrow J_f$) Experiment	Good $\bar{L}$	Good K
5/2 1/2	1/2 1/2	100 (9)	100	100
	3/2 1/2	135 (50)	86	29
	3/2 3/2	41 (21)	29	0
	5/2 3/2	24 (12)	21	0
5/2 3/2	1/2 1/2	65 (7)	29	0
	3/2 3/2	100 (37)	100	100
7/2 3/2	3/2 3/2	100 (12)	100	100
	5/2 3/2	122 (66)	75	150

[a]Taken from Warner and Van Isacker (1990).

validity of the pseudo-spin symmetry results may be enhanced in the superdeformed states of nuclei which have been discovered in the last few years.

5.7.3. O(6) Symmetry

The $O(6)$ limit is undoubtedly the best empirically established symmetry in even-even nuclei. The prime experimental example is that of ^{196}Pt, although the neighbouring Pt nuclei and, to a lesser extent, the heavier Os nuclei, are also strongly influenced by this type of structure. Further examples exist in the $A \approx 130$ region. The proton single-particle levels in the Pt region are shown in Fig. 5.21. Clearly, for nuclei with $Z > 76$, the obvious symmetry structure to study for positive-parity states would be $U^B(6) \times U^F(6)$, corresponding to fermion orbits with $j = \frac{1}{2}$ and $\frac{3}{2}$. Unfortunately, however, an $O(6)$ decomposition does not exist in this case. The studies of odd-proton nuclei in this region have therefore been compared with the structure $U^B(6) \times U^F(4)$, which includes only the $j = \frac{3}{2}$ orbit, and must rely on the assumption that the $j = \frac{1}{2}$ contributions play only a minor role in the low-lying structure.

Turning now to the odd-neutron nuclei in this region, the relevant shell model orbits are displayed in Fig. 5.22. Here it is clear that the $j = \frac{1}{2}, \frac{3}{2}, \frac{5}{2}$ negative-parity states are well separated from the rest, so that the odd Pt nuclei in particular should offer an excellent testing ground for the $O(6)$ decomposition of $U^B(6) \times U^F(12)$. Indeed, given the good core symmetry, the correct set of

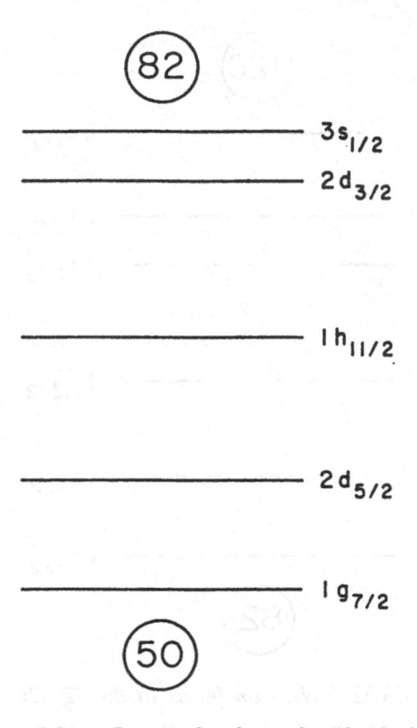

FIGURE 5.21. *Proton levels in the 50–82 shell.*

single-particle states and the ability to use the $U^{BF}(6)$ coupling, here there is really no obvious excuse left if the symmetry predictions fail!

Historically the $\text{Spin}^{BF}(6)$ limit which represents the $O(6)$ limit of the structure $U^{B}(6) \times U^{F}(4)$, was the first example of a boson–fermion symmetry to be proposed and tested. An example of a fit to low-lying states in ^{191}Ir is shown in Fig. 5.23. While we have not studied this type of symmetry explicitly in this Chapter, we can understand that all the states shown correspond to the lowest representation of $\text{Spin}^{BF}(6)$ which takes the place of $O^{BF}(6)$ and has quantum numbers $(\sigma_1, \sigma_2) = (\frac{17}{2}, \frac{1}{2})$. Similarly, the numbers in

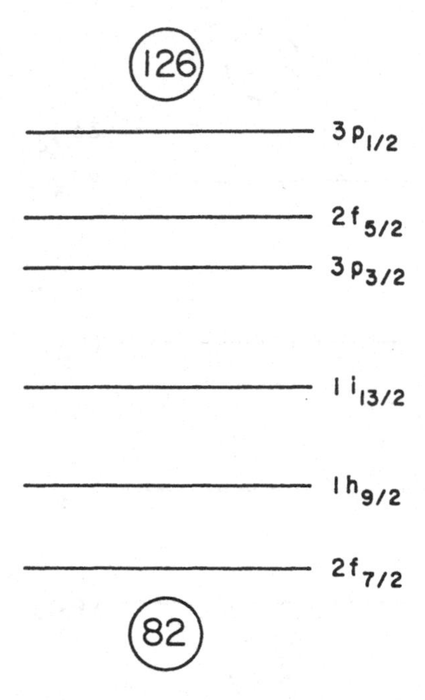

FIGURE 5.22. *Neutron levels in the* 82–126 *shell.*

parenthesis to the left of each level are the (τ_1, τ_2) quantum numbers of $\mathrm{Spin}^{BF}(5)$, (cf. $O^{BF}(5)$).

While the quantitative discrepancies between theory and experiment are by no means negligible, the $U^B(6) \times U^F(4)$ scheme does seem to describe the basic features of the energy spectra in each case. In particular, all the low-lying known states are accounted for and there are, as yet, no additional states known which do not appear in the theoretical picture. It is now necessary to go on to ask whether the associated classification of states, in terms of the $O(6)$ quantum numbers, can be confirmed. This can be done via the E2 transitions which probe the collective structure of the states.

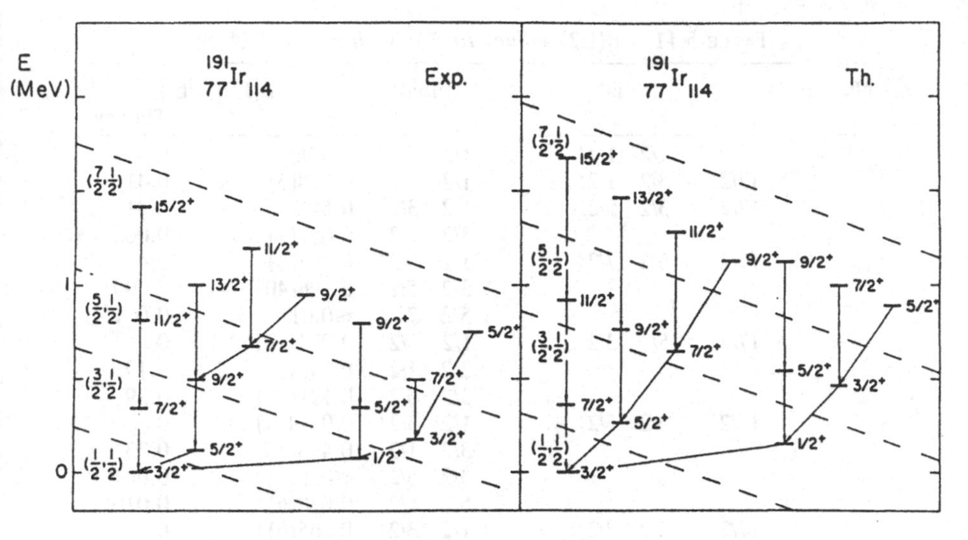

FIGURE 5.23 *Comparison of* ^{191}Ir *with the predictions of the O(6) symmetry from* $U^B(6) \times U^F(4)$. *See text for details. (Taken from Iachello and Van Isacker, 1991.)*

As discussed earlier, the use of a generator of $\mathrm{Spin}^{BF}(6)$ for the E2 operator will result in selection rules which are represented by the conditions $\Delta\sigma_1 = 0$, $\Delta\tau_1 = \pm 1$. The resulting pattern of strong "intra-band" E2 transitions is shown in Fig. 5.23. A quantitative comparison of measured absolute B(E2) values with the predictions is given in Table 5.11. Although, again, the quantitative level of agreement between theory and experiment is by no means ideal, it is clearly good enough to confirm the validity of the proposed classification of many of the states, in that the "allowed" transitions are indeed observed to be the strongest in each case. Thus these data support the postulate that the symmetry of the $O(6)$ scheme survives to an appreciable extent in the odd-even nuclei. In fact, the strong transitions originate principally from the boson part of the E2 operator, so that the agreement with the data is essentially confirming the predicted core-particle mixing in the final wave functions.

We return now to the $U^B(6) \times U^F(12)$ structure and its $O^{BF}(6)$ limit represented by the group chain of eq. (5.31). As pointed out earlier, an ideal empirical example should be ^{195}Pt and this does indeed turn out to be the case. Figure 5.24 illustrates the results for ^{195}Pt Quantum numbers of $U^{BF}(6)$ are in square brackets and

TABLE 5.11. *B*(E2) *values in* ^{191}Ir *with* e_{BF} = 0.14 eb.

Nucleus	Initial			Final[a]		B[E2](e^2b^2)	
	σ_1	τ_1	J^π	τ_1	J^π	Exp[b]	Theory
^{191}Ir	17/2	1/2	$1/2^+$	1/2	$3/2^+$	≤0.02	0
	17/2	3/2	$1/2^+$	1/2	$3/2^+$	0.130(3)	0.41
	17/2	3/2	$5/2^+$	1/2	$3/2^+$	0.64(3)	0.41
				3/2	$1/2^+$	0.057(11)	0.062
	17/2	3/2	$7/2^+$	1/2	$3/2^+$	0.293(6)	0.41
				3/2	$5/2^+$	0.196(40)	0.090
				5/2	$3/2^+$	≤0.011	0.031
	17/2	5/2	$3/2^+$	1/2	$3/2^+$	0.073(13)	0
				3/2	$1/2^+$	0.24(4)	0.19
				3/2	$5/2^+$	0.121(37)	0.29
	17/2	5/2	$5/2^+$	1/2	$3/2^+$	0.0111(4)	0
				3/2	$1/2^+$	0.448(70)	0.33
				3/2	$5/2^+$	≤0.1	0.084
				5/2	$3/2^+$	0.028(6)	0.010
	17/2	5/2	$7/2^+$	1/2	$3/2^+$	0.065(6)	0
				3/2	$5/2^+$	0.137(18)	0.28
				5/2	$3/2^+$	≤0.01	0.020
				5/2	$5/2^+$	≤0.02	0.037
	17/2	7/2	$(5/2^+)$	1/2	$3/2^+$	≤0.03	0
	15/2	1/2	$3/2^+$	1/2	$3/2^+$	0.024(6)	0
				3/2	$1/2^+$	0.0035(9)	0
				3/2	$5/2^+$	0.0094(9)	0
				5/2	$3/2^+$	0.0003(3)	0
	15/2	3/2	$(1/2^+, 3/2^+)$	1/2	$3/2^+$	0.024(4)	0
	15/2	3/2	$(5/2^+)$	1/2	$3/2^+$	0.014(2)	0
	15/2	5/2	$(3/2^+)$	1/2	$3/2^+$	0.03(1)	0

[a] σ_f = 17/2.
[b] Data are taken from Mundy (1984).

show that in this case, the symmetric representation [7, 0] is the ground state. However the Casimir operator of $U^{BF}(6)$ is still essential in that the next experimental states belong to [6, 1]. The σ quantum numbers are shown below the quasi-band structures, while (τ_1, τ_2)-L are given to the left of the levels. The agreement between experiment and theory is impressive, particularly when we remember that this represents the solution to the problem of coupling an odd neutron in one of three possible orbits to a γ-unstable core. Thus the structure here is not that of a vibrational or a rotational nucleus. Nevertheless, the pseudo-spin symmetry is still apparent, by virtue of the pseudo-spin couplets centred around each $\tilde{L}$-value.

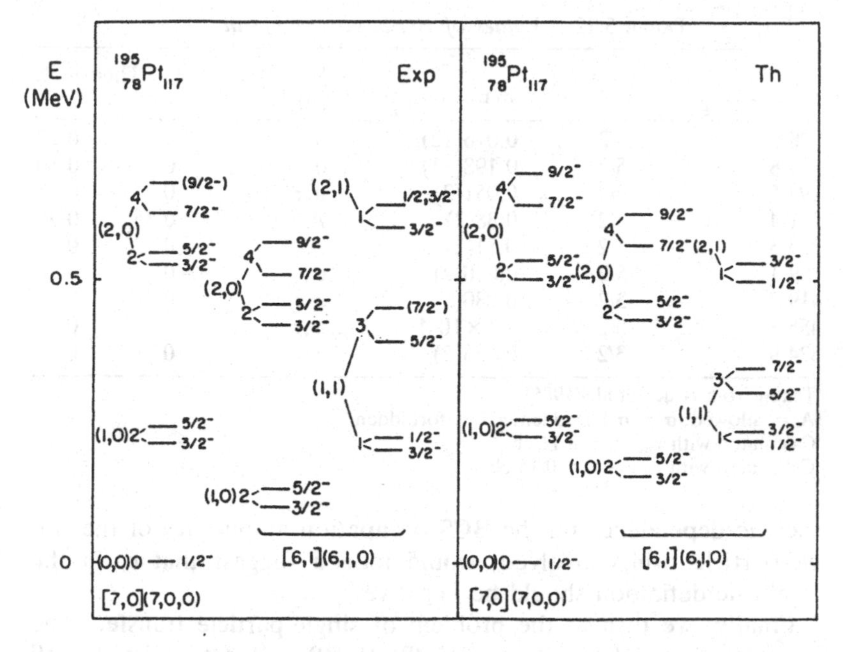

FIGURE 5.24. *Comparison of* ^{195}Pt *with the predictions of the O(6) symmetry of* $U^B(6) \times U^F(12)$. *See Fig. 5.10 for a definition of the quantum numbers.* (*Based on the fit to the data of Mauthofer et al., 1986.*)

The selection rules for E2 transitions given in Table 5.3 can be denoted by $\Delta(\sigma_1, \sigma_2) = (0, 0)$ and $\Delta(\tau_1, \tau_2) = (\pm 1, 0)$ or $(0, \pm 1)$. Thus, in the scheme of Fig. 5.24, only the states in the first σ representation with τ-labels (1,0) should be populated directly from the ground state in Coulomb excitation. This prediction is compared with experiment in Table 5.12. Clearly, while the states at 211 and 239 keV do indeed receive the largest E2 strength, there is nevertheless significant population of the states at 99 and 129 keV, which are forbidden in the symmetry scheme by the σ-selection rule. However, it was pointed out earlier that this rule arises specifically from a *cancellation* of the boson and fermion contributions to the E2 matrix element, and hence relies on the assumption of equal effective charges for the bosonic and fermionic sectors. If this constraint is relaxed, then the values in the last column of Table 5.12 can be obtained for $e_b = -e_F = 0.15$ e.b. In fact, the more general form of the IBFM E2 operator, which includes a

TABLE 5.12. *Values of $B(E2; g.s. \rightarrow J_f^\pi)$ in* ^{195}Pt.

Level energy	J_f^π	B(E2) [e^2b^2][a]	Selection rule[b]	Theory 1[c]	Theory 2[d]
98.9	3/2$^-$	0.076(12)	σ	0	0.07
129.8	5/2$^-$	0.198(13)	σ	0	0.10
199.5	3/2$^-$	0.051(3)	σ,τ	0	0
211.4	3/2$^-$	0.38(2)	A	0.35	0.35
239.3	5/2$^-$	0.51(2)	A	0.53	0
389.1	5/2$^-$	0.020(2)	σ,τ	0	0
419.7	3/2$^-$	0.030(2)	σ,τ	0	0
455.3	5/2$^-$	$<1\times10^{-4}$	σ,τ	0	0
524.9	3/2$^-$	0.033(2)	τ	0	0

[a]Taken from Bruce et al. (1985).
[b]A = allowed; σ = σ-forbidden; τ = τ-forbidden.
[c]Calculated with $e_B = e_F =$ 0.11 eb.
[d]Calculated with $e_B = -e_F =$ 0.15 eb.

specific dependence on the BCS occupation amplitudes of the single-particle orbits involved, would indeed suggest that e_F in the algebraic definition should be negative.

Finally, we turn to the problem of single-particle transfer. The operators are defined in eqs. (5.48)–(5.50) and data exists for all single-neutron transfer reactions leading to and from ^{195}Pt. The comparison with the data for the stripping and pick-up reactions leading to ^{195}Pt is shown in Table 5.13, and it is clear that the agreement is rather good. The parameters determined from these data can now be used to predict the spectroscopic strengths for the inverse reactions leading from ^{195}Pt, and the results are shown in Table 5.14. Here, it is evident that while the agreement for the pick-up reaction leading to ^{194}Pt is satisfactory, in the stripping, a number of states in ^{196}Pt which are forbidden by the selection rules of the transfer operator are populated with significant strength. A similar situation is found in the $U^B(6) \times U^F(4)$ case discussed earlier, in that the major disagreements between theory and experiment appear in the study of transfer from the odd-even to the neighbouring even-even nuclei. In the $U^B(6) \times U^F(12)$ case, however, the source of the discrepancies cannot be attributed to an inadequate single particle space, but must rather lie in the chosen form of the operator and/or in the treatment of the excited levels in the even-even nuclei in terms of a pure $O(6)$ symmetry.

TABLE 5.13. *Experimental spectroscopic strengths in* ^{195}Pt *versus* $U^B(6) \times U^F(12)$ *predictions.*[a]

	Final State				$^{196}Pt \rightarrow {}^{195}Pt$		$^{194}Pt \rightarrow {}^{195}Pt$	
E(keV)	(σ_1,σ_2)	(τ_1,τ_2)	$\tilde{L}$	J^π	S_{exp}	S_{th}[b]	S_{exp}	S_{th}[c]
0.0	(7,0)	(0,0)	0	$1/2^-$	0.60	0.60	0.54	0.50
99.0	(6,1)	(1,0)	2	$3/2^-$	0.72	0.73	0.68	0.73
129.8	(6,1)	(1,0)	2	$5/2^-$	1.56	1.46	1.52	1.57
199.5	(6,1)	(1,1)	1	$3/2^-$	0.12	0	0.04	0
211.4	(7,0)	(1,0)	2	$3/2^-$	0.20	0.19	0.18	0.02
239.3	(7,0)	(1,0)	2	$5/2^-$	0.30	0.38	0.08	0.05
389.1	(6,1)	(1,1)	3	$5/2^-$	not seen	0	not seen	0
419.7	(6,1)	(2,0)	2	$3/2^-$	not seen	0	not seen	0.08
455.3	(6,1)	(2,0)	2	$5/2^-$	0.04	0	not seen	0
524.8	(7,0)	(2,0)	2	$3/2^-$	weak	0	0.08	0.04

[a]Taken from Bijker (1984).
[b]Calculated using $\xi^2_{1/2} = 0.48$, $\xi^2_{3/2} = 0.24$, $\xi^2_{5/2} = 0.32$ in eq. (5.48).
[c]Calculated using $\theta^2_{1/2} = 0.036$ $\theta^2_{3/2} = 0.045$, $\theta'^2_{3/2} = 0.012$, $\theta^2_{5/2} = 0.065$, and $\theta'^2_{5/2} = 0$ in eq. (5.50).

TABLE 5.14. *Experimental single neutron strengths from* ^{195}Pt *versus* $U^B(6) \times U^F(12)$ *predictions.*[a]

E(keV)	Σ, τ, J^π	Transferred l, j	$^{195}Pt \rightarrow {}^{196}Pt$ S_{exp}	S_{th}
0	6, 0, 0$^+$	1, 1/2	0.54	0.50
356	6, 1, 2$^+$	1, 3/2	0.09	0.05
		3, 5/2	0.09	0.11
689	6, 2, 2$^+$	1, 3/2	0.10	0
877	6, 2, 4$^+$	3, 7/2	0.09	0
1015	6, 3, 3$^+$		not seen	0
1135	6, 3, 0$^+$	1, 1/2	0.11	0
1361	6, 4, 2$^+$		not seen	0
1402	4, 0, 0$^+$	1, 1/2	0.19	0
1604	4, 1, 2$^+$		not seen	0
1677	4, 1, 2$^+$	1, 3/2	0.22	0

E(keV)	Σ, τ, J^π	Transferred l, j	$^{195}Pt \rightarrow {}^{194}Pt$ S_{exp}	S_{th}
0	7, 0, 0$^+$	1, 1/2	0.27	0.25
328	7, 1, 2$^+$	1, 3/2	0.03	0.01
622	7, 2, 2$^+$	1, 3/2	0.08	0.02
1267	7, 3, 0$^+$	1, 1/2	0.02	0

[a]Taken from Bijker (1984). S_{th} has been calculated with the parameter values of Table 5.13.

5.8. SUPERSYMMETRY

We saw in Chapter 2 how the three dynamical symmetries $U(5)$, $SU(3)$ and $O(6)$ all stem from the single parent group structure $U^B(6)$, and we have seen in this chapter how these symmetries can also appear in odd-mass nuclei from a structure of the form $U^B(6) \times U^F(m)$. It is possible now to go one step further and ask if we can postulate an even higher group structure which gives rise to both the even-even and odd-even algebras. This involves the introduction of a superalgebra, denoted by $U(6/m)$ and the group chains which originate from it give rise to dynamical supersymmetries.

Superalgebras contain bilinear boson operators of the form $b_i^\dagger \tilde{b}_j$ and fermion operators $a_i^\dagger \tilde{a}_j$ but they also contain the boson–fermion products $b_i^\dagger \tilde{a}_j$ or $a_i^\dagger \tilde{b}_j$ which annihilate a fermion and create a boson, or vice-versa. In the nuclear physics applications which interest us here, we cannot tolerate the latter class of operator, since they would involve a change in nucleon number by one and hence mix states of different nuclei! The concept of supersymmetry thus seems inappropriate. However we know that the idea of dynamical symmetries is to find subalgebras which contain subsets of the original operators, still obeying the required closure rules with respect to commutation. Obviously, we are looking for a subalgebra which drops the nucleon number-changing operators. This is simply

$$U(6/m) \supset U^{B}(6) \times U^{F}(m). \qquad (5.61)$$

There are other subalgebras of $U(6/m)$ which do not conserve the nucleon number but they will not be considered here.

We now have to consider the representations of $U(6/m)$. Again, for nuclear physics applications, we consider only one, namely, the totally supersymmetric representation $[\mathfrak{N}\}$, where $\mathfrak{N} = N_B + N_F$, in which all the bosonic indices are symmetrized and the fermionic ones are antisymmetrized. The next problem is to find which representations of $U^{B}(6) \times U^{F}(m)$ are contained with $[\mathfrak{N}\}$. These can be described by the labels of $U^{B}(6)$ and $U^{F}(m)$ which are just N_B and N_F, respectively. The values for the first four representations are shown in Table 5.15.

The possible significance and applicability of supersymmetry in the nuclear domain now becomes more apparent. The treatment of an even-even nucleus with $\mathfrak{N}$ bosons, and an odd-even nucleus with $\mathfrak{N} - 1$ bosons and 1 fermion, can be viewed as the description of different representations of $U^{B}(6) \times U^{F}(m)$, contained within the same representation $\mathfrak{N}$ of $U(6/m)$. The representation with $\mathfrak{N} - 2$ bosons and 2 fermions describes, in principle, the two-quasiparticle states which lie above the pairing gap in the next nucleus in the multiplet. The specific Casimir operators of the groups in eq. (5.61) contribute only to the binding energy of the system and yield a quadratic dependence on $\mathfrak{N}$ which is consistent with other models and, indeed, with the semi-empirical mass formula. However, the other consequence of this approach is that the remainder of the group chain is common to both the even-even and odd-even members of the $U(6/m)$ representation. Thus both must display the

TABLE 5.15 *First four representations of $U^{B}(6) \times U^{F}(m)$ contained within $[\mathfrak{N}]$.*

N_B	N_F
$\mathfrak{N}$	0
$\mathfrak{N}$-1	1
$\mathfrak{N}$-2	2
$\mathfrak{N}$-3	3

properties of the assumed dynamical symmetry, and the energy spectrum of each must be described by the same parameters in the eigenvalue expression.

In fact, in his article (Iachello, 1980) proposing the existence of dynamical supersymmetries in nuclei, Iachello gave three conditions which must be satisfied simultaneously:

(i) The states of the combined system of bosons and fermions can be simultaneously classified with a complete set of group theoretical labels.

(ii) These states are split but not mixed by the Hamiltonian.

(iii) The *same* eivenvalue expression describes *both* bosonic and fermionic spectra.

Conditions (i) and (ii) demand the existence of identical symmetries in the bose and bose–fermi systems. This in turn mandates a certain single-particle structure for the fermion sector, as discussed in earlier sections. It is condition (iii) which describes the crucial additional step which implies a common origin for the two symmetry chains, in terms of a supersymmetric group. Thus experimental studies which prove the validity of selection rules for various observables are studying the symmetry aspects of the two nuclei in question, which in itself is a necessary but not sufficient condition for supersymmetry. The simultaneous description of the energy spectra of even-even and odd-even nuclei must also be possible.

Studies of single-particle transfer cross sections offer a good example to elucidate the distinction between these two concepts. At first glance, it would seem that these represent the ideal data with which to search for supersymmetry, since the relevant operators between the two members of a supermultiplet are precisely those which transform bosons to fermions and vice-versa. However, the matrix elements of these operators between states in the even-even and odd-even nuclei depend only on the wave functions assumed in each case, which in turn are purely a result of the assumed symmetry in each nucleus. Moreover, invoking the concept of supersymmetry sheds no light on the specific structure of the transfer operator. For example, in calculating the transfer between the ground state of the even-even nucleus with N bosons, and a state with spin j in its odd-even supersymmetric partner with $N - 1$ bosons and 1 fermion, the appropriate operator, to lowest order, would be T_j of eq. (5.49). The use of this operator thus involves a choice of the constants θ_j and $\theta_{jj'}$ but, unfortunately, supersymmetry offers no insights into their values.

Similar arguments can be made concerning electromagnetic transition rates in supersymmetric partners. While in this case, the form of the bosonic and fermionic operators is often specified by the symmetry chain under consideration, and the bosonic effective charge is normally assumed equal in the two nuclei, none of these features is mandated uniquely by supersymmetry rather than symmetry alone. More specifically, identical results would be obtained if the appropriate symmetry chains were assumed in the even-even and odd-even nuclei, *but with very different strengths for the Casimir operators in each case,* and hence very different eigenvalue expressions. It is indeed the constraint that the eigenvalue expressions be identical in the nuclei of a supermultiplet which provides the additional and essential observable test of supersymmetry, namely, the energy spectra of the two nuclei.

Some examples of supersymmetry partners are shown in Fig. 5.25. Note that, as shown, the supersymmetry partner of an odd-*A* nucleus is not the core nucleus, but rather the one with one additional boson. Detailed comparisons of these and other supermultiplets are available in several reviews. Here we will just point

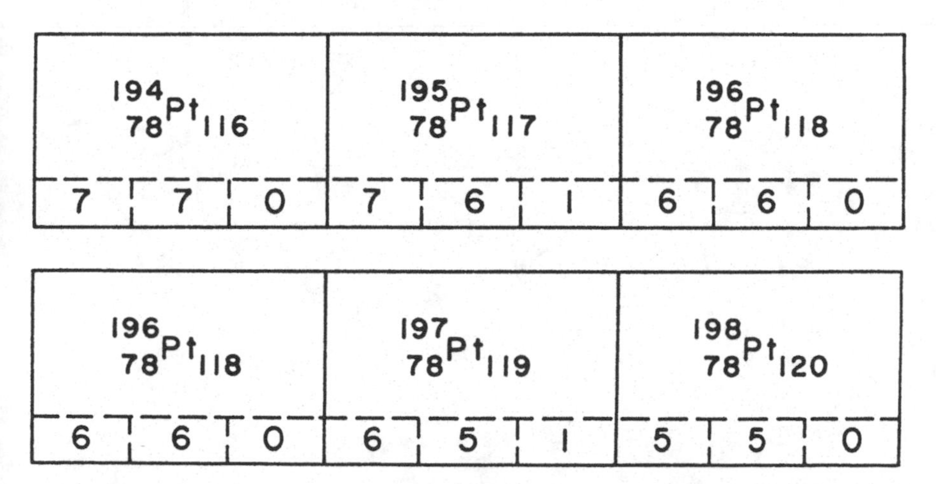

FIGURE 5.25. *Two examples of odd-neutron* Pt *nuclei which can be treated in the $O(6)$ symmetry scheme of $U^{B}(6) \times U^{F}(12)$. In each case, the even-even nucleus on the left is the* SUSY *partner, while the one on the right is the core nucleus. The numbers underneath each nucleus denote (from left to right) $\mathfrak{N}$, N_B and N_F.*

out that the quality of agreement obtained in comparing both the odd-even and even-even members of a supermultiplet with a single supersymmetry prediction is frequently similar to that shown earlier for the odd-even or even-even nucleus alone.

We may conclude by considering the significance and utility of the idea of dynamical supersymmetry in nuclear structure. The bosons entering the problem here are, of course, not fundamental entities but rather represent a mathematical representation of the underlying fermion pairs. Thus the concept of supersymmetry is an algebraic rather than a physical one. It can also be argued that the fact that neighboring even-even and odd-even nuclei can be described by the same Hamiltonian is hardly surprising and only reflects the fact that the addition of a particle has little effect on core properties, such as the moment of inertia. However, it is also clear from the discussion of the preceding sections that the constraints implied by supersymmetry link the structure of the even-even nucleus with the single particle Hamiltonian and with the strength and form of the core-particle interaction. The existence of this latter correlation in particular is not so easily dismissed, at least in systems which depart from spherical symmetry.

BIBLIOGRAPHY

R. Bijker, "Dynamical Boson–Fermi symmetries in nuclei," PhD Thesis, University of Groningen, 1984.

R. Bijker and V.K.B. Kota, *Ann. Phys.* **187**, 148 (1988).

A. Bohr, *Mat. Fys. Medd. Dan. Vid. Selsk* **26** (14) (1952) and

A. Bohr and B. R. Mottelson, *Mat. Fys. Medd. Dan. Vid. Selsk* **27** (16) (1953). The original papers on the Unified Model.

A. M. Bruce, W. Gelletly, J. Lukasiak, W. R. Phillips and D. D. Warner, *Phys. Lett.* **165B**, 43 (1985).

R. F. Casten, *Nuclear Structure from a Simple Perspective* (Oxford University Press, 1990). A good source of physical insight into various approaches to nuclear structure; the Nilsson Model is a good example.

R. F. Casten, P. Kleinheinz, P. T. Daly and B. Elbek, K. Dan, *Vidensk. Selsk. Mat. Fys. Medd.* **38** (13) (1972).

S. Das Gupta and N. de Takasy, *Am. J. Phys.* **44**, 47 (1976). A different derivation of the usual formalism for odd-A rotational nuclei, starting from a weak coupling basis in which H_{core} is diagonal.

F. Iachello, *Phys. Rev. Lett.* **44**, 772 (1980).

F. Iachello and P. Van Isacker, *The Interacting Boson–Fermion Model* (Cambridge University Press, Cambridge, 1991). A very complete treatment of the IBFM and its associated algebra, giving a large number of explicit formulae.

A. Mauthofer, K. Stelzer, J. Gerl, Th.W. Elze, Th. Happ, G. Eckert, T. Faestermann, A. Frank and P. Van Isacker, *Phys. Rev.* **C34**, 1958 (1986).

S. J. Mundy, J. Lukasiak and W. R. Phillips, *Nucl. Phys.* **A426**, 144 (1984).

O. Scholten and D. D. Warner, *Phys. Lett.* **142B**, 315 (1984).

A. de Shalit and H. Feshbach, *Theoretical Nuclear Physics, Vol.1: Nuclear Structure* (Wiley, New York, 1974). Excellent treatise on nuclear structure in general, with good sections on the Unified Model.

P. Van Isacker, A. Frank and H. Z. Sun, *Ann Phys.* **157**, 183 (1984). A valuable source of analytic results for the $U(6/12)$ group chains.

J. Vervier, *Rivista Nuovo Cimento* **10** (9) (1987). A comprehensive review of experimental tests of bose–fermi symmetry and supersymmetry schemes.

D. D. Warner and A. M. Bruce, *Phys. Rev.* **C30**, 1066 (1984).

D. D. Warner and P. Van Isacker, *Phys. Letts.* **247B**, 1 (1990).

CHAPTER 6

Extensions of the s–d Boson Model

KRIS HEYDE

6.1 INTRODUCTION

The Interacting Boson Model (IBM) has been developed during the last years to a level where it is possible to give a unified description of low-lying quadrupole collective excitations, using an interacting system of *s* and *d* bosons. Making no distinction between protons and neutrons, the IBM-1 approximation (see Chapters 2, 3) is obtained whereas, by explicitly taking into account both proton and neutron degrees of freedom, the more microscopically founded IBM-2 results (see Chapter 4). Unless stated explicitly otherwise, we shall denote with IBM in the present chapter, the IBM-1 model.

Recently, in many transitional nuclei near closed shells as well as in strongly deformed nuclei, low-lying extra states have been observed experimentally that cannot be accounted for within the $(sd)^N$ boson space. In particular, near the $Z = 50$ closed-shell mass region i.e., the Zr, Mo, Ru, Pd, Cd, Sn . . . nuclei, extra $J^\pi = 0^+$, 2^+ and 4^+ levels show up below an excitation energy needed for creating a two quasi-particle configuration. Moreover, some of these levels are characterized by peculiar E2 and E4 decay modes. In some well-studied strongly deformed nuclei, such as the doubly-even Gd nuclei, experimental evidence for the importance of the hexadecapole degree of freedom has resulted. Since low-lying excitations have mostly a rather simple structure and cannot be discarded as such, the main motivation of extending the *sd*-IBM can be found in the need to incorporate those new modes of motion and study their interactions with the monopole and quadrupole bosons.

Here we shall study, in the framework of the IBM, mainly the influence that incorporating the hexadecapole boson (or *g*-boson) will have. We do this, in a schematic but transparent way, for the different limits of the IBM ($U(5)$, $O(6)$ and $SU(3)$). We relate the Hamiltonian strength parameters and effective charges to the underlying shell-model structure. Applications of the schematic model to some specific regions of nuclei i.e. the ^{104}Ru nucleus for the $U(5)$, $O(6)$ limits and the ^{156}Gd nucleus for the $SU(3)$ limit, are also performed. We subsequently study some general properties that arise when we treat the *g*-boson on the same footing as the *s* and *d* bosons. Thereby we study and discuss a number of characteristics of the *sdg*-boson model, in particular, its $SU(3)$ limit.

Besides the *g*-boson, low-lying negative parity states of octupole and dipole structure have been determined near closed shells in strongly deformed nuclei. The consequences of introducing $l = 3$ and $l = 1$ bosons i.e. the *f* and *p*-boson are studied. The negative parity *p*-boson also allows for a discussion of high-lying dipole excitations in the giant dipole resonance region as well as for a description of certain phenomena related to α-clustering in nuclei. Both the *sdf*, *spd* and full *spdf* boson models are discussed in that respect.

After discussing the methods that relate algebraic models to a description of the nucleus where shape variables are used, we introduce cubic terms within the *sd* IBM in order to be able to describe deviations from axial symmetry. The effects of cubic anharmonicities in the IBM have been illustrated for the $U(5)$, $O(6)$, $SU(3)$ limits. Effects of these cubic terms have also been studied for the transitional nucleus ^{104}Ru and the $O(6)$ nucleus ^{196}Pt.

In most IBM studies, the number of bosons is obtained from the number of valence nucleons outside closed-shell configurations. It has been shown, though, that near closed shells, in particular in the $Z = 50$ (Cd, Sn, . . .) and $Z = 82$ (Hg, Pt, Pb, Po, . . .) regions, excitations of nucleon pairs across the closed shells can become energetically possible. The need for including these intruder excitations within the IBM was recognized, at the expense of working without a fixed boson number. So, a way to include intruder excitations in the IBM is presented with illustrations for the ^{114}Cd nucleus. Finally, effects of non-constant boson number in deformed nuclei can be shown to result in backbending properties for the rotational motion in the nucleus.

In the present chapter, we thereby present very natural extensions of the original IBM, motivated by the need to understand the low-lying excitations observed below the energy where nucleon pair breaking starts to complicate a simple description of nuclear, collective modes of motion. Most of the extensions retain much of the simplicity of the original *sd* IBM.

6.2 *g*-BOSONS IN THE INTERACTING BOSON MODEL

6.2.1 The *U*(5) and *O*(6) Limits

6.2.1.1. Introduction

In nuclei with low-lying anharmonic quadrupole vibrational excitations, in which the excitation energy of the two-phonon states is of the order of the energy needed to break a pair of nucleons and create a $(j)^2_{J=4,6,8,\ldots}$ state, mixing between the two-phonon and intrinsic excitations can result. These conditions are mainly fulfilled when one type of nucleon (protons or neutrons) deviates from a closed shell configuration by ± 4 nucleons i.e. the Pd, Cd, Sn, Cd, Te, Xe nuclei near $Z = 50$. Concentrating on the $J^\pi = 4^+$ coupled two-nucleon configurations, sometimes a collective enhancement can result so that one can speak of a collective *g*-boson excitation.

Here, we start from the situation drawn schematically in Fig. 6.1, which can be used to visualize the unperturbed spectrum in the vibrational limit (the *U*(5) dynamical symmetry). We indicate the levels, corresponding to the s^N, s^{N-1}, $s^{N-2}d^2$, . . . configurations on the left-hand side of this figure. On the right-hand side, we indicate the extreme weak-coupling limit of adding a hexadecapole boson (*g*-boson) to the harmonic quadrupole vibrational *U*(5) spectrum, resulting in the $s^{N-1}g$, $s^{N-2}dg$, $s^{N-3}d^2g$, . . . configurations.

Using the assumption that coupling the *sd* and *g* boson degrees of freedom will result from a quadrupole-quadrupole force, and using the general extended form for the quadrupole operator

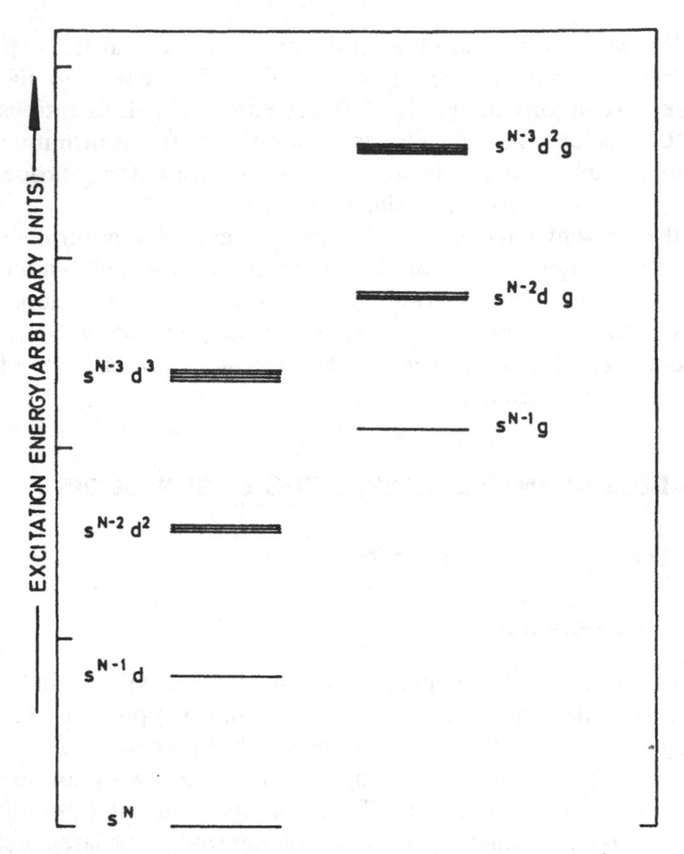

FIGURE 6.1. *Schematic representation of the* $|(sd)^N\rangle$ *and the* $|(sd)^{N-1} \otimes g\rangle$ *configurations in the U(5) limit (anharmonic quadrupole vibrational spectra).*

$$\begin{aligned} Q_{sdg} &= (d^\dagger s + s^\dagger \tilde{d})^{(2)} + \alpha(d^\dagger \tilde{d})^{(2)} + \beta(d^\dagger \tilde{g} + g^\dagger \tilde{d})^{(2)} \\ &\quad + \gamma(g^\dagger \tilde{g})^{(2)} \\ &= Q_{sd} + \beta Q_{gd} + \gamma Q_{gg}, \end{aligned} \tag{6.1}$$

one can write for the Hamiltonian describing a $|(sd)^N\rangle$ system, in interaction with a $|(sd)^{N-1} \otimes g\rangle$ system, the following form:

$$H = \epsilon_d d^\dagger \cdot \tilde{d} + \epsilon_g g^\dagger \cdot \tilde{g} + \kappa Q_{sdg} \cdot Q_{sdg}. \tag{6.2}$$

Taking out that part that is already contained in the *sd*-space H_{sd}, and which is given as

$$H_{sd} = \epsilon_d d^\dagger \cdot \tilde{d} + \kappa Q \cdot Q + \kappa' J \cdot J + \kappa'' P^\dagger \cdot P + q_3 T_3 \cdot T_3 + q_4 T_4 \cdot T_3, \tag{6.3}$$

using the definitions of the multipole operators as defined in the Dronten lecture notes of Iachello (Iachello, 1980) i.e.

$$Q = (d^\dagger s + s^\dagger \tilde{d})^{(2)} - \frac{\sqrt{7}}{2}(d^\dagger \tilde{d})^{(2)}, \qquad J = -\sqrt{10}\,(d^\dagger \tilde{d})^{(1)},$$

$$P = \tfrac{1}{2}(\tilde{d} \cdot \tilde{d}) - \tfrac{1}{2}(s \cdot s), \tag{6.4}$$

$$T_3 = (d^\dagger \tilde{d})^{(3)} \quad \text{and} \quad T_4 = (d^\dagger \tilde{d})^{(4)},$$

one obtains an approximate Hamiltonian

$$H \approx H_{sd} + \epsilon_g g^\dagger \cdot \tilde{g} + H_{\text{int}}(sd;\, g). \tag{6.5}$$

The parameters appearing in the multipole expansion of the IBM Hamiltonian (6.3) are slightly different from the ones used in Chapter 2. The relations will be outlined in Appendix 6.2. Moreover, some of the Racah algebra symbols (tensor product, scalar product of tensors, Clebsch–Gordan coefficients, . . .) are differing from the ones used in Chapter 2. Here too, we discuss the notation used in some more detail in Appendix 6.2. The part, called the interaction Hamiltonian H_{int}, follows from the $Q_{sdg} \cdot Q_{sdg}$ interaction and results from

$$H_{\text{int}}(sd;\, g) = \kappa(\beta Q_{gd} + \gamma Q_{gg}) \cdot Q_{sd}. \tag{6.6}$$

We write the approximate equal sign on (6.5) since terms, quadratic in the Q_{gd} and Q_{gg} parts of the quadrupole operator are not taken into account. Taking out those parts that are responsible for interactions between the g boson and the sd-bosons only, one can derive the following starting from eq. (6.6), i.e.,

$$H_{\text{int}}(sd;\, g) = \xi Q_{gg} \cdot Q_{sd} + \zeta H_{\text{mix}}(sd;\, g), \tag{6.7}$$

with $\xi = \kappa\gamma$; $\zeta = \kappa\beta$ and

$$H_{\text{mix}}(sd;\, g) = (g^\dagger s^\dagger)^{(4)} \cdot (\tilde{d}\tilde{d})^{(4)} + \text{h.c} + (g^\dagger d^\dagger)^{(2)} \cdot (\tilde{d} s)^{(2)} + \text{h.c} + \alpha \sum_L 5 \begin{Bmatrix} 4 & 2 & 2 \\ 2 & 2 & L \end{Bmatrix} [(g^\dagger d^\dagger)^{(L)} \cdot (\tilde{d}\tilde{d})^{(L)} + \text{h.c}]. \tag{6.8}$$

When studying, in a schematic way, the coupling of the $|(sd)^N\rangle$

configurations (called Σ states) with the $|(sd)^{N-1} \otimes g\rangle$ configuration (called Γ states), we use the simplest term from eq. (6.8), i.e., we reduce H_{mix} $(sd; g)$ to just one term.

$$H_{\text{mix}}(\Sigma, \Gamma) = [(d^{\dagger}d^{\dagger})^{(4)}(\tilde{g}s)^{(4)}]^{(0)} + \text{h.c.} \tag{6.9}$$

The effects of the other terms in eq. (6.8) relating to mixing are quite similar to the term we handle here. In the following discussion, we restrict ourselves to the inclusion of only one g-boson excitation.

In the same spirit, the electromagnetic transition operators are modified according to

$$\begin{aligned} T(\text{E2})_{sdg} = {} & e_{ds}^{(2)}(d^{\dagger}s + s^{\dagger}\tilde{d})^{(2)} + e_{dd}^{(2)}(d^{\dagger}\tilde{d})^{(2)} \\ & + e_{gd}^{(2)}(g^{\dagger}\tilde{d} + d^{\dagger}\tilde{g})^{(2)} + e_{gg}^{(2)}(g^{\dagger}\tilde{g})^{(2)}, \end{aligned} \tag{6.10}$$

and

$$T(\text{E4})_{sdg} = e_{dd}^{(4)}(d^{\dagger}\tilde{d})^{(4)} + e_{gs}^{(4)}(g^{\dagger}s + s^{\dagger}\tilde{g})^{(4)}. \tag{6.11}$$

At this point, the extra effective boson charges $e_{gd}^{(2)}$, $e_{gg}^{(2)}$, $e_{dd}^{(4)}$, and $e_{gs}^{(4)}$ are just a number of extra parameters. We shall relate these effective charges to the underlying shell-model structure as much as possible to determine both their sign and magnitude. Therefore, we have carried out most calculations of energy spectra and E2 and E4 decay rates in a schematic way.

6.2.1.2. Perturbation Theory

Starting from the mixing Hamiltonian of eq. (6.9), we can calculate, in the $U(5)$ limit, the wave function for the lowest yrast $J^{\pi} = 4^{+}$ level, using first-order pertubation theory. The resulting, perturbed wave functions, we denote with a horizontal bar on top of the basis vector i.e. $|\overline{\ }\rangle$. For the $J_i^{\pi} = 4_1^{+}$ level, we elaborate on the calculation in Appendix 6.1. We obtain as a result

$$|\overline{4_1^{+}}\rangle = |s^{N-2}d^{2}; 4^{+}\rangle - \zeta \sqrt{\frac{(N-1)2}{9}} \frac{1}{|\Delta E|} |s^{N-1}g; 4^{+}\rangle, \tag{6.12}$$

with $\Delta E = 2(\epsilon_d - \epsilon_s) - \epsilon_g$, the energy denominator in pertubation theory which has a negative value and is of the order of $\cong 1$ MeV for the $Z = 50$ mass region. Similarly, E2 reduced matrix elements between the 4_1^{+} and 2_1^{+} states are obtained as

$$\langle\overline{2_1^+}||T(\text{E2})_{sdg}||\overline{4_1^+}\rangle = e_{ds}^{(2)}3\sqrt{2(N-1)}\left[1-\frac{e_{gd}^{(2)}}{e_{ds}^{(2)}}\zeta\frac{1}{|\Delta E|}\frac{\sqrt{5}}{9}\right], \quad (6.13)$$

which, for a general yrast E2 transition, can be generalised into the expression

$$\begin{aligned}\langle\overline{J-2) = (2n_d-2)}||T(\text{E2})_{sdg}||\overline{J=2n_d}\rangle \\ = \sqrt{(N-n_d+1)n_d(4n_d+1)}\, e_{ds}^{(2)} \quad (6.14) \\ \times\left[1-\frac{e_{gd}^{(2)}}{e_{ds}^{(2)}}\zeta\frac{1}{|\Delta E|}\frac{\sqrt{5}}{9}(n_d-1)\right].\end{aligned}$$

Yrast means the particular set of states that occur at the lowest energy possible, for each given J^π value i.e., the 0_1^+, 2_1^+, 4_1^+, 6_1^+, . . . ground-state band. For the hexadecapole transitions, we obtain

$$\langle 0_1^+||T(\text{E4})_{sdg}||\overline{4_1^+}\rangle = -e_{gs}^{(4)}\zeta\frac{1}{|\Delta E|}\sqrt{2N(N-1)}\,3, \quad (6.15)$$

and the general yrast E4 matrix element becomes

$$\begin{aligned}\langle\overline{(J-4) = (2n_d-4)}||T(\text{E4})_{sdg}||\overline{J=2n_d}\rangle \\ = -e_{gs}^{(4)}\zeta\frac{1}{|\Delta E|}\frac{2J+1}{3} \quad (6.16) \\ \times\sqrt{n_d(n-1)(N-n_d+1)(N-n_d+2)}.\end{aligned}$$

From the expressions (6.13) and (6.14), we notice that the Σ and Γ contributions to the yrast $B(\text{E2})$ values result in positive interference if the product $e_{gd}^{(2)}/e_{ds}^{(2)}\,\zeta$ is negative. Taking $e_{ds}^{(2)}$ as a positive effective charge ($\cong$0.1 eb in the schematic calculations; this is also a typical value obtained from IBM-calculations in the $Z = 50$ mass region), and having $\zeta > 0$, this requires negative values of $e_{gd}^{(2)}$. Already from the lowest order perturbation analysis we notice that the yrast $B(\text{E2})$ values can become much larger than the standard IBM results. Experimental hexadecapole strengths obtained via particle inelastic scattering will then give the possibility to obtain further information about the effective charge $e_{gs}^{(4)}$ and $e_{dd}^{(4)}$, using eq. (6.15) and eq. (6.16).

The quadrupole moment for the yrast band members, however, will only deviate from their Σ values by a number proportional to $\zeta^2 e_{gg}^{(2)}$. This is easily observed when calculating diagonal matrix elements using the E2-operator from eq. (6.10) and the wave function of eq. (6.12), but then, second-order pertubation theory is needed. Besides the sd quadrupole moment, corrections quadratic in the mixing parameter ζ result. Therefore, one will need positive effective boson charges $e_{gg}^{(2)}$ in order to make the quadrupole moments less negative compared with the Σ values. This latter value (sign and magnitude) can again be related to the underlying shell-structure of the particular mass region under study (see Section 6.2.1.3).

6.2.1.3. Schematic Model Calculations

In the previous sections we have indicated in a rather general framework what the influence of including the hexadecapole boson will have on energy spectra and electromagnetic properties. We relied mainly on the simplicity of the vibrational limit (the $U(5)$ wave functions) to evaluate the various expressions in Section 6.2.1.2. One can extend this method to the $O(6)$ limit in a rather similar way, even though analytical results are much less transparent. To this end and to give a feeling of what influence the g-boson has on spectra and E2 properties, we carry out some calculations coupling a g-boson to an exact $U(5)$ and $O(6)$ core. We use the following parameters for the Hamiltonian of eq. (6.5):

(i) $U(5)$: ϵ_d = 0.4 MeV; c_0 = 0.17 MeV; c_2 = 0.10 MeV and c_4 = 0.07 MeV,

(ii) $O(6)$: κ'' = 0.10 MeV; κ' = 0.02 MeV and q_3 = 0.15 MeV.

In the $U(5)$ situation, the sd-boson Hamiltonian can also be written according to eq. (2.164) (see Chapter 2). Using the d-boson energy (ϵ_d) and the anharmonicity parameters (c_0, c_2, c_4), a large class of vibrational-like nuclei can be described. These parameters are adapted for the ^{104}Ru nucleus starting from the ^{104}Ru IBM-2 parameters. They can, however, be used as a standard set of parameters for the schematic $U(5)$ and $O(6)$ calculations. For the unperturbed g-boson energy, we take a typical value of $\epsilon_g \cong 2$ MeV; which is near to 2Δ, the energy needed to create a two-quasi particle excitation in the $Z \cong 50$ ($A \cong 100$) mass region.

The effective boson charges used are $e_{ds}^{(2)}$ = 0.1eb; $e_{dd}^{(2)}$ = -0.1 eb and we vary $e_{gg}^{(2)}$ and $e_{gd}^{(2)}$ from 0 to ± 0.3 eb in steps of 0.1 eb. For the E4 operator, we use the boson charge $e_{dd}^{(4)}$ = $e_{gs}^{(4)}$ = 1.0 eb^2. In the latter case, no experimental hexadecapole matrix elements are known and only an exploratory calculation could be carried out. All calculations discussed below are done with a mixing strength parameter of $\zeta = 0.2$ MeV.

In Fig. 6.2, we show part of the unperturbed (no mixing) $U(5)$ and $U(5) \otimes g$ spectra for a $(sd)^{N=8}$ boson system. In Figs. 6.3 and 6.4, we show the yrast B(E2) values and the static quadrupole moments for the yrast band members, respectively. In each case, we also compare with the results using the full IBM Hamiltonian (see eq. (6.3)), with parameters obtained from the projected IBM-2 parameters in ^{104}Ru. Some remarkable points result:

(i) In the yrast band, for negative $e_{gd}^{(2)}$ values, for both the $U(5)$ and the $O(6)$ limits, the B(E2) cut-off factor is shifted to higher angular momenta (see Fig. 6.3). Due to the constructive interference between the Σ and the Γ amplitudes, B(E2) values result which are doubled compared to the standard IBM cal-

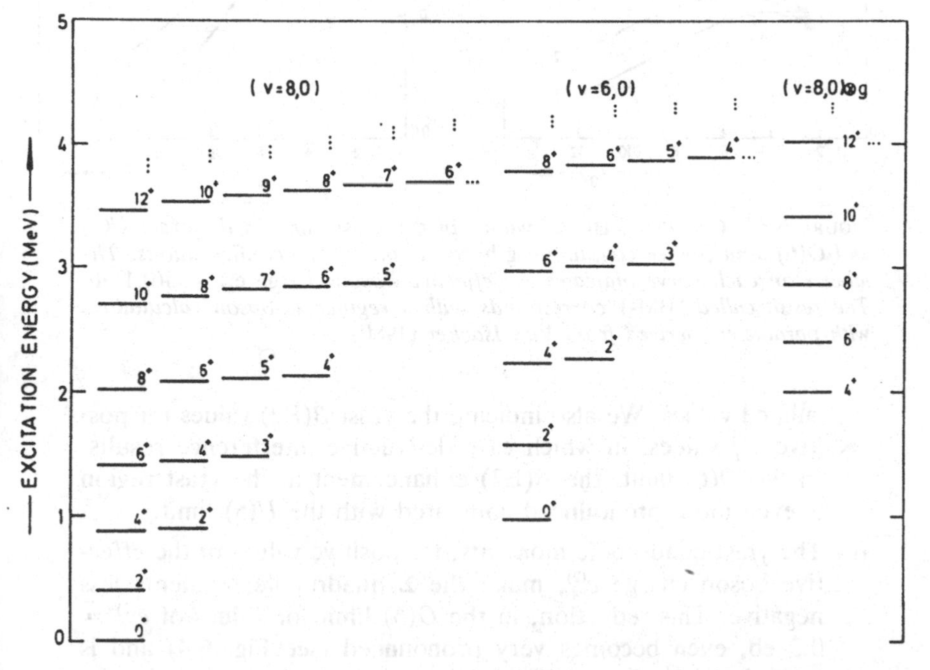

FIGURE 6.2. *Pure U(5) limit for N = 8 bosons. Also the lowest band for the* $|(sd)^7 \otimes g\rangle$ *configurations is given. Only the lowest bands are shown. The quantum numbers between brackets v and* n_Δ *denote the seniority quantum number and the number of boson triplets, coupled to zero angular momentum, respectively* (Casten, 1988).

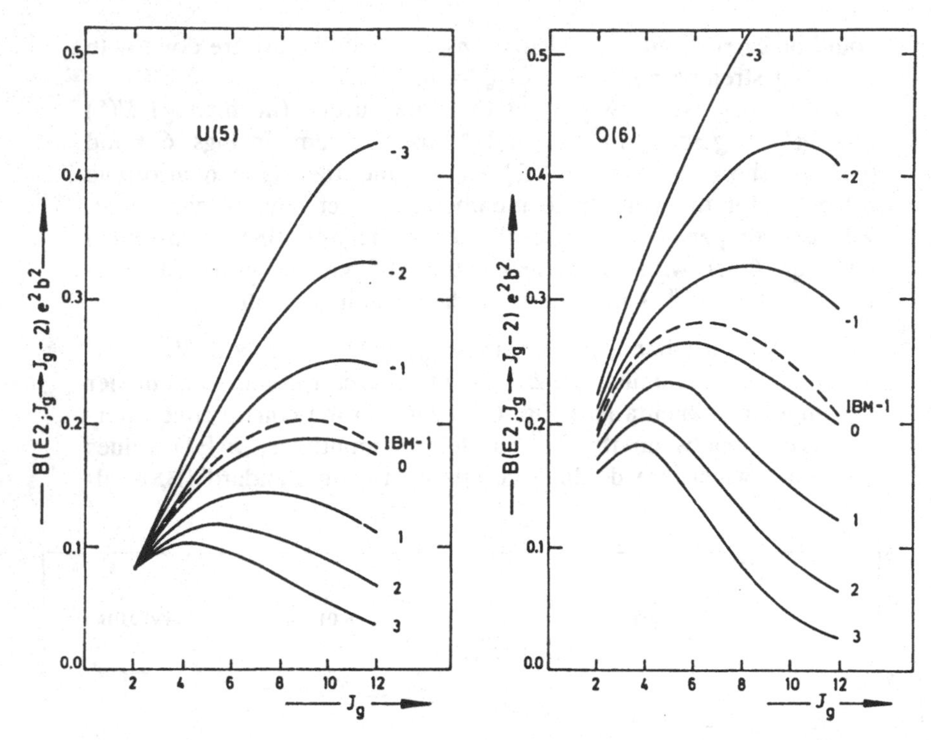

FIGURE 6.3. *Calculated* B(E2) *values in the* yrast *band in the exact* $U(5)$ *and* $O(6)$ *limits, when coupling a g-boson to the* $|(\mathrm{sd})^N\rangle$ *configurations. The label* i *on each curve indicates the effective boson charge* $e_{gd}^{(2)} = i(0.1)$eb. *The result called* IBM-1 *corresponds with a regular sd-boson calculations with parameters derived from Van Isacker* (1980).

culated values. We also indicate the yrast B(E2) values for positive $e_{gd}^{(2)}$ values, in which case destructive interference results. In the $O(6)$ limit, the B(E2) enhancement in the yrast region is even more pronounced compared with the $U(5)$ limit.

(ii) The yrast quadrupole moments, for positive values of the effective boson charge $e_{gg}^{(2)}$, make the Σ quadrupole moments less negative. This reduction, in the $O(6)$ limit for values of $e_{gg}^{(2)}$ = 0.3 eb, even becomes very pronounced (see Fig. 6.4) and is related to the fact that in the $O(6)$ limit description of the ground-state band, d-boson admixtures are present to a larger extent compared with the $U(5)$ limit.

The above modifications of E2 properties are quite important and

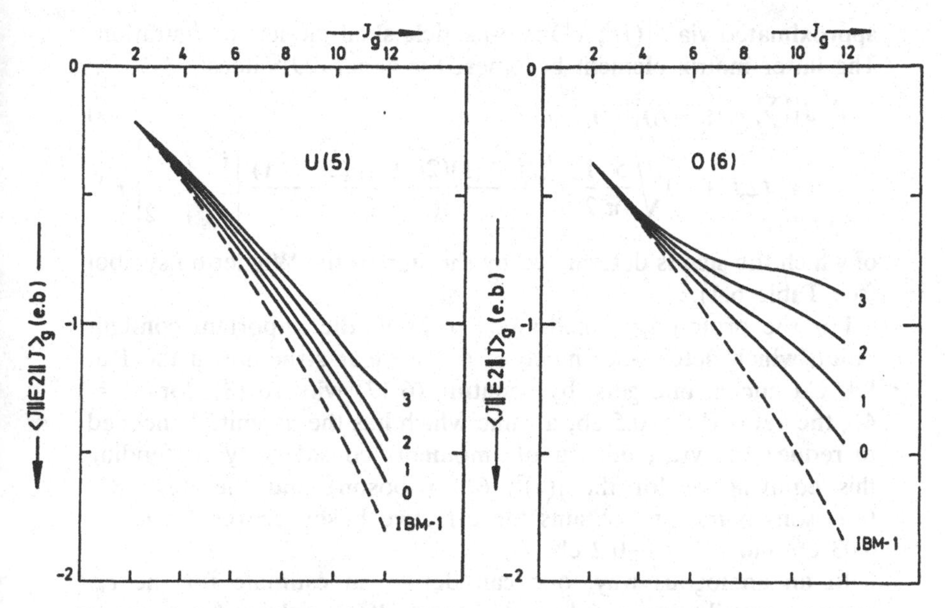

FIGURE 6.4. *Same caption as Fig. 6.3, but for the* yrast *quadrupole moments. Moreover, the label i on the different curves now gives the effective boson charge* $e_{gg}^{(2)}$ = i(0.1) eb.

will be discussed in Section 6.2.1.5. where we compare with the data in the even-even ^{104}Ru nucleus.

6.2.1.4. Shell-Model Estimates

We now try to relate, in light of the results obtained in Section 6.2.1.3., some of the boson effective charges from the underlying shell-model transition matrix elements in the $Z \cong 50$ mass region i.e. in the $N = 4$ harmonic oscillator shell.

One can estimate the magnitude of $e_{gg}^{(2)}$ by equating the boson matrix element

$$\langle s^{N-1}g;\, 4^+ \| e_{gg}^{(2)}(g^\dagger\tilde{g})^{(2)} \| s^{N-1}g;\, 4^+\rangle, \tag{6.17}$$

with the two-particle E2 matrix element

$$\langle (j)^{(2)};\, 4^+ \| \sum e_i r_i^2 Y_2(\hat{r}_i) \| (j)^2;\, 4^+\rangle. \tag{6.18}$$

Here, we tacitly make the assumption that the g-boson can be approximated via a$|(j)^2; 4^+\rangle$ two-particle shell-model configuration. The latter matrix element becomes, for general J values

$$\langle (j)^2; J || \sum e_i r_i^2 Y_2(\hat{r}_i) || (j)^2; J\rangle \tag{6.19}$$

$$= (2J+1)\sqrt{\frac{5}{4\pi}}\frac{1}{2}\sqrt{\frac{(2j-1)(2j+3)(2j+1)}{j(j+1)}}\begin{Bmatrix} j & J & j \\ J & j & 2 \end{Bmatrix}\langle r^2\rangle,$$

of which the sign is determined by the sign of the Wigner 6-j symbol (see Table 6.1).

For the proton $1g_{9/2}$ shell-model orbital, the important configuration which determines much of the nuclear structure in the Ru, Pd, Cd nuclei, one gets, by equating (6.17) with (6.18), for $J^\pi = 4^+$, the value $e_{gg}^{(2)} \cong 0.5$ eb, a value which has the magnitude needed to reduce the yrast quadrupole moments considerably. Extending this equivalence for the $|(j)^2; 6^+\rangle$ (i-boson) and the $|(j)^2; 8^+\rangle$ (k-boson) pairs, one obtains the effective boson charges of $e_{ii}^{(2)} \cong 0.03$ eb and $e_{kk}^{(2)} \cong -0.2$ eb.

In an analogous way, one can derive an estimate for the $e_{gd}^{(2)}$ charge, contributing mainly to the yrast B(E2) values, from equating the E2 matrix element

$$\langle s^{N-1}d; 2^+ || T(\text{E2})_{sdg} || s^{N-1}g; 4^+\rangle, \tag{6.20}$$

with the shell-model transition E2 matrix element

$$\langle (j)^2; 2^+ || \sum e_i r_i^2 Y_2(\hat{r}_i) || (j)^2; 4^+\rangle$$

$$= (-1)^{2j+1} 2\sqrt{5}\sqrt{9}\langle j || r^2 Y_2 || j\rangle \begin{Bmatrix} j & 2 & j \\ 4 & j & 2 \end{Bmatrix}. \tag{6.21}$$

The latter matrix element has a negative sign, and thus a negative $e_{gd}^{(2)}$ effective boson charge results. Again using, for j, the proton $1g_{9/2}$ proton shell-model configuration, one obtains the value $e_{gd}^{(2)} \cong$

TABLE 6.1 *The sign of the Wigner 6-j symbol* $\begin{Bmatrix} j & J & j \\ J & j & 2 \end{Bmatrix}$.

	$J = 2$	$J = 4$	$J = 6$	$J = 8$
$j = \frac{5}{2}$	+	−		
$j = \frac{7}{2}$	+	+	−	
$j = \frac{9}{2}$	+	+	−	−
$j = \frac{11}{2}$	+	+	+	−

-0.15 eb, a value which is of the order of magnitude used in the schematic calculations in order to enlarge the yrast $B(\mathrm{E2})$ values substantially.

Although the above arguments are only used in an approximate way and can therefore not serve as a quantitative calculation of boson effective charges, one can still obtain an idea of the sign and magnitude for the extra effective charges used in the E2 transition operator, when incorporating a *g*-boson in the IBM formalism. In a similar way, estimates for the effective hexadecapole charge $e_{gs}^{(4)}$ can be obtained.

6.2.1.5. Application to the Ru Nuclei: ^{104}Ru

In order to study a region of nuclei in which the schematic model calculations of coupling a *g*-boson to a $U(5)$ or $O(6)$ core can be applied, we should look for nuclei intermediate between both limits. It has been discussed (Chapter 3) that the Ru and Pd nuclei, in the mass $A \cong 100$ region, are probably good transitional nuclei between $U(5)$ and $O(6)$. Moreover, in recent experiments, particular results in ^{104}Ru have been obtained that point towards the necessity of incorporating extra degrees of freedom besides the *s*- and *d*-bosons:

(i) There is a sudden increase in the $B(\mathrm{E2})$ values between the $4_1^+ \rightarrow 2_1^+$ and the $6_1^+ \rightarrow 4_1^+$ transitions;

(ii) There is no clear evidence for a reduction in the yrast $B(\mathrm{E2})$ values, in contrast to the IBM predictions, which show a cut-off at spin $J^\pi = 6^+$.

Not only the IBM, but also calculations in the framework of the asymmetric rotor model (Davydov and Filippov, 1958) and in the collective model studies of Gneuss, Greiner and Hess (Gneuss, 1971; Hess et al., 1980) are unable to explain the above features (i) and (ii).

The schematic model calculations of Section 6.2.1.3. give hope to explain both salient features by including, besides the $|(sd)^N\rangle$ (Σ configurations), also $|(sd)^{N-1} \otimes g\rangle$ (Γ configurations) excitations. Although no unique evidence exists for the observation of low-lying hexadecapole states, near $E_x \cong 2$ MeV, in ^{104}Ru, a $J^\pi = 4^+$ level has been observed that cannot be explained easily within the standard descriptions of low-lying collective quadrupole excitations.

To treat this nucleus, we start from the full IBM Hamiltonian of eqs. (6.3)–(6.5) with a g-boson mixing Hamiltonian as described by eq. (6.9). The IBM parameters, obtained from IBM-2 parameters for ^{104}Ru (Van Isacker, 1980) are: $\epsilon_d = 0.6$ MeV and $\epsilon_g = 2.0$ MeV; $\kappa = -0.010$ MeV, $\kappa' = 0.010$ MeV, $\kappa'' = 0.098$ MeV, $q_3 = 0.050$ MeV; $\zeta = 0.2$ MeV.

The calculated spectrum is shown in Fig. 6.5 where both the results for $\zeta = 0$ MeV (unmixed calculation: U) and for $\zeta = 0.2$ MeV (mixed calculation using the g-boson: M) are indicated and compared with the experimental data.

Concentrating on the IBM calculations, including a one g-boson excitation, the overall agreement with experiment is good but for the γ-band, a very pronounced staggering effect $(3^+, 4^+)$, $(5^+, 6^+)$, . . . shows up which is not as strongly pronounced in the experimental data. This strong staggering in the γ-band is mainly a consequence of the fact that the corresponding IBM Hamiltonian potential energy surface, being close to the $O(6)$ limit (see the large pairing part (κ'')), shows the typical γ-soft (or even γ-unstable in

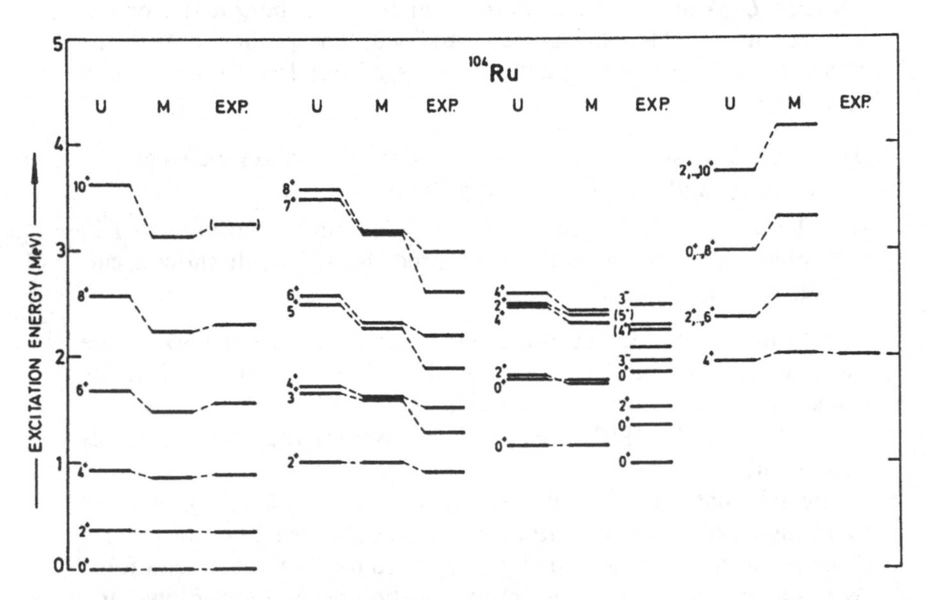

FIGURE 6.5. *Comparison of the experimental* ^{104}Ru *spectrum with* IBM *calculations without* g-bosons (U : *unmixed*) *and when coupling a* g-boson *to the same* IBM $|(sd)^N\rangle$ *core, interacting via the Hamiltonian of eq. (6.9)* (*mixed* : M).

the extreme $O(6)$ limit) potential in the classical (β, γ) plane. This staggering can be reduced, yielding better agreement with experiment, by introducing specific higher order terms in the Hamiltonian that give rise to cubic anharmonicities in the *sd*-boson model space (see Section 6.4).

Extensive comparison with the experimental data for B(E2) values and the static quadrupole moments is shown in Figs. 6.6–6.9. We have used the E2 operator of eq. (6.10) with the effective charges: $e_{ds}^{(2)} = 0.1$ eb, $e_{dd}^{(2)} = -0.05$ eb and the *g*-boson charges $e_{gg}^{(2)} = e_{gd}^{(2)} = 0, \pm 0.1, \pm 0.2$ and ± 0.3 eb. In each case, we also compare with the IBM ($\zeta = 0$). Here, we notice that for a value of $e_{gd}^{(2)} = -0.2$ eb, as used in the calculations of Section 6.2.1.4, both the order of magnitude of the yrast B(E2) values and the shift of the B(E2) cut-off to much larger angular momenta are observed (see Fig. 6.6). The same features are also observed and qualitatively reproduced for the γ-band B(E2) values (see Fig. 6.7). For the $J_\gamma \rightarrow J_{g-2}$ B(E2) values, which are normally very small, for a value of $e_{gd}^{(2)} = -0.2$ eb, the general trend is again well described (Fig. 6.8). Concerning the static quadrupole moments in the yrast band (Fig. 6.9), even for the larger $e_{gg}^{(2)}$ values, the experimental data are not well reproduced. This points to the need to introduce some triaxiality since, in the framework of the asymmetric rotor, the values and the systematics in the yrast band are well described.

Concluding, we can say that incorporating a *g*-boson degree of freedom in the IBM description of collective excitations extends the scope of this model. Moreover, we have noticed that, in the present example of ^{104}Ru, rather good agreement in both the energy spectra and the B(E2) values can be obtained.

6.2.2. The *SU*(3) Limit

In the discussion of Section 6.2.1, a *g*-boson was coupled to a $U(5)$ and $O(6)$ core. We now consider the coupling of a *g*-boson to a $|(sd)^N\rangle$ $SU(3)$ core, first in a schematic way. Afterwards, application to the even-even strongly deformed Gd nuclei is carried out.

6.2.2.1. Study of a Schematic Model

In nuclei, close to the rotational limit, a natural way to restrict H_{sd} of eq. (6.3) is to use its form in the $SU(3)$ limit. With this approx-

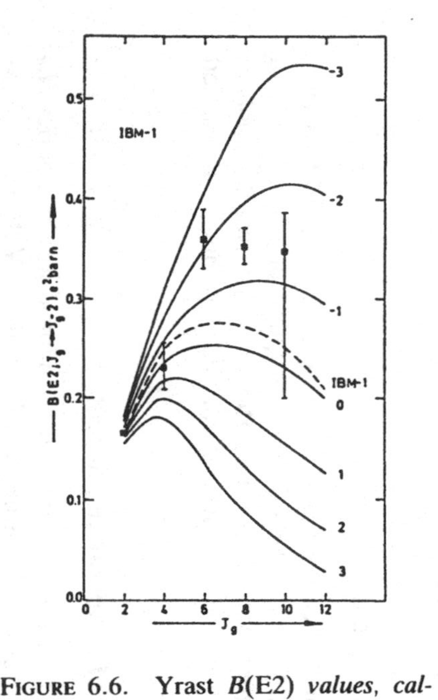

FIGURE 6.6. Yrast B(E2) *values, calculated using the* IBM *Hamiltonian coupled with one* g-*boson excitations. The curves correspond to different effective boson charges* $e_{gd}^{(2)}$ *and are labeled with a parameter* i ($e_{gd}^{(2)}$ = i(0.1) eb). *The experimental data for* ^{104}Ru *are taken from Stachel* (1982).

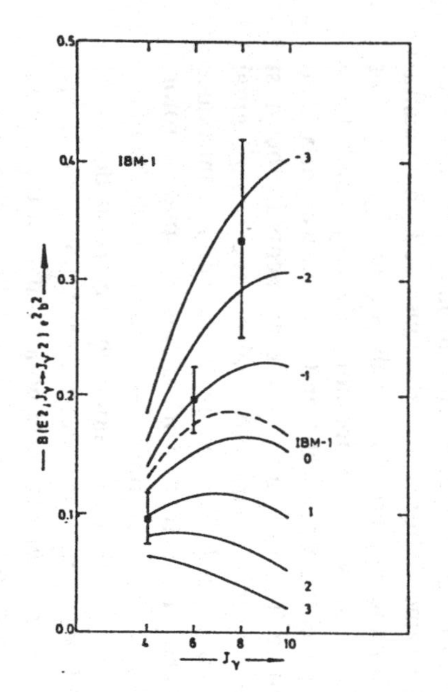

FIGURE 6.7. *Same as for Fig. 6.6, but for the γ-band*

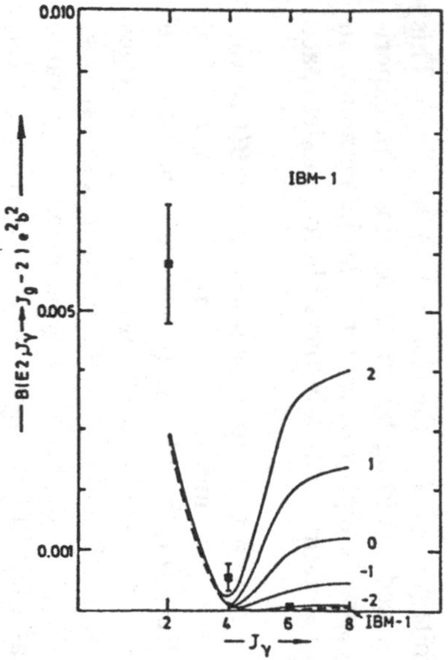

FIGURE 6.8. *Same as for Fig. 6.6, but for the* $J_\gamma \to J_{g-2}$ *transitions.*

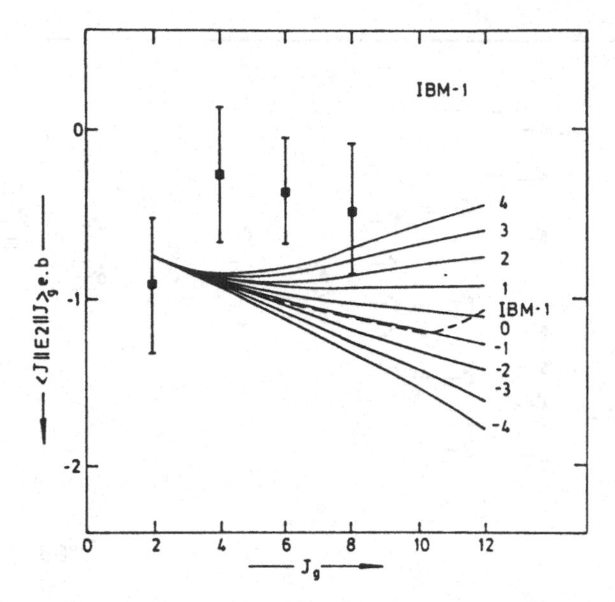

FIGURE 6.9. *Same as for Fig. 6.6, but for the static quadrupole moments in the* yrast *band. The curves are labeled with a parameter* i, *indicating the effective boson charge* $e^{(2)}_{gg}$ = i(0.1) eb.

imation, we are able to derive analytical formulae and gain insight in the energy spectra and their dependence on the different parameters appearing in the Hamiltonian.

A striking feature of the experimental spectra is the ordering of the different bands. For instance, in ^{156}Gd (Fig. 6.10), besides the ground-state, β and γ-bands; a number of extra bands show up, in particular a $K^\pi = 4^+$ band near $E_x \approx 1.5$ MeV, that are not easily obtained starting from the $SU(3)$ limit of the interacting *sd*-boson model. It is tempting to try to understand this feature without ad-hoc assumptions on the signs of the parameters in the Hamiltonian. In order to investigate this problem, we consider the following simplified form of the Hamiltonian:

$$H = -\kappa Q \cdot Q - \kappa_l Q \cdot Q_l, \tag{6.22}$$

where we use the definition $Q_l = (l^\dagger \tilde{l})^{(2)}$ and Q was defined in eq. (6.4). We leave the multipolarity l unspecified first. The appropriate basis to diagonalize the Hamiltonian (6.22) is of course the $SU(3)$ basis. Since we are interested here in the states arising from

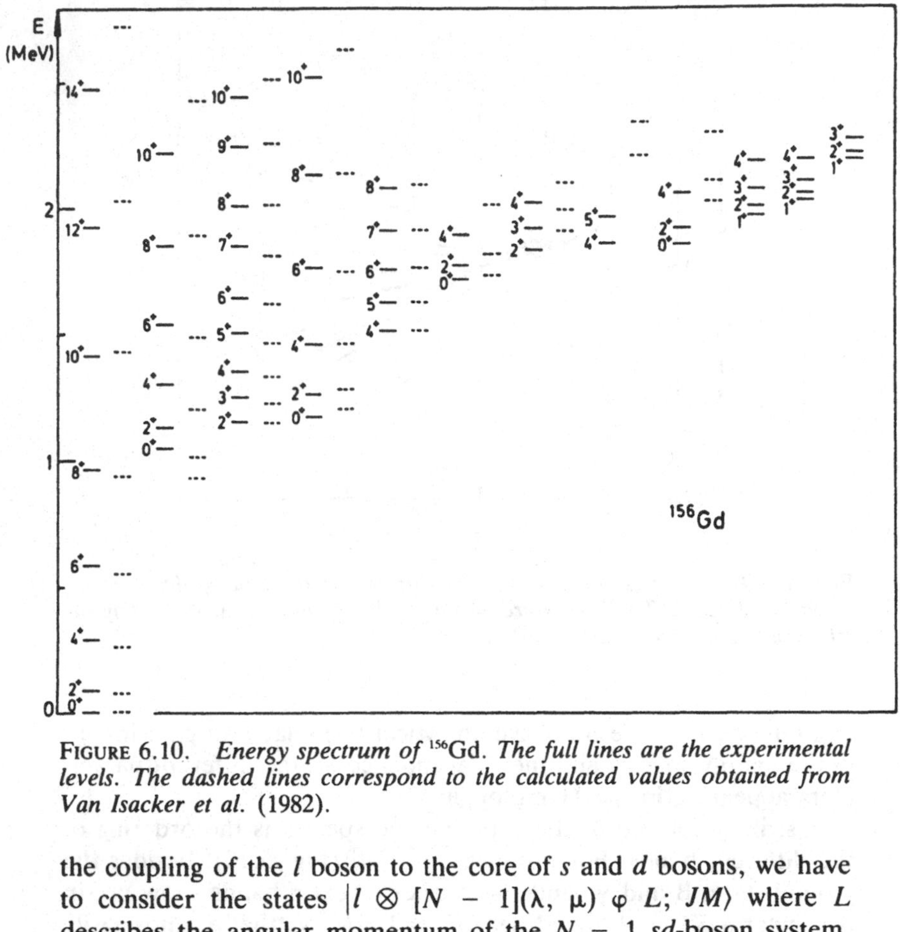

FIGURE 6.10. *Energy spectrum of* ^{156}Gd. *The full lines are the experimental levels. The dashed lines correspond to the calculated values obtained from Van Isacker et al.* (1982).

the coupling of the l boson to the core of s and d bosons, we have to consider the states $|l \otimes [N - 1](\lambda, \mu)\ \varphi\ L; JM\rangle$ where L describes the angular momentum of the $N - 1$ sd-boson system. The quantum number φ then distinguishes between the same L-values appearing in a given $(\lambda, \mu)SU(3)$ representation (Chapter 2). For the ground-state band $(\lambda, \mu) \equiv (2N - 2, 0)$, L determines uniquely the ground-state wave functions. Moreover, since Q is a generator of $SU(3)$, it suffices to consider, without making any approximation, the coupling of the l boson to the ground state band i.e. to consider the states $|l \otimes [N - 1](2N - 2, 0)L; JM\rangle$. The matrix elements of the Hamiltonian (6.22) in this basis are given by

$$\langle l \otimes [N-1](2N-2,0)L'; JM|H|l \otimes [N-1](2N-2,0)L; JM\rangle$$
$$= \tfrac{3}{8}\,\kappa L(L+1)\delta_{LL'} - \kappa_l\sqrt{5}(-1)^J$$
$$\times \begin{Bmatrix} l & L' & J \\ L & l & 2 \end{Bmatrix} \qquad (6.23)$$
$$\times \langle (2N-2,0)L'||Q||(2N-2,0)L\rangle.$$

Analytical energy formulae can be obtained in two limiting cases:

(i) $\kappa \gg \kappa_l$: In this limit no regular band structure develops (Fig. 6.11, left-hand part); the energies are given in first-order per-

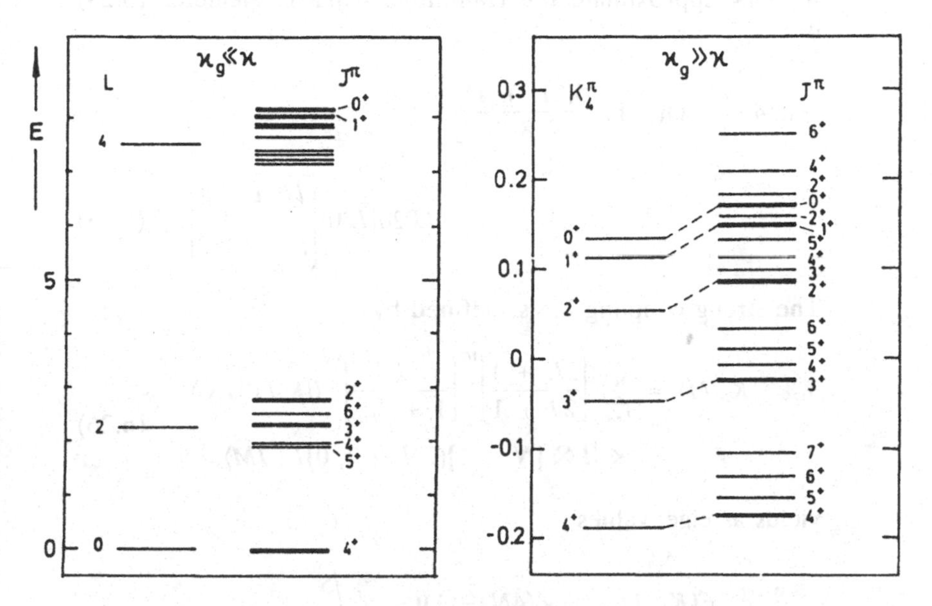

FIGURE 6.11. *Energy spectra resulting from the coupling of a g-boson to a SU(3) core both in the cases of weak coupling ($\kappa_g \ll \kappa$) and of strong coupling ($\kappa_g \gg \kappa$). In the case of weak coupling, we take $\kappa = \frac{1}{5}(4N-1)\kappa_g$ and apply eq. (6.24). For the strong coupling, we take $\kappa = (1/500(4N-1)\kappa_g$ and apply eqs. (6.27) and (6.28). The left columns result from perturbation theory in zeroth order, the right columns give results in first order. The energy is given in units κ in the weak coupling limit, and in units $(4N-1)\kappa_g$ in the strong coupling limit. For an explanation of the symbols* L, K_g *and* J *we refer to the text. The thickened lines represent* yrast *levels (weak coupling) or band heads (strong coupling).*

turbation theory by

$$E(J, L) \cong \frac{3}{8}\kappa L(L+1) - \kappa_l(-1)^J\left[\frac{5(2L+1)}{8}\right]^{1/2} \times (4N-1)\begin{Bmatrix} l & L & J \\ L & l & 2 \end{Bmatrix}\langle L0\,20|L0\rangle. \tag{6.24}$$

(ii) $\kappa \ll \kappa_l$: In this limit, by making a linear combination of the weak-coupling wave functions for different angular momenta L obtained after coupling a g-boson to the lowest $(2N-2, 0)$ $SU(3)$ representation, the Hamiltonian (6.22) can be shown to become diagonal (Arima and Iachello, 1978). For large N, we may approximate the Hamiltonian matrix elements (6.23) by

$$-\kappa_l(4N-1)(-1)^J\left[\frac{5(2L+1)}{8}\right]^{1/2} \times \langle L020|L'0\rangle\begin{Bmatrix} l & L' & J \\ L & l & 2 \end{Bmatrix}. \tag{6.25}$$

The strong coupling basis, defined by

$$|K_l JM\rangle = \sum_{L\,\mathrm{even}}\left[\frac{2L+1}{2J+1}\right]^{1/2}\left[\frac{2}{1+\delta_{K_l,0}}\right]^{1/2}\langle lK_l L0|JK_l\rangle \times |l\otimes[N-1](2N-2,0)L; JM\rangle, \tag{6.26}$$

yields as eigenvalues

$$E(K_l, J) \cong -\kappa_l(4N-1)(-1)^l\sqrt{\frac{5}{8}} \times \frac{3K_l^2 - l(l+1)}{[(2l-1)l(2l+1)(l+1)(2l+3)]^{1/2}}. \tag{6.27}$$

The quantum number K_l plays the role of an angular momentum projection quantum number on the symmetry axis of a deformed nucleus as discussed by Bohr and Mottelson (1975).

This formula can be developed further by considering a small perturbation term—$\kappa Q \cdot Q$. In first-order perturbation theory, we then obtain the following additional contribution to the energy (6.27) (see Fig. 6.11, right-hand part)

$$\Delta E(K_l, J) = \frac{3\kappa}{8}[J(J + 1) + l(l + 1) - 2K_l^2] - \kappa(2N^2 - N - 1). \tag{6.28}$$

Thus, whereas in the limit $N \to \infty$ and $\kappa = 0$ the energy $E(K_l, J)$ does not depend on the angular momentum J, this degeneracy is lifted by considering a small $Q \cdot Q$ perturbation. In this way, for each band-head characterized by a K_l value, a rotational-band with $J(J + 1)$ spectrum develops.

In Fig. 6.11 we show, in a schematic way, the results for both limits, for a *g*-boson. We have chosen a positive sign for the different constants κ, κ_g. It is clear that with this positive sign, in the case of the coupling of a *g*-boson, the $J^\pi = 4^+$ state is the lowest for the two limits and consequently for the region in between. In Fig. 6.12 we show the behaviour of the energy of the states $|g \otimes [N - 1](2N - 2, 0)L; JM\rangle$ as a function of the ratio κ_g/κ. From this figure, the emergence of the different bands $K^\pi = 4^+, 3^+, \ldots, 0^+$ can be inspected. Also, the characteristic feature that the bands with higher K value are more regular than the bands built on the low K values can be noted. This is a simple consequence from the fact that fewer levels with the same J^π are present and thus configuration mixing disturbs the regular band structure in a minimal way.

In analogy with the coupling of a fermion particle to the core of *s* and *d*-bosons (see Chapter 5), one may consider in the *g*-boson part of the Hamiltonian of eq. (6.5), besides the single *g*-boson energy ϵ_g, also a contribution (exchange term) of the form

$$\Lambda : [(g^\dagger \tilde{d})^{(4)}(d^\dagger \tilde{g})^{(4)}]^{(3)} : , \tag{6.29}$$

where the colons denote that the normal ordering of the operators has to be taken. The microscopic origin of the so-called exchange term has been discussed in the case of the coupling of a fermion. However, for a *g*-boson this microscopic interpretation has not yet been given. In order to evaluate the effect of this exchange contribution, we can calculate its matrix element in a rotational basis

$$\langle g \otimes [N-1](2N-2,0)L'; JM | :[(g^\dagger \tilde{d})^{(4)}(d^\dagger \tilde{g})^{(4)}]^{(0)} : | g \otimes [N-1](2N-2,0)L; JM\rangle$$
$$= 3(N-1)[(2L+1)(2L'+1)]^{1/2} \tag{6.30}$$
$$\times \sum_R \begin{Bmatrix} 2 & L & R \\ J & 4 & 4 \end{Bmatrix} \begin{Bmatrix} 2 & L' & R \\ L & 4 & 4 \end{Bmatrix}$$
$$\times \langle (2N-4,0)R; (2,0)2 \| (2N-2,0)L'\rangle$$
$$\times \langle (2N-4,0)R; (2,0)2 \| (2N-2,0)L\rangle,$$

where the double-barred matrix elements are reduced $SU(3) \supset O(3)$ Wigner coefficients (Vergados, 1968). However, since the exchange contribution 6.29 cannot be written in terms of the generators of $SU(3)$, it will mix the different representations (λ, μ) and a calculation within the restricted basis $|g \otimes [N-1](2N-2, 0)L; JM\rangle$ will give a sensible approximation only for small values of Λ, the coupling strength of the exchange term in eq. (6.29). In Fig. 6.13, we show the variation of the energy spectrum as a function of the parameter Λ. It can be noted that for large values of Λ, the $J^\pi = 3^+$ level will become the lowest state. However, for such large values of Λ, a calculation within the restricted basis does not give reliable results anymore and a full diagonalization, taking into account all core states, will be required.

6.2.2.2. Application to the Even-Even Gd Nuclei

In this section, we briefly discuss some of the most important results of the g-boson calculations as carried out for the doubly-even Gd nuclei with $90 < N < 96$, using a more general Hamiltonian than the schematic $SU(3)$ Hamiltonian of eq. (6.22). The latter Hamiltonian only splits the states originating from the same core coupled to a g-boson but does not mix the different $|(sd)^N\rangle$ (Σ states) and $|(sd)^{N-1} \otimes g >$ (Γ states) configurations. Therefore, we have introduced the mixing Hamiltonian, that was also used in Section 6.2.1, to study the $U(5)$ and $O(6)$ limiting situations. The total Hamiltonian used here is the same as given in eq. (6.5). The resulting parameters, which give a best fit for the ^{156}Gd nucleus are: $\kappa = -0.015$ MeV, $\kappa' = 0.00645$ MeV; $\kappa'' = -0.0054$ MeV; $\kappa_g = 0.075$ MeV and $\epsilon_d = 0.100$ MeV; $\epsilon_g = 1.415$ MeV. The value κ_g equals $-\kappa\gamma$ in eq. (6.7). The coefficient ζ, determining the mixing

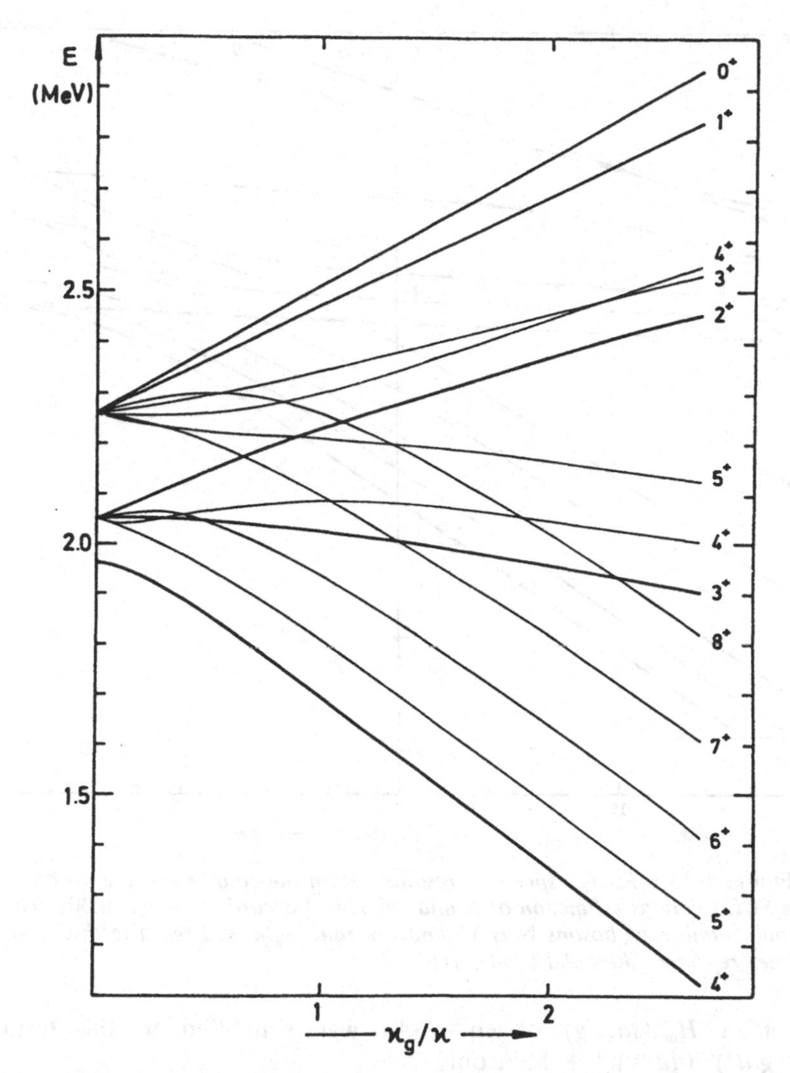

FIGURE 6.12. *Energy spectrum resulting from the coupling of a g-boson to a SU(3) core as a function of the ratio κ_g/κ. The total number of bosons N is 12. The thickened lines represent band head levels.*

between the *sd* and *g* bosons, is not very important in determining the energy spectra. A value of $\zeta = 0.1$ was determined from E2 decay properties connecting Γ states to Σ states. In the application to the even-even Gd nuclei, the mixing Hamil-

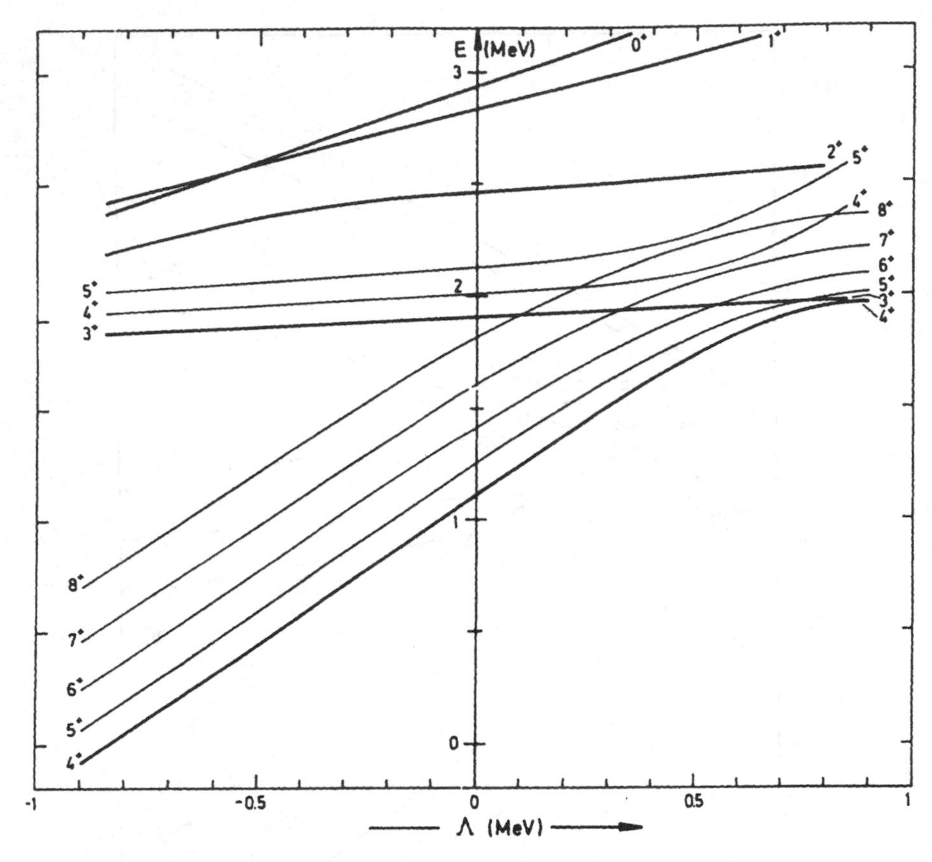

FIGURE 6.13. *Energy spectrum resulting from the coupling of a g-boson to a SU(3) core as a function of Λ and calculated according to eq.* (6.30). *The total number of bosons* N *is* 12 *and the ratio* $\kappa_g/\kappa \cong 2.68$. *The thickened lines represent the band head levels.*

tonian $H_{mix}(sd;\ g)$ of eq. (6.8) was simplified to the term $[(g^{\dagger}d^{\dagger})^{(2)}(\tilde{d}\tilde{d})^{(2)}]^{(0)}$ + h.c., only.

A detailed discussion of ^{156}Gd, however, needs the inclusion of even more degrees of freedom then just the hexadecapole boson. The *s*-boson was defined as a coherent 0^+ pair state starting from the underlying shell structure (see Chapter 4). At an energy close or somewhat lower than twice the pairing gap energy, non-coherent 0^+ pairs can occur. The lowest of these non-coherent 0^+ pairs is often introduced as an *s'* boson. In the detailed study of the even-even Gd nuclei, this *s'* boson was incorporated besides the *g* boson.

A detailed discussion of this extended set of calculations can be found in (Van Isacker et al., 1982).

Keeping to the most important extra states, a $K^\pi = 4^+$ band reproduces the experimental $K^\pi = 4^+$ at $E_x = 1.611$ MeV rather well (See Fig. 6.10, dashed lines). The lowest $K^\pi = 4^+$ band obtained within the *sd*-boson model, appears near $E_x \approx 2.2$ MeV and thus indicates the importance of the *g*-boson in reproducing the lower-lying $K^\pi = 4^+$ band.

A stringent test for the structure of the *g*-boson $K^\pi = 4^+$ band is obtained from a study of its E2 decay properties. In Fig. 6.14, we present E2 decay of the $J^\pi = 5^+$ and 4^+ members of this band into the ground, β, γ band mainly. For that purpose we use a simplified form of the general $T(\mathrm{E2})_{sdg}$ operator of eq. (6.10), resulting in the expression

$$T(\mathrm{E2})_{sdg} = e_{ds}^{(2)}(d^\dagger s + s^\dagger \tilde{d})^{(2)} + e_{dd}^{(2)}(d^\dagger \tilde{d})_{(2)} + e_{gg}^{(2)}(g^\dagger \tilde{g})^{(2)}. \quad (6.31)$$

Here, it is the $(g^\dagger \tilde{g})^{(2)}$ operator which almost fully accounts for the reduced transition probabilities. Most features of the E2 γ-decay can be explained: weak transitions from the $K^\pi = 4^+$ band to the ground band, stronger to the β-band and still stronger to the γ-band and a very strong $5_\Gamma^+ \to 4_\Gamma^+$ intraband transition.

Experimentally, the best known E4 transition in the Gd isotopes is the $0_1^+ \to 4_1^+$ transition. The experimental matrix element $\langle 0_1^+ || T(\mathrm{E4}) || 4_1^+ \rangle$ decreases smoothly with increasing neutron number (see Fig. 6.15). A calculation with the E4 operator $(d^\dagger \tilde{d})^{(4)}$ and considering only *s* and *d* bosons, yields matrix elements increasing with neutron number. On the other hand, a calculation including a *g*-boson and using the full E4 operator of eq. (6.11) can reproduce the experimental trend. However, one must bear in mind that the E4 operator (6.11) contains two coefficients, which have to be determined from the experiment, and only four experimental points are known. Thus, it is certainly worthwhile to repeat a similar analysis of the $\langle 0_1^+ || T(\mathrm{E4}) || 4_1^+ \rangle$ matrix element for a wider class of nuclei.

6.2.3. Group Theoretical Structure of Interacting *s*, *d* and *g*-Bosons

Instead of carrying out a phenomenological study of the interacting *sdg*-boson model, we can, in a more formal and general way, accen-

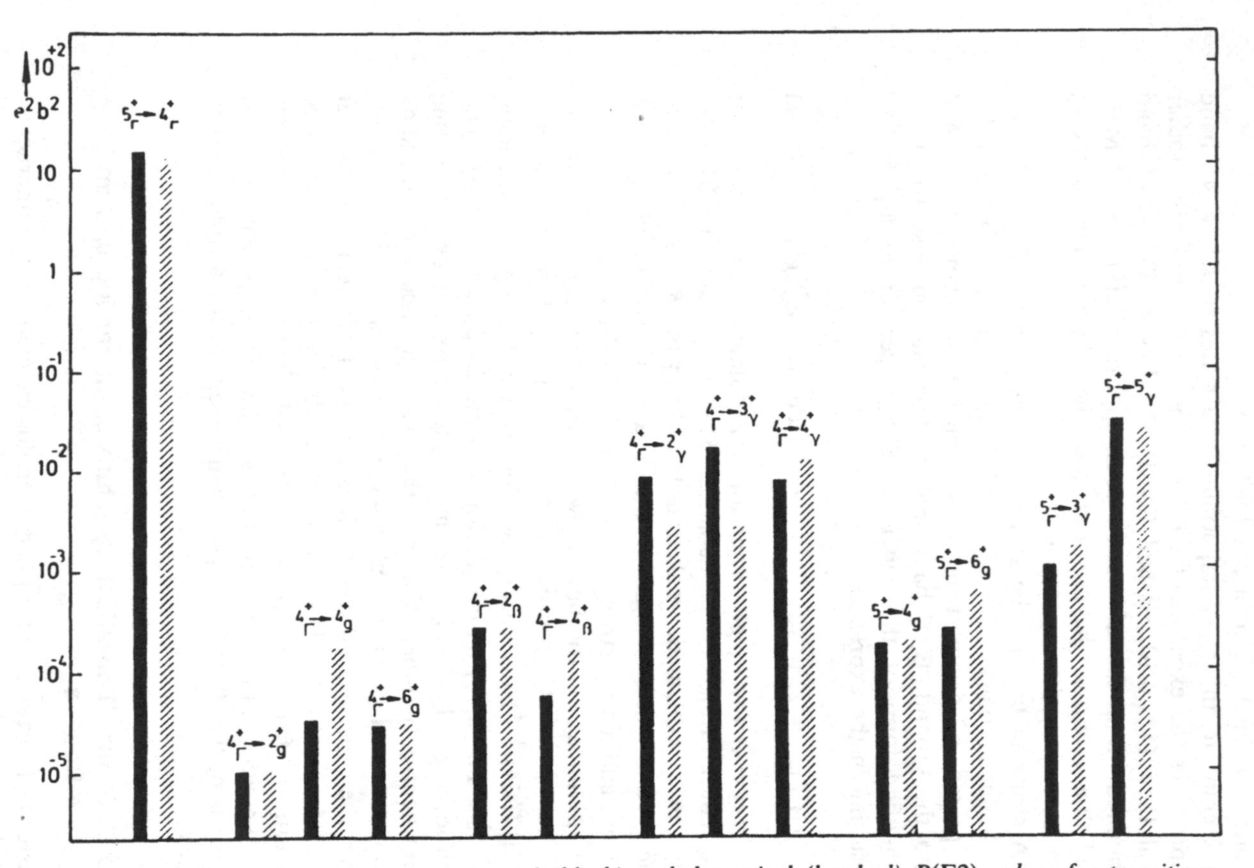

FIGURE 6.14. *Comparison of the experimental (black) and theoretical (hatched)* B(E2) *values for transitions from the* $J^\pi = 4^+$ *and* 5^+ *members of the hexadecapole band (denoted by* 4^+_Γ *and* 5^+_Γ*) to the members of the ground,* β- *and* γ-*bands.*

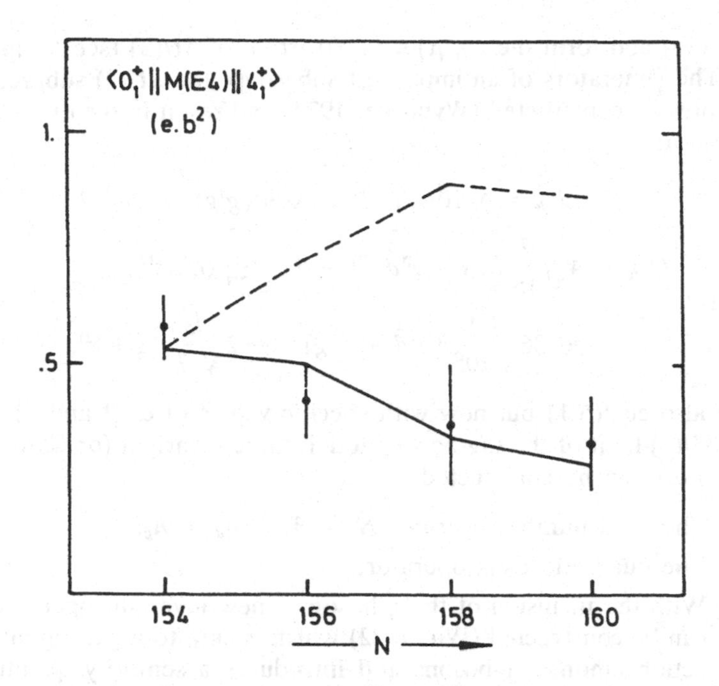

FIGURE 6.15. *The matrix element* $\langle 0_1^+ || T(E4) || 4_1^+ \rangle$ *as a function of the neutron number N in the* Gd *isotopes. The dashed curve represents the* IBM *calculation with* s *and* d*-bosons only. The full curve is a calculation including a g-boson.*

tuate dynamical symmetries that are present in this enlarged system.

Here, in contrast to the $U(6)$ symmetry related to the 6 components $s^\dagger$, $d^\dagger_\mu$ ($\mu = -2, \ldots, +2$), the g-boson $g^\dagger_\nu$ ($\nu = -4, \ldots, +4$) introduces 9 extra components resulting in the $U(15)$ symmetry. In a system with many s, d and g-bosons, and going beyond perturbation theory, we can also study sdg-boson models (with n_g small) in a weak-coupling approach. Here, low-lying hexadecapole vibrations will still dominate the energy spectra and one has $\epsilon_g \gg \epsilon_s, \epsilon_d$. In a strong-coupling limit, however, with many g-bosons present and treating the g-bosons on the same footing with the s and d bosons ($\epsilon_g \approx \epsilon_s \approx \epsilon_d$), new realizations of dynamical symmetries do show up.

The extended group structure contains the operators

$$b^\dagger_\alpha \equiv s^\dagger, d^\dagger_\alpha, g^\dagger_\alpha. \tag{6.32}$$

The 15 operators $b^\dagger_\alpha$ constitute an elementary representation [1] of $U(15)$ and form the $(\lambda, \mu) \equiv (4, 0)$ irrep. of $SU(3)$ (see Chapter 2). The generators of an important subgroup, the $SU(3)$ subgroup, can now be constructed (Wyborne, 1974) and result in the extended operators

$$J_{sdg} = \sqrt{10}(d^{\dagger}\tilde{d})^{(1)} + \sqrt{60}(g^{\dagger}\tilde{g})^{(1)}, \tag{6.33}$$

$$\begin{aligned} Q_{sdg} = {} & 4\sqrt{\frac{7}{15}}(d^{\dagger}s + s^{\dagger}\tilde{d})^{(2)} - 11\sqrt{\frac{2}{27}}(d^{\dagger}\tilde{d})^{(2)} \\ & + 36\sqrt{\frac{1}{105}}(g^{\dagger}\tilde{d} + d^{\dagger}\tilde{g})^{(2)} - 2\sqrt{\frac{33}{7}}(g^{\dagger}\tilde{g})^{(2)}, \end{aligned} \tag{6.34}$$

(see also eq. (6.1) but now with specific values of α, β and γ). In the $SU(3)$ limit of the *sdg*-boson model, three invariant (or Casimir) operators can be constructed:

(i) The total number operator $\hat{N} = \hat{n}_s + \hat{n}_d + \hat{n}_g$.
(ii) The quadratic C_{2SU3} operator,
(iii) With the inclusion of the g-boson, a new invariant operator $\hat{S}$ can be constructed (Wu, 1982) which is akin to a pairing interaction amongst g-bosons and introduces a seniority quantum number ω counting the number of $(\lambda, \mu) \equiv (0, 4)$ coupled pairs. The eigenvalue in a $U(15) \supset SU(3) \supset O(3) \supset O(2)$ basis reads $\langle\hat{S}\rangle = \omega(2N - 2\omega + 3)$. The latter operator causes a splitting of the degeneracy for (λ, μ) representations that occur more than once for a given N-value.

In Fig. 6.16, we illustrate a typical $SU(3)$ dynamical symmetry, resulting for a $N = 8$ boson system, comparing the *sd*- and full *sdg*-boson model. Three major differences in the extended model show up, compared to the *sd*-model.

(i) For a given boson number N, in all the representations, $2N$ is replaced by $4N$ indicating a maximal spin in the ground-band $J_{max} = 4N$ instead of $2N$. So, the major, low-lying representations (λ, μ) now become

$$\begin{aligned} (\lambda, \mu) \equiv {} & (4N, 0), (4N - 4, 2), \\ & (4N - 6, 3), (4N - 8, 4)^2, (4N - 6, 0), \ldots \end{aligned} \tag{6.35}$$

where the superscript 2 indicates that two $(4N - 8, 4)$ representations, distinguished by their ω values, occur.

(ii) The number of representations increases rapidly and, more

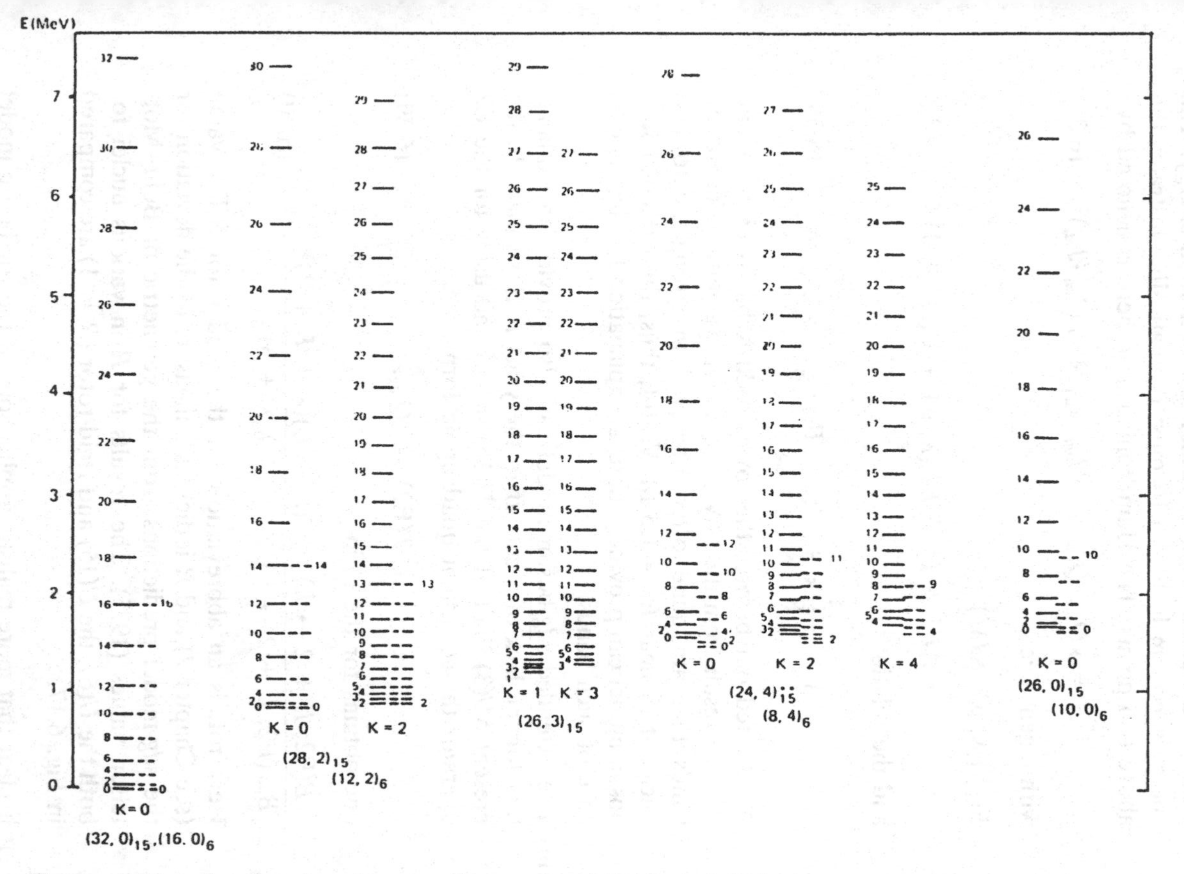

FIGURE 6.16. *A comparison between the spectra of U(15) and U(6) dynamical symmetries for N = 8. All energy levels of the figure are of positive parity. Taken from Wu (1982).*

importantly, representations e.g., $(4N - 6, 3)$ containing $K =$ 1 and 3 bands can result at rather low excitation energy. The energy spectrum in Fig. 6.16 most dramatically illustrates the above two points. The Hamiltonian used here is denoted by

$$H = \kappa Q_{sdg} \cdot Q_{sdg} + \kappa' J_{sdg} \cdot J_{sdg} + \delta\,(J_{sdg} \cdot J_{sdg})^2, \quad (6.36)$$

with eigenvalues

$$E([N](\lambda, \mu), KJM) = AJ(J + 1) + BC(\lambda, \mu) + C(J(J + 1))^2, \quad (6.37)$$

and the relations

$$A = -\frac{3}{8}\kappa + \kappa'; \qquad B = \frac{\kappa}{2}; \qquad C = \delta. \quad (6.38)$$

As pointed out before, the lowest odd-K bands ($K = 1$ and $K = 3$) result from the $(4N - 6, 3)$ representation. These are bands that are quite often observed in strongly deformed nuclei at or near $E_x \approx 1.5$ MeV and, thus, the extended *sdg*-boson model can provide a natural explanation for the appearance of such bands.

(iii) The extended *sdg*-boson model can also provide an explanation for the lack of a cut-off in the yrast B(E2) values. In the present $SU(3)$ limit of the $U(15)$ model and using for the E2 operator the consistent quadrupole form

$$T(\mathrm{E2}) = eQ_{sdg}, \quad (6.39)$$

one obtains for the ratio

$$\frac{B(\mathrm{E2}; L + 2 \rightarrow L)}{B_{\mathrm{rot.}}(\mathrm{E2}; L + 2 \rightarrow L)} = \frac{(\lambda - L)(\lambda + L + 3)}{\lambda(\lambda + 3)} = R. \quad (6.40)$$

Here rot. is an abbreviation for the rigid rotor B(E2) value (see Chapter 2) and R indicates a measure of the deviation for the *sdg*-model predictions from the geometric or Bohr–Mottelson values (1975). The results for R in various nuclei for both the $U(6)$, the $U(15)$ and rigid-rotor ($R = 1$) are compared in Fig. 6.17.

It is clear that more realistic applications of the sdg-boson model to strongly deformed nuclei must allow for the consideration of the other $SU(3)$ non-trivial invariant operator $\hat{S}$ and for some symmetry

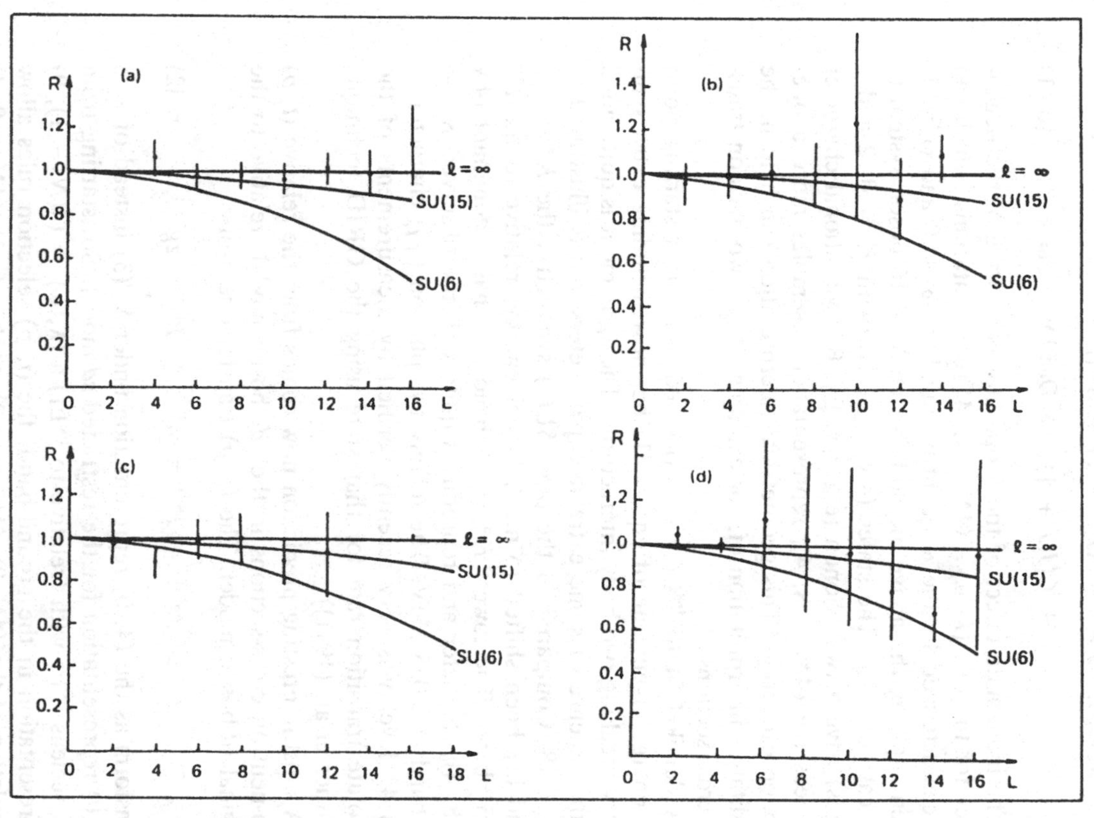

FIGURE 6.17. *The comparison in the ground-state band B*(E2) *values for U*(15), *the U*(6) *symmetries and the rigid-rotor* ($l \rightarrow \infty$) system. The different cases are: (a) ^{232}Th *with* $N = 12$; (b) ^{168}Hf *with* $N = (12)$; (c) ^{170}Hf *with* $N = 13$ *and* (d) ^{164}Yb *with* $N = 12$. *Taken from Wu* (1982).

breaking. The energy eigenvalues, including the $\hat{S}$ operator, become

$$E([N](\lambda, \mu)\omega, KJM) = AJ(J + 1) + BC(\lambda, \mu) + C(J(J + 1))^2 + D\omega(2N - 2\omega + 3). \quad (6.41)$$

The immediate effect of the $\hat{S}$ operator is a lifting of the degeneracy for the two $(\lambda, \mu) \equiv (4N - 8, 4)$ representations, which can be characterized by the ω quantum number, $\omega = 0$ and $\omega = 1$, when applying the above method for the $N = 16$ boson system in the case of ^{168}Er. The states $(4N - 8, 4)_0$ (with $K = 0$, 2 and 4 bands) have a two phonon (e.g. $\gamma\gamma$, $\beta\gamma$, $\beta\beta$) vibrational character whereas the $(4N - 8, 4)_1$ representation resembles more a one-phonon excitation. This geometrical interpretation can help in the qualitative interpretation of, for example, (t, p) two-nucleon transfer cross-sections.

A detailed fit to ^{168}Er was performed using as a starting point the energy eigenvalue of eq. (6.41) but now including extra symmetry breaking two-body interactions. The agreement is quite good at the expense of some extra free parameters and is illustrated in Fig. 6.18. Compared to the pure $SU(3)$ symmetry, the $K^\pi = 1^+$ band has been shifted to much higher energy, relative to the $K^\pi = 3^+$ band. The lowest $K^\pi = 4^+$ band now has a dominant $(4N - 8, 4)_0$ character and reproduces quite well the observed $K^\pi = 4^+$ band at 2.055 MeV. The mainly two-phonon $(\gamma\gamma)$ character of this 4^+ level was very recently verified by measurements of the absolute transition rate for that state using the GRID technique (Börner et al. (1991)).

A most interesting prediction now comes from the relative (t, p) two-neutron cross-sections in the *sdg*-boson model, relative to the original *sd*-boson model. The (t, p) transfer operators

$$P^{(0)} = xs^\dagger \ ; \qquad P^{(2)} = yd^\dagger \ ; \qquad P^{(4)} = zg^\dagger, \qquad (6.42)$$

transform as the (4, 0) representation under $SU(3)$ instead of as a (2, 0) representation for the restricted *sd*-model. So, starting from the nucleus ^{166}Er with (relative to ^{168}Er) a $(\lambda, \mu) = (4(N - 1), 0)$ representation in the ground-band, the (t, p) selection rules allow population of the $(4N, 0)$, $(4N - 4, 2)$, $(4N - 6, 3)$, $(4N - 8, 4)$ representations i.e. of levels in the $K^\pi = 0^+$ ground band, β-band and $(4N - 8, 4)_1$ one-phonon band members. In the *sd*-boson model, however, only the $(2N, 0)$ and $(2N - 4, 2)$ representations

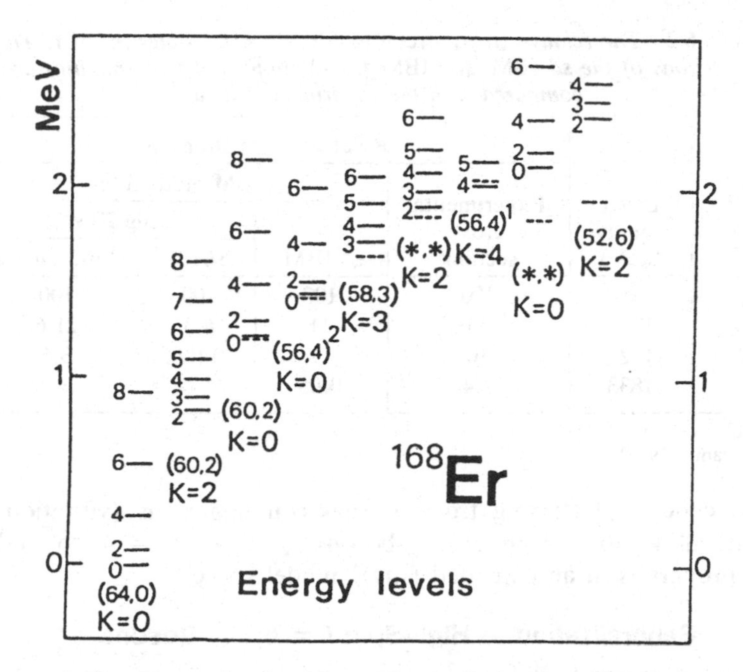

FIGURE 6.18. *Energy levels with positive parity bands in* ^{168}Er *below* E_x = 2.4 MeV. *Solid lines correspond to the theoretical values and dashed lines to the experimental band-head energies. The labels below each band-head repesent the SU*(3) (λ, μ) *quantum numbers of its main component and the K quantum number. States with an asterisk cannot be assigned to a single SU*(3) *representation because of strong configuration mixing. The SU*(3) *labels* $(56, 4)^1$ *and* $(56, 4)^2$ *indicate the* ω = *0 and* ω = 1 ($4N$ − 8, 4) *representations, respectively. Taken from Akiyama et al.* (1986).

are strongly excited. Because of the distinction in the ($4N$ − 8, 4) $\omega = 0$ and $\omega = 1$ bands, relative to a geometrical interpretation, in ^{168}Er, mainly the one-phonon $\omega = 1$ states will be selectively excited in (*t*, *p*) transfer. This is well supported from the results shown in Table 6.2 where the *SU*(3) limit of the *U*(6), *U*(15) groups is compared with the data for (*t*, *p*) transfer into $J^\pi = 0^+$ levels and with the results of the more realistic calculation of Akiyama et al. (1986).

From these examples, the extension into an *sdg*-boson model and the *SU*(3) dynamical symmetry considerations have proven to be a highly successful extension.

Recently a computer code was constructed, in order to carry out expensive *sdg*-boson model calculations, by Y. Devi et al. (1990).

TABLE 6.2 *The relative* (t, p) strengths for $K^{\pi} = 0^+$ *states in* ^{168}Er. *The predictions of the sd* IBM, *sdg* IBM *(SU(3) limit and band-mixing) are compared with the experimental data.*

Band K_i^{π}	Level energy (keV)	Relative (t, p) strength: Experimental cross section[c]	IBM predictions: sd IBM[c]	sdg IBM[a]: $SU(3)$	sdg IBM[a]: Band mixing
0_1^+	0	100	100	100	100
0_2^+	1217	15	11	8.3	21.6
0_3^+	1422	10	2	14.7	3.6
0_4^+	1833	2.4	0.05	–	7.6

[c] D. G. Burke et al. (1985).
[a] Akiyama (1985).

This code, still starting from a weak-coupling basis (vibrational limit) allows for a number of g-bosons n_g, up to $n_g = 4$, to study E4 properties in an extended $(sdg)^N$ model space.

6.2.4. Generalization to High-Spin $l = 6, \ldots$ Bosons

We have shown in Section 6.2.3 that attempts to handle the full *sdg*-boson model space with the *g*-boson treated on the same footing with the *s* and *d* bosons, the $SU(3)$ dynamical symmetry, could be studied in quite some detail. Serious modifications of the *sd*-boson model appeared that gave rise to a better reproduction of the data in strongly deformed nuclei.

A particular interesting result was a significant modification in the ground-state $B(\text{E2})$ cut-off behaviour. Generalizations with inclusion of even higher-spin even parity $l = 4, 6 \ldots$ bosons can be included too. Finite boson number effects are much weakened by replacing N by $2N$ in all $SU(3)$ expressions (see Chapter 2) when incorporating the $g - (l = 4)$ boson. In the generalized situation, with $\eta = l_{max}$, the boson degeneracy becomes $d = \frac{1}{2}(\eta + 1)(\eta + 2)$. In the $SU(3)$ limit, the (λ, μ) ground-band representation then becomes $(\eta N, 0)$ with $J_{max} = \eta.N$. This modifies the yrast $B(\text{E2})$ values into the expression (using the E2 operator of eq. (6.39))

$$B(\text{E2}; L + 2 \rightarrow L)$$
$$= e^2 \frac{3}{4} \frac{(L + 2)(L + 1)(\eta N - L)(\eta N + L + 3)}{(2L + 3)(2L + 5)}, \qquad (6.43)$$

and the quadrupole moment becomes

$$Q(L) = e\frac{1}{2}(-1)^{L+1}(2\eta N + 3)\frac{L}{2L + 3}. \qquad (6.44)$$

Equation (6.43) includes the cut-off on B(E2) values, which stems from the last two factors in the numerator, and reduces to the *sd*-boson $SU(3)$ model result after the substitution of $\eta = 2$. So, the cut-off is largely reduced when incorporating the *g*-boson. Including higher *l*-bosons ($l = 6$, $l = 8$) makes the yrast B(E2) values approach the rigid rotor predictions as is illustrated for the case of $N = 15$ bosons (^{238}U) in Fig. 6.19.

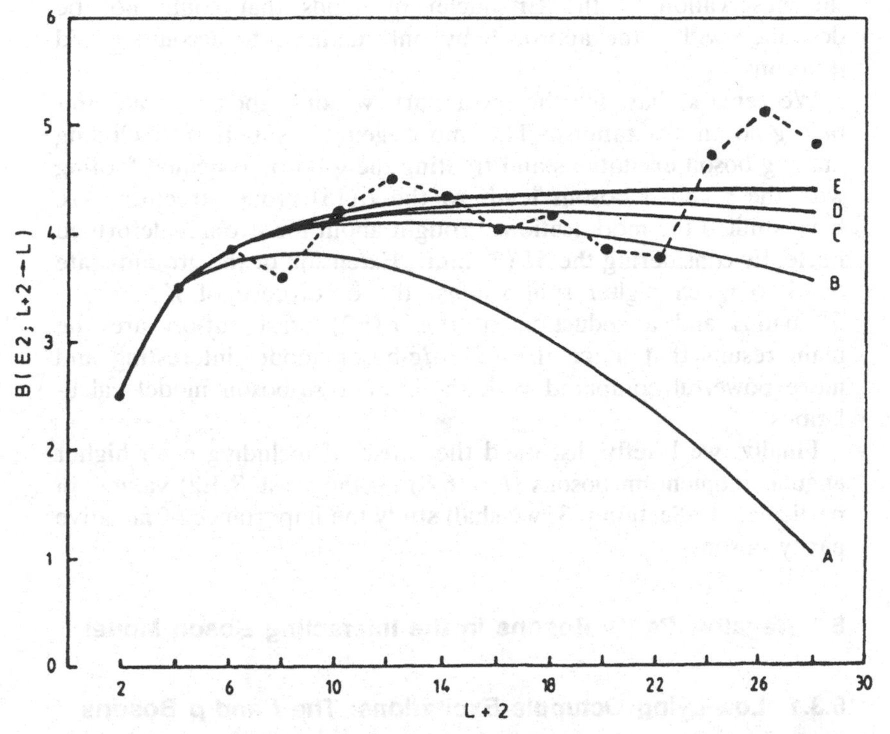

FIGURE 6.19. *Schematic behaviour of eq. (6.43) for various η values. The curves have the following correspondence: A, $\eta = 2$; B, $\eta = 4$, C, $\eta = 6$; D, $\eta = 8$ and curve E corresponds with the rigid-rotor predictions. The full circles, joined by broken lines represent the data in ^{238}U with $N = 15$ bosons. Taken from Ratna Raju (1982).*

6.2.5. Conclusions

We have studied, in a schematic approach, the influence on the standard IBM model description of energy spectra and E2 and E4 transitions of coupling a g-boson to the configurations $|(sd)N\rangle$. We have carried out model studies in the $U(5)$, $O(6)$ and $SU(3)$ limits as well for some specific nuclei (^{104}Ru, even-even Gd nuclei).

The most pronounced features of this g-boson configuration show up in the yrast B(E2) values. Much larger values ($\approx$ factor of 2) can result as compared with the standard IBM approach. Moreover, the yrast cut-off is shifted to higher angular momenta. In the deformed nuclei, we have indicated in an almost parameter free study how extra $K^\pi = 4^+, 3^+, 2^+, 1^+, 0^+$ low-lying hexadecapole bands can result. This particular result can also shed new light on the observation in the Er nuclei of bands that could not be described within the approach by only taking into account s and d-bosons.

We remark that, for the most part, we only took into account one g-boson excitations. The more general situation, including many g-boson excitations and treating the g-boson on equal footing with the s and d bosons leads to the $U(15)$ group structure. We have studied the modifications brought about for strongly deformed nuclei by considering the $SU(3)$ limit. Extension of the ground-state band to much higher spin values, the occurrence of $K^\pi = 1^+$, 3^+ bands and a reduction of the B(E2) yrast cut-off are the main results that make the full sdg-boson model interesting and more powerful compared with the former sd-boson model calculations.

Finally, we briefly discussed the effect of including even higher angular momentum bosons ($l = 6,8$) on the yrast B(E2) values, in particular. In Section 6.3, we shall study the importance of negative parity bosons.

6.3 Negative Parity Bosons in the Interacting Boson Model

6.3.1. Low-Lying Octupole Excitations: The *f* and *p* Bosons

Negative parity levels have been observed throughout the nuclear chart. In particular, low-lying octupole (3^-) excitations near closed shells and octupole deformation in deformed nuclei are well known. Those states are most often characterized by enhanced E3 transitions. In order to incorporate those states within an interacting boson model description, extensions to include the f-boson (and possibly the p-boson) degree of freedom are necessary.

6.3.1.1. f-Boson Group Structure and Dynamical Symmetries

The sdf-boson model is quite well described in the early articles of Arima and Iachello (Arima and Iachello, 1978, 1976; Scholten et al., 1978). Here, we discuss the basic elements of this extended model. The model, using the 13 creation operators

$$b_\alpha^\dagger \equiv s^\dagger, \qquad d_\alpha^\dagger, \qquad f_\alpha^\dagger, \tag{6.45}$$

constitutes a $U(13)$ group structure (compared to the smaller sd $U(6)$ group structure) which is a complex system, in general. In obtaining tractable limits, we start by including only one f-boson configurations, $(sd)^{N-1}\cdot f$, besides the $(sd)^N$ states, very much along the same lines as the g-boson was introduced and discussed in Section 6.2.

The total Hamiltonian then has the general form

$$H = H_{sd} + H_f + V_{sd,f}, \tag{6.46}$$

where H_{sd} describes the sd-interacting boson system, H_f the f-bosons and $V_{sd,f}$ the coupling term. Even including a single f-boson excitation, many terms can be constructed that would contribute to H_f and $V_{sd,f}$. Therefore, it is very natural to study some of the limiting cases: the $U(5)$ and $SU(3)$ dynamical symmetries.

(i) $U(5)$ limit: Near closed shells, where vibrational spectra are observed, the lowest negative parity state is a $J^\pi = 3^-$ level, followed by a grouping of other negative-parity states at higher energies. This is reminiscent of a $U(5)$ scheme, as shown on the left-hand side in Fig. 6.20. Using one f-boson only, $n_f = 1$, spectra are obtained by coupling this f-boson to the exact $U(5)$ sd-interacting boson system. In the weak-coupling limit, relatively simple and analytic results can be obtained (Arima and Iachello, 1976).

Electromagnetic transitions, in particular the E1 and E3 transitions, are described by the operators in the $U(5)$ limit as

$$T(\mathrm{E1}) = e_1(f^\dagger\tilde{d} + d^\dagger\tilde{f})^{(1)} + \text{higher-order terms}, \tag{6.47}$$

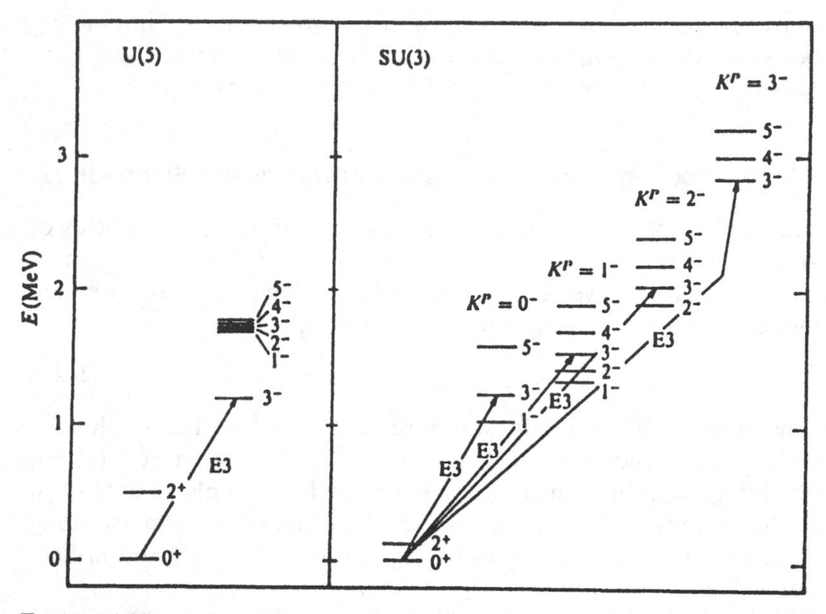

FIGURE 6.20. *Schematic representation of the octupole low-lying energy spectrum. On the left-hand side the U*(5) *limit is given, on the right-hand side, the SU*(3) *limit is presented. The* E3 *excitation is indicated via the upgoing arrows in both limits. Taken from Scholten et al.* (1978).

and

$$T(\text{E3}) = e_3(f^\dagger s + s^\dagger \tilde{f})^{(3)}. \tag{6.48}$$

in the $U(5)$ limit, only the first excited $J^\pi = 3^-$ state with $n_s = N - 1$, $n_d = 0$, $n_f = 1$ is connected to the $n_s = N$ $J^\pi = 0^+$ ground-state.

(ii) $SU(3)$ limit: In deformed nuclei, a series of strongly coupled negative-parity bands results. Here, starting from the $SU(3)$ quadratic Casimir operator describing the sd-boson system, a rather simple expression for the interaction Hamiltonian $V_{sd,f}$ can be obtained which has the advantage of giving analytical results. In this $SU(3)$ limit, we use the form

$$H = H_{sd} + \epsilon_f \hat{n}_f + \theta_f \hat{N} + \kappa_1 J_f \cdot J + \kappa_2 Q_f \cdot Q, \tag{6.49}$$

with $\hat{n}_f$ and $\hat{N}$ the f-boson and total boson number operators and

$$J_f = 2\sqrt{7}\,(f^\dagger\tilde{f})^{(1)}, \qquad Q_f = -2\sqrt{7}(f^\dagger\tilde{f})^{(2)}. \quad (6.50)$$

If we consider the states, obtained by coupling a single f-boson to an $SU(3)$ wave-function, describing the remaining $N-1$ sd-boson system, $|f[N-1](\lambda, \mu)\varphi L; JM\rangle$, then the first four terms of eq. (6.49) are diagonal in this basis. The last term is non-diagonal but since Q is a generator of $SU(3)$, the matrix elements can easily be calculated (see also Section 6.2.2.). If we now construct the basis for this f-boson system, along the lines of Section 6.2.2.1, eqs. (6.26) and (6.27), the energy eigenvalues can be written as

$$E(f[N-1](\lambda, \mu)K_f; JM) = AJ(J+1) + B_f K_f^2 + C_f, \quad (6.51)$$

where A, B_f and C_f are particular combinations of the parameters ϵ_f, θ_f, κ_1 and κ_2. The resulting spectrum is shown in the right-hand side of Fig. 6.20 where $K_f^\pi = 0^-, \ldots, 3^-$ bands are present.

In the $SU(3)$ limit, the $J^\pi = 3^-$ states of all four bands have rather large E3 octupole transitions, described with a more general T(E3) operator

$$T(\mathrm{E3}) = e_3\,[(f^\dagger s + s^\dagger\tilde{f})^{(3)} + \chi_3(f^\dagger\tilde{d} + d^\dagger\tilde{f})^{(3)}]. \quad (6.52)$$

The relative B(E3) values for decay from the $K_f^\pi = 0^-, 1^-, 2^-, 3^-$ bands are in the ratio 1:2:2:2. In general, E3 transitions are rather well explained but E1 transitions are not so well described and will be discussed later.

(iii) $O(6)$ limit: Using the same methods as under (i) and (ii), the possiblity exists of coupling an f-boson to an $O(6)$ core. In this case, the quadrupole operator appearing in both H_{sd} and $V_{sd,f}$, the coupling Hamiltonian, is taken with its $O(6)$ value i.e. putting $\chi = 0$ in the quadrupole operator of eq. (6.4) instead of the standard value of $-\sqrt{7}/2$. Numerical studies have been carried out by Engel (1986) for the even-even $^{190-196}$Pt nuclei.

6.3.1.2. Numerical Applications

Even though the $U(5)$ and $SU(3)$ dynamical symmetries illustrate many of the salient features associated with the f-boson degree of freedom, more realistic situations have to be studied too. A still very transparent way to perform such a calculation with a transition between the $U(5)$ and $SU(3)$ symmetries was carried out making use of the Hamiltonian in eq. (6.49), though without the $J_f \cdot J$ term. Some typical examples for the change in energy spectra (Fig. 6.21) and in B(E3) values (Fig. 6.22) are presented for the case in which H_{sd} is allowed to undergo a transition from $U(5)$ to $SU(3)$. In these illustrations one observes the smooth variation from the more vibrational pattern in spectra (near $N = 82$) into a $K^\pi = 0^-$ band as lowest negative parity band near $N = 94$.

Recently, extensive calculations were performed by Barfield et

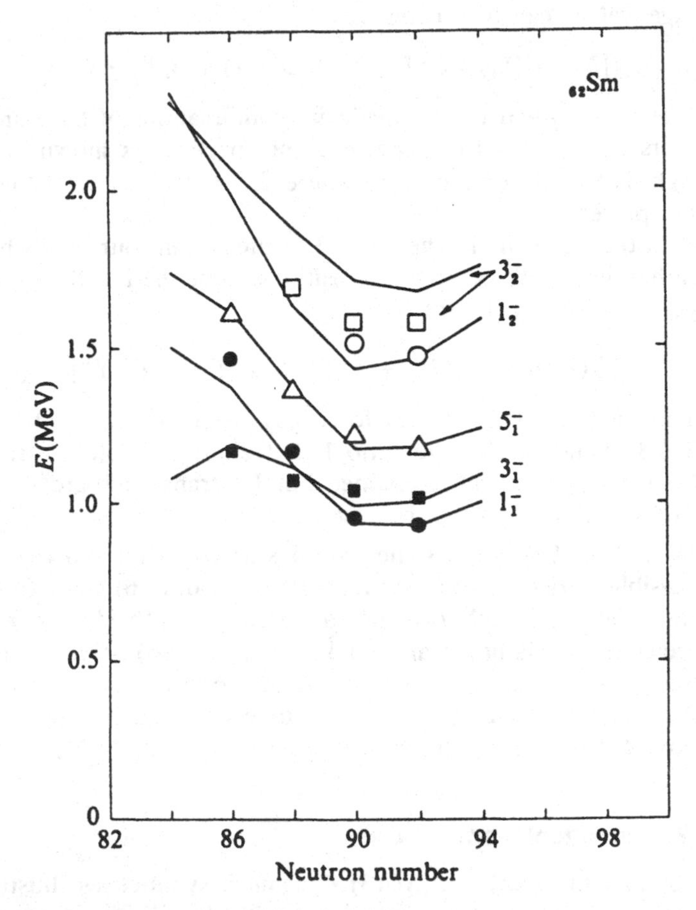

FIGURE 6.21. *Comparison of the calculated (lines) and experimental* (●, ■, △, ○, □) *energy levels of negative-parity states in the even-even* Sm *nuclei. Taken from Scholten et al.* (1978).

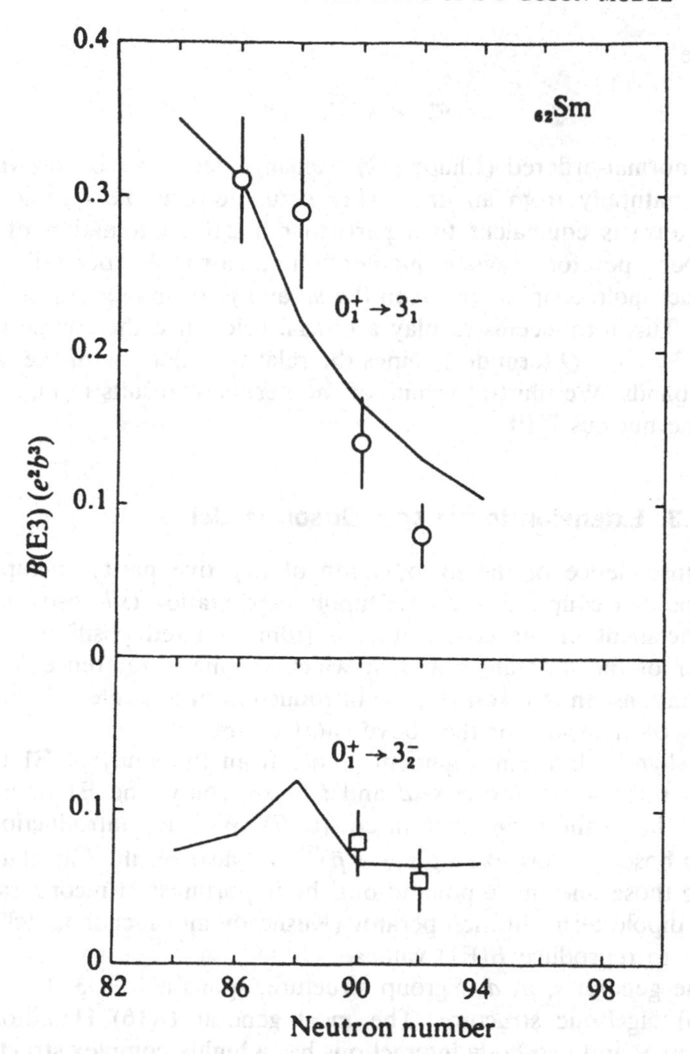

FIGURE 6.22 *Comparison of the calculated and experimental* (○, □) $B(E3)$ *values for the* $0_1^+ \rightarrow 3_1^-$ *and* $0_1^+ \rightarrow 3_2^-$ *E3 transitions. Taken from Scholten et al.* (1982).

al. (1988). The formalism used is very near to the Hamiltonian described in eq. 6.49. Here now, an extra "exchange" term has been added

$$H_{\text{exch.}} = \xi : E_{df}^{\dagger} \cdot E_{df} : \tag{6.53}$$

where

$$E^{\dagger}_{df} = \sqrt{5}(d^{\dagger}\tilde{f})^{(3)}. \tag{6.54}$$

This normal-ordered (Chapter 2) exchange term can be shown to arise naturally from an underlying octupole-octupole interaction. This term is equivalent to a particular linear combination of the number operator, angular momentum, quadrupole, octupole and hexadecapole coupling between the *sd* and *f*-boson degrees of freedom. This term seems to play a crucial role since the competition with the $\kappa_2 Q_f \cdot Q$ term determines the relative orderning of the octupole bands. We illustrate some of the pertinent results in Fig. 6.23 for the nucleus ^{168}Er.

6.3.1.3. Extension to the *spdf*-Boson Model

A consequence of the introduction of negative parity (octupole) bosons that couple to the quadrupole deformation (*sd*-bosons) is a displacement of the center-of-mass from its fixed position at the center of the laboratory axis in which we make reference for all calculations. In that respect, the introduction of a dipole or *p*-boson might compensate for the above shortcomings.

A slightly different argument comes from the study of E1 transition rates. Considering *s*, *d* and *f* bosons only, the E1 operator has to be of the form given in eq. (6.47). With the introduction of the *p*-boson, terms like $(p^{\dagger}s + s^{\dagger}\tilde{p})^{(1)}$ can also result. Calculations along those lines have pointed out the importance of incorporating such dipole terms in the operator (Kusnezov and Iachello, 1988) in order to reproduce $B(\mathrm{E1})$ values.

The general *s*, *p*, *d*, *f* group structure, spans a 1 + 3 + 5 + 7 $U(16)$ algebraic structure. The most general $U(16)$ Hamiltonian with one- and two-body interactions has a highly complex structure. A simplified Hamiltonian

$$H = \epsilon_p \hat{n}_p + \epsilon_d \hat{n}_d + \epsilon_f \hat{n}_f + \kappa Q_{spdf} \cdot Q_{spdf} + \kappa' J_{spdf} \cdot J_{spdf} + \kappa'' P^{\dagger}_{spdf} \cdot P_{spdf}, \tag{6.55}$$

where ϵ_l denotes the l-boson energy and Q_{spdf}, J_{spdf} and $P^{\dagger}_{spdf}$ are related to the $SU(3)$ Casimir operator, the angular momentum operator and the pairing operator (related to $O(16)$), respectively.

Some of the crucial properties related to this *spdf* model are E1

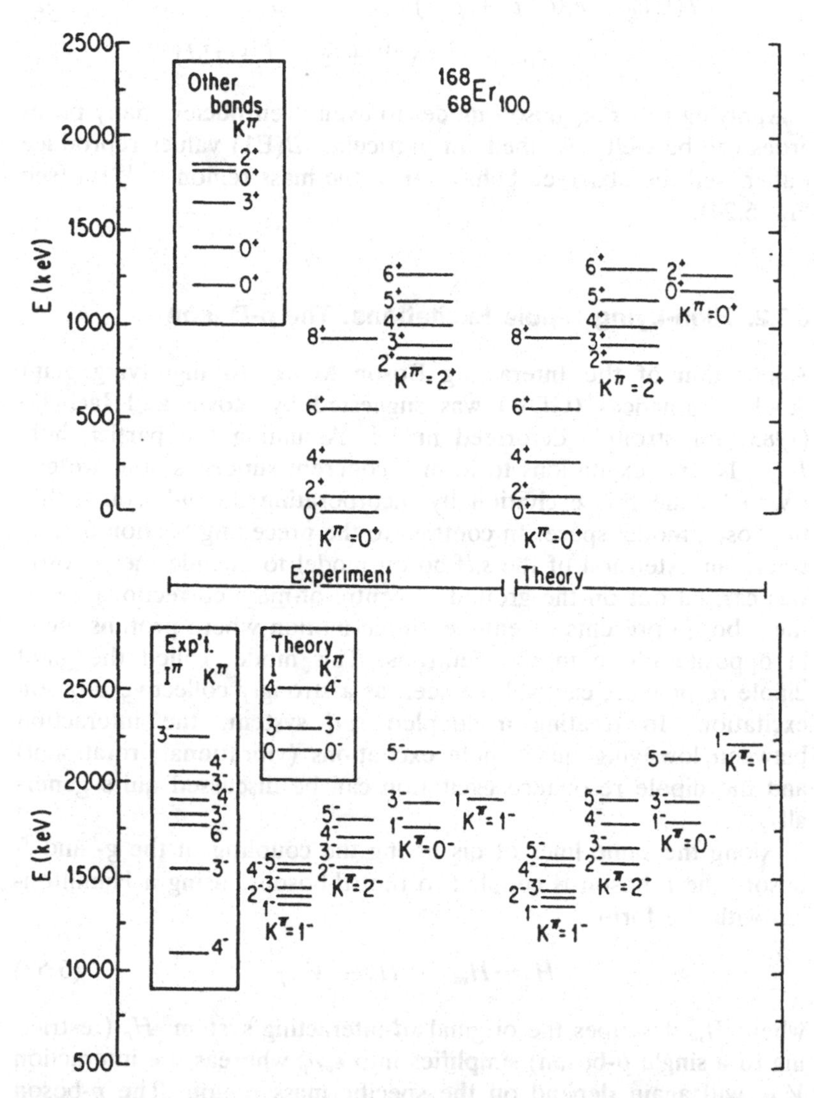

FIGURE 6.23. *Comparison of calculated and experimental negative-parity states in* ^{168}Er. *The boxes contain experimental band-heads thought to represent dominant two-quasi particle excitations and the next-higher-lying theoretical bands. Taken from Barfield et al.* (1988).

transitions. Here, one has to use an extended form of eq. (6.47) with the result

$$T(\mathrm{E1}) = e_1\{(d^\dagger\tilde{p} + p^\dagger\tilde{d})^{(1)} + \chi_1(s^\dagger\tilde{p} + p^\dagger s)^{(1)} + \chi_1'(d^\dagger\tilde{f} + f^\dagger\tilde{d})^{(1)}\}. \tag{6.56}$$

Applying this *spdf*-boson model to even-even nuclei, many properties can be well described. In particular, $B(\mathrm{E1})$ values reproduce rather well the observed behaviour in the mass region $^{144-148}$Ba (see Fig. 6.24).

6.3.2. High-Lying Dipole Excitations: The *p*-Boson

Application of the Interacting Boson Model to high-lying giant dipole resonances (GDR) was suggested by Rowe and Iachello (1983) for strongly deformed nuclei. Assuming the particle-hole $J^\pi = 1^-$ $1\hbar\omega$ excitations to form a coherent superposition, we can try to handle this excitation by incorporating the *p*-boson within the boson model space. In contrast to the preceding Section 6.3.1.3 where an extension of the *sdf* boson model to include the *p*-boson was carried out on the ground of centre-of-mass corrections, here, the *p*-boson presents a genuine dipole motion where protons move in opposite phase to the neutrons. This mode, called the giant dipole resonance, can well be seen as a strongly collective *p*-boson excitation. In treating a coupled *spd* system, the interaction between low-lying quadrupole excitations (vibrational, rotational) and the dipole resonance excitation can be discussed quite generally.

Along the same lines of discussing the coupling of the *g*- and *f*-boson, the *p*-boson is coupled to the *sd*-bosons, using a Hamiltonian with the form

$$H = H_{sd} + H_p + V_{sd,p}, \tag{6.57}$$

where H_{sd} describes the original *sd*-interacting system, H_p (restricting to a single *p*-boson) simplifies into $\epsilon_p\hat{n}_p$ whereas the interaction $V_{sd,p}$ will again depend on the specific mass region. The *p*-boson energy ϵ_p follows from the empirical giant-dipole energy behaviour as $\epsilon_p = 77.5\mathrm{A}^{-1/3}$ MeV. For vibrational or transitional nuclei, we use

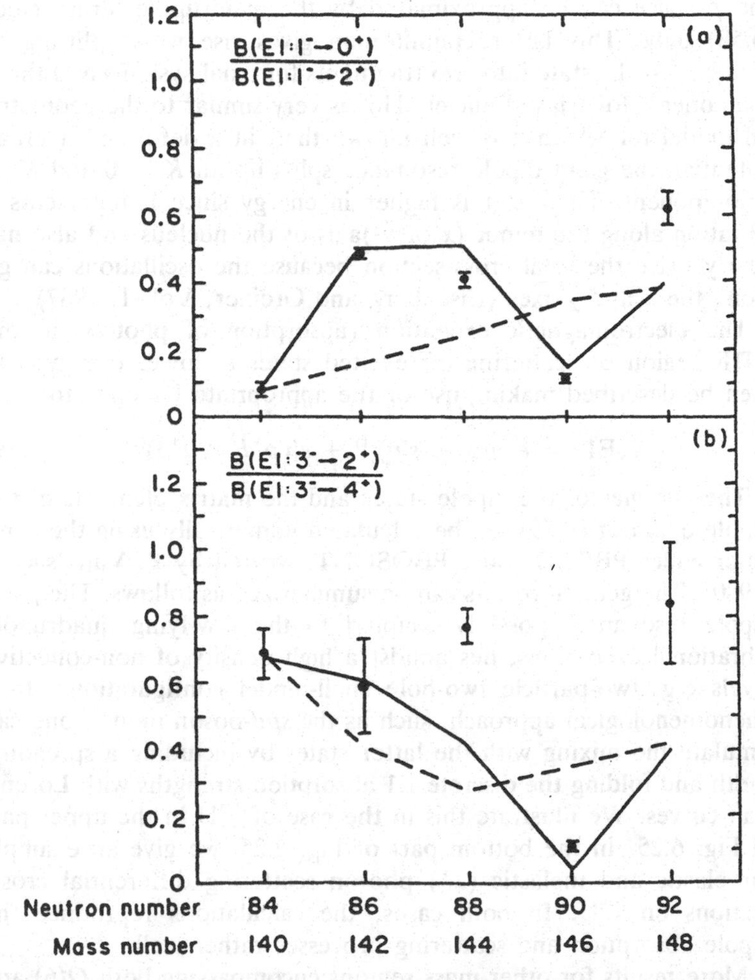

FIGURE 6.24. *The ratios for certain* B(E1) *values* (B(E1; $1^- \rightarrow 0^+$)/B(E1; $1^- \rightarrow 2^+$) *in part* (a) *and* B(E1; $3^- \rightarrow 2^+$)/B(E1; $3^- \rightarrow 4^+$) *in part* (b) *for the even-even* Ba *nuclei with mass number* $140 \leq A \leq 148$. *Calculations have been carried out within an extended spdf boson model. The dashed line represents the use of a* T(E1) *operator with fixed parameters while the solid line allows for variation of the* E1 *charges in eq.* (6.56). *Taken from Kusnezov and Iachello* (1988).

$$V_{sd,p} = b_0(d^\dagger\tilde{d})^{(0)}\cdot(p^\dagger\tilde{p})^{(0)} + b_1(d^\dagger\tilde{d})^{(1)}\cdot(p^\dagger\tilde{p})^{(1)} \\ + b_2[s^\dagger\tilde{d} + d^\dagger s + \chi\,(d^\dagger\tilde{d})^{(2)}]\cdot(p^\dagger\tilde{p})^{(2)}, \tag{6.58}$$

whereas for strongly deformed nuclei, the coupling between the *sd* and *p* space can be approximated by the quadrupole term of eq. (6.58) only. This latter Hamiltonian gives rise to a splitting of dipole $J^\pi = 1^-$ state into two fragments for axial systems and three components for triaxial nuclei. This is very similar to the geometrical model for which it is well known that, in a deformed average potential, the giant dipole resonance splits into a $K = 0$ and $K = 1$ component. The latter is higher in energy since it represents a vibration along the minor (x or y-)axis of the nucleus and also has nearly twice the total cross-section because the oscillations can go along the x and y-axes (Eisenberg and Greiner, Vol. 1, 1987).

The electromagnetic excitation (absorption of photons in the GDR region or scattering off excited states at lower energy) can then be described making use of the appropriate E1 operator

$$T(\mathrm{E1}) = e_1(p^\dagger s + s^\dagger\tilde{p})^{(1)} + e_1'(p^\dagger\tilde{d} + d^\dagger\tilde{p})^{(1)}. \tag{6.59}$$

The energies of the dipole states and the matrix elements of the dipole operator (6.59) can be calculated numerically using the computer codes PBOSON and PBOSONT, written by P. Van Isacker (1980). The general results can be summarized as follows. The giant dipole resonance, possibly coupled to the low-lying quadrupole vibrational excitations, lies amidst a high density of non-collective levels e.g. two-particle two-hole shell-model configurations. In a phenomenological approach, such as the *spd*-boson model, one can simulate the mixing with the latter states by including a spreading width and folding the discrete E1 absorption strengths with Lorentzian curves. We illustrate this in the case of ^{238}U in the upper part of Fig. 6.25. In the bottom part of Fig. 6.25, we give an example for elastic and inelastic (2_1^+) photon scattering differential cross-sections on ^{238}U. In both cases, the calculations reproduce the dipole absorption and scattering processes rather well.

More results for other mass regions encompassing both $O(6)$ and $U(5)$-like nuclei have been discussed by Maino et al. (1985, 1986).

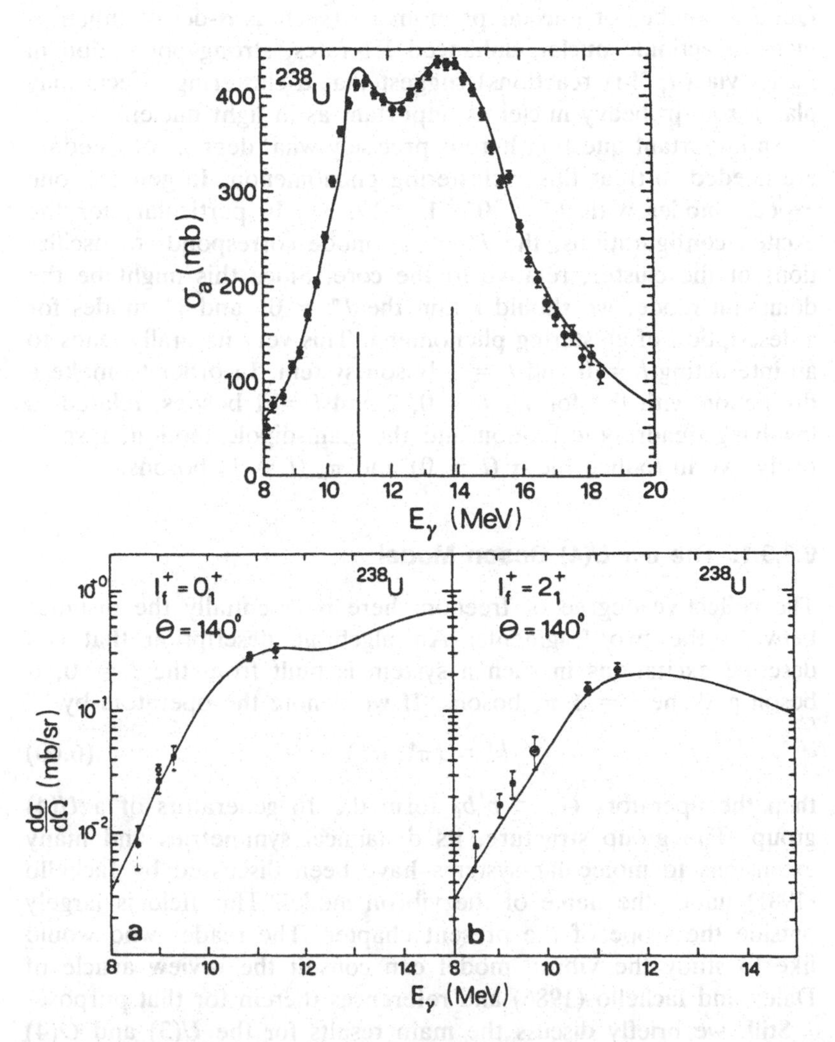

FIGURE 6.25. *Comparison of calculated and experimental total photo-absorption cross-sections in the strongly deformed nucleus* ^{238}U. *The fragmentation of* E1 *strength is shown in the vertical lines (upper part). In the lower-part, the elastic* (0_1^+) *and inelastic* (2_1^+) *differential photon scattering cross-sections* $d\sigma/d\Omega$ *for the same nucleus* ^{238}U *are shown, as a function of the photon energy* ($E\gamma$), *and this at a scattering angle of* $\theta = 140°$ *(calculations are performed in the* SU(3) *limit). Taken from Zuffi et al.* (1987).

6.3.3. Clustering in Nuclei: Interplay of Dipole and Quadrupole Excitations

Quite a number of nuclear phenomena (such as α-decay enhancement in actinide nuclei, enhanced E1 rates, strong population in nuclei via (d, ^{6}Li) reactions) suggest that α-clustering effects may play a role in heavy nuclei as important as in light nuclei.

An important question here is precisely what degrees of freedom are needed to treat this α-clustering phenomenon. In general, one expects modes with $J^\pi = 0^+, 1^-, 2^+, 3^-$. In particular, for the excited configurations, the $J^\pi = 1^-$ mode corresponds to oscillations of the cluster, relative to the core. Since this might be the dominant mode, we should retain the $J^\pi = 0^+$ and 1^- modes for a description of clustering phenomena. This very naturally leads to an interacting $l = 0$ and $l = 1$ boson system. In order to make a distinction with the former $l = 0, 2$ and $l = 1$ bosons, related to low-lying quadrupole motion and the giant-dipole motion, respectively, we introduce the σ ($l = 0$) and π_μ ($l = 1$) bosons.

6.3.3.1. The $\sigma\pi$ $U(4)$ Boson Model

The collective degree of freedom here is essentially the distance between the two fragments. An algebraic description that can describe excitations in such a system is built from the $l = 0$, σ boson and the $l = 1$ π_μ bosons. If we denote the operators by

$$b_\alpha^\dagger \equiv (\pi_\mu^\dagger, \sigma^+), \tag{6.60}$$

then the operators $G_{\alpha\beta} \equiv b_\alpha^\dagger b_\beta$ form the 16 generators of a $U(4)$ group. The group structure, its dynamical symmetries and many extensions to molecular systems have been discussed by Iachello (1981) under the name of the vibron model. This field is largely outside the scope of the present chapter. The reader who would like to study the vibron model can consult the review article of Daley and Iachello (1986) and references therein for that purpose.

Still, we briefly discuss the main results for the $U(3)$ and $O(4)$ dynamical symmetries, contained in the (σ, π) $U(4)$ model.

(i) One chain, (I), has the structure

$$U(4) \supset U(3) \supset O(3) \supset O(2). \tag{6.61}$$

Here, the most general Hamiltonian diagonal in this chain

reads

$$H^{(I)} = H_0 + x_1 C_{1U3} + x_2 C_{2U3} + x_3 C_{2O3}, \qquad (6.62)$$

with corresponding eigenvalues

$$E^{(I)}(N, n_\pi, L, M) = E_0 + x_1' n_\pi + x_2' n_\pi(n_\pi + 3) + x_3' L(L + 1). \qquad (6.63)$$

Here the x_i''s are linear combinations of the x_i coefficients.

(ii) A second chain (II) has the structure

$$U(4) \supset O(4) \supset O(3) \supset O(2). \qquad (6.64)$$

The most general dynamical symmetry Hamiltonian reads

$$H^{(II)} = H_0 + x_1 C_{2O4} + x_2 C_{2O3}, \qquad (6.65)$$

with corresponding eigenvalues

$$E^{(II)}(N, \omega, L, M) = E_0 + x_1' \omega(\omega + 2) + x_2' L(L + 1). \qquad (6.66)$$

Instead of ω, one can use the vibrational quantum number v, related to ω via the expression

$$v = (N - \omega)/2; \qquad v = 0, 1, \ldots, N/2 \text{ or } (N - 1)/2. \qquad (6.67)$$

Typical spectra for the $U(4)$ limits (i) and (ii) are presented in Fig. 6.26 on the left-hand and right-hand side, respectively. Whereas the (i) limit looks more like a vibrational system, the limit (ii) is more reminiscent of a rotational-like structure.

The electric dipole operator, appropriate to calculate E1 transitions in the cluster system, is

$$T(E1) = e_1(\pi^\dagger \sigma + \sigma^\dagger \tilde{\pi})^{(1)}. \qquad (6.68)$$

6.3.3.2. Clustering at Low Excitation Energy

In studying clustering in heavy nuclei, one (or both) of the clusters could be deformed. Algebraic techniques that treat the complex interacting system where the quadrupole deformation degree of freedom is important, and is incorporated within the *sd* $U(6)$ model, as well as the clustering effects, described by the $U(4)$ group structure, form the basis for a detailed study (see Fig. 6.27) of the interplay of clustering and quadrupole collectivity. Here, one studies the product group structure

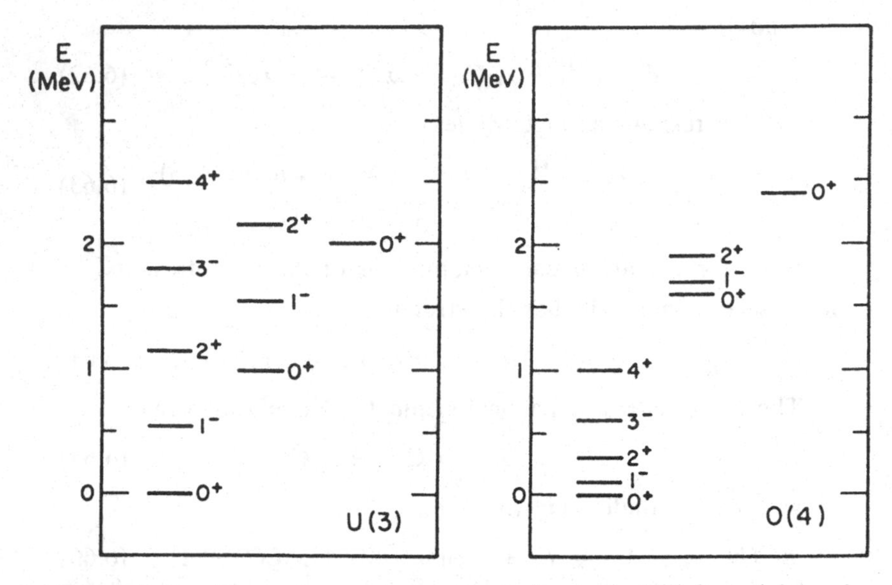

FIGURE 6.26. *The energy spectra, corresponding to the* $U(3)$ *limit (left-hand part) and* $O(4)$ *limit (right-hand part) within the* $U(4)$ $\sigma\pi$ *boson model. The energy spectra are expressed by the eigenvalues in eqs.* (6.63) *and* (6.66), *respectively.*

$$G \equiv U_{sd}(6) \otimes U_{\sigma\pi}(4), \tag{6.69}$$

with a Hamiltonian, coupling the (sd) and $(\sigma\pi)$ degrees of freedom.

In looking for applications to nuclear physics and by concentrating on the various limiting cases in both the $U(6)$ and the $U(4)$ groups, a very rich variety of energy spectra results. One application to the α-clustering in heavy nuclei, where the $SU(3)$ limit of $U(6)$ is taken and where the coupling term is restricted to a quadrupole-plus-dipole term gives a rather good reproduction of the energy spectra in $Z = 88$ (Ra) and $Z = 90$ (Th) nuclei and is illustrated in Fig. 6.28. More applications can be found (Daley and Iachello, 1986).

6.3.4. Conclusions

The above discussions have illustrated a number of results obtained by incorporating the octupole and dipole degree of freedom into the Interacting Boson Model. The data on negative parity states

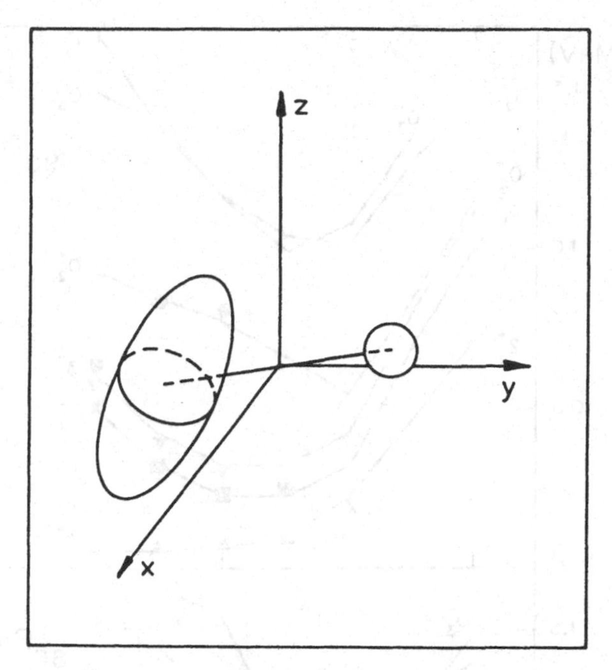

FIGURE 6.27. *The relative translational modes of motion of a two-cluster system, described by the U(4) group, coupled with the quadrupole modes of motion of the larger cluster, described by the U(6) group.*

(energy, E3 and E1 decay properties, α-decay, . . .) are quite extensive on deformed rare-earth and actinide nuclei. On the theoretical side, the *f*-boson has been incorporated in a number of numerical studies. The group theoretical structure, implied by the *f*-boson i.e., $U(13)$ and, more extensively, by also taking into account the *p* boson, the $U(16)$ group structure, have been presented.

Besides, α-cluster excitations could also be incorporated within an interacting boson approximation, on equal footing with the quadrupole modes of motion through coupling the *sd* $U(6)$ and $\sigma\pi$ $U(4)$ groups. Here, quite a number of existing data on α-decay widths, α-transfer spectroscopic factors and E1 decay probabilities can be tested using the above ideas.

The E1 mode is a very appropriate means of studying the large variety of nuclear processes ranging from octupole vibrations and octupole deformation effects to clustering and the giant dipole res-

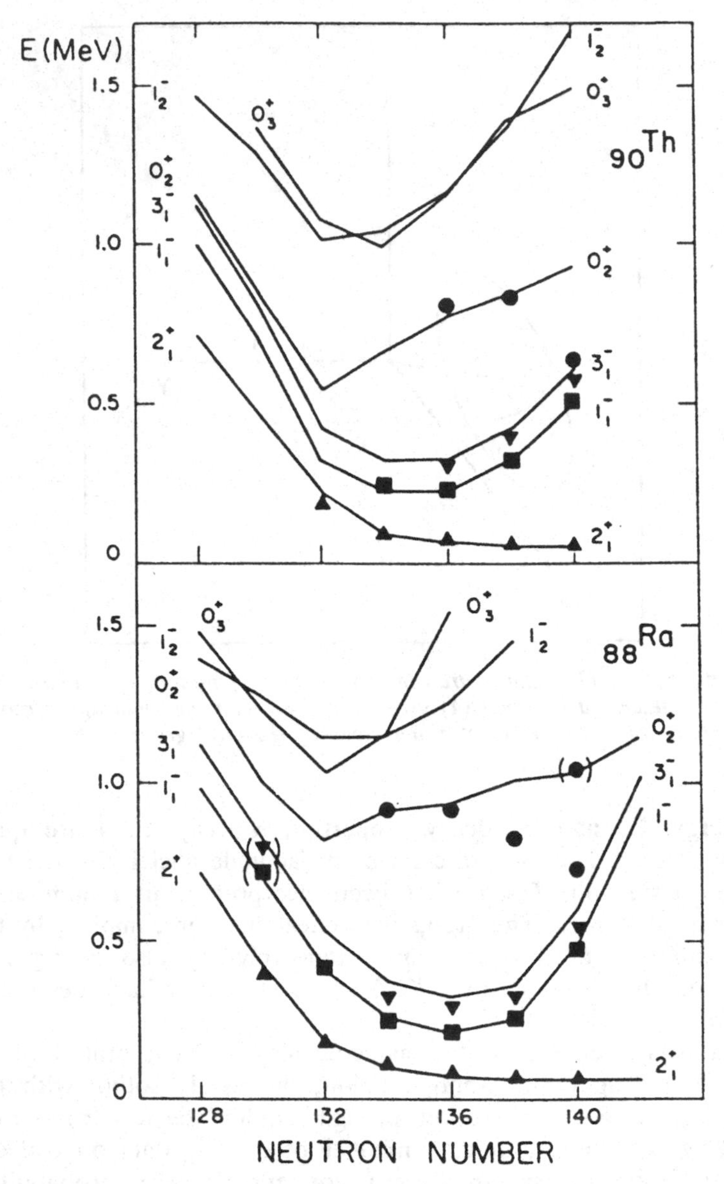

FIGURE 6.28. *Low-lying collective bands in the even-even* Ra (Z = 88) *and* Th (Z = 90) *nuclei. In this calculation, alpha-clustering and quadrupole degrees of freedom are treated starting from the* $U(6) \otimes U(4)$ *group. The coupling Hamiltonian is restricted to quadrupole and dipole terms. Taken from Daley and Iachello* (1986).

onance region. This very wide spectrum is schematically presented in Fig. 6.29.

This entire study field of negative parity excitations has only been discussed in a rather compact way and is clearly amenable for a more extensive and systematic investigation.

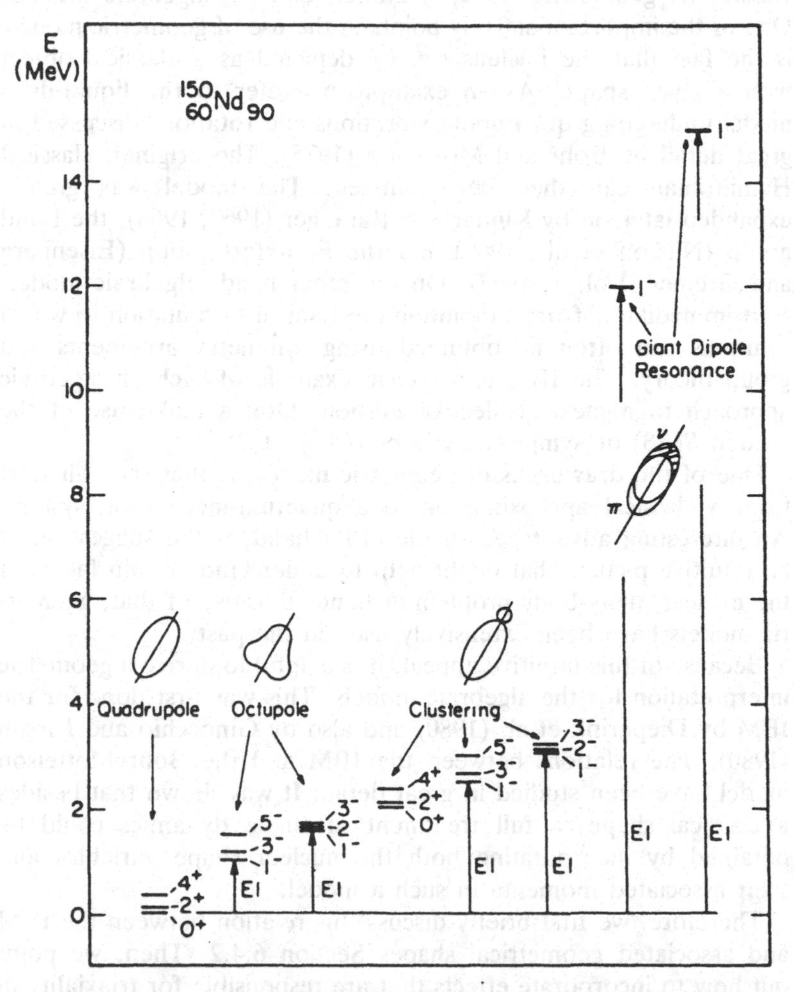

FIGURE 6.29. *Various possibilities of exciting the nucleus with the* E1 *dipole operator. Starting from octupole vibrations and octupole deformed bands, passing through alpha-clustering within different limiting situations, one can end up, at higher energy, using the* p*-boson in describing the giant-dipole resonance. Taken from Iachello* (1985).

6.4 TRIAXIALITY AND CUBIC BOSON INTERACTIONS

6.4.1. Introduction

Macroscopic nuclear collective models can be divided in two large classes: (i) geometrical (shape) models and (ii) algebraic models. One of the important starting points in the use of geometric models is the fact that the nucleus can be depicted as a classical object with a given shape. As an example we refer to the liquid-drop model undergoing quadrupole vibrations and rotations discussed in great detail by Bohr and Mottelson (1975). The original, classical Hamiltonian can then be quantized. The model was greatly expanded later on by Kumar and Baranger (1967, 1968), the Lund group (Nilsson et al., 1969) and the Frankfurt group (Eisenberg and Greiner, Vol. 1, 1987). On the other hand, algebraic models start immediately from a quantum mechanical formulation to which solutions can often be obtained using symmetry arguments and group theory. The IBM is a typical example of such an algebraic approach to nuclear, collective motion. Others make use of the pseudo $SU(3)$ or symplectic groups (Chapter 7).

One of the drawbacks of geometric models is that they all start from a classical approximation to a quantum-mechanical system. An interesting advantage, on the other hand, is the suggestion of an intuitive picture that might help to understand certain facets of the nuclear many-body problem at hand. Because of that, geometric models have been extensively used in the past.

Because of this intuitive appeal, it is helpful to derive a geometric interpretation for the algebraic models. This was first done for the IBM by Dieperink et al. (1980) and also by Ginocchio and Kirson (1980). The relations between the IBM and the Bohr–Mottelson model have been studied in great detail. It was shown that besides a classical shape, a full treatment of shape dynamics could be obtained by incorporating both the nuclear shape variables and their associated momenta in such a model.

Therefore, we first briefly discuss this relation between the IBM and associated geometrical shapes Section 6.4.2. Then, we point out how to incorporate effects that are responsible for triaxiality in the IBM via the use of cubic anharmonicities Section 6.4.3. We finally discuss some specific applications to the transitional nucleus ^{104}Ru that we have treated earlier in terms of g-bosons in Section 6.4.4.

6.4.2. Geometrical Shapes—the Classical Limit

The mapping of quantum observables into c-number functions has always been an important problem in quantum mechanics. The precise procedures for performing this task have been formulated by Glauber (1963) using the coherent-state representation. The step of connecting a quantum description with the corresponding classical motion is best known in the study of electromagnetism and quantum optics. The same methods can now be used in order to obtain a correspondence between the algebraic interacting boson model results and obtained using the quantization of shape dynamics (quadrupole vibrations, rotations) for the atomic nucleus.

The method described is based on the use of coherent states (See Chapter 2 and 3) for a boson system. Thereby, one starts from the general boson creation operator (boson condensate)

$$B^{\dagger} = \sum_i \alpha_i^* b_i^{\dagger}, \tag{6.70}$$

where $b_i^{\dagger}$ creates a boson i and the α_i are general variables. They are supposed to be real variables (in a Cartesian basis) although for the most general study of this problem complex α_i have to be used. For a given boson Hamiltonian H, the expression

$$\frac{\langle B^N|H|B^N\rangle}{\langle B^N|B^N\rangle}, \tag{6.71}$$

is a function both of N and of the α_i. It was shown by Gilmore (1978, 1979) that, in the limit $N \rightarrow \infty$ (classic limit) the minimum of the expression (6.71) as a function of the α_i equals the quantum mechanical ground-state energy for the Hamiltonian H. So, one obtains an energy surface in the variables α_i for which the minimum corresponds to the ground state. The calculation of the matrix elements in (6.71) can be simplified very much using the commutation relations

$$[b_i, f(b)] = \frac{\partial}{\partial b_i^{\dagger}} f(b), \tag{6.72a}$$

$$[f(b), b_i^{\dagger}] = \frac{\partial}{\partial b_i} f(b), \tag{6.72b}$$

where $f(b)$ is a polynomial expression in the creation and annihilation operators $b_i^{\dagger}$ and b_i. One obtains

$$\begin{aligned}
\langle B^N | B^N \rangle &= \langle B^{N-1} | \sum_i \alpha_i b_i | B^N \rangle \\
&= \langle B^{N-1} | \sum_i \alpha_i \frac{\partial}{\partial b_i^\dagger} | B^N \rangle \\
&= N \langle B^{N-1} | \sum_i \alpha_i^2 | B^{N-1} \rangle \\
&= N\alpha^2 \langle B^{N-1} | B^{N-1} \rangle,
\end{aligned} \tag{6.73}$$

where we have defined $\alpha^2 = \Sigma_i \alpha_i^2$. This process of reducing the number of quanta from N to $N - 1$ can be performed, until one obtains the final result $N!\alpha^{2N}$.

For a general one-body operator

$$A_1 = \sum_{ij} \xi_{ij} b_i^\dagger b_j \tag{6.74}$$

one finds for the normalized matrix element

$$\begin{aligned}
\langle A_1 \rangle &\equiv \frac{\langle B^N | A_1 | B^N \rangle}{\langle B^N | B^N \rangle} \\
&= \frac{N}{\alpha^2} \sum_{i,j} \alpha_i \alpha_j \xi_{i,j},
\end{aligned} \tag{6.75}$$

and for a two-body boson operator

$$A_2 = \sum_{ijkl} \eta_{ijkl} b_i^\dagger b_j^\dagger b_k b_l \tag{6.76}$$

one obtains as a result

$$\langle A_2 \rangle = \frac{N(N-1)}{\alpha^4} \sum_{ijkl} \eta_{ijkl} \alpha_i \alpha_j \alpha_k \alpha_l. \tag{6.77}$$

When we restrict ourselves to the study of monopole and quadrupole degrees of freedom in the IBM, using $s^\dagger$ ($l = 0$) and $d_\mu^\dagger$ ($\mu = -2, \ldots +2$) where μ refers to the projection of $l = 2$ on a space-fixed axis, the condensate of eq. (6.70) becomes

$$B^\dagger = s^\dagger + \sum_m \beta_m d_m^\dagger. \tag{6.78}$$

Here, the expansion coefficients β_m denote the five collective quadrupole variables when describing quadrupole vibrations of a classical object defined in the laboratory system. If the nucleus we describe undergoes axially symmetric shape changes, one can transform the five β_m into the two intrinsic variables (β, γ) and three Euler angles defining the position of the nucleus relative to the lab system. One thus obtains

$$\beta_0 = \beta \cos\gamma, \quad \beta_{\pm 1} = 0, \quad \beta_{-2} = \beta_2 = \frac{1}{\sqrt{2}} \beta \sin\gamma. \tag{6.79}$$

We can evaluate the classical limit expressions for the different terms that appear in a multipole expansion of the *sd*-interaction Hamiltonian using eq. (6.78). As an example, we evaluate one of the terms in some detail. For the $\hat{n}_d$ operator, a one-body operator, which, written out in detail, reads $\hat{n}_d = \Sigma_\mu d^\dagger_\mu d_\mu$, the intrinsic coefficients of ξ_{ij} of eq. (6.75) become diagonal with contributions only from the *d*-boson. The expression $\Sigma_{ij}\, \alpha_i \alpha_j \xi_{ij}$ in eq. (6.75) then simplifies to β^2 and $\alpha^2 = \Sigma_i\, \alpha_i^2 = 1 + \beta^2$. Combining these intermediate results, we obtain finally

$$\langle \hat{n}_d \rangle = N \frac{\beta^2}{1 + \beta^2}, \tag{6.80a}$$

$$\langle Q.Q \rangle = N \frac{1}{1 + \beta^2} \left[5 + \frac{11}{4} \beta^2 \right] + \frac{N(N-1)}{(1 + \beta^2)^2} \left[\frac{\beta^4}{2} + 2\sqrt{2}\, \beta^3 \cos 3\gamma + 4\beta^2 \right], \tag{6.80b}$$

$$\langle J \cdot J \rangle = 6N \frac{\beta^2}{1 + \beta^2}, \tag{6.80c}$$

$$\langle P^\dagger \cdot P \rangle = \frac{N(N-1)}{4} \left(\frac{1 - \beta^2}{1 + \beta^2} \right)^2, \tag{6.80d}$$

$$\langle T_3 \cdot T_3 \rangle = \frac{7}{5} N \frac{\beta^2}{1 + \beta^2}, \tag{6.80e}$$

$$\langle T_4 \cdot T_4 \rangle = \frac{9}{5} N \frac{\beta^2}{1 + \beta^2} + \frac{N(N-1)}{(1 + \beta^2)^2} \frac{18}{35} \beta^4. \tag{6.80f}$$

It is now quite instructive to study the energy surface as a function of β and γ. For the $U(5)$ limit one finds that

$$H = \epsilon_d \hat{n}_d + \sum \frac{1}{2} \sqrt{2l + 1}\, c_l ((d^\dagger d^\dagger)^{(l)} (\tilde{d}\tilde{d})^{(l)})^{(0)} \tag{6.81a}$$

has the classical limit (see also eq. (2.164) in Chapter 2)

$$\langle H \rangle = \epsilon_d \frac{N\beta^2}{1 + \beta^2} + x_1 N(N - 1) \frac{\beta^2}{(1 + \beta^2)^2}, \tag{6.81b}$$

with

$$x_1 = \frac{1}{10} c_0 + \frac{1}{7} c_2 + \frac{9}{35} c_4.$$

For the $SU(3)$ Hamiltonian

$$H = -\kappa_1 Q \cdot Q - \kappa_2 J \cdot J, \tag{6.82a}$$

where

$$Q \equiv (s^\dagger \tilde{d} + d^\dagger s)^{(2)} + \chi\, (d^\dagger \tilde{d})^{(2)}, \tag{6.83}$$

we get for the classical limit

$$\langle H \rangle = -\kappa_1 \left[\frac{N}{1 + \beta^2}(5 + (1 + \chi^2)\beta^2) + \frac{N(N-1)}{(1 + \beta^2)^2} \left(\frac{2}{7}\chi^2 \beta^4 - 4\sqrt{\frac{2}{7}}\chi \beta^3 \cos 3\gamma + 4\beta^2 \right) \right] - \kappa_2 \frac{6N\beta^2}{1 + \beta^2}. \tag{6.82b}$$

Finally, the $O(6)$ Hamiltonian

$$H = AP^\dagger \cdot P + BC_{2SO5} + CJ \cdot J, \tag{6.84a}$$

where

$$P^\dagger = \frac{1}{2} (s^\dagger s^\dagger - d^\dagger \cdot d^\dagger), \tag{6.85}$$

has the classical limit

$$\langle H \rangle = A \frac{N(N-1)}{4} \left(\frac{1 - \beta^2}{1 + \beta^2} \right)^{(2)} + \kappa_3 \frac{N\beta^2}{1 + \beta^2}, \tag{6.84b}$$

in which $\kappa_3 = 2B + 6C$. The $U(5)$, $SU(3)$ and $O(6)$ Hamiltonians are slightly different from some choices made in the literature but the present parametrizations are most often used in phenomenological studies. For the relation between various representations, see Chapter 2.

In Figs. 6.30 and 6.31 we show a typical example for each limit.

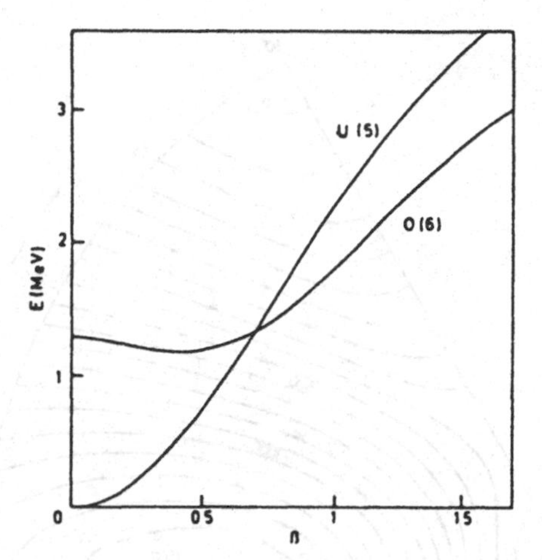

FIGURE 6.30. *Potential energy, as a function of the deformation parameter β, in the U(5) and O(6) limit, for the nuclei* ^{102}Ru *and* ^{196}Pt, *respectively, according to eqs.* (6.81b), (6.84b).

The $U(5)$ and $O(6)$ limits are both independent of γ and a γ-dependence only appears in the $SU(3)$ limit. We point out that, for deformed nuclei, the β values cannot directly be compared to the deformation values β_{BM} used in the Bohr–Mottelson model (Bohr and Mottelson, 1975). The difference stems from the fact that in the IBM, β describes the deformation associated with the valence nucleons ($2N$ nucleons). In the collective model, β_{BM} refers to the deformation of all nucleons. An approximate relation between both has been derived by Ginocchio and Kirson (1980) and reads

$$(\beta)_{BM} \cong 1.18 \left(\frac{2N}{A}\right)\beta.$$

The energy surface in the $U(5)$ limit always has, for realistic values for ϵ_d and c, a minimum at $\beta = 0$. Still it does not correspond to harmonic oscillations in the β direction because

(i) the β^4 factor coming from the $[(d^\dagger d^\dagger)^{(l)}(\tilde{d}\tilde{d})^{(l)}](0)$ term, and
(ii) the factors $(1 + \beta^2)$ and $(1 + \beta^2)^2$ in the denominator.

These factors are not present in the potentials used, for example, in the Bohr–Mottelson (Bohr and Mottelson, 1975) model and give rise to a different behavior for $\beta \rightarrow \infty$: in the Bohr–Mottelson

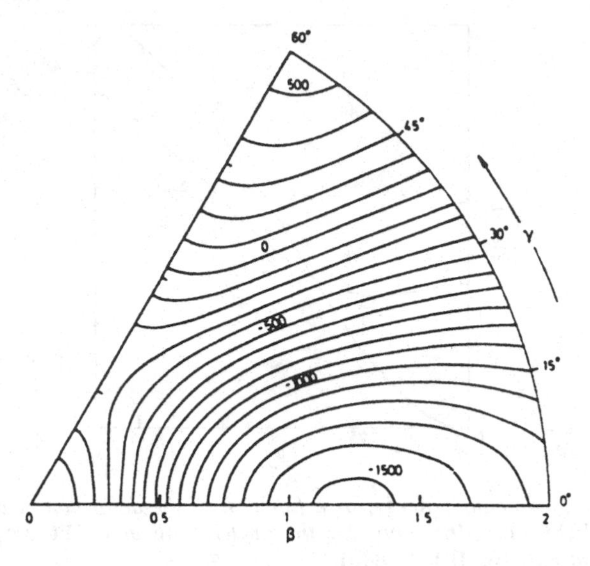

FIGURE 6.31. *Potential energy surface for the SU(3) limit, applied to the nucleus* ^{156}Gd, *using eq.* (6.82b). *The energies are given in between the contour lines.*

model, the potential goes to infinity whereas in the IBM, the potential approaches a constant value.

For the $SU(3)$ limit, and using $\chi = \sqrt{7}/2$ (or $-\sqrt{7}/2$) one obtains the energy surface corresponding to eq. (6.82b). These two choices lead to different shapes: the first is an oblate shape where the minimum in the potential energy surface is at $\gamma = 60°$, while the second is a prolate nucleus with minimum at $\gamma = 0°$. In each case, the minimum corresponds to an axial shape. Even the most general IBM Hamiltonian will never lead to a triaxial minimum in the (β, γ) plane.

In the $O(6)$ limit, one has a competition between the two terms in eq. (6.84b): the second term tries to put the minimum at $\beta = 0$; the first term prefers a stable, β deformed shape. Since in IBM fits to actual nuclei, $A \gg \kappa_3$ and the N-dependence multiplying the A-coefficient is quadratic, the $O(6)$ limit corresponds to a deformed γ-unstable rotor with minimum for the energy at $\beta = 1$.

6.4.3. Cubic Terms: The Three Limiting Cases

Since the work of Wilets and Jean (1956) and of Davydov, Filippov and Chaban (1958, 1960), it is known that the γ-degree of freedom plays an important role in many nuclei where static (quadrupole) deformation occurs. These models adopt a geometrical picture of the nucleus to describe its collective excitations, and emphasize the importance of deviations from axial symmetry. In the model of Wilets and Jean (1956) the potential energy is assumed to be independent of the γ-degree of freedom, hence the name γ-unstable rotor. In the model of Dayvdov et al. (1958, 1960) on the other hand, a rigid triaxial shape is used for the nucleus. Although these models are, in general, successful in reproducing the experimental data, they have as a disadvantage the lack of a unified theory, since each is based on a different hypothesis concerning triaxiality.

The IBM has provided us with an alternative description of nuclear collective excitations, which, in contrast to geometrical models, is of an algebraic nature. One of the most attractive features of the IBM is certainly its unifying capacity and one might try to use this property to establish a coherent theory of triaxiality in the framework of the IBM. However, even the most general Hamiltonian of the IBM cannot give rise to stable, triaxial shapes. In contrast, triaxiality does occur in particular dynamic symmetries of the IBM-2 that take into account the distinct character of the proton and neutron bosons.

We show that by incorporating cubic terms in the Hamiltonian of the IBM, one may obtain nuclei that exhibit a minimum associated with the potential in the γ-direction i.e., $V(\gamma)$ for γ-values that are differing from $\gamma = 0°, 60°$. The word "triaxial" indicates a rigid axial asymmetric shape (Davydov et al., 1958 and 1960) and expresses the fact that the γ-variable is taking on a single value ($\gamma \neq 0°, 60°$). We study the influence of such terms on the energy spectrum in each of the three dynamic symmetries. Our analysis will be restricted to the IBM-1, that is, making no distinction between protons and neutrons.

Introducing the cubic terms as

$$d(l,k,r) = ((d^\dagger d^\dagger)^{(l)} d^\dagger)^{(r)} \cdot ((\tilde{d}\tilde{d})^{(k)} \tilde{d})^{(r)}, \qquad (6.86)$$

one finds five linear combinations of the type (6.86) that are uniquely characterized by r ($r = 0, 2, 3, 4, 6$). For $r = 2, 3$ and 4, there are different choices in l or k. In particular, one can always choose $l = k$. For $r = 2$, 3 and 4, the different choices of l and k can be related making use of the relations that exist between them:

$$
\begin{aligned}
((d^\dagger d^\dagger)^{(2)} d^\dagger)^{(2)} &= \frac{2\sqrt{5}}{7} ((d^\dagger d^\dagger)^{(0)} d^\dagger)^{(2)}, \\
((d^\dagger d^\dagger)^{(4)} d^\dagger)^{(2)} &= \frac{6}{7} ((d^\dagger d^\dagger)^{(0)} d^\dagger)^{(2)}, \\
((d^\dagger d^\dagger)^{(4)} d^\dagger)^{(3)} &= -\sqrt{\frac{2}{5}} ((d^\dagger d^\dagger)^{(2)} d^\dagger)^{(3)}, \\
((d^\dagger d^\dagger)^{(4)} d^\dagger)^{(4)} &= \sqrt{\frac{10}{11}} ((d^\dagger d^\dagger)^{(2)} d^\dagger)^{(4)}.
\end{aligned}
\tag{6.87}
$$

So, one can always choose $l = k$. The relations for the $d(l, k, r)$ if $l \neq k$ are easily obtained using eqs. (6.87).

The classic limit for the $d(l, k, r)$ then becomes (using eqs. (6.72))

$$
\langle d(l,k,r) \rangle = N(N-1)(N-2) \frac{\beta^6}{(1+\beta^2)^3} (X_0 + X_1 \cos^2 3\gamma), \tag{6.88}
$$

with X_0, X_1 as given in Table 6.3. Because of the $\cos^2 3\gamma$ dependence, the cubic term can, in combination with the general H_{sd} Hamiltonian, lead to minima corresponding to a value $\gamma \neq 0°, 60°$.

So, the Hamiltonian we consider is of the form

$$
H = H_{sd} + \sum_L \theta_L [d^\dagger d^\dagger d^\dagger]^{(L)} \cdot [\tilde{d}\tilde{d}\tilde{d}]^{(L)}, \tag{6.89}
$$

where H_{sd} is the standard Hamiltonian of the IBM,

TABLE 6.3. *The coefficients* X_0 *and* X_1 *[see eq. (6.88)] that appear in the expressions for the classical limit* $d(l,k,r)$ *for the cubic terms defined in eq.* (6.86).

r	0	2	3	4	6
$l = k$	2	0	2	2	4
X_0	0	$\frac{1}{5}$	$-\frac{1}{7}$	$\frac{3}{49}$	$\frac{14}{55}$
X_1	$\frac{2}{35}$	0	$\frac{1}{7}$	$\frac{3}{35}$	$-\frac{8}{385}$

$$H_{sd} = \epsilon_d \hat{n}_d + \kappa Q \cdot Q + \kappa' J \cdot J + \kappa'' P^\dagger \cdot P + q_3 T_3 \cdot T_3 + q_4 T_4 \cdot T_4. \tag{6.90}$$

For each of the three limits of the IBM [$U(5)$, $O(6)$ and $SU(3)$] the influence of the various cubic terms of eq. (6.89) on the energy spectrum has been studied. In Figs. (6.32a–c) we show typical spectra in the three limits. The parameters of eqs. (6.89) and (6.90) are specified in the caption. Figs. (6.32a–c) illustrate that in both the $U(5)$ and $O(6)$ limits the ground-state band (gsb) and the first excited 2^+ band are not changed by the inclusion of cubic terms in the Hamiltonian. On the other hand, the 3^+, 4^+, 5^+, . . . bands are lowered in energy since for these levels the cubic term with $L = 3$ is the most simple and straightforward term to be used and gives a diagonal energy contribution. One could accomplish almost the same effect using a general, linear combination of various terms with different L values. In the $SU(3)$ limit the ground-state band

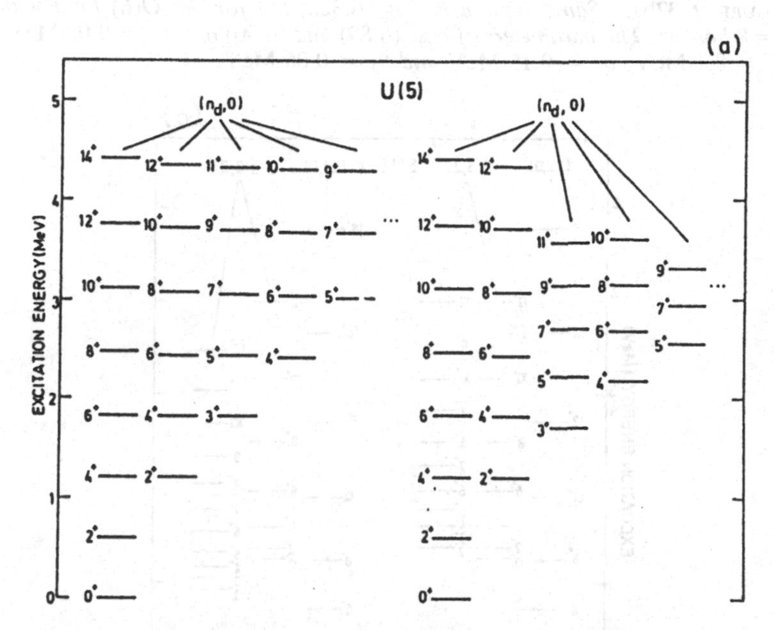

FIGURE 6.32(a). *The lower band members in the $U(5)$ limit for $N=7$ bosons. The left part of the figure shows the spectrum in the exact $U(5)$ limit and the right part shows the spectrum when the $L=3$ cubic term is added. The parameters of eqs.* (6.89) *and* (6.90) *are:* $\epsilon_d = 0.6$ MeV, $\kappa' = 0.001$ MeV, $\theta_3 = 0.06$ MeV.

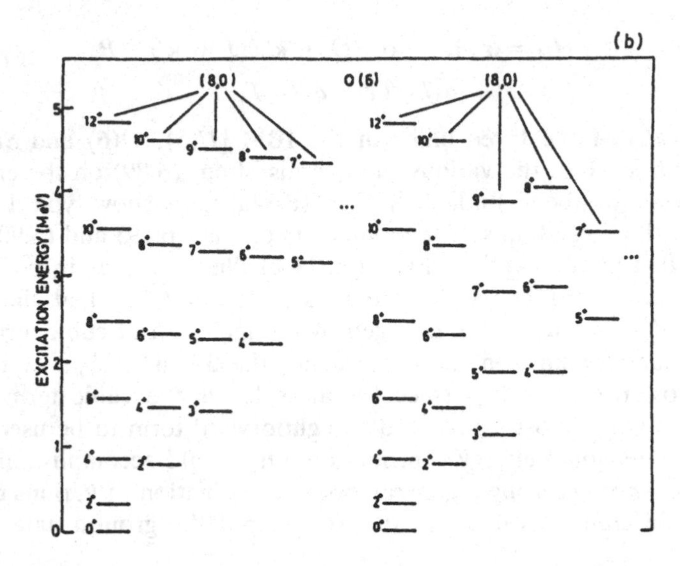

FIGURE (6.32b). *Same caption as Fig.* 6.32a, *but for the* $O(6)$ *limit with* $N=8$ *bosons. The parameters of eqs.* (6.89) *and* (6.90) *are:* $\kappa' = 0.02$ MeV, $\kappa'' = 0.1$ MeV, $q_3 = 0.15$ MeV *and* $\theta_3 = 0.06$ MeV.

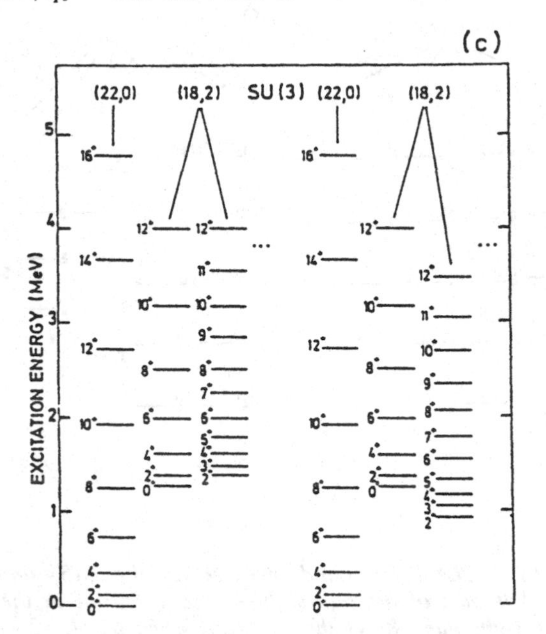

FIGURE 6.32(c). *Same captian as Fig. 6.32*(a), *but for the* $SU(3)$ *limit with* N = *11 bosons, the parameters of eqs.* (6.89) *and* (6.90) *are:* $\kappa = -0.02$ MeV, $\kappa' = 0.01$ MeV *and* $\theta_3 = 0.02$ MeV.

and the β-band are unaltered but the γ-band is shifted in energy with respect to the $SU(3)$ calculation.

In order to obtain a more intuitive insight into the problem of triaxial shapes, the classical limit of the Hamiltonian of eq. (6.89) can be calculated. In the intrinsic frame of reference one obtains a potential energy surface dependent on β and γ, which for a Hamiltonian with an $L = 3$ cubic term reads

$$\begin{aligned} E(\beta,\gamma) = {} & \epsilon_d \frac{N\beta^2}{1+\beta^2} + \kappa\left[\frac{N}{1+\beta^2}\left(5+\frac{11}{4}\beta^2\right)\right. \\ & \left. + \frac{N(N-1)}{(1+\beta^2)^2}\left(\frac{\beta^4}{2} + 2\sqrt{2}\,\beta^3\cos 3\gamma + 4\beta^2\right)\right] \\ & + \kappa'\frac{6N\beta^2}{1+\beta^2} + \kappa''\frac{N(N-1)}{4}\left[\frac{1-\beta^2}{1+\beta^2}\right]^2 \\ & + q_3 N\frac{7}{5}\frac{\beta^2}{1+\beta^2} \\ & + q_4\left[N\frac{9}{5}\frac{\beta^2}{1+\beta^2} + N(N-1)\frac{18}{35}\frac{\beta^4}{(1+\beta^2)^2}\right] \\ & + \theta_3 N(N-1)(N-2)\frac{1}{7}\frac{\beta^6}{(1+\beta^3)^2} \\ & \times(-1+\cos^2 3\gamma). \end{aligned} \tag{6.91}$$

This expression allows us to visualize the influence of the $L = 3$ cubic term in a (β, γ)-potential energy plot, as is illustrated in Figs. (6.33a–c) in the case of the $U(5)$, $O(6)$ and $SU(3)$ limits, respectively. One indeed, observes that, in particular the $O(6)$ limit, a stable triaxial minimum results at $\gamma = 30°$ and $\beta \cong 0.7$. The need for cubic terms in order to have a minimum in $E(\beta,\gamma)$, different from $\gamma = 0°$ or 60°, is clear. Without the cubic terms, the only dependence on γ in eq. (6.91) resides in the cos3γ term, which implies always either a prolate or an oblate minimum (provided κ is negative), depending on the sign one considers in the quadrupole operator Q. Only by including terms proportional to $\cos^2 3\gamma$ does a minimum in $E(\beta,\gamma)$ occur with $\gamma \neq 0°, 60°$.

The above discussion is based upon a study of the minimum values (β, γ) of the expression $E(\beta, \gamma)$ of eq. (6.91) corresponding to a given IBM Hamiltonian. Thereby, one concentrates on the static (potential energy) part of the Hamiltonian. In the full quantum

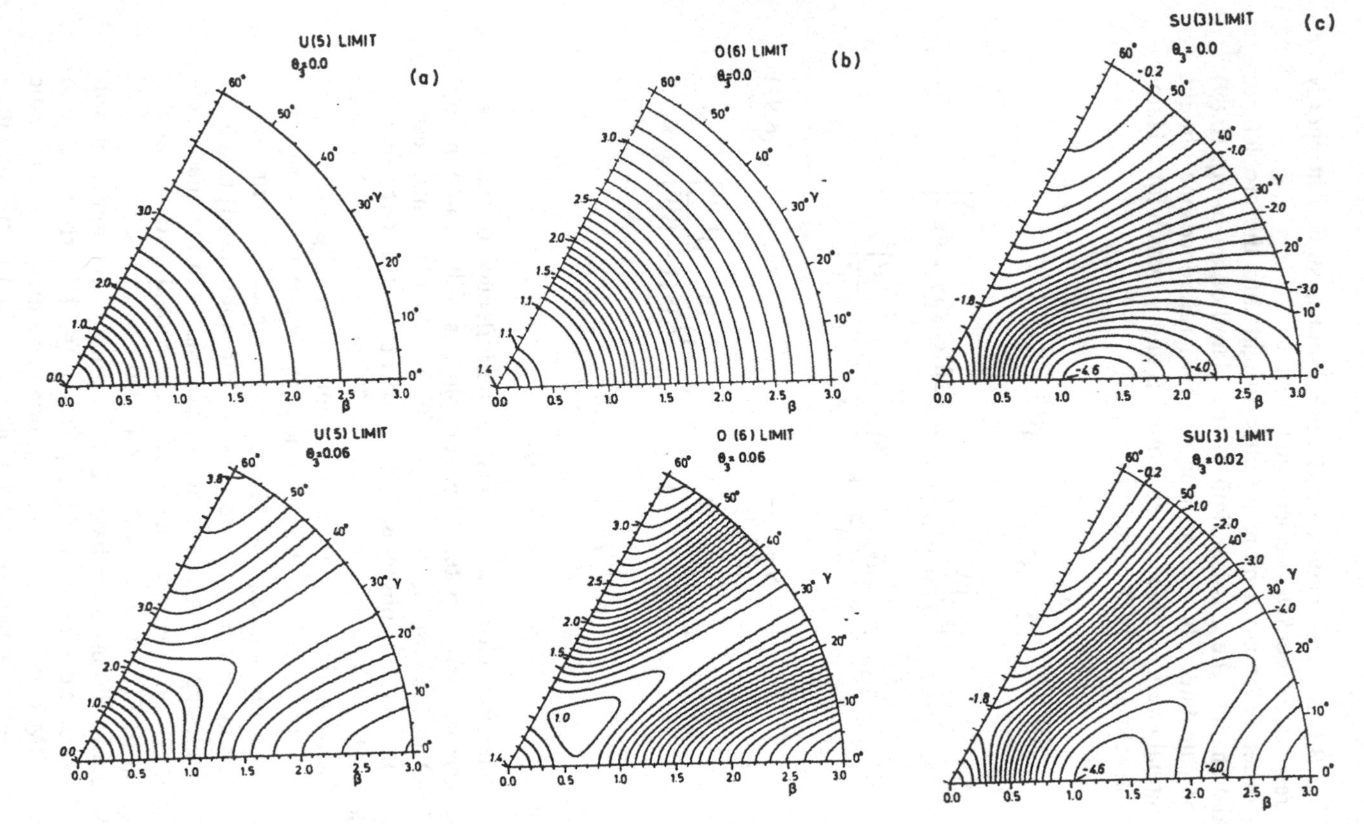

FIGURE 6.33(a). *Potential energy surface, according to eq.* (6.91) *for the* $U(5)$ *limit with the parameters given in the caption of Fig.* 6.32(a).

FIGURE 6.33(b). *Potential energy surface, according to eq.* (6.91) *for the* $O(6)$ *limit with the parameters given in the caption of Fig.* 6.32(b).

FIGURE 6.33(c). *Potential energy surface, according to eq.* (6.91) *for the* $SU(3)$ *limit with the parameters given in the caption of Fig.* 6.32(c).

mechanical problem dynamics (kinetic energy) has to be taken into account properly which implies deviations from the minimum γ-value. In this way, an effective γ-value can be determined as the expectation value of γ, using the quantum mechanical wave function, allowing for vibrational excitations, rotational excitations being excluded.

A slightly different procedure to relate an effective γ-value to the IBM χ parameter in the quadrupole operator has been discussed in Chapter 3.

6.4.4. Application to ^{104}Ru

The nucleus ^{104}Ru has been extensively studied in the framework of the IBM. We have discussed in Section 6.2. the hexadecapole degree of freedom via a *g*-boson and this was shown to be of crucial importance for the description of many detailed properties of transitional nuclei such as the E2 decay properties within the ground-state band, E2 transitions connecting the γ- and ground-state bands as well ground-state quadrupole moments. However, the excitation energies of the levels of the γ-band are not substantially altered by the addition of a *g*-boson. They still deviate strongly from the experimental data (Fig. 6.34). These remaining discrepancies support the hypothesis of ^{104}Ru being an axially asymmetric but not γ-unstable nucleus and suggest the use of the $L = 3$ cubic term in the IBM Hamiltonian. Its effect is illustrated in Figs. 6.34 and 6.35. The levels of the γ-band are changed from a $(3^+, 4^+)$, $(5^+, 6^+)$, $(7^+, 8^+)$ odd-even sequence into a band with more regular spacings, which is typical for a nucleus which is neither γ-soft or rigid triaxial, but has a shape somewhere in between. This example illustrates that the use of cubic terms in the IBM enables one to describe nuclei ranging from γ-soft to rigid triaxial, hence covering a wide range of possible shapes.

When including cubic terms in the Hamiltonian, one should at the same time extend the E2 operator to higher order,

$$
\begin{aligned}
T(\mathrm{E2}) = {} & e_{ds}(d^\dagger s + s^\dagger \tilde{d})^{(2)} + e_{dd}(d^\dagger \tilde{d})^{(2)} \\
& + \sum_{L=\mathrm{even}} e_L\{[(d^\dagger d^\dagger)^{(L)}(\tilde{d}s)^{(2)}]^{(2)} + \mathrm{h.c.}\} \qquad (6.92) \\
& + \sum_{L,L'=\mathrm{even}} e_{L,L'}[(d^\dagger d^\dagger)^{(L)}(\tilde{d}\tilde{d})^{(L')}]^{(2)}.
\end{aligned}
$$

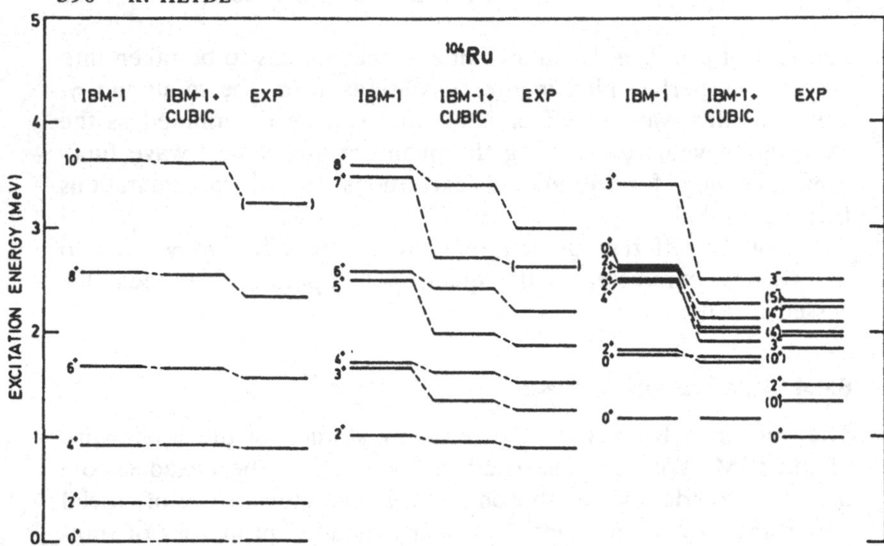

FIGURE 6.34. *Comparison of the experimental spectrum of* ^{104}Ru *taken from* Stachel *et al.* (1982) *with a* IBM *calculation* (ϵ_d = 0.6 MeV, κ = −0.01 MeV, κ' = 0.01 MeV, κ'' = 0.098 MeV *and* q_3 = 0.05 MeV) *and with a* IBM *calculation with the same parameters and a* L = 3 *cubic term* (θ_3 = 0.11 MeV). *Note that if* κ' *were zero, the* γ*-band doublets* $(3^+_\gamma, 4^+_\gamma)$, $(5^+_\gamma, 6^+_\gamma)$ *would become degenerate at* $\theta_3 = 0$ *giving the results of the* γ*-unstable system.*

This is still not the most general one- and two-body transition operator for s- and d-bosons. We will later indicate with arguments based on perturbation theory that the operator (6.92) is appropriate for describing ground-state band E2 properties (see Fig. 6.36). The extension (6.92) has the disadvantage of increasing the number of effective boson charges significantly. However, one can show that in the pure $U(5)$ limit the calculation of ground-state band B(E2) values and diagonal reduced matrix elements needs only two of these effective boson charges.

In the $U(5)$ limit the gsb wave functions read

$$|s^N; 0^+\rangle, |s^{N-1}d; 2^+\rangle, |s^{N-2}d^2; 4^+\rangle, |s^{N-3}d^3; 6^+\rangle, \ldots, \quad (6.93)$$

and the new expressions for $B(\mathrm{E2}; J \to J-2)$ and $\langle J||T(\mathrm{E2})||J\rangle$ become

$$B(\mathrm{E2}; J \to J-2) = (N - n_d)(n_d + 1)\left[e_{ds} + e_4 \frac{\sqrt{5}}{3} n_d\right]^2, \quad (6.94)$$

only (L = 4 contributes), and

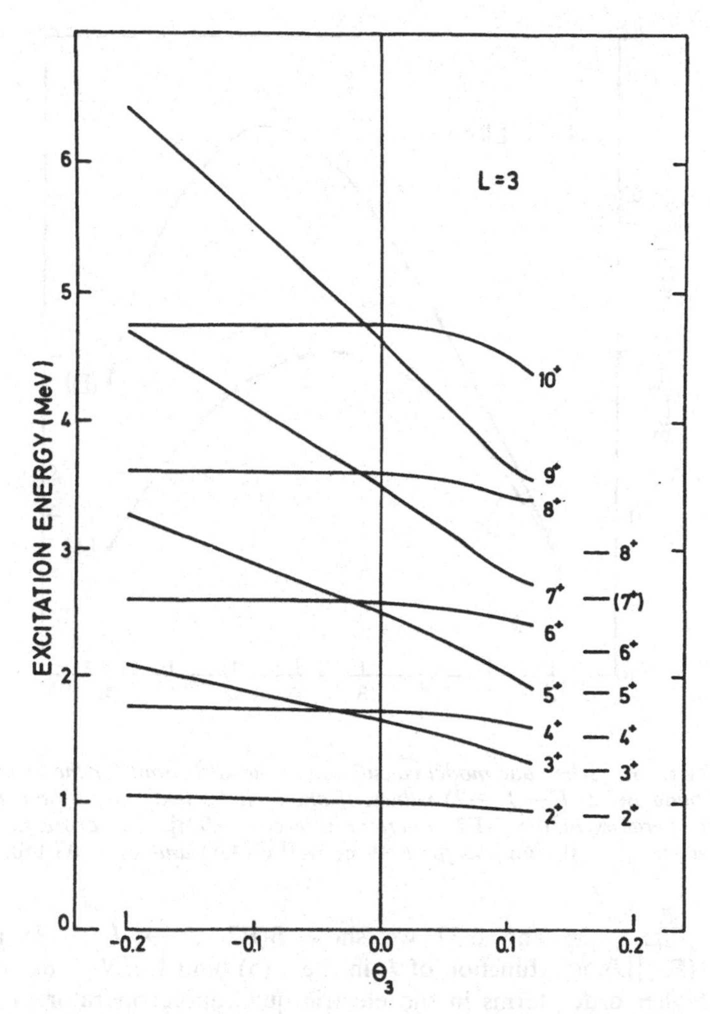

FIGURE 6.35. *Variation of the levels of the γ-band as a function of the strength parameter θ_3. The parameters of the* IBM *Hamiltonian of eqs. (6.89) and (6.90) are given in the caption of Fig. 6.34. At the extreme right, the experimental γ-band in* ^{104}Ru *is also drawn for comparison.*

$$\langle J||T(E2)||J\rangle = \left[\frac{(2J+1)(2J+3)(2J+2)}{7J(2J-1)}\right]^{1/2} \times \left[e_{dd}n_d + e_{4,4}\frac{7}{3}\sqrt{\frac{2}{11}}(n_d-1)n_d\right], \tag{6.95}$$

(only $L = L' = 4$ contributes), respectively.

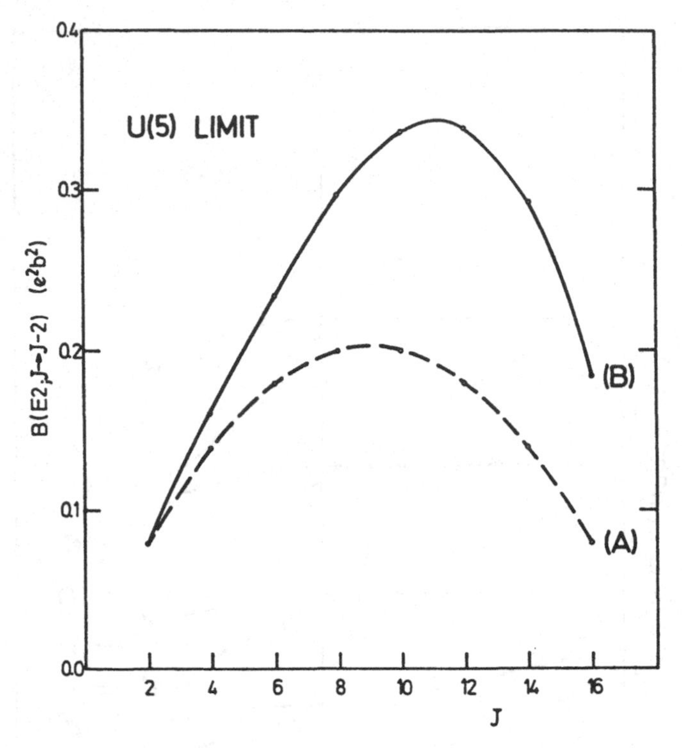

FIGURE 6.36. *Schematic model calculation in the U(5) limit for the ground-state band B(E2; J → J − 2) values, if one includes two-body boson transition operators in the T(E2) operator [see eq. (6.92)]. The expression is drawn for* e_{ds} = 0.1 eb *and for both* e_4 = 0 eb (A) *and* e_4 = 0.01eb(B).

In Figs. 6.36 and 6.37 we show $B(E2, J \rightarrow J - 2)$ and $\langle J||T(E2)||J\rangle$ as a function of J, in the $U(5)$ limit for $N = 8$. Since the higher order terms in the electric quadrupole operator of eq. (6.92) will generally be of the order $e_L \cong e_{L,L'} \cong e^2_{1-body}$, we use in the schematic $U(5)$ calculations the values $e_4 = 0.01$ eb and $e_{4,4} = 0.01$ eb. The ground-state band $B(E2; J \rightarrow J - 2)$ values grow to much larger values for the $6^+ \rightarrow 4^+$, $8^+ \rightarrow 6^+$ and $10^+ \rightarrow 8^+$ transitions and the diagonal reduced E2 matrix elements no longer show the pure $U(5)$ behavior. Both features appear in ^{104}Ru.

If one studies more complicated situations [other than $U(5)$] and other E2 transitions, the above simplifications for the $U(5)$ limit are not valid any more. One way to keep some simplicity is to introduce cubic anharmonicities in a consistent-Q framework (Cas-

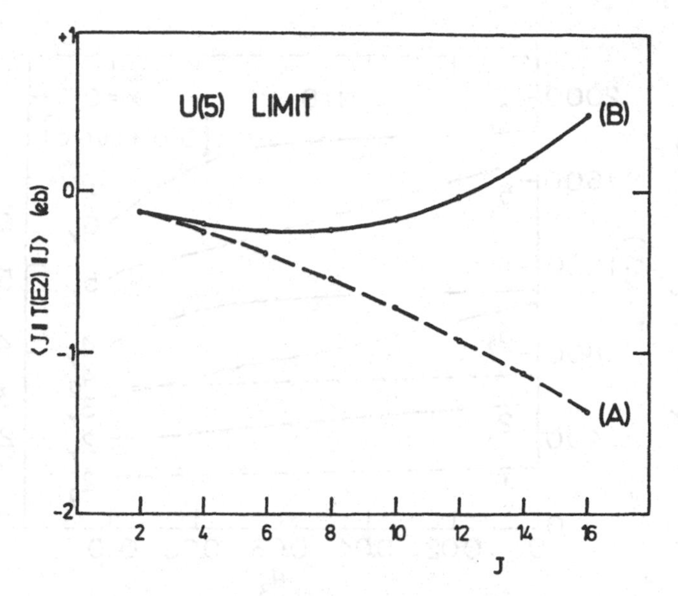

FIGURE 6.37. *Schematic model calculation in the* U(5) *limit for the ground-state band diagonal reduced matrix element* $\langle J||T(\mathrm{E2})||J\rangle$ *if one includes two-body transition operators in the* T(E2) *operator* [*see eq.* (6.92)]. *The expression is drawn for* $e_{dd} = -0.05$ eb *and for both* $e_{4,4} = 0$ eb(A) *and* $e_{4,4} = 0.01$ eb(B).

ten and Warner (1988)), which is an interacting boson model description in which the parameter χ appearing in the quadrupole operator for the Hamiltonian and in the T(E2) operative, is identical. Our example suggests we use the simple Hamiltonian

$$H = \kappa Q \cdot Q + \theta_3 (d^\dagger d^\dagger d^\dagger)^{(3)} \cdot (\tilde{d}\tilde{d}\tilde{d})^{(3)}, \tag{6.96}$$

where the quadrupole operator in the Hamiltonian is identical to the E2 operator used in order to calculate E2 transitions. Using this scheme, we can investigate, in particular, the evolution of the γ-band energy staggering as a function of χ and θ_3. It can be expected that, when θ_3 is small, the potential $V(\beta, \gamma)$ is quite γ-soft and the energy levels of the γ-band will behave as in the usual $O(6) \rightarrow SU(3)$ transition when χ is varied. For larger θ_3 values, the cubic term may dominate, and, even for small χ, one will no longer obtain an $O(6)$-like staggering. Some typical examples are illustrated in Fig. 6.38. Near the $O(6)$ limit (upper part), a small θ_3 value leaves the $O(6)$-like staggering intact but a more uniform

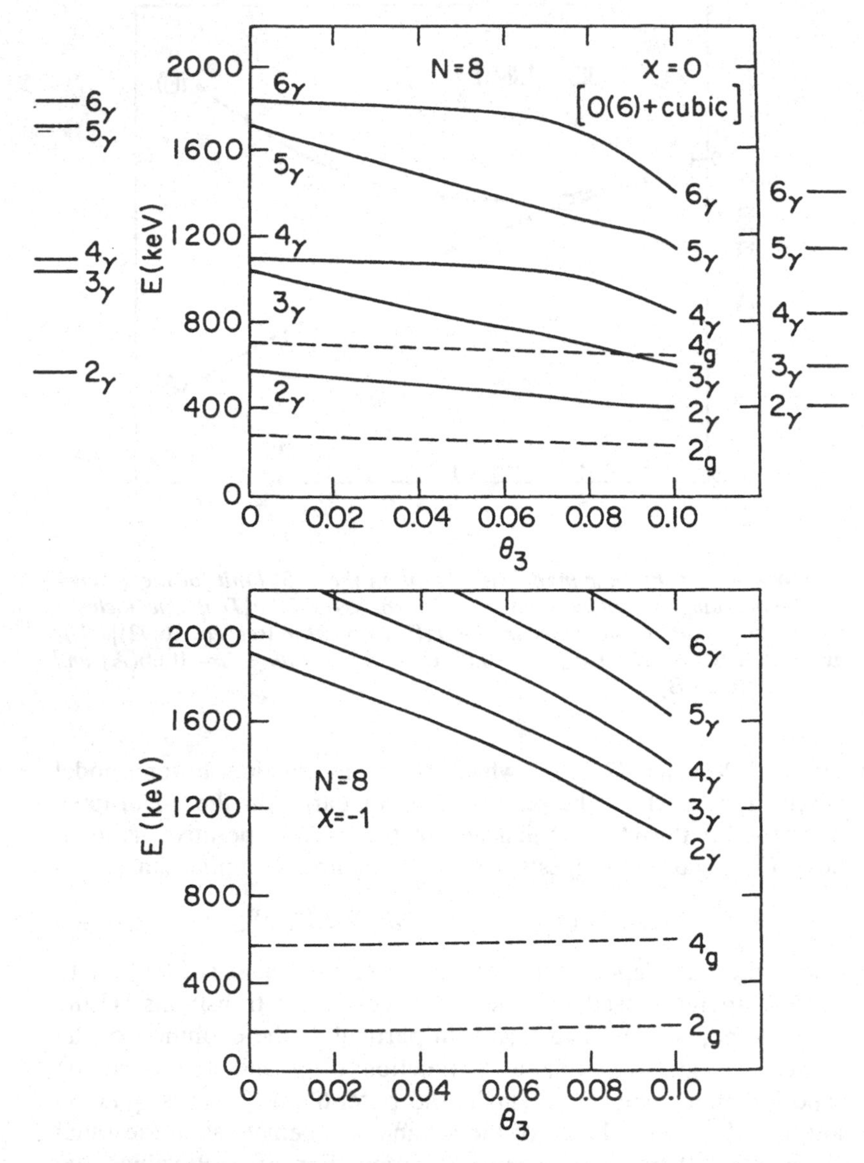

FIGURE 6.38. *Ground (dashed) and γ-band (solid) energy levels as a function of θ_3 for two values of χ, one corresponding to O(6) (χ=0) and the other to nuclei near SU(3)(χ = −1.0), for N=8 bosons. In the upper part, the γ-band energies for θ_3 = 0 and 0.10 are extracted and displayed on the left and right, respectively.*

spacing is achieved for larger θ_3 values. This change in structure mainly results from a potential minimum developing at $\gamma = 30°$ as θ_3 becomes larger (Casten et al. (1985)). In the bottom part of the figure, the IBM potential $V(\beta, \gamma)$, already exhibits a well-defined minimum at $\gamma = 0°$ for $\chi = -1.0$ and the effect of the θ_3 term is then similar to the introduction of an increasing $V(\gamma)$ potential in the asymmetric rotor model i.e., it rapidly lowers the γ-band without much affecting the staggering. Thus, the effect of the cubic term is most significant when the $V(\beta, \gamma)$ potential is nearly independent on γ i.e., near the $O(6)$ limit.

Concluding this Section 6.4.4, we again emphasize that the $O(6)$ limit corresponds geometrically to a γ-unstable potential, implying wave functions with a mean γ-value $\langle\gamma\rangle = \gamma_{\text{eff.}} = 30°$ together with large γ-fluctuations around this value. For the rigid, triaxial rotor with $\gamma = 30°$, the wave functions again have $\gamma_{\text{eff.}} = 30°$ but rather small γ-fluctuations. Many predictions of the γ-soft and triaxial potentials with the same effective γ-asymmetry are identical. A crucial difference centers on the γ-band energy staggering. In the triaxial rotor, levels appear in doublets $(2^+_\gamma, 3^+_\gamma)$, $(4^+_\gamma, 5^+_\gamma)$, . . . , whereas the γ-unstable $O(6)$ limit leads to a 2^+_γ, $(3^+_\gamma, 4^+_\gamma)$, $(5^+_\gamma, 6^+_\gamma)$, . . . staggering. The experimentally observed energy spacing in the γ-band is in many cases (see Fig. 6.34 for ^{104}Ru and the spectrum of ^{196}Pt in Fig. 6.39) more regular. From the above studies one can conclude that for nuclei with a large $\gamma_{\text{eff.}}$, the amount of γ-dependence needed in the potential $V(\gamma)$ is very little (at most 50–200 keV). Despite the interest in this issue for decades, it is only within the IBM (at the $O(6)$ analytic starting point) that this question of γ-softness vs. triaxiality in low-energy nuclear spectra could be resolved (Zamfir and Casten, 1991).

6.4.5. Conclusions

We have shown that nuclei with triaxial features can be described in the framework of the IBM by considering cubic terms in the model Hamiltonian. The effect of this addition to the Hamiltonian was studied in the three different limits of the IBM ($U(5)$, $O(6)$ and $SU(3)$) and visualized by taking the classical limit of this extended Hamiltonian. This discussion pointed out the occurrence of a stable minimum in the γ-degree of freedom if one considers the $L = 3$ cubic term. As an illustration of the usefulness of this approach, we presented applications to the axially symmetric

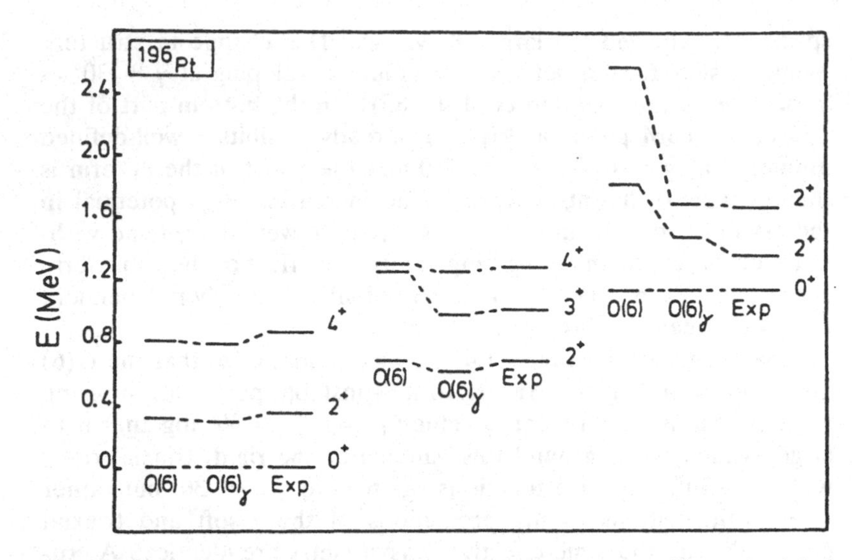

FIGURE 6.39. *Calculated and observed low-lying levels of* ^{196}Pt. *The label* $O(6)$ *denotes levels calculated without a cubic interaction* ($\theta_3 = 0$) *in the pure* $O(6)$ *limit (taken from Casten et al.* (1985). *Levels marked by* $O(6)_\gamma$, *indicate the results of a calculation including the cubic interaction.*

nucleus ^{104}Ru and the $O(6)$ nucleus ^{196}Pt. With regard to electromagnetic properties of the ground-state band, we showed that these are not changed drastically by cubic terms in the IBM, unless these higher order terms are treated consistently in Hamiltonian and transition operators.

6.5 NUCLEON EXCITATIONS ACROSS CLOSED SHELLS: INTRUDER EXCITATIONS

6.5.1. Intruder Excitations Near Closed Shells

6.5.1.1. Introduction: Microscopic Basis for Pair Excitations

In nuclei with few valence neutrons (-protons), one observes collective quadrupole vibrational type of excitations. Using the IBM, this means that the energy spectra could be described within the $U(5)$ limit. However, on quite a number of occasions, the above simple spectra are perturbed by the presence of extra levels that often show rotational-like characteristics in medium-heavy and heavy nuclei with a large neutron excess. Many such levels have been observed since, in many even-even nuclei as well as in odd-A nuclei. For a review of the status of such excitations, in particular near or at the closed shells we refer the reader to Heyde et al. (1983) and Wood et al. (1992). One of the earliest and very clear-cut examples was the observation of a rotational-like band structure in the Sn nuclei with A = 112, 114, 116, 118 and a 0^+ band head energy of $E_x \approx 2$ MeV together with the observation of the intruder 0^+ systematics in the Pb even-even nuclei.

Theoretically, one explains these intruder states as particle-hole (p-h) excitations across the closed shell (Fig. 6.40) i.e., as 2p–2h in even-even nuclei (0^+ intruder excitations) and 1p–1h in odd-A nuclei. Particle-hole excitations across closed-shell configurations as the microscopic basis for nuclear shape coexistence were originally discussed by G.E. Brown for light even-even nuclei such as ^{16}O, ^{40}Ca.

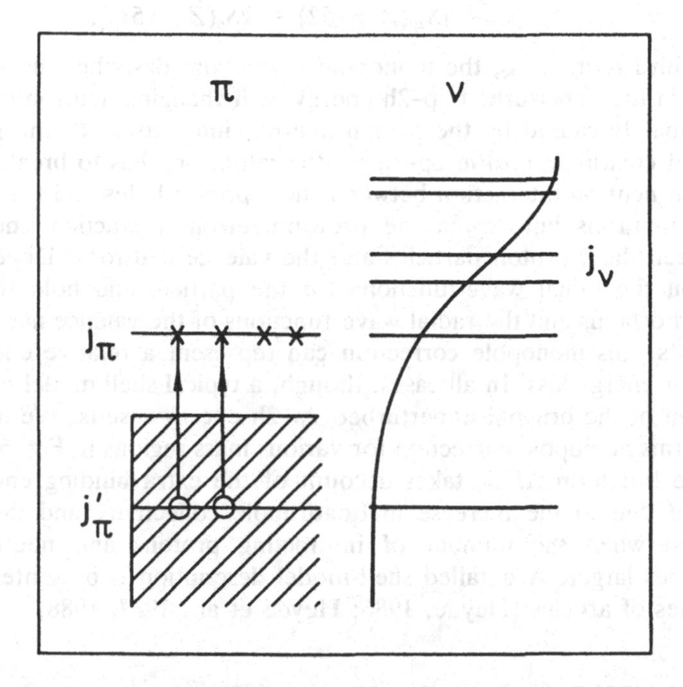

FIGURE 6.40. *Schematic illustration of a proton intruder* 0^+ 2p–2h *excitation where* j_π *denotes the regular orbital,* j'_π *the intruder orbital and* j_ν *the neutron orbitals filled with a Bardeen–Cooper–Schrieffler* (BCS)*-like pair distribution* (ν).

The excitation energy of the lowest intruder state, described as a 2p–2h configuration in even-even nuclei, is described by the expression

$$E_{\text{intr.}}(0^+) = 2(\epsilon_p - \epsilon_h) - \Delta E_{\text{pairing}} + \Delta E_M + \Delta E_{\text{coll.}}. \quad (6.97)$$

The first term denotes the unperturbed energy for the 2p–2h excitation and, for a proton 2p–2h excitation, can be related to experimental one-proton separation energies (for $Z = 50$):

$$2(\epsilon_p - \epsilon_h) \cong 2[S_p(Z = 50) - S_p(Z = 51)]. \quad (6.98)$$

The second term describes the gain in pairing energy for a 0^+ coupled particle (2p) and hole (2h) pair, since the 2p–2h excitation is, of course, not composed of independent particles. This number can also be estimated from one- and two-proton separation energies:

$$\begin{aligned} \Delta E_{\text{pairing}} = {} & [2S_p(Z = 50) - S_{2p}(Z = 50)] \\ & + [S_{2p}(Z = 52) - 2S_p(Z = 51)]. \end{aligned} \quad (6.99)$$

The third term, ΔE_M, the monopole correction, describes the variation in the unperturbed 2p–2h energy with changing neutron number, mainly caused by the proton-neutron interaction. In the process of creating a proton 2p–2h configuraton, one has to break the proton-neutron interaction between the 2 proton holes and the valence neutrons but regain the proton-neutron interaction energy between the 2 proton particles and the valence neutrons. Depending on the radial wave-functions for the particle and hole shell-model orbitals and the radial wave functions of the valence neutron orbitals, this monopole correction can represent a relative energy gain or energy loss. In all cases, though, a typical shell-model modulation of the original unperturbed 2p–2h energy results. We illustrate this monopole correction for various mass regions in Fig. 6.41.

The last term $\Delta E_{\text{coll.}}$ takes account of the extra binding energy gained due to the increase in quadrupole collectivity and deformation when the number of interacting protons and neutrons becomes larger. A detailed shell-model description is presented in a series of articles (Heyde, 1986; Heyde et al., 1987, 1988).

6.5.1.2. IBM-2 Configuration Mixing: A Consistent κ Approach

In order to evaluate the above collective contribution $\Delta E_{\text{coll.}}$, we can use the IBM-2 approach to describe intruder excitations near

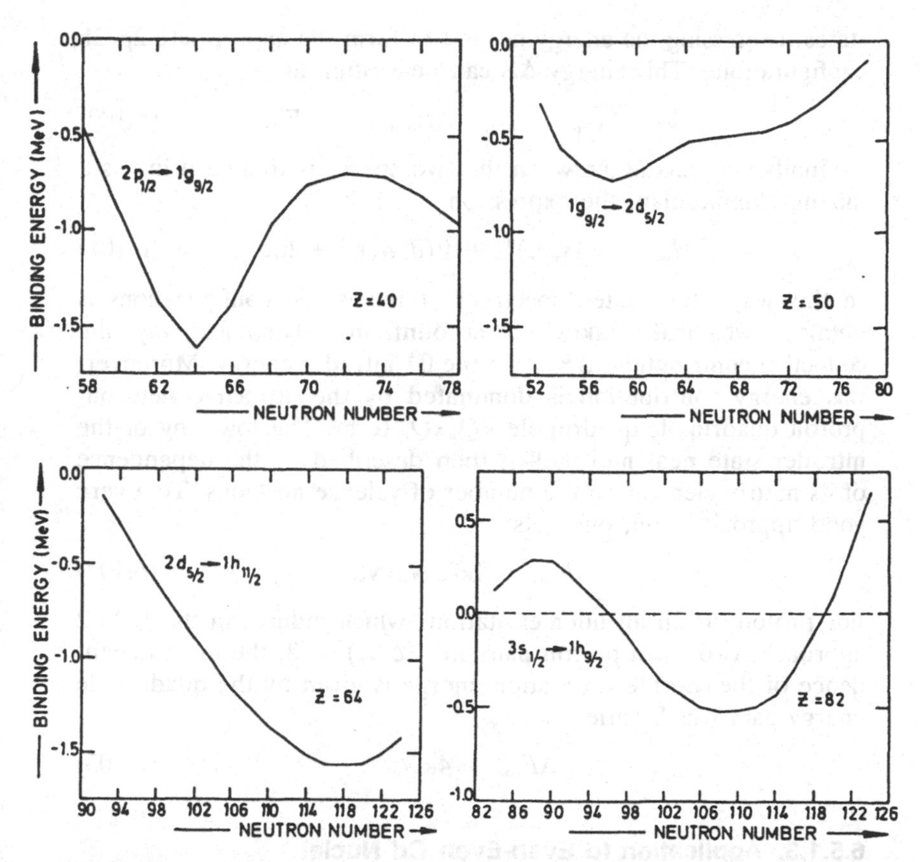

FIGURE 6.41. *The monopole binding energy ΔE_M correction (see Heyde, (1987, 1988)) for the Z = 40,64 subshell and Z = 50,82 shell closures. In each case, the type of excitation across the shell (subshell) closure $j_\pi \rightarrow j'_\pi$ is indicated.*

closed shells. Then, as originally discussed by Duval and Barrett (1982), the normal configuration space is described as an interacting system of N_ν neutron bosons and N_π proton bosons. The intruder configuration is then described by N_ν neutron bosons interacting with $N_\pi + 2$ proton bosons, the extra two bosons describing the 2p–2h core excited configuration. In this approximation, no distinction is made between the particle and hole configurations that form bosons in different major shells. To obtain the final energy spectrum the IBM-2 Hamiltonian is diagonalized for both configurations, separately. An energy ΔE has to be added for the intruder states, expressing the energy needed to form the appropriate 2p–2h configurations. This energy ΔE can be written as

$$\Delta E \equiv 2(\epsilon_p - \epsilon_h) - \Delta E_{\text{pairing}} + \Delta E_M. \qquad (6.100)$$

Finally, the mixing between the two spaces is treated using as a mixing Hamiltonian, the expression

$$H_{\text{mix.}} = \alpha(s_\pi^\dagger s_\pi^\dagger)^{(0)} + \beta(d_\pi^\dagger d_\pi^\dagger)^{(0)} + \text{h.c.} \qquad (6.101)$$

In this way, the excited spectrum of the mixed configurations is obtained which also takes into account, in a dynamical way, the collective contribution, $\Delta E_{\text{coll.}}$, to the 0^+ intruder energy. Moreover, this energy contribution is dominated by the attractive neutron-proton quadrupole-quadrupole $\kappa Q_\pi \cdot Q_\nu$ term. The lowering of the intruder state near mid-shell is then described by the dependence of its matrix element on the number of valence nucleons. To a very good approximation, one gets

$$\Delta E_{\text{coll.}} \cong 2\kappa(\Delta N_\pi)N_\nu. \qquad (6.102)$$

For proton 2p–2h intruder excitations which induce, in the IBM-2 approach, two extra proton pairs i.e. $(\Delta N_\pi) = 2$, the main dependence of the intruder excitation energy is given by the quadrupole energy gain which varies as

$$\Delta E_{\text{coll.}} \cong 4\kappa N_\nu. \qquad (6.103)$$

6.5.1.3. Application to Even-Even Cd Nuclei

We shall apply the above method to describe the even-even Cd nuclei in which compelling evidence for the observation of 0^+ intruder states has been obtained. Besides the regular quadrupole phonon triplet of states (0^+, 2^+, 4^+) extra 0^+ and 2^+ states show up, pointing towards the importance of proton 2p–2h intruder configurations.

Since the matrix element of the quadrupole force shows a very specific and strong dependence (see eq. (6.103)) on the number of interacting nucleons, the large gain in binding energy $4\kappa N_\nu$ for the 2p–2h intruder configuration can supply the necessary explanation for the very low energy of these intruder states.

We shall now show that when one uses expression (6.97), which is derived from shell-model arguments together with a consistent determination of the IBM-2 parameters for the normal and intruder configuration, a good description of the mixing of both configurations can be obtained. The consistency is based on the fact that the IBM-2 parameters, when going from one nucleus to another within a shell, are approximately known on the basis of microscopic studies and phenomenological studies of chains of isotopes. Some parameters, such as κ, are nearly constant in a shell, while others, such as χ_ν, χ_π are rapidly changing.

The nucleus ^{114}Cd has been investigated by many authors and an almost complete set of E2 reduced matrix elements for the low-lying normal and intruder states became available recently. We therefore concentrated on this nucleus. In ^{114}Cd the normal configuration is described as an interacting system of $N_\nu = 8$ neutron bosons and one $N_\pi = 1$ proton boson with respect to the $Z = 50$ shell. The intruder configuration is described by $N_\nu = 8$ neutron bosons and $N_\pi = 3$ proton bosons. The large set of parameters occuring in such a mixing calculation gives great freedom. We determine these parameters as follows. For $H_{\text{IBM-2}}(N_\pi = 1)$ the parameters of Heyde et al. (1982) are taken with the exception of the values of χ_ν and χ_π as will be discussed below. For the $N_\pi = 3$ Hamiltonian, expression 6.100 is first used to determine the value of ΔE. The value of the 2p–2h unperturbed single-particle energy, $2(\epsilon_p - \epsilon_h) = 8.780$ MeV is determined from the experimental one-proton separation energies using the prescription given before. The pairing correction is obtained from the experimental one- and two-proton separation energies as outlined before and yields $\Delta E_{\text{pairing}} = 4.366$ MeV. Finally, the monopole correction is calculated (see Fig. 6.41 for the $Z = 50$ region) and is $\Delta E_M = -0.412$ MeV. Hence, we get $\Delta E = 2(\epsilon_p - \epsilon_h) - \Delta E_{\text{pairing}} + \Delta E_M \cong 4.0$ MeV. The last correction in expression (6.97) is due to the p-n quadrupole collectivity and will be given by the extra binding energy of the $N_\pi = 3$ configuration with respect to the $N_\pi = 1$ configuration. This contribution is essentially dominated by the neutron-proton quadrupole interaction energy, although the other interaction terms also contribute. Having determined ΔE in a precise way, we now start from the parameter set as obtained by Heyde et al. (1982). In order to move the intruder state to the right energy the value of κ has been chosen as -0.15 MeV. Finally, we have adjusted the d-boson energy ϵ_d in order to reproduce the B(E2) values as discussed below.

As mentioned before, we modified the values of χ_ν and χ_π with respect to those used in (Heyde, 1982), for the normal as well as for the intruder configuration. This is motivated by the recently measured quadrupole moments of the three lowest 2^+ states: $Q(2_i^+)$ = −0.27 eb (+0.01, −0.02), +0.69 eb (+0.03, −0.04), +0.22 eb (+0.08, −0.20) (Jolie and Heyde, 1990). The large and positive quadrupole moment of the second excited 2^+ state is in contradiction to all previous calculations where a negative quadrupole moment of −0.31 eb was obtained. Since the quadrupole moments in IBM-2 are mainly determined by the sum $\chi_\nu + \chi_\pi$, the experimental values of $Q(2_1^+)$ and $Q(2_2^+)$ are used to determine the sum $\chi_\nu + \chi_\pi$ as −0.95 for $N_\pi = 1$ and 0.25 for $N_\pi = 3$. This was done before mixing the two configurations. Furthermore, using the value for χ_ν from Heyde et al. (1982), we obtain the values $\chi_\pi = -0.05$ for the normal and $\chi_\pi = 0.75$ for the intruder states. Here, we should mention that the resulting spectra and electromagnetic properties are not very sensitive to the fact that χ_π is different in the $N_\pi = 1$ and $N_\pi = 3$ system, as long as the above sums are fulfilled.

In order to calculate the E2 transitions for the mixed states we use the quadrupole transition operator

$$T(\mathrm{E2}) = e_1(Q_\pi + Q_\nu)_1 + e_3(Q_\pi + Q_\nu)_3, \tag{6.104}$$

where $(Q_\pi + Q_\nu)_i$; $i = 1, 3$ is the quadrupole operator acting in each subsystem. The effective charge e_3 was taken to be 0.103 eb which is the value obtained in the description of the Ru isotopes. The charge e_1 was fitted in order to reproduce the $B(\mathrm{E2}; 0_1^+ \to 2_1^+)$ giving $e_1 = 0.086$ eb.

Using $\alpha = \beta = 0.08$ MeV in eq. (6.101), we finally adjust the boson energy ϵ_d for the intruder configuration taking into account experimental energies and the B(E2) values. The parameters so determined are listed in Table 6.4. When varying around $\epsilon_d = 0.3$ MeV, all features of the quintuplet of states around 1.2 MeV can

TABLE 6.4. *The parameters used in the calculations for* ^{114}Cd. *All quantities are in units of* MeV, *except for* χ_π *and* χ_ν *which are dimensionless. The parameters FS and FK are related to the Majorana strength parameters* $\xi_1 = \xi_3$ = *FS–FK,* ξ_2 = *FS where* ξ_1, ξ_2 *and* ξ_3 *can be found in Iachello and Arima* (1987).

N_π	ϵ	κ	χ_ν	χ_π	$c_{0\nu}$	$c_{2\nu}$	$c_{0\pi}$	$c_{2\pi}$
1	0.83	−0.14	−0.9	−0.05	−0.2	−0.05	0.0	0.0
3	0.30	−0.15	−0.5	0.75	−0.2	−0.05	0.0	0.2

FS = 0.06, FK = 0.12, $\alpha = \beta = 0.08$, $\Delta = 4.00$.

be reproduced although not simultaneously for the same value of ϵ_d. This is due to the fact that all five states are mixing at the same time. To show this we give in Fig. 6.42 the experimental diagonal E2 matrix elements and the theoretical values for $\epsilon_d = 0.28$, 0.30 and 0.32 MeV. In Fig. 6.43 we also show the corresponding energy spectra obtained with these three values. Of the three values the last one best reproduces the excitation energies. In general the E2 matrix elements are described in a qualitative way. Finally, we mention that in our calculation we can make the ratio $B(\mathrm{E2}; 0_3^+ \rightarrow 2_2^+)/B(\mathrm{E2}; 0_3^+ \rightarrow 2_1^+)$ become quite large due to cancelations between the normal and intruder contributions which make the numerator very small. Such very large ratios have been observed experimentally.

6.5.2. Multiple Pair Excitations and Backbending

As discussed before, the fact that the IBM only describes excitations of valence nucleons is an important short-coming of the model. The fact that one needs to incorporate fermion excitations across a major shell closure by increasing the boson number by two units in order to describe low-lying intruder excitations was illustrated in the case of ^{114}Cd in Section 6.5.1. A problem related to the finite number of bosons in a given valence space is the appearance of a cut-off on various observables, in particular in $B(\mathrm{E2})$ values as J increases. One normally obtains analogous expressions as in the geometrical models but corrected by a cut-off function which is boson number and spin dependent, i.e.,

$$B(\mathrm{E2}; J \rightarrow J-2)_{\mathrm{IBM}} = g(N, J)\, B(\mathrm{E2}; J \rightarrow J-2)_{\mathrm{BM}}. \quad (6.105)$$

There has been a lot of discussion on this point and it has become clear that the high-spin properties of nuclei cannot just be described by s and d bosons alone in the IBM. So, we discuss here an extension of the method discussed under Section 6.5.1. but now for a system where one considers $N, N+2, N+4, \ldots$ bosons. We present some applications in the deformed rare-earth region (the Dy isotopes) (see Fig. 6.44). All calculations are performed using the IBM-1.

The Hamiltonian we start from has the form

$$H = \epsilon_d(N)\hat{n}_d + \kappa''(N)P^\dagger \cdot P + \kappa'(N)J \cdot J + \kappa(N)Q \cdot Q. \quad (6.106)$$

In applying this Hamiltonian to a series of deformed nuclei, the

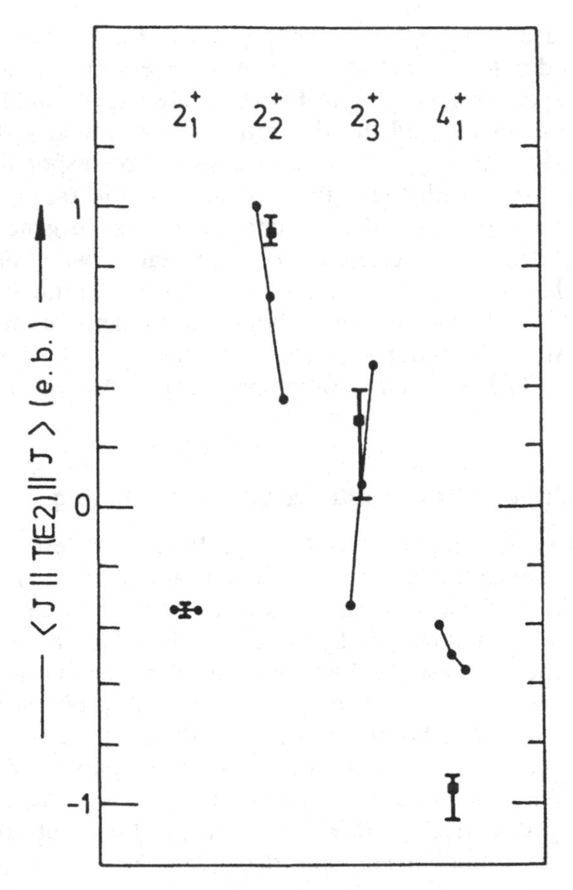

FIGURE 6.42. *The experimental* E2 *reduced matrix elements (black squares with error bar) compared with the theoretical calculations with* $\epsilon_d(N_\pi = 3)$ *equal to* 0.28 MeV *(left),* 0.30 MeV *(middle),* 0.32 MeV *(right). Results are for* ^{114}Cd *and are taken from Jolie and Heyde* (1990).

Dy isotopes, the first two terms only contribute a small perturbation to the rotational-like spectrum. The Hamiltonian (6.106) is diagonalized in a basis with non-constant boson number

$$|(sd)^N\rangle \oplus |(sd)^{N+2}\rangle \oplus |(sd)^{N+4}\rangle \oplus \dots . \qquad (6.107)$$

In practice, this basis has to be truncated which is done in most cases at $N_{max} = N + 6$. This truncation is sufficient in describing levels with spin up to $J \lesssim 30$. The Hamiltonian coupling the different parts of the basis (6.107) has the form as (6.101). Moreover,

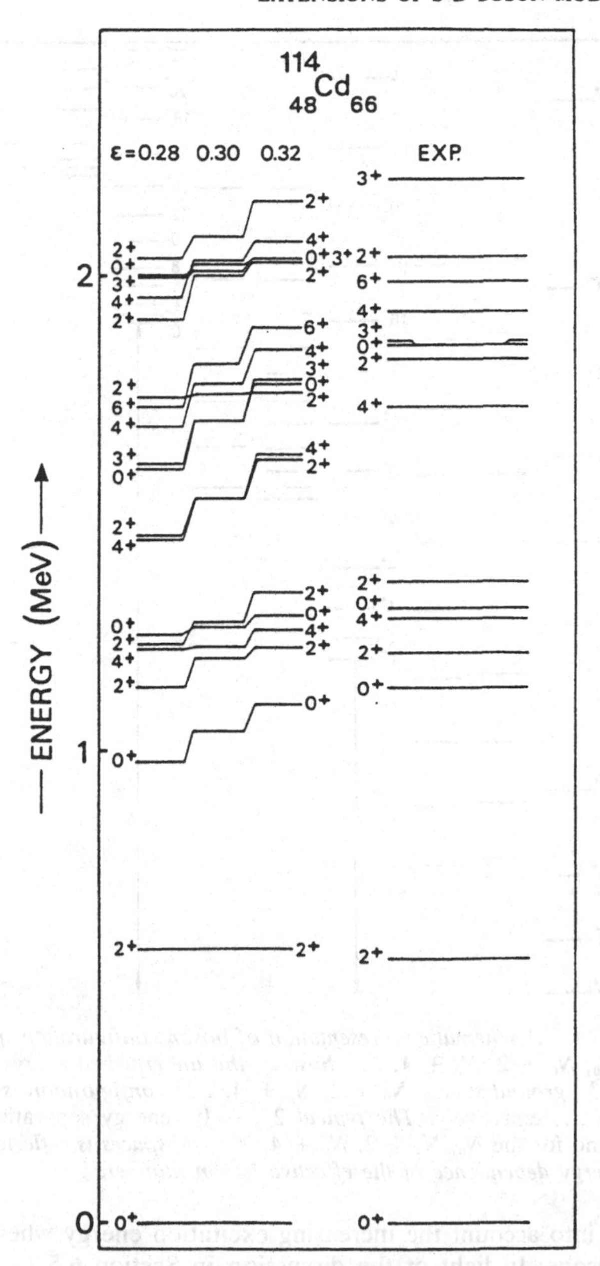

FIGURE 6.43. *Comparison between the calculated energy levels (left) with* $\epsilon_d(N_\pi = 3)$ *equal to* 0.28 MeV *(left),* 0.30 MeV *(middle),* 0.32 MeV *(right) and the experimental ones. Taken from Jolie and Heyde* (1990).

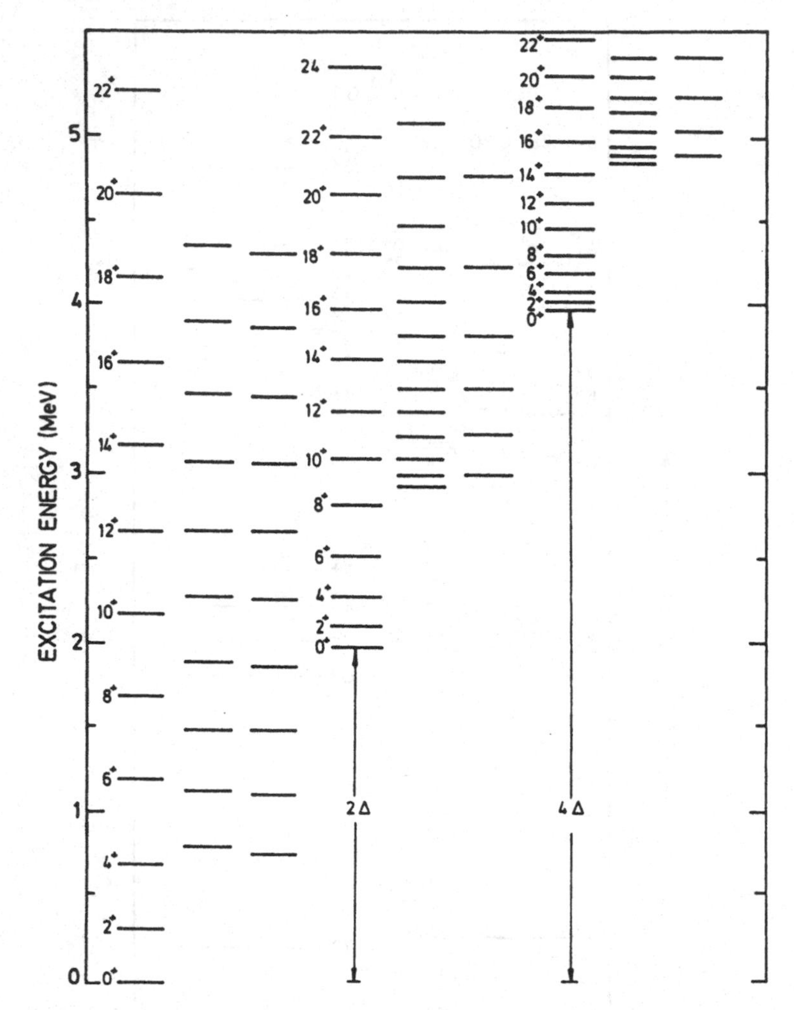

FIGURE 6.44. *A schematic representation of boson configuration space containing N_0, $N_0 + 2$, $N_0 + 4$, . . . bosons. the unperturbed energy to excite from the N_0 ground state a $N_0 + 2$, $N_0 + 4$, . . . configuration is given by* 2Δ, 4Δ, . . . *respectively. The typical* $2^+ - 0^+$ energy separation in the lowest band for the N_0, $N_0 + 2$, $N_0 + 4$, . . . *subspaces is reflected via the boson energy dependence of the effective boson number.*

we take into account the increasing excitation energy when adding extra bosons. In light of the discussion in Section 6.5.1., this can be written as $\Delta\hat{N}$ (with $\hat{N}$ the total boson number operator).

Performing the calculations, one gets in principle a large number of parameters because of the N-dependence. Applying this to the study of the $^{154-164}$Dy isotopes, a number of simplifying assumptions can be used

(i) κ, κ', κ'': independent of N (with the exception of ^{154}Dy)

(ii) we choose a linear variation in N for ϵ_d

$$\epsilon_d = \theta(N_{\max} - \hat{N}). \tag{6.108}$$

The latter dependence expresses the fact that a $U(5) \rightarrow SU(3)$ transition results as the number of active bosons increases. For a more detailed discussion of the parameters, and how Δ and α, β were determined, we refer to (Heyde, 1984).

In Fig. 6.45, we compare the x-component of the total spin (J_x) versus the rotational frequency $\hbar\omega$ for the ground-band both for the data in the Dy isotopes and for the calculations. For large values of J, one has the approximate relations

$$\begin{aligned} J_x &= [J(J+1) - K^2]^{1/2} \approx J + \tfrac{1}{2} \quad \text{for } J^2 \gg K^2, \\ \hbar\omega &\cong \tfrac{1}{2}\,(E(J) - E(J-2)). \end{aligned} \tag{6.109}$$

For the lighter Dy nuclei, one clearly observes the backbending around spin $J = 16$ which disappears for the heavier nuclei. In Fig. 6.46, we show the B(E2) values in the ground-band as a function of spin, using the standard E2 operator. The strength e and the χ value are determined to obtain optimal agreement both in magnitude and in shape. Up to spin 16, a good reproduction results. The dip in the curve also goes away with increasing mass of the Dy-isotopes.

We finally illustrate the diagonal matrix element $\langle J || T(\text{E2}) || J \rangle$ (proportional to the intrinsic quadrupole moment) in Fig. 6.47. The variations of both the $(d^\dagger s + s^\dagger \tilde{d})^{(2)}$ and the $(d^\dagger \tilde{d})^{(2)}$ terms are given as a function of spin J. The results show the importance of the value of χ in producing a change in sign for the intrinsic quadrupole moment.

So, one observes that the phenomena related to backbending, B(E2) cut-off and the like can be accommodated within the IBM by no longer using a strict boson number N. Still, one has to remark that a large number of parameters must then be used in the calculations. The results have to be taken with some caution in comparing with the experimental data and phenomena like band-

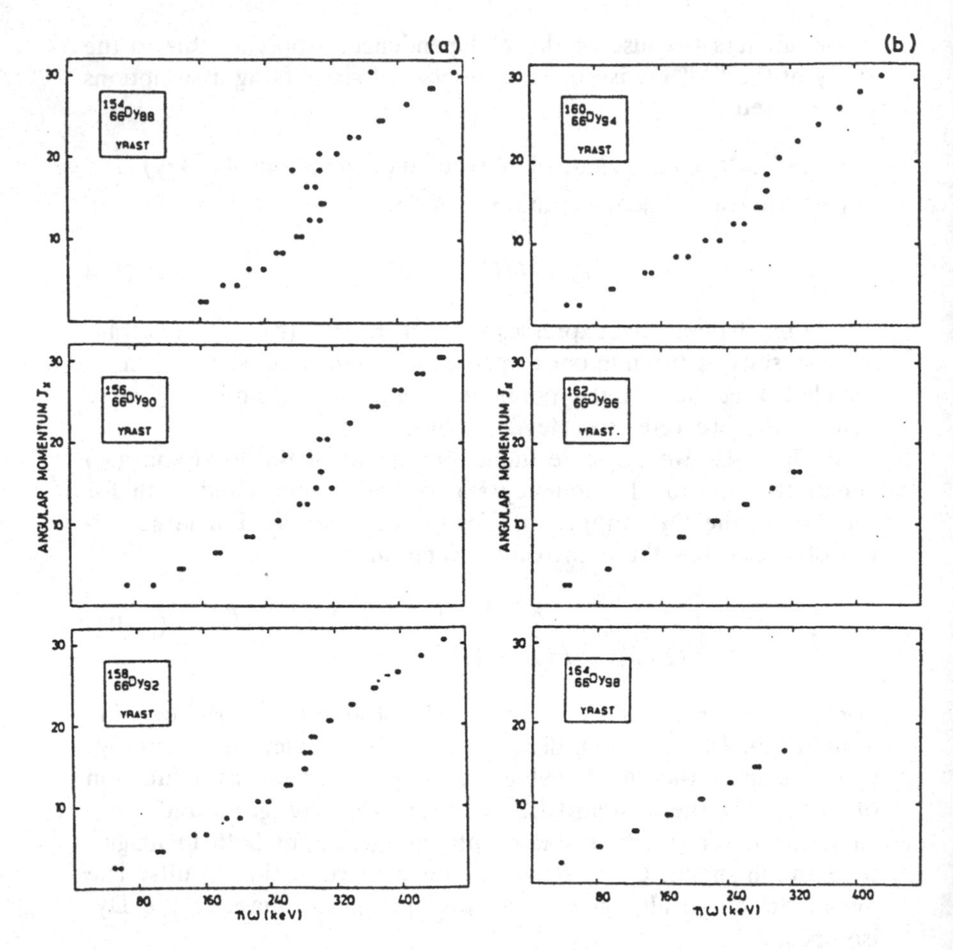

FIGURE 6.45(a),(b). *Alignment plots, displaying the "x-component" of the total angular momentum* [*according to eq.* (6.109)] *versus the "rotational" frequency* $\hbar\omega$, *deduced from the excitation energies using relation* (6.109) *for* $^{154\text{-}164}$Dy. *The theoretical values are given by the open circles whereas the experimental data are denoted by the filled circles.*

crossing frequencies, since ground-state band B(E2) values appear to be rather sensitive to the parameter choices.

6.6 Conclusion

In the present chapter, we have discussed possible extensions to the highly successful *sd*-interacting boson model. The main reason

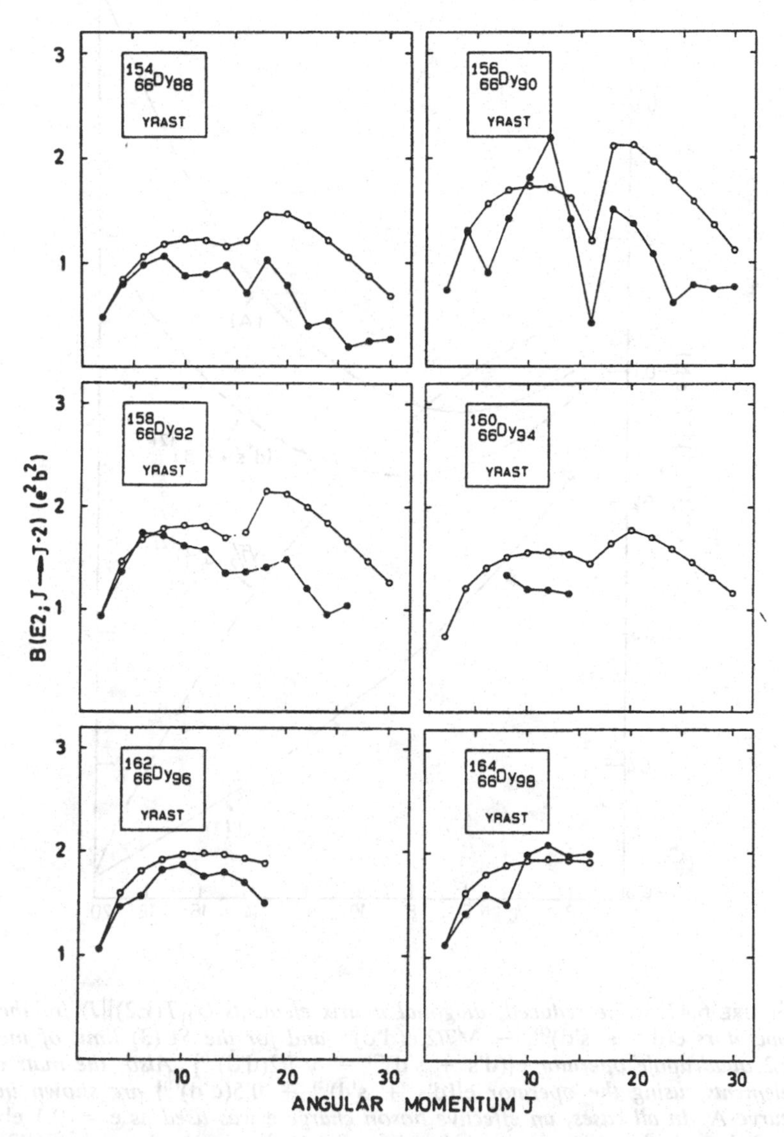

FIGURE 6.46. *Detailed comparison for the* yrast *B*(E2) *values in* $^{154-164}$Dy *between calculated results (open circles) and experimental data (filled circles). The* $B(\mathrm{E2}; J \rightarrow J-2)$ *values are given in units of* e^2b^2.

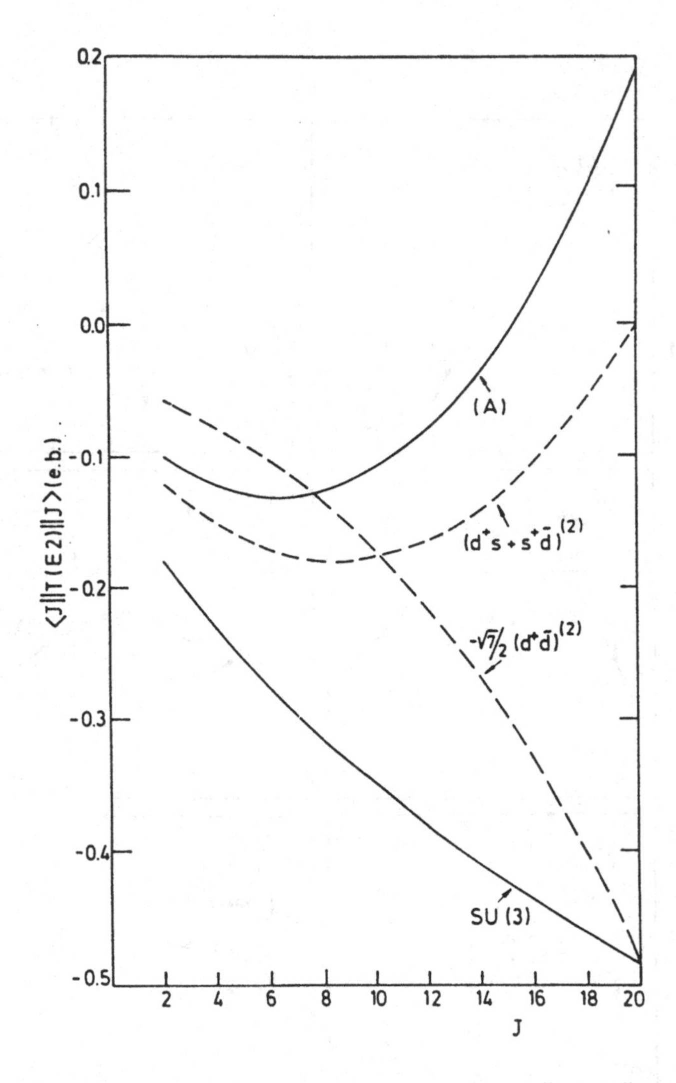

FIGURE 6.47. *The reduced, diagonal matrix elements* $\langle J\|T(E2)\|J\rangle$ *for the operators* $e(d^{\dagger}s + s^{\dagger}\tilde{d})^{(2)}$, $-\sqrt{7}/2.e(d^{\dagger}\tilde{d})^{(2)}$ *and for the SU(3) limit of the E2 quadrupole operator* $e\{(d^{\dagger}s + s^{\dagger}\tilde{d})^{(2)} - \sqrt{7}/2(d^{\dagger}\tilde{d})^{(2)}\}$. *Also, the matrix elements, using the operator* $e\{(d^{\dagger}s + s^{\dagger}\tilde{d})^{(2)} + 0.5(d^{\dagger}\tilde{d})^{(2)}\}$ *are shown in curve* A. *In all cases, an effective boson charge* e *was used as* e = 0.1 eb *and the calculations were carried out for* $N=10$ *bosons in the exact* $SU(3)$ *limit.*

for including the g-boson, or the hexadecapole degree of freedom, is the indication that in nuclei near closed shells ($U(5)$, $O(6)$ type nuclei), pair breaking and the appearance of 4^+ coupled nucleon pairs are well known. In strongly deformed rare-earth nuclei ($SU(3)$ type nuclei), ground-state stable hexadecapole deformation is exhibited and so points towards including the g-boson in the interacting boson model description of nuclear collective motion.

We have studied the effects of including one g-boson in the different dynamical symmetries. Analytical results have been obtained that give insight in how the addition of the g-boson modifies both the energy spectra and the electromagnetic E2 and E4 decay properties. More detailed numerical studies were carried out for the Ru nuclei and the Gd nuclei.

In the discussions of Sections 6.2.1 and 6.2.2 we concentrated on modifications implied by a single g-boson. Extensions to many g-boson excitations were also discussed, in particular using a group-theoretical approach in Section 6.2.3.

In Section 6.3, the introduction of negative parity bosons within the IBM was discussed. The octupole and dipole degrees of freedom (f and p bosons) are instrumental for the description of low-lying octupole excitations (vibrations, octupole $K^\pi = 0^-, 1^-, 2^-, 3^-$ bands) and high-lying dipole states (giant-dipole resonances), respectively. Besides a number of numerical studies, the group structure and dynamical symmetries were highlighted. Here, we also mentioned the rich spectrum of possible excitation modes, resulting from clustering (e.g., α-clustering, . . .), in particular in its applications to heavy nuclei. The nuclear vibron or $U(4)$ model was outlined as well as its two dynamical symmetries. The mutual interaction between the sd bosons ($U(6)$) and the dipole motion ($U(4)$) was discussed within the coupled $U(6) \otimes U(4)$ algebraic structure.

The lack of shapes within the sd-interacting boson model with potential minima for $\gamma \neq 0°$ and $60°$ (a result that is obtained using methods that relate given algebraic structures to a geometrical shape) represents a significant limitation. Using the method of coherent states, the relation between algebraic models and geometrical shapes has been discussed at some length. It was then pointed out that the extension of the sd-IBM Hamiltonian to include cubic terms allows for the introduction and description of potentials with minima for $\gamma \neq 0°$ or $60°$. Besides a detailed study of modifications in the dynamical $U(5)$, $O(6)$ and $SU(3)$ symmetries through the cubic terms, a detailed numerical study for ^{104}Ru was performed. A rather close correlation with results, studied in Section 6.2, extending the IBM by including the g-boson, stands out.

Finally, in Section 6.5, the importance of nucleon pair excitations across closed shells is discussed. In particular, the question of 0^+ intruder states near (or at) closed shells ($Z = 50$, $Z = 82$) was presented, within the context of the IBM. We discussed as consistent an approach as possible, connected to the underlying shell-model description of intruder 2p–2h excitations. An attempt was also made, using a non-constant boson number or, equivalently, including mp–mh excitations ($m = 0, 2, 4, 6, \ldots$) in the boson model space, to handle problems like back-bending and B(E2) cut-offs in ground-state band E2 transitions.

Even though extending a highly successful and elegant algebraic structure, the sd-IBM boson model, might seem disturbing, we have pointed out that some extensions (the hexadecapole degree of freedom explicitly or, implicitly using cubic interactions, and the need for nucleon pair excitations, across closed shells) keep most of the simplicity and elegance of the IBM and at the same time, extend the scope and range of applications significantly.

APPENDIX 6.1

In Section 6.2.1.2, we use perturbation theory to study the effects of g-boson admixtures in the IBM where the $U(5)$ limit is considered. Here, we evaluate in detail the perturbed $|4_1^+\rangle$ wave function of eq. (6.12), starting from the mixing Hamiltonian of eq. (6.9).

The first-order perturbation theory result gives

$$a_{\text{mix.}} = \langle s^{N-2}d^2, 4^+|\zeta H_{\text{mix.}}(\Sigma, \Gamma)|gs^{N-1}, 4^+\rangle \frac{1}{\Delta E}, \qquad (A.1)$$

where ΔE is the energy denominator i.e. $\Delta E = 2(\epsilon_d - \epsilon_s) - \epsilon_g$, and has a negative value. In a first step, the H_{mix} coupling Hamiltonian is recoupled in order to separate the g-boson from the sd-operators. One then finds

$$a_{\text{mix.}} = \frac{\zeta}{\Delta E}\frac{1}{\sqrt{9}}\langle s^{N-2}d^2, 4^+|\tilde{g}\cdot(s(d^\dagger d^\dagger)^{(4)})^{(4)}|gs^{N-1}, 4^+\rangle. \quad (A.2)$$

Now, we use Racah-algebra reduction rules (see Heyde (1990)) to separate the matrix element (A.2) into a part depending on the g-bosons and another part depending on the sd-bosons. We obtain

$$a_{\text{mix.}} = \frac{\zeta}{\Delta E}\frac{1}{\sqrt{9}}\begin{Bmatrix}0 & 4 & 4\\ 0 & 4 & 4\end{Bmatrix} \times \langle s^{N-2}d^2, 4^+||(s(d^\dagger d^\dagger)^{(4)})^{(4)}||s^{N-1}, 0^+\rangle\langle 0^+||\tilde{g}||g\rangle. \tag{A.3}$$

Inserting the values of the Wigner 6j-symbol and the g-boson reduced matrix element $\langle 0||\tilde{g}||g\rangle$, where $|0\rangle$ is the vacuum for g-bosons, the mixing amplitude is

$$a_{\text{mix.}} = \frac{\zeta}{\Delta E}\frac{1}{9}\langle s^{N-2}d^2, 4^+||(s(d^\dagger d^\dagger)^{(4)})^{(4)}||s^{N-1}, 0^+\rangle. \tag{A.4}$$

Again, using Racah-algebra reduction rules we separate the s-boson part from the d-boson part (see Heyde (1990)) and find

$$a_{\text{mix.}} = \frac{\zeta}{\Delta E}\begin{Bmatrix}0 & 4 & 4\\ 0 & 0 & 0\\ 0 & 4 & 4\end{Bmatrix}\langle s^{N-2}||s||s^{N-1}\rangle\langle d^2, 4^+||(d^\dagger d^\dagger)^{(4)}||d^0, 0^+\rangle. \tag{A.5}$$

Inserting the values of the Wigner 9-j symbol (1/9), the s-boson reduced matrix element $\sqrt{N-1}$, the mixing amplitude can be written as

$$a_{\text{mix.}} = \frac{\zeta}{\Delta E}\frac{1}{9}\sqrt{N-1}\langle d^2, 4^+||(d^\dagger d^\dagger)^{(4)}||d^0, 0^+\rangle. \tag{A.6}$$

Using reduction rules from Racah-algebra, by inserting a complete set of d-boson states between the two $d^\dagger$ creation operators in (A.6), we finally obtain the expression

$$a_{\text{mix.}} = \frac{\zeta}{\Delta E}\frac{1}{9}\sqrt{N-1}\sqrt{9}\langle d^2, 4^+||d^\dagger||d, 2^+\rangle \times \langle d, 2^+||d^\dagger||d^0, 0^+\rangle\begin{Bmatrix}2 & 2 & 4\\ 4 & 0 & 2\end{Bmatrix}. \tag{A.7}$$

Filling in the remaining reduced d-boson matrix elements $\sqrt{2}$, $\sqrt{9}$ and $\sqrt{5}$, respectively, as well as the value of the Wigner 6-j symbol $1/\sqrt{5}\cdot 1/\sqrt{9}$, the final expression for $\alpha_{\text{mix.}}$ reads

$$a_{\text{mix.}} = \frac{\zeta}{\Delta E}\sqrt{\frac{2}{9}}\sqrt{N-1}. \tag{A.8}$$

APPENDIX 6.2

In this Appendix, we give a short outline of the notation used in the present chapter, mainly relating to the Racah-algebra representing operators (fermion, boson) and the various tensor products of these operators. In particular, we shall indicate where the present notation differs from similar quantities as used throughout this book but mainly in Chapter 2.

6.2.1. Operators and Racah-Algebra

$b^{\dagger}_{l,m}$: creation-operator of a boson with angular momentum l and projection m, (B.1)

$b_{l,m}$: Hermitian conjugate to $b^{\dagger}_{l,m}$, (B.2)

$$\tilde{b}_{l,m} = (-1)^{l-m} b_{l,-m}, \tag{B.3}$$

$s^{\dagger}$, $d^{\dagger}$, $g^{\dagger}$: $b^{\dagger}_{l}$ $(l = 0, 2, 4)$, (B.4)

$s^{\dagger}_{\rho}(d^{\dagger}_{\rho})$: $(\rho = \pi, \nu)$: proton- or neutron $s(d)$ boson, (B.5)

$a^{\dagger}_{j,m}$: fermion creation operator for a state with angular momentum j, projection m, (B.6)

$a_{j,m}$: Hermitian conjugate to $a^{\dagger}_{j,m}$, (B.7)

$\tilde{a}_{j,m} = (-1)^{j+m} a_{j,m}$ (the phase factor $(-1)^{j-m}$ is used in Chapter 2 by Lipas), (B.8)

$$B^{l'm'}_{lm} = b^{\dagger}_{l,m} b_{l',m'}, \tag{B.9}$$

$$[T^{(k_1)}T^{(k_2)}]^{(k)}_{\kappa} = \sum_{\kappa,\kappa'} \langle k_1\kappa, k_2\kappa' | k\kappa\rangle T^{(k_1)}_{\kappa} T^{(k_2)}_{\kappa'}. \tag{B.10}$$

The tensor and tensor coupling notation used here differs from the notation used by Lipas in Chapter 2 where tensors are denoted as $T_{k\kappa}$ and a tensor product by $[T_{k_1}T_{k_2}]_{k\kappa}$. The scalar (or dot) product

$$\begin{aligned} T^{(k)} \cdot T^{(k)} &= (-1)^k \sqrt{2k+1}\,[T^{(k)}T^{(k)}]_0^{(0)} \\ &= (-1)^\kappa \sum_\kappa T_\kappa^{(k)} T_{-\kappa}^{(k)}, \end{aligned} \tag{B.11}$$

has the same definition. The notation of reduced matrix elements, denoted by

$$\langle aJ||T^{(k)}||\alpha'J'\rangle \quad \text{(Heyde, (1990))}, \tag{B.12}$$

is the same as in Chapter 2.

We can now define the angular momentum coupled boson operators

$$B_\kappa^{(k)}(ll') = [b_l^\dagger \tilde{b}_{l'}]_\kappa^{(k)}, \tag{B.13}$$

$$F_\kappa^{(k)}(ll') = (1 + \delta_{ll'})^{-1}[B_\kappa^{(k)}(ll') + B_\kappa^{(k)}(l'l)], \tag{B.14}$$

with the boson number operator $\hat{n}_l$ defined as

$$\hat{n}_l = \sqrt{2l+1}(b_l^\dagger b_l)_0^{(0)} = b_l^\dagger \cdot \tilde{b}_l. \tag{B.15}$$

6.2.2. Boson Hamiltonians—Casimir Invariants

Various expressions, denoting a general IBM-1 Hamiltonian, have been used in the literature (see Chapter 2). A very useful form is the multipole decomposition,

$$\begin{aligned} H = {} & \epsilon_d d^\dagger \cdot \tilde{d} + \kappa'' P^\dagger \cdot P + \kappa' J \cdot J \\ & + \kappa Q \cdot Q + q_3 T_3 \cdot T_3 + q_4 T_4 \cdot T_4, \end{aligned} \tag{B.16}$$

which is used in Chapter 2 but with a slightly different parameter notation i.e. eq. (2.167)

$$\begin{aligned} H_{\text{Mult.}} = {} & \epsilon'' d^\dagger \cdot \tilde{d} + a_0 P^\dagger \cdot P + a_1 J \cdot J \\ & + a_2 Q \cdot Q + a_3 T_3 \cdot T_3 + a_4 T_4 \cdot T_4. \end{aligned} \tag{B.17}$$

The relations, connecting the two Hamiltonians, read

$$
\begin{aligned}
\epsilon'' &= \epsilon_d && \text{EPS}, \\
a_0 &= \kappa'' && 2\ \text{PAIR}, \\
a_1 &= \kappa' && \tfrac{1}{2}\ \text{ELL}, \\
a_2 &= \kappa && \tfrac{1}{2}\ \text{QQ}, \\
a_3 &= q_3 && 5\ \text{OCT}, \\
a_4 &= q_4 && 5\text{HEX},
\end{aligned}
\tag{B.18}
$$

where the corresponding input name for these parameters in the code PHINT is also given.

The Casimir operators (invariant operators) and their corresponding eigenvalues used in the present chapter are also given, using the angular momentum coupled $B_\kappa^{(k)}$ operators of eq. (B.13). There are some differences with the corresponding Casimir operators as used in Chapter 2, in particular for the C_{2SU3}, C_{2O6}, C_{2O5} and C_{2O3} operators.

$$
\begin{aligned}
C_{1U6} &= B_0^{(0)}(00) + \sqrt{5}B_0^{(0)}(22) && N, \\
C_{2U6} &= B^{(2)}(02)\cdot B^{(2)}(20) \\
&\quad + B^{(2)}(20)\cdot B^{(2)}(02) \\
&\quad + B^{(0)}(00)\cdot B^{(0)}(00) \\
&\quad + \sum_{k=0}^{4} B^{(k)}(22)\cdot B^{(k)}(22) && N(N+5), \\
C_{1U5} &= \sqrt{5}B_0^{(0)}(22) && n_d, \\
C_{2U5} &= \sum_{k=0}^{4} B^{(k)}(22)\cdot B^{(k)}(22) && n_d(n_d+4), \\
C_{2SU3} &= 2Q\cdot Q + \tfrac{3}{4}J\cdot J && \lambda^2+\mu^2+\lambda\mu+3(\lambda+\mu), \\
C_{2O6} &= (B^{(2)}(02) + B^{(2)}(20)) \\
&\quad \times (B^{(2)}(02) + B^{(2)}(20)) \\
&\quad + 2B^{(1)}(22)\cdot B^{(1)}(22) \\
&\quad + 2\,B^{(3)}(22)\cdot B^{(3)}(22) && \sigma(\sigma+4),
\end{aligned}
$$

$$C_{2O5} = 2B^{(1)}(22) \cdot B^{(1)}(22) + 2B^{(3)}(22) \cdot B^{(3)}(22) \qquad \tau(\tau + 3),$$

$$C_{2O3} = J \cdot J \qquad J(J + 1). \tag{B.19}$$

The relations connecting both representations are given below where the Casimir operators in Chapter 2 are given with the index L (Lipas) i.e., $C_{nOk}(L)$, $C_{2SU3}(L)$

$$C_{2O5} \Leftrightarrow \tfrac{1}{2}\, C_{2O5}(L), \qquad C_{2O3} \Leftrightarrow \tfrac{1}{2}\, C_{2O3}(L),$$
$$C_{2SU3} \Leftrightarrow \tfrac{3}{2}\, C_{2SU3}(L), \qquad C_{2O6} \Leftrightarrow \tfrac{1}{2}\, C_{2O6}(L). \tag{B.20}$$

At the same time, we give in the tables various transformation coefficients relating the IBM-1 Hamiltonian using Casimir operators with, (i) a Hamiltonian expressed in terms of the operators $\hat{N}$, $\hat{N}^2$, $\hat{n}_d$, $\hat{D}_0$, $\hat{D}_2$, $\hat{D}_4$ characterizing the vibrational (or $U(5)$) limit and, (ii) with a Hamiltonian using the multipole expansion of eq. (B.17).

TRANSFORMATION RELATIONS

H	C_{1U6}	C_{2U6}	C_{1U5}	C_{2U5}	C_{2O5}	C_{2O3}
$\hat{N}$	1	5				
$\hat{N}^2$		1				
$\hat{n}_d$			1	5	4	6
$\hat{D}_0$				2	-8	-12
$\hat{D}_2$				2	2	-6
$\hat{D}_4$				2	2	8

(B.21)

$$\hat{D}_j = \tfrac{1}{2}\sqrt{2J+1}\,[(d^\dagger d^\dagger)^{(J)}(\tilde{d}\tilde{d})^{(J)}]_0^{(0)}$$

H	C_{2SU3}	C_{2O6}	C_{2O5}	C_{2O3}
$\hat{N}$		4		
$\hat{N}^2$		1		
$P^+ \cdot P$		-4		
$J \cdot J$	$\frac{3}{4}$		$\frac{1}{5}$	1
$Q \cdot Q$	2			
$T_3 \cdot T_3$			2	

(B.22)

6.2.3. Special Relations

Here, we give a number of very useful relations.

In the $SU(3)$ limit where the quadrupole operator eigenvalue in the $SU(3)$ basis is quite often needed, we give the precise relation between this quadrupole operator and the quadratic C_{2SU3} and $J \cdot J$ operators, as well as the eigenvalues.

$$Q \cdot Q = \tfrac{1}{2} C_{2SU3} - \tfrac{3}{8} J \cdot J, \tag{B.23}$$

$$-\kappa\langle Q \cdot Q\rangle = -\frac{\kappa}{2}[\lambda^2 + \mu^2 + \lambda\mu + 3(\lambda+\mu)] + \frac{3\kappa}{8} J(J+1), \tag{B.24}$$

$$-\kappa'\langle J \cdot J\rangle = -\kappa' J(J+1). \tag{B.25}$$

In the $O(6)$ limit, a form which is slightly different from the Hamiltonian used in Chapter 2 (eq. (2.176)) to express this particular dynamical symmetry, has been used quite often,

$$H = \frac{A}{4}\hat{N}(\hat{N}+4) - \frac{A}{4} C_{2O5} + \frac{B}{6} C_{2O5} + C C_{2O3}, \tag{B.26}$$

with corresponding eigenvalue

$$\langle H\rangle = \frac{1}{4} A\,[(N-\sigma)(N+\sigma+4)] + \frac{B}{6}\tau(\tau+3) + CJ(J+1). \tag{B.27}$$

Using a multipole expansion, we can also rewrite the Hamiltonian of eq. (B.26) as

$$\begin{aligned} H &= AP^{\dagger} \cdot P + \left(\frac{B}{30} + C\right) L \cdot L + \frac{B}{3} T_3 \cdot T_3 \\ &= a_0 P^{\dagger} \cdot P + a_1 J \cdot J + a_3 T_3 \cdot T_3. \end{aligned} \tag{B.28}$$

Two other interesting forms are

$$H = \kappa Q \cdot Q = \kappa(d^{\dagger}s + s^{\dagger}\tilde{d})^{(2)} \cdot (d^{\dagger}s + s^{\dagger}\tilde{d})^{(2)} \quad (\text{with}\, \chi = 0), \tag{B.29}$$

and

$$H = \kappa(C_{2O6} - C_{2O5}), \tag{B.30}$$

with the following expression as eigenvalue:

$$\langle H\rangle = \kappa[\sigma(\sigma+4) - \tau(\tau+3)]. \tag{B.31}$$

ACKNOWLEDGEMENTS

I am most indebted to P. Van Isacker, J. Jolie, A. Sevrin who were at the origin of the extensions discussed here (*g*-boson, cubic interactions) and to Ph. Duval and B. R. Barrett relating to including pair excitations across closed shells within the IBM. Much help was obtained from J. Moreau and G. Wenes relating to the detailed, numerical studies on triaxiality in nuclei. I would like to express my sincere thanks to P. Van Isacker without whose efforts the present chapter could not have been written. Many aspects were discussed in detail in his "Hoger Aggregaat" thesis. IBM-1 codes have been extended by. him, including (i) *g*, *s'*, *d'* bosons; (ii) *p*-bosons; (iii) triaxial terms; (iv) multi-boson excitations.

The author would like to thank NATO for research grant CRG 920011 which helped support the final stages of this work. The author would also like to thank the IIKW and the NFWO for constant support which made the present work possible. Last, but not least, he is most grateful to D.dutré-Lootens and L. Waerniers-Schepens for their expert typing and transforming my handwriting into a most elegant text, and to R. Verspille for the first-class graphics and drawings.

BIBLIOGRAPHY

Y. Akiyama, *Nucl. Phys.* **A433,** 369 (1985). Detailed discussion of the *SU*(3) limit within the *sdg*-boson model.

Y. Akiyama, K. Heyde, A Arima and N. Yoshinaga, *Phys. Lett.* **B173**, 1 (1986).

A. Arima and F. Iachello, *Ann. Phys.* (N.Y.) **99**, 253 (1976). Basic IBM article on the vibrational or *U*(5) limit. Contains many analytical results.

A. Arima and F. Iachello, *Ann. Phys.* (N.Y.) **111**, 201 (1978). Basic IBM article on the rotational or *SU*(3) limit. Contains many analytical results.

A. F. Barfield, B. R. Barrett, J. L. Wood and O. Scholten, *Ann. Phys.* (N.Y.), **182**, 344 (1988). In-depth study of the IBM model, including one *f*-boson excitations. Systematics of E3 properties in deformed, rare-earth nuclei.

A. Bohr and B. Mottelson, *Nuclear Structure,* Vol. 2 (W. A. Benjamin, Reading, 1975). Part 2 of a monumental work on nuclear structure. This volume concentrates on collective properties of the nucleus.

H. G. Börner, J. Jolie, S. J. Robinson, B. Krusche, R. Piepenbring, R. F. Casten, A. Aprahamian and J. P. Draayer, *Phys. Rev. Lett.*, **66**, 691

(1991). Evidence for the existence of two-phonon collective excitations in deformed nuclei.

G. E. Brown and A. M. Green, *Nucl. Phys.* **75**, 401 (1966); *Ibid* **85**, 87 (1966). Seminal articles discussing the multi-particle multi-hole origin to shape coexisting states in light even-even nuclei.

D. G. Burke, W. F. Davidson, J. A. Cizewski, R. E. Brown, E. R. Flynn and J. W. Sunier, *Can. J. Phys.* **63**, 1309 (1985).

R. F. Casten et al., *Nucl. Phys.* **A439**, 289 (1985).

R. F. Casten and D. D. Warner, *Revs. Mod. Phys.* **60**, 389 (1988). IBM-review article emphasizing symmetries and simplicity more than detailed analytical and numerical applications.

H. J. Daley and F. Iachello, *Ann. Phys.* (N.Y.) **167**, 73 (1986). Study of clustering in nuclei, including quadrupole deformation at the same time. In particular, the $SU(3)$ limit is used and analytic expressions for energies, transition rates and α-decay widths are discussed.

A. S. Davydov and G. F. Filippov, *Nucl. Phys.* **8**, 237 (1958). Basic article on rigid triaxial rotor model.

A. S. Davydov and A. A. Chaban, *Nucl. Phys.* **20**, 499 (1960). Studies the coupling of rotational- and vibrational excitations in non-axial nuclei.

Y. D. Devi and V.K.B. Kota, TN-90-68; Physical Research Laboratory Fortran programmes for spectroscopic calculations in (*sdg*)-boson space: the package SDG-IBM1. Fortran program manual explaining how to obtain analytical and numerical results in the *sdg*-boson model.

A.E.L. Dieperink, O. Scholten and F. Iachello, *Phys. Rev. Lett.* **44**, 1747 (1980). Relation between IBM and geometrical models, using the method of coherent states.

P. D. Duval and B. R. Barrett, *Nucl. Phys.* **A376**, 213 (1982). Basic article, introducing the methods to incorporate 2 particle–2 hole excitations within the framework of the IBM.

J. M. Eisenberg and W. Greiner, *Nuclear Theory,* Vol. 1 (North-Holland, 1987).

J. Engel, *Phys. Lett.* **171B**, 148 (1986).

R. Gilmore and D. H. Feng, *Phys. Lett.* **76B**, 26 (1978); *Nucl. Phys.* **A301**, 189 (1978); R. Gilmore, *J. Math. Phys.* **20**, 891 (1979). General description of mathematical methods to evaluate the classical limit expression for the ground-state energy in an interacting boson system.

J. N. Ginocchio and M. W. Kirson, *Phys. Rev. Lett.* **44**, 1744 (1980); *Nucl. Phys.* **A350**, 31 (1980). Classical limit of the algebraic IBM model is studied in some detail.

R. J. Glauber, *Phys. Rev.* **C130**, 2529(1963); **C131**, 2766(1963). Basic article on the introduction of coherent states. Gives detailed calculations with various applications to scattering, quantum optics, and so on.

G. Gneuss and W. Greiner, *Nucl. Phys.* **A171**, 449 (1971). Extension of the standard Bohr–Mottelson model.

P. O. Hess, M. Seiwert, J. Maruhn and W. Greiner, *Z. Phys.* **A296**, 147 (1980). Numerical studies within the extended Bohr–Mottelson model, with applications to the rare-earth region.

K. Heyde, P. Van Isacker, M. Waroquier, G. Wenes and M. Sambataro,

Phys. Rev. **C25**, 3160 (1982). Study within the IBM and within a particle-core coupling description of intruder states, with application to ^{114}Cd.

K. Heyde, P. Van Isacker, M. Waroquier, G. Wenes, Y. Gigase and J. Stachel, *Nucl. Phys.* **A398**, 235 (1983). Study of the *g*-boson, coupled to an *sd*-boson system using the $U(5)$ and $O(6)$ symmetries. Application to the transitional nucleus ^{104}Ru.

K. Heyde, P. Van Isacker, M. Waroquier, J. L. Wood and R. A. Meyer, *Phys. Repts.* **102**, 291 (1983). Review article on properties of intruder states, mainly concentrating on odd-mass nuclei.

K. Heyde, J. Jolie, P. Van Isacker, J. Moreau and M. Waroquier, *Phys. Rev.* **C29**, 1428 (1984). Backbending and high-spin states in the IBM, starting from a system with non-constant boson number. Application to the even-even Dy nuclei.

K. Heyde et al., *Nucl. Phys.* **A466**, 189 (1987); *ibid.* **A484**, 275 (1988). Shell-model description of intruder states in even-even, odd-mass and odd-odd nuclei.

K. Heyde, *The Nuclear Shell Model,* Springer series in Nuclear and Particle Physics, (Springer-Verlag, 1990). Text-book on the nuclear shell model, including chapters on angular momentum coupling and Racah-algebra.

F. Iachello, *Chem. Phys. Lett.* **78**, 581 (1981). Introduction of the vibron $U(4)$ model in the context of molecular structure, using the $\sigma(l=0)$ and $\pi(l=1)$ bosons.

F. Iachello in: *Nuclear Structure*, K. Abraham, K. Allaart and A.E.L. Dieperink, eds. (Plenum Press, New York and London, 1980), p. 53. One of the first review lecture notes on the Interacting Boson Model.

F. Iachello, *Phys. Lett.* **160B**, 1 (1985).

F. Iachello and A. Arima, *The Interacting Boson Model* (Cambridge Univ. Press, Cambridge, 1987). Complete monograph on the IBM and its various versions. Mainly accentuating analytic results.

J. Jolie and K. Heyde, *Phys. Rev.* **C42**, 2034(1990).

K. Kumar and M. Baranger, *Nucl. Phys.* **A92**, 608 (1967); *ibid.* **A110**, 490 (1968); *ibid.* **A122**, 273 (1968). Article series discussing a possible microscopic basis to the Bohr–Mottelson description of collective motion.

D. Kusnezov and F. Iachello, *Phys. Lett.* **209B**, 420 (1988); D. Kusnezov, Ph.D. Thesis, University of Princeton, (1988), unpublished. Study of the spdf IBM. The article concentrates on E1 properties in the Ba nuclei; the thesis of D. Kusnezov gives detailed derivations of the dynamical symmetries contained in $U(16)$.

G. Maino, A. Ventura, L. Zuffi and F. Iachello, *Phys. Lett.* **152B**, 17 (1985).

G. Maino, A. Ventura, P. Van Isacker and L. Zuffi, *Phys. Rev.* **C33**, 1089 (1986).

S. G. Nilsson et al., *Nucl. Phys.* **A131**, 1 (1969). Classic article on the Nilsson model: single-particle motion in a deformed field.

R. D. Ratna Raju, *J. Phys.* **G8**, 1663 (1982). Extension of the IBM to include high-spin bosons.

D. J. Rowe and F. Iachello, *Phys. Lett.* **130B**, 231 (1983). Effect of the dipole degree of freedom (*p*-boson), using the $SU(3)$ limit of the *sd*-model Hamiltonian, on the splitting of the giant dipole resonance.

O. Scholten, F. Iachello and A. Arima, *Ann. Phys.* (N.Y.) **115**, 325 (1978). Basic article, studying various transitional regions between the dynamical $U(5)$, $O(6)$ and $SU(3)$ symmetries, including the *f*-boson.

J. Stachel et al., *Nucl. Phys.* **A383**, 429 (1982).

P. Van Isacker, *Computer Codes* PBOSON and PBOSONT, unpublished, 1980. Description of the spd boson computer codes. PBOSON and PBOSONT.

P. Van Isacker, K. Heyde, M. Waroquier and G. Wenes, *Nucl. Phys.* **A380**, 383 (1982). Study of the *g*-boson, coupled to an *sd*-boson system using the $SU(3)$ symmetry. Application to the even-even Gd nuclei.

P. Van Isacker and G. Puddu, *Nucl. Phys.* **A348**, 125 (1980). IBM-2 (proton-neutron boson model) study of the even-even transitional Ru and Pd nuclei.

J. D. Vergados, *Nucl. Phys.* **A111**, 681 (1968). Extensive discussion of $SU(3) \supset O(3)$ Wigner coupling and re-coupling coefficients. Large set of tables with numerical values.

L. Wilets and M. Jean, *Phys. Rev.*, **102**, 788 (1956). Basic article on the γ-unstable rotor, within a geometrical framework.

J. L. Wood, K. Heyde, W. Nazarewicz, M. Huyse and P. Van Duppen, *Phys. Repts.* **215**, 101 (1992). Basic review article on coexistence in even–even nuclei.

H. C. Wu, *Phys. Lett.* **110B**, 1 (1982). $SU(3)$ limit of the *sdg* $U(15)$ boson model, containing a discussion on (λ,μ) representations in the $U(15)$ model.

B. G. Wyborne, *Classical Groups for Physicists* (Wiley, New York, 1974). Classical textbook on groups for physicists.

N. V. Zamfir and R. F. Casten, *Phys. Lett.* **270**, 265 (1991). A study about the possible best signatures, relating to the question on γ-softness versus triaxiality in low-energy nuclear spectra.

L. Zuffi, P. Van Isacker, G. Maino and A. Ventura, *Nucl. Instr. and Meth.* **A255**, 46 (1987). Study of photo-absorption and photon scattering in strongly deformed nuclei, using the $SU(3)$ limit of the spd IBM.

CHAPTER 7

Fermion Models

JERRY P. DRAAYER

7.1. FERMIONS VERSUS BOSONS

In this chapter we switch to a description of atomic nuclei wholly in terms of their fermion degrees of freedom. Specifically, the nucleus ${}^{A}_{Z}X_{N}$ will be considered to be a collection of Z protons and N neutrons (atomic number $A = Z + N$) that are spin $\frac{1}{2}$ fermions. These nucleons satisfy the Pauli Exclusion Principle, so spurious configurations like those that can arise in boson applications due to the neglect of the antisymmetry requirement among particles are ruled out. If pairs of nucleons couple to form objects that behave as bosons, as we held to be the case in previous chapters, it will be a consequence of the dynamics of the system of interacting fermions, and not an externally imposed condition. For light nuclei ($A \lesssim 28$) it is convenient to think of the nucleons as members of an isospin doublet ($t = \frac{1}{2}$ with $t_z = +\frac{1}{2}$ for neutrons and $t_z = -\frac{1}{2}$ for protons) while for heavy nuclei with sizable neutron excesses ($A \gtrsim 100$) an identical particle neutron-proton formalism is more appropriate. In the former case, a nucleon can exist in any one of four possible spin-isospin states, while in the latter case, only two spin states are allowed per particle type. Under either scenario, however, the many-particle shell-model picture uses basis states that change sign under an odd number of nucleon permutations. As we will see, the antisymmetry requirement of a fermion description is more difficult to deal with than a boson formulation of the dynamics. Nonetheless, since nucleons are fermions this approach affords the possibility of a truly microscopic theory of nuclear structure.

All fermion models of nuclear structure in use today are variations of the many-particle shell-model that evolved from the single-particle picture independently introduced in the late 1940s by Mayer, as well as by Haxel, Jensen and Suess. From the earliest days, this development was driven by the inability of the single-particle picture to account for observed collective phenomena in nuclei, particularly enhanced E2 transition rates between members of rotational bands. In fact, this failure of the simplest shell-model scheme also spurred the development of the highly successful and complementary collective model. While a collective model description—which parameterizes nuclei in terms of shape variables rather than individual nucleon degrees of freedom—may be too simple for an in-depth understanding of a multi-modal nuclear system, it is unquestionably a very successful theory because it provides an elementary, first-level understanding of a large volume of nuclear data.

An algebraic fermion model is a many-particle shell-model theory which treats nucleons within a nucleus as fermions (in contrast with the IBM which treats nucleon pairs as bosons) and invokes special group symmetries associated with collective behavior (like the IBM) in an attempt to achieve a tractable microscopic description of nuclear phenomena. Several different schemes will be described in this chapter; all will be identified by the generic Algebraic Fermion Model (AFM) label. One scheme, the so-called Fermion Dynamical Symmetry Model, exploits a different realization of the single-particle orbital structure than the others and will therefore be identified with the FDSM acronym used for it in the literature. The others have the oscillator $SU(3)$ symmetry (realized in different ways) as a common ingredient. For light nuclei, there is the basic Elliott model and its multi-shell extension called the Symplectic Shell Model (SSM). The latter, which is sometimes also referred to in the literature as the Microscopic Collective Model (MCM), has a rich and convoluted history ranging from work on collective-coordinate models to investigations into complementary group structures. For heavy nuclei, the pseudo-spin concept can be exploited to gain another realization of $SU(3)$—the so-called pseudo-$SU(3)$ model—and its multishell extension called the pseudo-symplectic scheme.

The main difficulty in developing a many-particle shell-model theory stems from the fact that while many properties of the interaction between nucleons are known from nucleon-nucleon scattering data, their renormalization in the nuclear medium is not at all well understood. Furthermore, even if a suitable nucleon-nucleon interaction was known, further renormalization is required due to basis truncation because typical dimensionalities—which go as the binomial $\binom{n}{m} \approx n^m$ for m identical nucleons in n levels—are too large for the best modern or presently foreseeable computers. The end result is that neither the interaction nor the model space can be separately or sharply specified. This being the case, one might question the feasibility of gaining an in-depth understanding of nuclear phenomena, however, it is this very dilemma which makes the nuclear problem so intriguing. Because of these uncertainties, one has license to be creative in seeking solutions to what remains one of nature's most challenging puzzles—systems that display ordered motion demonstrated by single-particle and many-particle collective phenomena as well as chaotic behavior. A pleasant surprise is that the nucleus itself seems to filter out much of the noise generated by the nucleon-nucleon interaction so only the strongest features, such as rotational collectivity, survive. Algebraic models exploit the fact that certain modes normally dominate and the associated set of operators usually define special group symmetries. These symmetries can be used to partition the full space into subspaces which can be ordered by relative importance, thereby allowing a reasonable criterion for basis truncation.

7.1.1. Statistics and the Fatal Flaw

The boson approximation works well for many nuclei for a very simple reason that goes beyond the fact that nucleons like to form s, d and even g pairs: the number of valence nucleons is small compared to the number of available levels. For example, if ^{20}Ne is considered to be a system of 4 spin $\frac{1}{2}$ particles outside a closed ^{16}O core, then the number of available levels, ignoring all symmetries except the one associated with a permutation of the 4 valence particles, is simply $\binom{12}{4} = 495$. Hence, an estimate for the probability of a individual nucleon feeling the effect of the exclusion principle is 4/495, or about 1 part in 125, which is simply a statistical prediction for the fractional occupancy of the levels. However, if other symmetries such as total angular momentum and isospin are taken into account, the number of distinguishable arrangements is reduced significantly. For example, in the ^{20}Ne case the number of $(J^\pi, T) = (0^+, 0)$ states is only 21, so the probability of seeing effects due to the Pauli Principle for this configuration is increased from 1 part in 125 to about 1 part in 5. An extreme case is the ^{20}Ne $(J^\pi, T) = (8^+, 0)$ configuration, which occurs only once.

Since the simplest s and d boson picture, with no broken pairs taken into account, predicts 2 states for ^{20}Ne with $J^\pi = 0^+$ and none for $J^\pi = 8^+$, one might conclude that the boson approximation underestimates the number of available levels. This follows because the grouping of nucleons into s and d pairs is overly restrictive, and other allowed two-nucleon configurations, like spin-triplet pairs, are ruled out. Because of the pair restriction, it is easy to understand why the number of distinct boson configurations is less than the number of fermion states when the fractional occupancy is small. On the other hand, another problem occurs (the boson ansatz predicts new configurations that are forbidden by the Pauli Principle) when the fractional occupancy approaches the mid-shell $\frac{1}{2}$ value. Specifically, while the simplest boson picture predicts a leading $SU(3)$ irrep with $(\lambda,\mu) = (2N,0)$ for a system of $2N$ particles (N bosons), for $2N > k\Omega/3$ (where $k\Omega$ is the *total* shell degeneracy) this irrep is not an allowed configuration. To see how this comes about, we note that for the leading $SU(3)$ symmetry $\lambda = n_z - n_x$ and $\mu = n_x - n_y$ where n_α counts the total number of quanta, with two per boson, in the α-th direction. Up to $2N = k\Omega/3$ the quanta can all go into the "z" direction, so $(\lambda, \mu) = (2N, 0)$. However, when the number of quanta exceeds one-third the maximum $(k\Omega/3)$ the balance up through $2k\Omega/3$ must go into the "x" direction, and hence $\lambda = k\Omega/3 - (2N - k\Omega/3) = 2k\Omega/3 - 2N$ and $\mu = 2N - k\Omega/3$ which at the half-shell $2N = k\Omega/2$ point take on the values $(\lambda, \mu) = (k\Omega/6, k\Omega/6)$. For $2N > 2k\Omega/3$ the "y" direction plays a role, and the leading irrep is $(0, 2\bar{N})$ where $2\bar{N} = k\Omega - 2N$ is the hole equivalent of the particle configuration. The full story is laid out in Table 7.1. While this has become known as the *fatal flaw* of the $SU(3)$ limit of the IBM theory, it is only fatal in the sense that the bosons are interpreted as paired fermions. For the mid-shell region $(k\Omega/3 < 2N < 2k\Omega/3)$ the leading representation is $(\lambda, \mu) = (2k\Omega/3 - 2N, 2N - k\Omega/3)$, and not $(2N, 0)$. To avoid spurious results, the unallowed representations should be excluded from the basis.

The fact that the boson picture underestimates the multiplicity of configurations when the number of pairs is small compared to the

TABLE 7.1. *Leading representations of* SU(3) *for a coupling scheme in which a total of* N $l = 0$ *and* $l = 2$ *fermion pairs are represented by* s *and* d *bosons, respectively. If the total shell degeneracy is* $k\Omega$, *then the maximum number of oscillator quanta is also* $k\Omega$ *since there are two quanta per boson. The maximum number of quanta in any one of the three cartesian directions must be* $k\Omega/3$; *and for the leading irrep,* $\lambda = n_z - n_x$ *and* $\mu = n_x - n_y$ *where* n_α *counts the number of quanta in the* α*-th direction* ($n = n_x + n_y + n_z$). *A particle-hole relationship,* $2\tilde{N} = k\Omega - 2N$, *applies for* $2N > k\Omega/2$. *General analytic expressions for the* (β, γ)*-shape variables are:* $\gamma = \tan^{-1}[\sqrt{3}(\mu + 1)/(2\lambda + \mu + 3)]$ *and* $(\beta/\beta_0)^2 = (\lambda^2 + \lambda\mu + \mu^2 + 3\lambda + 3\mu + 3)$.

$2N(2\tilde{N})$	λ	μ	Shape	γ	$(\beta/\beta_0)^2$
$0 \le 2N \le k\Omega/3$	$2N$	0	prolate	$0°$	$4N^2 + 6N + 3$
$k\Omega/3 \le 2N \le k\Omega/2$	$2k\Omega/3 - 2N$	$2N - k\Omega/3$		$0° < \gamma < 30°$	$4N^2 + k^2\Omega^2/3 - 2Nk\Omega + k\Omega + 3$
$2N = k\Omega/2 = 2\tilde{N}$	$k\Omega/6$	$k\Omega/6$	asymmetric	$30°$	$k^2\Omega^2/12 + k\Omega + 3$
$k\Omega/3 \le 2\tilde{N} \le k\Omega/2$	$2\tilde{N} - k\Omega/3$	$2k\Omega/3 - 2\tilde{N}$		$30° < \gamma < 60°$	$4\tilde{N}^2 + k^2\Omega^2/3 - 2\tilde{N}k\Omega + k\Omega + 3$
$0 \le 2\tilde{N} \le k\Omega/3$	0	$2\tilde{N}$	oblate	$60°$	$4\tilde{N}^2 + 6\tilde{N} + 3$

shell degeneracy, and introduces spurious ones when the fractional occupancy approaches the mid-shell value, should neither come as a surprise nor be considered its Achilles' heel. After all, the IBM is an approximation to the full shell-model scheme, and with its elegant simplicity comes some inherent weaknesses. As with all models, it is essential to appreciate its strengths (which in the IBM case are many) while at the same time understanding its limitations. The limitations can be dealt with in a number of ways. Missing configurations, for example, can be added by taking broken pairs into account. The simplest way to handle spurious states is to devise a means—algebraic or numeric—to remove them prior to carrying out a calculation. This requires special extra-model considerations, such as those put forward above and those which are now also a part of the FDSM (see below). A better choice of action (although much more demanding) is adopting fermion dynamics rather than a boson approximation. Unfortunately, the spurious state problem is the greatest near mid-shell where the boson description is needed most and a fermion approach leads to model spaces too large to handle on even the most sophisticated modern computational facility. This difficult region is the domain targeted by the AFMs.

It is important to appreciate that bosonic degrees of freedom still play important roles even when the occupancies are such that a fermionic calculation is required. An example, the so-called contraction limit of the symplectic model (a hybrid theory which treats particle excitations within the lowest valence shell as fermionic but intershell couplings as bosonic modes), will be presented in greater detail below. In this case the bosonic excitations are of two types, those associated with lifting nucleons out of a nearly filled core, and the scattering of valence nucleons into virtually unoccupied higher-lying spaces. The boson approximation applies in these cases, even though the fractional occupancy of the lowest set of valence orbitals may be high, because the affected spaces are either nearly full (core states so the hole occupancy is low) or nearly empty (higher-lying valence states so the probability of finding a level occupied is small) so blocking due to the Exclusion Principle is minimized. Therefore, boson dynamics are important, even within the framework of an extended AFM like the symplectic model.

7.1.2. Particle Permutation and Unitary Symmetries

The many-particle shell-model scheme starts with a single-particle picture and adds nucleons to orbitals of the valence space subject to the constraint of the Pauli Principle. For example, this means that if a space-spin wavefunction $\Phi = \psi\chi$ is used and the spatial part ψ of the product is symmetric under identical particle interchange, then the spin part χ must be antisymmetric, or vice versa (see Chapter 2). Recall that the particle-permutation symmetry of a wavefunction and its transformation properties under a basis change are directly related. As a simple example, note that a two-particle wavefunction $\psi_\alpha(1)\psi_\beta(2)$ can be transformed into $\psi_\alpha(2)\psi_\beta(1)$ either by interchanging particles 1 and 2, or by a transformation that carries state α and β and β into α. The first of these is a permutation group operation, while the second is a unitary transformation. The same is true for any linear combination of the two, like the symmetric (+) and antisymmetric (−) forms, $[\psi_\alpha(1)\psi_\beta(2) \pm \psi_\alpha(2)\psi_\beta(1)]/\sqrt{2}$, where the $\sqrt{2}$ factor is included to insure a normalized result, assuming the ψ's themselves are normalized.

The particle-permutation, unitary-transformation complementarity illustrated above for 2 particles extends to the general case, so a classification of m-particle wavefunctions into irreps of the unitary group in n dimensions ($U(n)$, where n is the dimensionality of the space) simultaneously specifies the irrep of the permutation group on m objects (S_m) to which each of the m-particle wavefunctions belongs. This $U(n)$ and S_m complementarity means that it is sufficient to give further serious consideration to only one of the two. $U(n)$ is simpler to deal with because it is a continuous group—involving familiar basis transformation and matrix concepts—so one normally focuses on it. Nonetheless, it must be understood that the permutation symmetry requirements are being satisfied. This complementarity of $U(n)$ and S_m applies to bosonic as well as fermionic schemes.

As indicated in Section 2.4, representations of $U(n)$ are labelled by a symbol $[f] = [f_1, f_2, \ldots, f_n]$ which has a simple pictorial representation called a Young diagram consisting of placing f_1 boxes directly adjacent to one another in a horizontal row, with $f_2 \leq f_1$ boxes left justified in a second row below and adjacent to the first set, etc., see eq. (2.62). As each box corresponds to one particle, $\Sigma_i f_i = m$ for an m-particle configuration. Now the actual construction of an m-particle wavefunction with a specific $[f]$ symmetry out of a set of n single-particle basis states is, in general, very complicated, and requires a much deeper understanding of the relationship between $U(n)$ and S_m than given even in Chapter 2. Fortunately, this can all be avoided by using a second quantization fermion picture, see Chapter 2. Indeed, practitioners of modern shell-model theories map the second quantized picture onto a bit-string representation of basis states, letting the computer deal with the antisymmetrization problem through elementary logical operations and rules.

If the dimensionality of the shell-model valence space is $k\Omega$ (k = 2 for spin and 4 for a spin-isospin picture) then an m-particle wavefunction must belong to the antisymmetric irrep of the unitary group in $k\Omega$ dimensions, $U(k\Omega)$. As noted above, this guarantees that it belongs to the antisymmetric irrep of the permutation group. For m particles this antisymmetric irrep is the one with all the f_i's equal to 1, $[f] = [1, 1, \ldots, 1] \equiv [1^m]$. However, the full shell-model space can be further partitioned into space and spin, or space and spin-isospin parts, with spatial dimensionality Ω. When this is done each part can have any allowed m-particle symmetry, so long as its complement has the conjugate symmetry to insure that the product remains totally antisymmetric under particle permutation (see Chapter 2).

The unitary symmetries associated with a factorization of the valence space into a direct product of two are denoted as $U(k\Omega) \supset U(\Omega) \otimes U(k)$ where $k = (2, 4)$ for the (spin, spin-isospin) cases. Overall antisymmetry in $U(k\Omega)$ requires that the $U(k)$ irrep be conjugate to the irrep of $U(\Omega)$. These irreps—$[f]$ of $U(\Omega)$ and the conjugate $[f^c]$ of $U(k)$—are related to one another by row-column interchange. This means that the $[f]$ and $[f^c]$ patterns can be determined from one another by reflection across a downward sloping 45° diagonal through the first box of the Young diagram. The direct product $U(k\Omega) \supset U(\Omega) \otimes U(k)$ decomposition is shown schematically in Fig. 7.1. For two particles, a space symmetric ($[f] = [2]$) configuration must be coupled to a spin or spin-isospin antisymmetric one ($[f^c] = [1^2]$). The first mixed symmetry case is encountered for three particles, with $[f] = [2, 1]$, which is self conjugate, $[f^c] = [2, 1]$. For 4 particles, in addition to the symmetric and antisymmetric combinations $[f] = [4]$ ($[f^c] = [1^4]$) and $[f] = [1^4]$ ($[f^c] = [4]$), there are 3 mixed symmetries: $[f] = [3, 1]$; $[2^2]$; and $[2, 1^2]$ with conjugates $[f^c] = [2, 1^2]$; $[2^2]$; and $[3, 1]$, respectively. Because $[f]$ determines $[f^c]$, and vice versa, and because the sym-

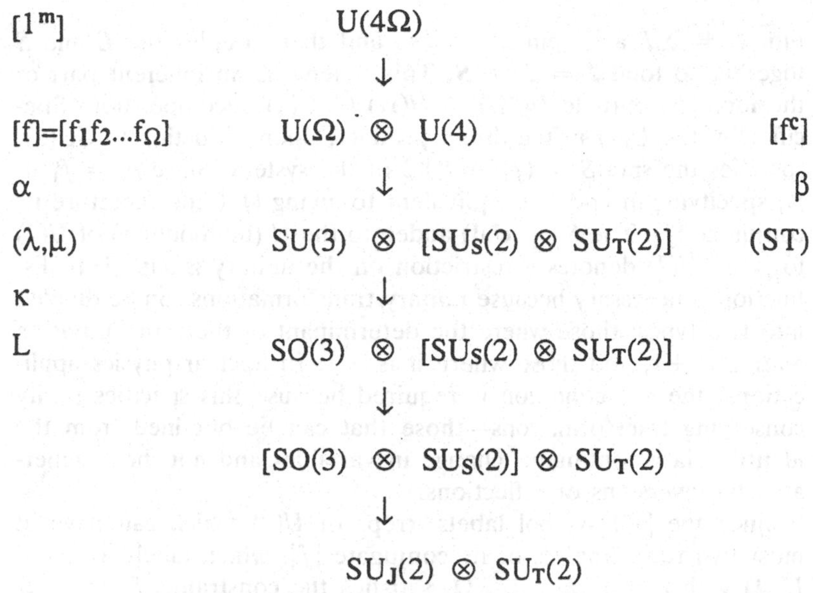

FIGURE 7.1. *Diagram showing the direct product* U(kΩ) ⊃ U(Ω) ⊗ U(k) *and further reductions of* U(Ω) *and* U(k) *for the* k = 4 *(space and spin-isospin) case. The irrep labels for groups in the* U(Ω) *and* U(k) *chains are given on the far left and right, respectively. Overall antisymmetry in* U(kΩ) *requires that* [f] *and* [f^c] *be related to one another by row-column interchange. The* α *and* κ *are integer labels for resolving multiple occurrences of* SU(3) *and* SO(3) *irreps in the* U(Ω) ⊃ SU(3) *and* SU(3) ⊃ SO(3) *reductions, respectively, while* β *is the* U(4) ⊃ [SU$_S$(2) ⊗ SU$_T$(2)] *multiplicity label. In the next to the last step, the* [SO(3) ⊗ SU$_S$(2)] ⊃ SU$_J$(2) *association and reduction is to gain the total angular momentum,* **L** + **S** = **J**, *as a good quantum number.*

metry under $U(k\Omega)$ is completely trivial ($[f] = [1^m]$), shell-model wavefunctions are usually labelled with m and $[f]$ only, $|m[f] \ldots\rangle$. The redundant $[1^m]$ and $[f^c]$ labels are suppressed. Actually, the particle number m is also redundant because $m = \Sigma_i f_i$ but it is normally not suppressed simply for the sake of clarity. The ellipses $(\ldots)$ in the ket refer to additional labels associated with irreps of subgroups of the unitary groups and their multiplicities, which are discussed next.

7.1.3. Traditional Schemes: LS and jj Coupling

The LS-coupling scheme first separately couples the orbital angular momenta and spins of the particles to total orbital angular momentum $\boldsymbol{L} = \Sigma_i \boldsymbol{l}_i$ and spin $\boldsymbol{S} = \Sigma_i \boldsymbol{s}_i$, and then couples the $\boldsymbol{L}$ and $\boldsymbol{S}$ together to total $\boldsymbol{J} = \boldsymbol{L} + \boldsymbol{S}$. This scheme is an inherent part of the identical particle $U(2\Omega) \supset U(\Omega) \otimes U(2)$ decomposition. Specifically, the $U(2)$ in the direct product, upon reduction to $SU(2)$, specifies the spin $S = (f_1^c - f_2^c)/2$ of the system. Since $m = f_1^c + f_2^c$, specifying m and S is equivalent to giving $[f^c]$ and therefore its conjugate $[f]$. The S (special) added to the U (unimodular) of $U(2)$ to give $SU(2)$ denotes a restriction on the unitary group. This distinction is necessary because unitary transformations can be divided into two types: those where the determinant of the transformation matrix is $+1$, and those where it is -1. In nuclear physics applications, the $+1$ condition is required because this specifies parity conserving transformations—those that can be obtained from the identity via a continuous change in variables and not those generated by inversions or reflections.

Since the $[f^c]$ symbol labels irreps of $U(2)$ which can have at most two rows, the f_i's of its conjugate $[f]$, which labels irreps of $U(\Omega)$ with $i = 1, 2, \ldots, \Omega$, satisfies the constraints $f_{i+1} \leq f_i \leq 2$. A filled shell is one with $m = 2\Omega$ and $f_1^c = \Omega = f_2^c$ (so $S = 0$) and f_i's that are all equal to 2. Also, there is always a particle-hole symmetry across mid-shell; the complement of an m-particle system has $m^{p\text{-}h} = 2\Omega - m$ holes. This particle-hole complementarity extends to the unitary irreps as $[f^{c,p\text{-}h}] = [\Omega - f_2^c, \Omega - f_1^c]$ (so $S^{p\text{-}h} = S$) and for $[f^{c,p\text{-}h}]$, $f_2^{p\text{-}h} = 2 - f_{r+1-i}$. The usual procedure is to use the description which involves the least number of either particles or holes. For example, relative to an ^{16}O core, ^{38}Ca is a 22 particle *ds*-shell nucleus, but relative to ^{40}Ca, it is a 2 hole *ds*-shell system. The structural complexity of these two descriptions is the same and equivalent to that of the 2 particle *ds*-shell nucleus ^{18}O.

While the spin $S = (f_1^c - f_2^c)/2$ of a system of m identical particles is fixed once $[f^c]$ is specified, the set of orbital angular momentum L values in an $[f]$ of $U(\Omega)$ is not given by a simple general rule. The reason is that L labels the irrep of an unnatural (or non-canonical in group theory jargon) subgroup $SO(3)$ of $U(\Omega)$. Recall that the rotation group in three dimensions, which we denoted as $SO(3)$ but which is also sometimes called R_3, is physically important because space is isotropic so observables must be $SU_J(2)$ scalars—operators that are independent of the frame of reference in which they are measured—and $SU_J(2) \supset SO(3) \otimes SU_S(2)$. It is mathematically an unnatural subgroup since it cannot be obtained from $U(\Omega)$ by eliminating one of the original basis states. The latter corresponds to a reduction from $U(\Omega)$ to $U(\Omega - 1)$ which is a mathematically natural (canonical) reduction but one that does not include $SO(3)$. From a practical point of view this means that a particular L value may occur many times in a given $[f]$, so care must be exercised in taking proper account of this multiplicity. Each occurrence of a given L corresponds to a separate and independent m-particle configuration. The total angular momentum J is obtained by coupling L to S, $\mathbf{J} = \mathbf{L} + \mathbf{S}$. The multiplicity of J states derives from both the multiplicity of L values and the coupling of those allowed L values to S.

As an example, consider the nontrivial ^{19}O case where ^{16}O is taken to be a closed core and the 3 extra-core neutrons are constrained to the ds-shell. In this case $k\Omega = 12$, and the unitary group decomposition is $U(12) \supset U(6) \otimes U(2)$. There are only 2 allowed $U(6)$ $[f]$ symmetries, $[2, 1]$ and $[1^3]$, since $[3]$ is ruled out because its conjugate $[\tilde{f}] = [1^3]$ has 3 rows, and 2 rows is the limit for $U(2)$. The $[f] = [2, 1]$ symmetry has L values $0, 1^2, 2^4, 3^2, 4$ while the $[1^3]$ has L values $1^2, 3^2$; the corresponding S values are $\frac{1}{2}$ and $\frac{3}{2}$, respectively. This means, for example, that 10 independent $J = \frac{3}{2}$ states can be built – 6 from the $[f] = [2, 1]$ symmetry and 4 from the $[1^3]$. The enumeration is straightforward once the $[f] \rightarrow L$ reduction is known. Procedures for determining the latter are known, and in fact a simple code giving the result for any combination of orbits is available. As an illustration of how quickly shell-model dimensionalities grow, we note that the $J = \frac{3}{2}$ dimensionality for the fp-shell nucleus ^{43}Ca is 28 while for ^{45}Ca it is 253 and for ^{47}Ca it is 1005.

In contrast with LS-coupling, the jj-coupling scheme starts by coupling the orbital angular momentum $\boldsymbol{l}$ and spin $\mathbf{s}$ of the individual particle to $\mathbf{j} = \boldsymbol{l} + \mathbf{s}$ and then afterwards coupling to the total angular momentum, $\mathbf{J} = \Sigma_i \mathbf{j}_i$. In this case for identical particles, $2\Omega = \Sigma_j(2j+1)$ includes both the space and spin degrees of freedom so there is no direct product decomposition. The reduction is from $U(2\Omega)$ directly to $SU_J(2)$, rather than through the LS-coupling scenario which goes from $U(2\Omega)$ to $U(\Omega) \otimes U(2)$ and then down to $SO(3) \otimes SU_S(2) \supset SU_J(2)$ with L of $SO(3)$ and S of $SU_S(2)$ coupling to give the J of $SU_J(2)$, $\mathbf{J} = \mathbf{L} + \mathbf{S}$. The subscript added to $SU(2)$ distinguishes the spin S and total angular momentum J unitary groups.

There are various other jj-coupling alternatives. One is to first decompose the 2Ω-dimensional space into orbits, $2\Omega = \Sigma_i 2\Omega_i$ normally with $2\Omega_i = (2j_i + 1)$; next reduce each of these spaces which have unitary symmetry $U(2\Omega_i)$ with respect to $SU_{J_i}(2)$; finally couple the J_i's to the total J of $SU_J(2)$. In this case, the group structure is $U(2\Omega) \supset U(2\Omega_1) \oplus U(2\Omega_2) \oplus \cdots \oplus U(2\Omega_k)$ if there are k orbits, with $\mathbf{J} = \Sigma_i \mathbf{J}_i$ where J_i is the angular momentum of the m_i particles in the $2\Omega_i$ subspace and $m = \Sigma_i m_i$. This scheme is conceptually the simplest of all, because it reduces the problem to one of dealing with the distribution of identical particles in a single j-shell and angular momentum coupling, nothing more. The basis is built by considering allowed distributions of m particles across k levels. This is called a configuration-j scheme. If $(m_\alpha) = (m_1, m_2, \ldots, m_k)$ denotes the α-*th* distribution of the m particles across the k levels, and $(J_\alpha) = (J_1, J_2, \ldots, J_k)$ the corresponding total angular momenta, then a complete labelling of basis states consists of specifying (m_α) and (J_α) plus all intermediate angular momentum couplings like $\mathbf{J}_{12} = \mathbf{J}_1 \times \mathbf{J}_2$, $\mathbf{J}_{12,3} = \mathbf{J}_{12} \times \mathbf{J}_3$, etc. Early versions of a general purpose shell-model code were built around the logic of this scheme. In addition to the reduction to a sum of single-j shells plus angular momentum coupling and recoupling, it has the further advantage of allowing for a very natural truncation through specifying a subset of allowed particle configurations, the (m_α) that are included.

As a jj-coupling example, consider the ^{19}O case again. The ds-shell space has three j-orbits, $(d_{5/2}, s_{1/2}, d_{3/2})$. There are 9 distinguishable 3-particle configurations: (3, 0, 0); (2, 1, 0); (2, 0, 1); (1, 2, 0); (1, 1, 1); (1, 0, 2); (0, 2, 1); (0, 1, 2); (0, 0, 3); the (0, 3, 0) does not occur as the $s_{1/2}$ orbital will hold at most 2 identical particles. To determine the dimensionality of the model space, for example for $J = \frac{3}{2}$, the allowed J's for each configurations must be worked out. This is where the antisymmetry requirement enters. For example, for the $(d_{5/2})^3(s_{1/2})^0(d_{3/2})^0$ configuration denoted by (3, 0, 0), only 1 of the 4 possible $J = \frac{3}{2}$ coupling belongs to an antisymmetric arrangement of the 3 particles $[f] = [1^3]$. For the (2, 0, 1) configuration there are 2, one from the coupling of $(d_{5/2})^2$, $J = 0$ to the $(d_{3/2})^1$ and one from the coupling of the $(d_{5/2})^2$, $J = 2$ to the $(d_{3/2})^1$. For the 9 configurations listed above, the number of times $J = \frac{3}{2}$ occurs in each is respectively: 1; 1; 2; 0; 2; 1; 1; 1; 1; for a total dimensionality of 10 as already given in the LS-coupling example. Since the $d_{5/2}$ orbital lies below the $s_{1/2}$ and the $s_{1/2}$ below the $d_{3/2}$, a natural way to truncate the basis is to eliminate configurations, starting with the (0, 0, 3) and working backwards.

The assumption underlying the *jj*-coupling approach is that the residual two-body interaction is less important than the one-body mean-field term that generates the single-particle orbital splitting. While this simple assumption may be reasonable for near closed-shell nuclei, away from closed shells it may or may not always be so. The reason for this ambiguity is that the magnitude of the one-body term is proportional to the number of particles, m, whereas for the two-body interaction it grows as the number of possible pairs, $m(m - 1)/2$. So if the one-body part is twice as strong on the average as the two-body part for $m = 2$, the two interactions will be comparable for $m = 5$ and for $m = 9$ the effect of the two-body part will be more than twice that of the one. In actual fact the situation is even worse than this indicates, because the two-body interaction can build in additional coherence due to the coupling to higher shells, as happens when rotations dominate.

7.1.4. Pairing Interaction and the Seniority Scheme

The first question one normally asks when seeking a solution to a physics problem is, "What are the constants of the motion?" For classical systems the answer to this question usually dictates a preferred coordinate system, like cylindrical geometry if there is axial symmetry. In quantum physics, constants of the motion are conserved quantities—like the parity of a system and its angular momentum (J^π) or (for light nuclei) its angular momentum and isospin (J^π, T)—corresponding to eigenvalues of invariant operators that commute with each other and with the Hamiltonian. In the classical case, the constants of the motion label orbits defining equivalent states of the system; in the quantum case, the eigenvalues of the invariant operators are the quantum numbers of the system. For bound systems like nuclei, each distinct set of quantum numbers defines a matrix eigenvalue problem, and accordingly, a finite number of distinct eigenstates corresponding to unique system configurations.

A long-standing problem in nuclear physics is finding additional invariant or near invariant operators in order that the dimensionality of a model space—the size of the matrix problem and the number of eigenstates associated with each set of quantum numbers—can be used to reduce the space to tractable size. Clearly, if there are additional true invariant operators, isospin (T) in light nuclei or F-spin in the IBM, for example, these should be incorporated into any solution. However, identifying *near* invariants which are associated with the approximate symmetries of a system (as opposed to *true* ones which correspond to exact symmetries) and giving a quantitative measure for the *goodness* of these symmetries is a difficult challenge. These non-exact (but important) symmetries exist when the states of a system can be further grouped into subsets with the matrix elements of H connecting the subsets small relative to the matrix elements of H within each subset. A quantitative measure for this requires a detailed knowledge of matrix elements of H.

There is a simpler global measure, however, which is statistical in nature because it depends only on traces of operators and operator products, which may be used to predict the goodness of a symmetry. Specifically, if K is a Hamiltonian-like operator built out of products and sums of generators of a symmetry group, then that symmetry will be good if K and H are strongly correlated. An operator correlation coefficient, ζ, that measures this is defined by

$$\zeta_{H,K} = \frac{\langle\langle(H - \overline{H})(K - \overline{K})\rangle\rangle}{\sqrt{\langle\langle(H - \overline{H})^2\rangle\rangle\langle\langle(K - \overline{K})^2\rangle\rangle}} \tag{7.1}$$

when in this expression, the double bracket $\langle\langle\ \rangle\rangle = d^{-1}\langle\ \rangle$ denotes the trace and a bar over an operator means its average, $\overline{O} = \langle O\rangle = d^{-1}\langle\langle O\rangle\rangle$. Clearly, if $K = \pm H$ the correlation coefficient $\zeta_{H,K} = \pm 1$, but in spaces with large dimensions d even a $|\zeta| \approx 0.5$ indicates a strong correlation.

The pairing operator P, which counts the number of pairs of nucleons in a many-particle wavefunction that are coupled to total angular momentum $J = 0$, plays a very important role in nuclear structure. The reason for its importance follows because the nucleon-nucleon interaction is basically attractive, and as a consequence favors the $(j)^2J = 0$ configuration which maximizes the spatial overlap of two-particle states. For a single j-shell geometry, the pairing operator is an invariant of the symplectic symmetry group $Sp(2j + 1)$ that sits between $U(2j + 1)$ and $SU_J(2)$: $U(2j + 1) \supset Sp(2j + 1) \supset SU_J(2)$. In a multi-orbit space with $2\Omega = \Sigma_j(2j + 1)$ the concept can obviously be extended. The additional quantum number gained by adding $Sp(2j + 1)$ to the reduction of $U(2j +$

1) is called the seniority and is usually labelled ν for the corresponding Hebrew word *vethek* used by Racah when first introducing the concept in atomic physics. In a many-particle scheme, the seniority is the number of particles *not* coupled to $J = 0$ pairs. An identical m-particle, single-j shell configuration can be labelled $|j^m \alpha \nu J M_J\rangle$ where α is again a running index that distinguishes the possible multiple configurations within the $(m\nu J)$ set. The α label is only needed for $j \geq 9/2$ since for $j \geq 7/2$ the $(m\nu J)$ labels suffice because there is no more than a single configuration with those quantum labels.

If the Hamiltonian is taken to be the harmonic oscillator with pairing,

$$H = H_0 - GP, \tag{7.2}$$

the eigenstates of the system can be organized by the seniority quantum number according to a simple J-independent result,

$$E(m,\nu) = -\frac{G}{4}(m-\nu)(2\Omega-\nu-m+2). \tag{7.3}$$

The pairing operator in eq. (7.2) can be given most simply in terms of a quasi-spin formulation of theory: $P = S_+S_- = S^2 - S_0^2 + S_0$ where $S_+ = \Sigma_{m>0}(-1)^{j-m}(a_m^+ \; a_{-m}^+)$ and $S_- = \Sigma_{m>0}(-1)^{j+m}(a_m a_{-m})$ are pair creation and annihilation operators, respectively, and $S_0 = \frac{1}{2}(N - \Omega)$ with $N = \Sigma_m (a_m^+ a_{-m})$ the number operator and $2\Omega = 2j + 1$ the shell degeneracy. The S_+, S_-, and S_0 operators satisfy the usual angular momentum commutation relations and define the $SU_Q(2)$ quasi-spin group. For an even number of particles, the $(\nu, J) = (0, 0)$ state lies lowest, the $\nu = 2$ configurations with $J = 2, 4, \ldots, 2j - 1$ come next—all at an excitation energy ΩG above the $(\nu, J) = (0,0)$ state—and so on. For an odd number of particles, the same expression applies: the $(\nu, J) = (1, j)$ state lies lowest and the set of $\nu = 3$ states next at a relative excitation energy $(\Omega - 1)G$, etc. While it is true that for identical particle configurations and reasonable interactions the lowest calculated eigenstate has $J = 0$ and a large overlap with the $(\nu, J) = (0, 0)$ configuration, there are no really good examples of nuclei where next one finds a degenerate or nearly degenerate set with $J = 2, 4, \ldots, 2j - 1$. An example that is often cited are the ${}^{90+m}_{40}Zr_{50+m}$ nuclei with neutrons filling a $j = \frac{5}{2}$ shell. In this case, the $J = 2$ and 4 states lie at excitation energies 0.93 and 1.49 (Mev), respectively, above the J

= 0 ground state. So the splitting of the $J = 2$ and $J = 4$ states, ~0.5 (Mev), is roughly half the pairing gap, ~1.0 (Mev).

This failure to find examples of nuclei with a pure pairing spectrum is indicative of the fact that typically, for realistic interactions, $|\zeta_{H,P}| \lesssim 0.2$. For example, the *ds*-shell Kuo–Brown interaction gives $|\zeta_{H,P}| = 0.14$ for a $(ds)^4$ configuration and 0.19 in the $(ds)^8$ space. This is in contrast with the quadrupole-quadrupole interaction $(Q \cdot Q)$ which typically has a stronger correlation with realistic interactions, $|\zeta_{H,Q\cdot Q}| \approx 0.5$. In considering these numbers, it is also important to realize that the P and $Q \cdot Q$ interactions are not orthogonal, $|\zeta_{P,Q\cdot Q}| \approx 0.2$, so a pairing spectrum displays some rotational features and vice versa. Nonetheless, for even-even nuclei near closed shells, light-ion stripping reactions like ${}^{A}_{Z}X_N({}^3\mathrm{He}, n){}^{A+2}_{Z+2}Y_N$ and ${}^{A}_{Z}X_N(t, p){}^{A+2}_{Z}Y_{N+2}$ or pickup reactions like ${}^{A}_{Z}X_N(t, {}^5\mathrm{Li}){}^{A-2}_{Z-2}Y_N$ and ${}^{A}_{Z}X_N(p, t){}^{A-2}_{Z}Y_{N-2}$ show enhanced strengths which indicates the transfer of a correlated pair as opposed to two uncorrelated nucleons.

7.1.5. Quadrupole-Quadrupole Interaction and *SU*(3) Symmetry

The LS-coupling scheme also has an intermediate group structure. Just as $Sp(2j + 1)$ lies between $U(2\Omega)$ and $SU_J(2)$ and is associated with pairing, the special unitary group in three dimensions, $SU(3)$, lies between $U(\Omega)$ and $SO(3)$ and is associated with rotations. In addition to the usual three angular momentum operators L_μ ($\mu = -1, 0, 1$), $SU(3)$ is generated by five special *algebraic* quadrupole operators, Q^a_μ ($\mu = -2, -1, 0, 1, 2$). These operators—$Q^a_\mu = \frac{1}{2}\sqrt{16\pi/5}[\Sigma_i(p_i^2 b^2 Y_{2\mu}(\hat{p}_i) + r_i^2 Y_{2\mu}(\hat{r}_i))]$, where b is the oscillator length parameter and the sum extends over all valence particles—are symmetric in the coordinates and momenta, thereby insuring they reproduce matrix elements of their real *collective* quadrupole counterparts, $Q^c_\mu = \sqrt{16\pi/5}[\Sigma_i(r_i^2 Y_{2\mu}(\hat{r}_i))]$, within a shell ($n' = n$), but have vanishing matrix elements—which distinguishes them from the Q^c_μ—between different shells ($n' \neq n$). It is important to emphasize that the Q^c have nonzero matrix elements between major shells differing by two oscillator quanta ($n' = n \pm 2$). These intershell couplings are responsible for the buildup of quadrupole collectivity in strongly deformed nuclei, and are the motivation behind the symplectic model discussed in a later section. Since these intershell couplings are neglected when Q^a is used in place of Q^c, one expects and finds that Q^a must be renormalized whenever it is used in place of Q^c.

The $SU(3)$ structure is the symmetry group of the three-dimensional oscillator. The existence of a symmetry higher than $SO(3)$ can be anticipated from fact that the eigenenergies of the oscillator are $E_n = (n + \frac{3}{2})\hbar\omega$, while the radial wavefunctions (R_{nl}) include the orbital angular momentum label l with values $l = n, n - 2, \ldots, 1$ or 0 for fixed n. This degeneracy in l for fixed n is a result of the embedding of $SO(3)$ in $SU(3)$, the symmetry group of the oscillator in three-space. The oscillator Hamiltonian does not distinguish between quanta in the z, x or y directions. Of course, adding an l^2 term to H_0 breaks this symmetry, as does the spin-orbit interaction, $l \cdot s$. Since these two interactions are a fundamental part of the shell-model picture, the question that naturally arises is whether or not this breaking of the $SU(3)$ symmetry destroys its usefulness. For nuclei up through the middle of the ds-shell, the answer is no; the level splitting generated by the one-body l^2 and $l \cdot s$ is small relative to the major shell separation distance, $\hbar\omega$. However for heavier nuclei, the splitting generated by the spin-orbit interaction even changes shell closures. Indeed this is a necessary feature to get shell and subshell closures to reproduce the magic numbers. We will return to this in a later section when the pseudo-spin scheme is introduced.

An appreciation for the connection between $SU(3)$ and rotations can be gained by noting, just as for the boson case, the eigenvalue of the quadrupole-quadrupole interaction is given in terms of the Casimir invariants of $SU(3)$ and $SO(3)$: $Q^a \cdot Q^a = 4C_2 - 3L^2$. So the simple Hamiltonian

$$H = H_0 - \tfrac{1}{2}\chi Q^a \cdot Q^a, \tag{7.4}$$

gives rise to a $L(L + 1)$ rotational spectrum. (Note: In the present context χ denotes the strength of the quadrupole-quadrupole interaction—it is not the χ introduced earlier to define the internal structure of the IBM quadrupole operator.) The various allowed representations of $SU(3)$ in $U(\Omega)$ lie at a relative excitation energy determined by the $2\chi C_2$ factor in H. Of course, a realistic Hamiltonian includes additional interaction terms; in particular, the one-body l^2 and $l \cdot s$ terms are not diagonal in the $U(\Omega) \supset SU(3) \supset SO(3)$ chain. Specifically, although the l^2 term does not mix representations of $U(\Omega)$, the $l \cdot s$ does, and neither conserve $SU(3)$. Since the one-body parts of the Hamiltonian are known to be very important, in order for $SU(3)$ and rotations to survive as an important symmetry, the $Q^a \cdot Q^a$ component in realistic interactions must dominate. The value quoted above for the correlation coefficient, namely $|\zeta_{H,Q\cdot Q}| \approx 0.5$, tells one that this is indeed the case. Looking ahead to the pseudo-spin and its pseudo-$SU(3)$ extension, we note that the correlation between H and $Q \cdot Q$ is even higher, approaching close to 0.8. Although the scalar correlation coefficient is a crude measure, it is quite reliable, and one can be certain that when it is large the corresponding symmetry will be good. For example, for the yrast band of ^{20}Ne—considered to be $(ds)^4$, $T = 0$ nucleus—the calculated eigenstates are between 60% and 80% pure $(\lambda, \mu) = (8, 0)$, the leading $SU(3)$ irrep for this nucleus, depending on which of the many available microscopic interactions one chooses to use.

7.2. QUANTUM ROTOR AND THE *SU*(3) MODEL

The *macroscopic* and *microscopic* options for interpreting nuclear phenomena have existed since the earliest period in the history of nuclear physics, and we continue to use these labels in a general way for characterizing models in use or under development today. Of course, there is no longer a macroscopic-microscopic dichotomy, as most modern theories fit solidly into the intermediate region between the extremes. For example, the simplest Bohr–Mottelson rotation-vibration model for even-A systems, and its Nilsson extension for strongly deformed odd-A nuclei have evolved into a variety of self-consistent theories, both simple and sophisticated. We will refer to these as *geometric* models, because they use (β, γ)-shape variables for describing quadrupole deformation in an intrinsic principal-axis frame, and a quantum top dynamics for a description of the accompanying rotational motion. From the shell-model perspective, there are now a variety of *algebraic* models exploiting group symmetries in order to focus more sharply on nuclear collective degrees of freedom. As applications to superdeformed phenomena are explored, both approaches are coming under very sharp and critical testing, because this regime of strong deformation focuses on the unique situation of collective rotational phenomena in isolation from other normally important effects. The challenge facing nuclear structure physicists is the need to adequately incorporate *microscopic* features in a *geometric* theory or *macroscopic* features in an *algebraic* one.

Significant progress towards bridging this gap between the macroscopic and microscopic pictures of nuclear structure has recently been made. In this section, we will review this progress for rotational motion; extensions along similar lines for vibrations and the coupling of rotations and vibrations are still under development. Specifically, a simple one-to-one mapping between the triaxial quantum rotor problem and an $SU(3)$ shell-model theory will be established. In order to accomplish this, the quantum rotor problem will be reformulated to identify the corresponding many-particle, shell-model theory. An explicit connection between the (β, γ)-shape variables of a macroscopic approach and the (λ, μ)-irrep labels of the microscopic theory follows from this development. This $(\beta, \gamma) \leftrightarrow (\lambda, \mu)$ mapping establishes an unambiguous link between the geometric and algebraic approaches to the structure of nuclei. Several examples—covering excitation energies and transition strengths—will be shown to demonstrate the quality of agreement between the two theories under the mapping. Because this matter is so important, we will show how the equivalence of these approaches arises at the level of the symmetry algebras of the two theories, even though our intent is to avoid using advanced group theory arguments when possible. We will also show how the theory can be used to identify a specific linear combination of shell-model operators expected to play a significant role in nuclear K-band splitting.

7.2.1. Triaxial Quantum Rotor

The triaxial rotor was one of the first problems considered after the introduction of the quantum theory. Interest in it peaked during the late 1920's and early 1930's as it was used as a model for studying molecular rotations. The Hamiltonian is given by

$$H_{\mathrm{ROT}} = A_1 I_1^2 + A_2 I_2^2 + A_3 I_3^2 \tag{7.5}$$

where $I_\alpha(\alpha = 1, 2, \text{and } 3)$ is the projection of the the total angular momentum of the rotor on the α–th body-fixed symmetry axis and A_α is the corresponding inertia parameter: $A_\alpha = \frac{1}{2\mathcal{I}_\alpha}$ where $\mathcal{I}_\alpha$ is the moment of inertia of the system about the α–th principal axis. Here we will deviate from the practice of using $\mathbf{J}$ for the angular momentum and use $\mathbf{I}$ and $\mathbf{L}$ instead, to denote body-fixed and lab-frame angular momentum operators, respectively. Later, when spin is included in the formulation, we will return to the use of $\mathbf{J} = \mathbf{I} +$

S = **L** + **S**. Note that **I** (vector) = **L** (vector); the use of different symbols (rather than primed and unprimed operators, for example) for the body-fixed and lab-frames is for clarity only.

The actual values for the A_α in eq. (7.5) depend on details of the model. Two extremes can be identified: rigid rotation and irrotational flow. A system is rigid if all of its constituent particles are locked in place and participated equally in the rotational motion—as for a simple top or a molecule. It is irrotational if the rotation is simply a manifestation of a rearrangement of the constituent particles—as happens when a water balloon is tossed in the air. Nuclear rotations are somewhere between these two limits. The more deformed the system, the more rigid-rotor-like it seems to become. As an example, recall that superdeformed bands suggest the underlying configurations have rigid-rotor characteristics.

The familiar principal-axis form for the rotor Hamiltonian eq. (7.5) can be rewritten in a less unfamiliar but nonetheless important and very revealing frame-independent representation by introducing three special scalar operators:

$$L^2 = \sum_\alpha L_\alpha L_\alpha = \sum_\alpha I_\alpha^2, \qquad X_3^c = \sum_{\alpha\beta} L_\alpha Q_{\alpha,\beta}^c L_\beta = \sum_\alpha \lambda_\alpha I_\alpha^2,$$
$$X_4^c = \sum_{\alpha,\beta,\gamma} L_\alpha Q_{\alpha,\beta}^c Q_{\beta,\gamma}^c L_\gamma = \sum_\alpha \lambda_\alpha^2 I_\alpha^2. \tag{7.6}$$

The L_α and $Q_{\alpha,\beta}^c$ in eq. (7.6) are Cartesian forms of the total angular momentum and collective quadrupole operators, respectively. (Remember the superscript c is appended to Q to denote the collective quadrupole operator which has non-vanishing matrix elements between major shells ($n' = n, n \pm 2$), in contrast with the algebraic quadrupole operator (Q^a) which has non-vanishing matrix elements only within a major shell, $n' = n$.) The last expression given for each scalar in eq. (7.6) is the form these operators take in the body-fixed, principal-axis system: $\langle Q_{\alpha,\beta}^c\rangle^{\mathrm{BF/PA}} = \lambda_\alpha \delta_{\alpha,\beta}$. This result presumes the eigenvalues of the quadrupole operator are sharp. The expression given in eq. (7.6) can be inverted to yield an expression for the I_α^2 in terms of L^2 and the X^c's:

$$I_\alpha^2 = [(\lambda_1\lambda_2\lambda_3)L^2 + (\lambda_\alpha^2)X_3^c + (\lambda_\alpha)X_4^c]/D_\alpha$$
$$\text{where } D_\alpha \equiv 2\lambda_\alpha^3 + \lambda_1\lambda_2\lambda_3. \tag{7.7}$$

Substituting this result for the I_α^2 into eq. (7.5) yields

$$H_{\mathrm{ROT}} = aL^2 + bX_3^c + cX_4^c, \tag{7.8}$$

where a, b and c depend on the inertia parameters and the eigenvalues of Q^c,

$$a = \sum_\alpha a_\alpha A_\alpha, \qquad b = \sum_\alpha b_\alpha A_\alpha, \qquad c = \sum_\alpha c_\alpha A_\alpha;$$
$$a_\alpha = \lambda_1\lambda_2\lambda_3/D_\alpha, \qquad b_\alpha = \lambda_\alpha^2/D_\alpha, \qquad c_\alpha = \lambda_\alpha/D_\alpha. \tag{7.9}$$

The eigenvalues and eigenstates of Hamiltonians eq. (7.5) and eq. (7.8) with parameters related through eq. (7.9) are the same.

7.2.2. Rotor Dynamics and *SU*(3) Symmetry

The reason for casting the rotor Hamiltonian into a frame-independent form—when its principal-axis representation is clearly simpler—is so its shell-model image can be readily identified. To achieve this, note that the many-body form for the L_α and $Q^c_{\alpha,\beta}$ operators are simply their single-particle forms summed over all particles,

$$L_\alpha = \sum_i l_\alpha(i) \quad \text{and} \quad Q^c_{\alpha,\beta} = \sum_i q^c_{\alpha,\beta}(i). \tag{7.10}$$

It would seem from this that the many-body, shell-model image of the rotor Hamiltonian is eq. (7.8), with the L_α and $Q^c_{\alpha,\beta}$ operators interpreted as in eq. (7.10). However, this interpretation ignores the underlying shell structure and the fermion character of the many-body system. In particular, it is important to remember that while the L_α have non-vanishing matrix elements only within a major oscillator shell, the $Q^c_{\alpha,\beta}$ couple shells differing by two quanta ($n' = n, n \pm 2$). Indeed, the off-diagonal ($n' = n \pm 2$) matrix elements of the $Q^c_{\alpha,\beta}$ are about equal in size to the diagonal ($n' = n$) ones. It follows from this that operators like $Q^c \cdot Q^c$ and the X^c's (even if used only as residual interactions) can actually destroy the shell structure. One way around this dilemma is to set the matrix elements of Q^c coupling to different major shells to zero, which has the effect of transforming the collective model quadrupole operators ($Q^c_{\alpha,\beta}$) into the algebraic counterparts ($Q^a_{\alpha,\beta}$). As noted above, these $Q^a_{\alpha,\beta}$ operators, along with the L_α's, generate $SU(3)$, the symmetry algebra of the isotropic harmonic oscillator Hamiltonian. In this case, the shell-model Hamiltonian (which reproduces rotor results, as shown below) is given by

$$H_{SU3} = H_0 + aL^2 + bX_3^a + cX_4^a. \tag{7.11}$$

Before determining shell-model values for the λ_αs, which will also establish a connection between the (β, γ)-shape variables of the collective model and the (λ, μ)-irrep labels of the shell-model approach, thereby completing the $H_{\text{ROT}} \leftrightarrow H_{SU3}$ correspondence, it is instructive to point out the relationship of these two theories at a more fundamental level. This can be accomplished most simply by comparing the algebras of their symmetry groups. The symmetry group of the quantum rotor is the semi-direct product $T_5 \wedge SO(3)$ where T_5 is the group generated by the five components of the collective quadrupole operator (Q^c_μ); as usual, $SO(3)$ is generated by the angular momentum operators (L_μ). (The semi-direct product symbol $\wedge$ denotes the fact that the generators of this special group structure divide into two sets, each separately closed under commutation, but the commutator of an element of one with an element of the other yielding an element of only one (not both) of the sets, symbolically: $[T_5, T_5] \to T_5$ and $[SO(3), SO(3)] \to SO(3)$ but $[T_5, SO(3)] \to T_5$, see below.) The generators of $SU(3)$, on the other hand, are the Q^a_μ's and the L_μ's. If Q^x denotes a generic quadrupole operator, then the commutation relations of the L's and the Q^x's are given by

$$[L_\mu, L_\nu] = -\sqrt{2}\langle 1\mu, 1\nu | 1, \mu + \nu \rangle L_{\mu+\nu}, \tag{7.12a}$$

$$[L_\mu, Q^x_\nu] = \sqrt{6}\langle 1\mu, 2\nu | 2, \mu + \nu \rangle Q^x_{\mu+\nu}, \tag{7.12b}$$

$$[Q^x_\mu, Q^x_\nu] = c\langle 2\mu, 2\nu | 1, \mu + \nu \rangle L_{\mu+\nu}, \tag{7.12c}$$

where $c = +3\sqrt{10}$ for $SU(3)[Q^x = Q^a]$ and $c = 0$ for $T_5 \wedge SO(3)[Q^x = Q^c]$. In eq. (7.12) the $\langle\, ,\, | \rangle$ symbol denotes an ordinary $[SO(3)]$ Clebsch–Gordan (Wigner) coefficient. Actually, there is one additional case, $c = -3\sqrt{10}$ for $Sl(3, R)[Q^x = Q^b \sim (x_i p_j + p_j x_i)]$ which identifies the symmetry associated with shear degrees of freedom. All three groups, $SU(3)$, $T_5 \wedge SO(3)$, and $Sl(3, R)$, are subgroups of the symplectic group $Sp(3, R)$, which we will discuss later at greater length. One can see from these commutation relations how the $SU(3)$ algebra of the isotropic oscillator goes over into $T_5 \wedge SO(3)$ of the rotor. Specifically, if Q^a is renormalized by dividing it by the square root of the second-order invariant of $SU(3)[Q^a \leftarrow Q^a/\sqrt{C_2}$ where by definition the invariant $C_2 = \frac{1}{4}(Q^a Q^a + 3L^2)$ commutes with both the Q^a's and the L's] the first (a) and second (b) commutators given in eq. (7.12) remain unchanged, while the $L_{\mu+\nu}$ on the right-hand side of the third (c) goes over into $L_{\mu+\nu}/C_2$. For low L values in large $SU(3)$ irreps, $L_{\mu+\nu}/C_2 \rightarrow 0$ and the $SU(3)$ algebra reduces to $T_5 \wedge SO(3)$. This renormalization of the Q^a is called a group contraction process; the arguments presented show the $SU(3)$ algebra reduces to the algebra of $T_5 \wedge SO(3)$ in the contraction limit, and consequently, the $SU(3)$ theory reduces to that of the rotor. Examples demonstrating this for small but finite L/C_2 values are given below.

7.2.3. Shape Variables $(\beta, \gamma) \leftrightarrow (\lambda, \mu)$ Irrep Labels

The parameters a, b and c of H_{SU3} eq. (7.11) are related to the A_α of H_{ROT} through eq. (7.9). However, to complete the mapping it is necessary to have shell-model values for the λ_α's. The simplest and most direct way this can achieved is by requiring a correspondence between invariants of the rotor and shell-model theories. This is a very natural requirement, since two theories describing the same physical phenomena should have the same constants of the motion. Because $SU(3)$ is a rank two group it has two invariants: C_2, with eigenvalue $[\lambda^2 + \lambda\mu + \mu^2 + 3(\lambda + \mu)]$; and C_3, with eigenvalue $[(\lambda - \mu)(\lambda + 2\mu + 3)(2\lambda + \mu + 3)]$, where λ and μ are the usual $SU(3)$ irrep labels—$(\lambda + \mu)$ and μ specify respectively the number of boxes in the first and second rows of a Young diagram labelling of the $SU(3)$ irreps. Note that C_2 is of degree two in the generators of $SU(3)$ while C_3 is of degree three. The symmetry group of the rotor $[T_5 \wedge SO(3)]$ also has two invariants: traces of the square $\{\text{Trace}[(Q^c)^2]\}$ and the cube $\{\text{Trace}[(Q^c)^3]\}$ of the collective quadrupole matrix. The eigenvalues of these invariant operators are $\lambda_1^2 + \lambda_2^2 + \lambda_3^2 \rightarrow (k\beta)^2$ and $\lambda_1\lambda_2\lambda_3 \rightarrow (k\beta)^3\cos(3\gamma)$, respectively. The requirement of a linear correspondence between these two sets of invariants, one for $SU(3)$ and the other for $T_5 \wedge SO(3)$, leads to the following relations:

$$\lambda_1 = -(\lambda - \mu)/3, \qquad \lambda_2 = -(\lambda + 2\mu + 3)/3, \qquad \lambda_3 = (2\lambda + \mu + 3)/3. \tag{7.13}$$

This correspondence, in turn, sets up a direct relationship between the (β, γ) shape variables of the collective model and the (λ, μ) irrep labels of $SU(3)$,

$$\beta^2 = (4\pi/5)(A\overline{r^2})^{-2}[\lambda^2 + \lambda\mu + \mu^2 + 3(\lambda + \mu) + 3], \quad (7.14)$$
$$\gamma = \tan^{-1}\left(\frac{\sqrt{3}\,(\mu + 1)}{2\lambda + \mu + 3}\right).$$

The (β, γ) and (λ, μ) relationship is shown schematically in Fig. 7.2 on a traditional (β, γ) or polar plot with β the radius vector and γ the azimuthal angle. The Cartesian components of β are given by

$$k\beta_x = k\beta\cos(\gamma) = (2\lambda + \mu + 3)/3, \quad (7.15)$$
$$k\beta_y = k\beta\sin(\gamma) = (\mu + 1)/\sqrt{3},$$

where $k^2 = (5/9\pi)(A\overline{r^2})^2$. These relations show that each (λ, μ)-

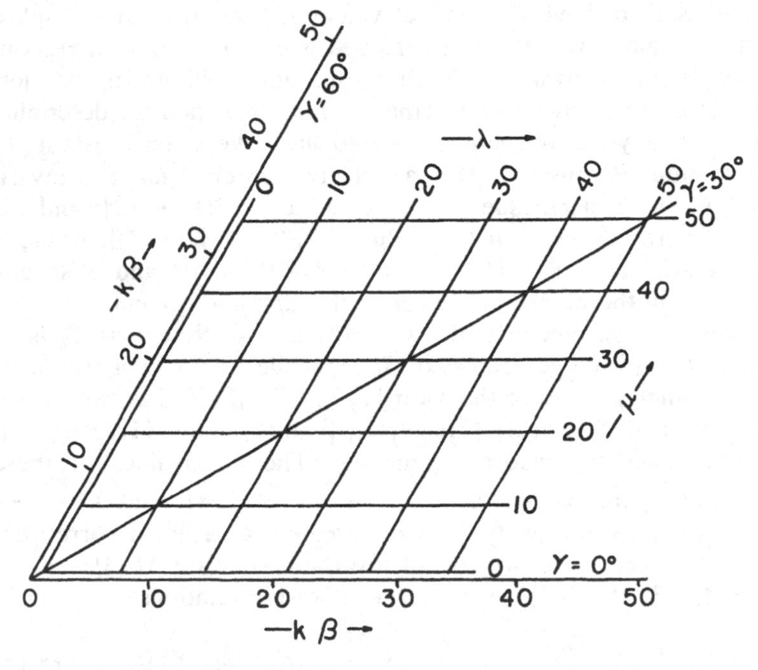

FIGURE 7.2. *A traditional (β,γ) or polar plot with β the radius vector and γ the azimuthal angle which shows the relationship between the collective model shape parameters (β,γ) and the* SU(3) *representation labels (λ,μ). The linear relationship holds for the special mapping defined by eq.* (7.13) *in the text. The magnitudes of β^2 and γ are given in terms of λ and μ by eq.* (7.14). *The cartesian components of β are given in eq.* (7.15). *(From Castaños, Draayer and Leschber,* 1988.)

irrep corresponds to a unique value for the pair (β, γ). Because the λ and μ are representation labels of $SU(3)$ they can only be nonnegative integers. Furthermore, for any particular nucleus the group structure, $U(\Omega) \supset SU(3)$ dictates a set of allowed (λ, μ) irreps for each $[f]$ value. The $[f] \rightarrow (\lambda, \mu)$ branching rule is called a group plethysm; tables as well as a computer code for determining the set of (λ, μ) irreps in $[f]$ are available. For example, for the *ds*-shell nucleus ^{24}Mg, the most important $U(\Omega = 6)$ symmetry is $[f] = [44]$ and this contains the following $SU(3)$ irreps: $(\lambda, \mu) = (8, 4)$; $(7, 3)$; $(8, 1)$; $(4, 6)$; $(5, 4)$; $(6, 2)^2$; $(3, 5)$; $(4, 3)$; $(5, 1)^2$; $(0, 8)$; $(2, 4)^2$; $(3, 2)$; $(4, 0)^2$; $(1, 3)$; and $(0, 2)$ where the superscript denotes the number of times the (λ, μ) irrep occurs (its multiplicity). The $[f] \rightarrow (\lambda, \mu)$ reduction means the microscopic fermion structure as determined through group plethysms places constraints on the allowed values for (β, γ)—in sharp contrast with the macroscopic liquid drop picture which takes β and γ to be continuous variables. This is then a very clear and significant difference in the macroscopic and microscopic approaches—the microscopic structure, together with the branching rules this implies, dictates the (λ, μ) values that can occur; and this implies constraints that should be placed on the allowed (β, γ) shapes of a macroscopic theory.

This correspondence between (λ, μ) and (β, γ) means that the potential energy surface concept, so much a part of the macroscopic picture, can be given a shell-model interpretation. The usual procedure used for determining the potential energy surface is to assume the Hamiltonian is a 5-dimensional oscillator based on the quadrupole shape excitations, plus an anharmonic residual interaction that is a polynomial function of maximum phonon order k in the scalars $[\beta^2$ and $\beta^3\cos(3\gamma)]$ of the theory,

$$H = H_0 + \sum_{p,q}^{2p+3q \leq k} C_{p,q}\beta^{2p+3q}[\cos(3\gamma)]^q. \tag{7.16}$$

This Hamiltonian is diagonalized in a basis built by considering all possible phonon states up through some maximum number (n) of quanta. The coefficients $C_{p,q}$ of the residual interaction are determined by fitting to known experimental data on excitation spectra and transition strengths. This approach is insensitive to k and n, provided both are sufficiently large; results for rare earth and actinide nuclei are given in the literature for k and n values up to 6 and 32, respectively. The $k = 6$ terms are needed to get a quadratic dependence on $\cos(3\gamma)$, and hence the possibility of a nonzero triaxial equilibrium shape. The energy surface is the expectation value of H in the calculated eigenstates plotted as a function of β and γ. If this procedure is revisited from the shell-model perspective suggested above, the allowed β and γ values must (at a minimum) be restricted by the $[f] \rightarrow (\lambda, \mu)$ constraint to a subplane of the full (β, γ)-plane. Some (β, γ) values are ruled out by the Pauli Principle. Indeed, since within a shell there is always a leading (λ, μ) for which C_2 has a maximum, the shell structure places a constraint on the maximum β deformation. The only way around this dilemma is to allow for the excitation of nucleons into higher lying shells, therefore including particle-hole configurations. The symplectic model, which we will study later, allows this to be done in a very controlled and structured manner.

7.2.4. Excitation Spectra and Transition Strengths

The group contraction argument given above shows that for relatively low L values in a (λ, μ) irrep with large C_2, one expects results for H_{ROT} and H_{SU3} to agree. The L content in a (λ, μ) irrep is given by the following rule:

$$L = K, K+1, \ldots, \lambda + \mu + 1 - K \quad \text{if } K \neq 0$$
$$\text{and } L = \lambda + \mu, \lambda + \mu - 2, \ldots, \text{or } 0 \quad \text{when } K = 0 \qquad (7.17)$$
$$\text{where } K = \min\{\lambda, \mu\}, \min\{\lambda, \mu\} - 2, \ldots, 1 \text{ or } 0.$$

The maximum L value in the (λ, μ) irrep is therefore $L = \lambda + \mu$, and since the eigenvalue of C_2 can be written as $(\lambda + \mu)(\lambda + \mu + 3) - \lambda\mu$, the ratio $L/C_2 \sim 1/(\lambda + \mu)$—which is a measure for the closeness of the $SU(3)$ and $T_5 \wedge SO(3)$ algebras—is certainly small for most shell-model applications indicating similar results for the macroscopic and microscopic descriptions. One can actually compare rotor and $SU(3)$ expressions for the matrix elements of their respective Q's, and such a comparison shows that the matrix elements of H_{ROT} and H_{SU3} typically agree to within 10% or better. Differences in the two theories grow with both K and L; but with significant deviations only for L values near the top of each K band, $L \sim \lambda + \mu + 1 - K$.

Since rotational behavior is a characteristic feature of many nuclei, it is worthwhile to provide an example of the close agreement between the rotor and $SU(3)$ theories. Before doing so, however, it is important to note that H_{ROT} has an additional symmetry, associated with rotations of the system by 180° about its principal axes. These rotations are generated by three special operators $T_\alpha = \exp(i\pi I_\alpha)$ where I_α is (as before) the projection of the total angular momentum on the α-th body-fixed axis. The set $\{E, T_1, T_2, T_3\}$ where E denotes the identity, closes on itself ($T_\alpha T_\beta = \epsilon_{\alpha\beta\gamma} T_\gamma$ and $T_\alpha^2 = E$) to form the point group D_2, known as the Vierergruppe from early work on the quantum rotor. This means that the Hamiltonian matrix can be brought into block diagonal form with the four classes of D_2—usually called A and B_α with $\alpha = 1, 2,$ and 3—labelling the distinct blocks. Specifically, a basis for the rotor which takes this symmetry into account is given by

$$\Psi^{(\lambda,\ \mu)|K|LM} = \frac{1}{\sqrt{2(1 + \delta_{K,0})}} (D^L_{KM} + (-1)^{\lambda+\mu+L} D^L_{-KM}) \quad (7.18)$$

where the $D^L_{K,M}$ are the usual rotation matrices, and the odd-even character of the integer labels λ and μ determine the symmetry class of the wavefunction. For example, if λ and μ are both even, then $\Psi^{(\lambda,\mu)|K|LM}$ transforms under D_2 according to its symmetry class A. On the other hand, if (λ, μ) is (even, odd), (odd, odd) or (odd, even), then the symmetry class is B_1, B_2, or B_3, respectively. This feature is important because it carries over to the shell-model case. The corresponding $SU(3)$ shell-model basis states—starting with the spatial configuration $|m[f]\alpha(\lambda, \mu)KLM\rangle$—are specifically given by

$$|[f]\alpha(\lambda, \mu)|K|LM\rangle = \frac{1}{\sqrt{2(1 + \delta_{K,0})}} (|[f]\alpha(\lambda, \mu)KLM\rangle + (-1)^{\lambda+\mu+L} |[f]\alpha(\lambda, \mu) - KLM\rangle). \quad (7.19)$$

It is now possible to turn this argument around and learn additional information about the structure of nuclei. Since there is no rule specifying that the $[f] \rightarrow (\lambda, \mu)$ reduction yields only even λ and μ values—including even-even cases—microscopic shell-model configurations with other than D_2 symmetry type A are likely to occur. Indeed, one should expect to find several rotational sequences within a nucleus, and these can carry any of the four D_2 symmetry types. This differs from what is traditionally accepted for a collective model description of even-even nuclei.

In Fig. 7.3, comparisons of excitation spectra for H_{ROT} and H_{SU3} in the $(\lambda, \mu) = (30, 8)$ irrep of $SU(3)$ are given. This (30, 8) irrep is the leading one in a pseudo-$SU(3)$ description of the ^{168}Er rare

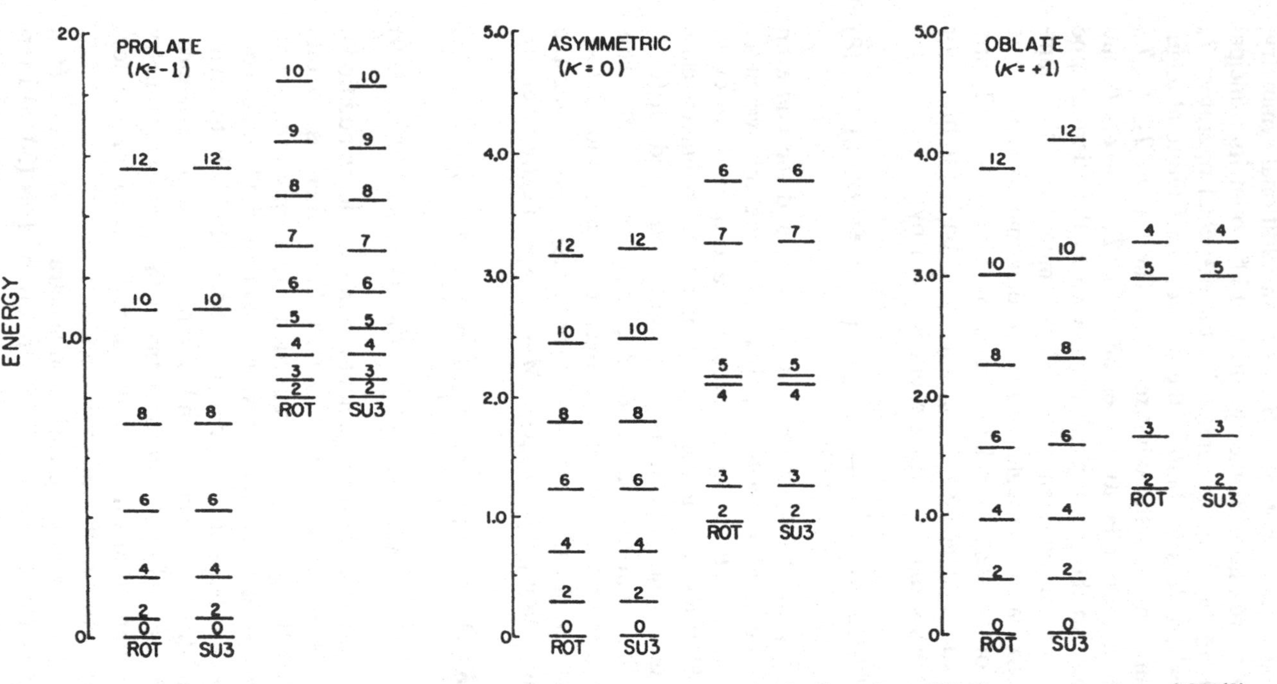

FIGURE 7.3. *Comparison of the excitation spectra for* H_{ROT} *and* H_{SU3} *in the* $(\lambda, \mu) = (30,8)$ *representation of* SU(3) *for three values of the asymmetry parameter,* $\kappa = -1$ *(prolate),* $\kappa = 0$ *(asymmetric) and* $\kappa = +1$ *(oblate), defined by* $\kappa = (2A_1 - A_3 - A_2)/(A_3 - A_2)$. *The parameters of* H_{SU3} *were determined from* H_{ROT} *through eq.* (7.9) *with the* λ_α *values determined through eq.* (7.13). *(From Leschber,* 1987.)

earth nucleus. Results are shown for three rotor geometries: $\kappa = -1$ (prolate, $A_2 = A_1 < A_3$); $\kappa = 0$ (asymmetric, $A_2 < A_1 = (A_2 + A_3)/2$); and $\kappa = +1$ (oblate, $A_2 < A_1 = A_3$). The κ measure, which is part of the molecular spectroscopy rotor jargon, is known as the asymmetry parameter, and is defined by the simple expression $\kappa = (2A_1 - A_3 - A_2)/(A_3 - A_2)$ where by convention $A_2 \leq A_1 \leq A_3$. It is not necessary to consider other A_α ranges since they can all be obtained from this choice by a simple permutation of the axis labels. The match-up is extremely good; even the spin inversions in the side bands for the asymmetric and oblate cases are accurately reproduced. Note in particular that under the mapping eq. (7.9) with the $\lambda_\alpha \leftrightarrow (\lambda, \mu)$ connection given by eq. (7.13), the same (λ, μ) reproduces the excitation spectra of the full range of rotor geometries. Since the (λ, μ) quantum numbers imply a specific distribution of oscillator quanta, which is the same regardless of the κ values chosen, the connection between the dynamics and the concept of a geometrical shape is not necessarily as straightforward as a liquid drop picture might suggest.

There are various other measures one can use to compare the theories; one which also involves excitation energies uses sum rules. The simplest and best known of these rules is one for the A type D_2 rotor symmetry: the sum of the two 2^+ energies is equal to the energy of the 3^+ state, $E_{2_1^+} + E_{2_2^+} = E_{3^+}$. This is an example of the more general exact result, $\text{Trace}(H_{\text{ROT}})_L = \text{Trace}(H_{\text{ROT}})_{L+1}$, which applies for any A type D_2 rotor geometry. The sum rule result is different for the four D_2 symmetry types, and as for the excitation spectra, the $SU(3)$ results reproduce the rotor sum rules. This is shown for traces of the two Hamiltonians in Fig. 7.4 for all four symmetry types, and this correspondence extends to quadratic and higher-order sums. For example, a comparison of the energy centroids $\{\epsilon = d^{-1}\text{Trace}(H)\}$ and variances $\{\sigma^2 = d^{-1}\text{Trace}[(H - \epsilon)^2]\}$ are shown in Fig. 7.5. Here one sees that the ratios $\epsilon_{SU3}/\epsilon_{\text{ROT}}$ and $(\sigma_{SU3}/\sigma_{\text{ROT}})^2$ begin to deviate significantly from unity when $L \sim (\lambda + \mu)/4$, which is consistent with the group contraction measure.

In Fig. 7.6, results are given comparing E2 transition strengths between eigenstates of H_{ROT} and H_{SU3} for different geometries; the example shown is a cross-over transition between bands. Since each interband transition matrix elements are typically weaker by an order of magnitude than the matrix elements of intraband transitions, this is a very tell-tale test, however as before, these results

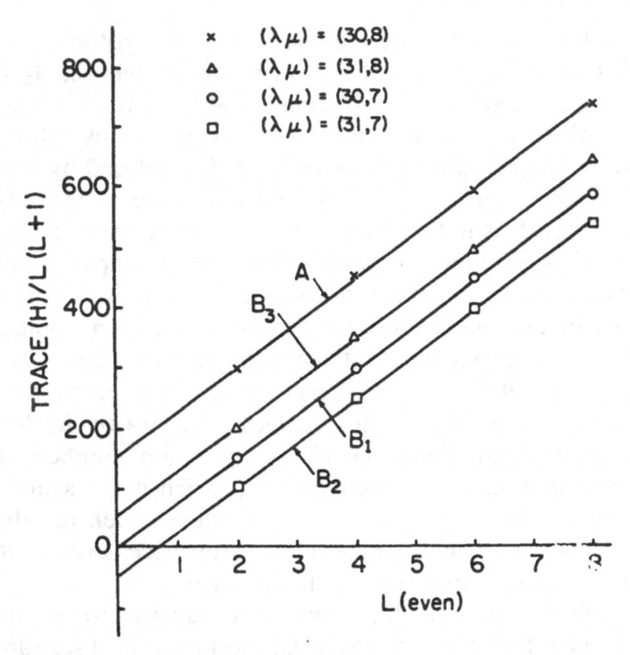

FIGURE 7.4. *Comparison of results for traces of* H_{ROT} *and* H_{SU3} *divided by* L(L+1) *for even* L *values in the* A *and* $B_\alpha(\alpha = 1, 2, 3)$ *symmetry types of the point group* D_2. *The same set of inertia parameters was used in all four cases with* (λ,μ) *values chosen as indicated. The parameters of* H_{SU3} *were determined from* H_{ROT} *through eq.* (7.9) *with the* λ_α *values determined through eq.* (7.13). *Whereas algebraic results exist for* H_{ROT}, *the* H_{SU3} *numbers were determined by explicit construction of the various L-submatrices and numerically determining the trace of each. (From Leschber,* 1987.)

confirm the near equivalence of the two theories. To demonstrate that a difference exists, a comparison for $(\lambda, \mu) = (8, 4)$—the leading irrep for the *ds*-shell nucleus ^{24}Mg—is shown in Fig. 7.7 using the most asymmetric ($\kappa = 0$) rotor geometry. In this case, the dimensionality of the *SU*(3) irrep is relatively small, and as stated above, differences can be easily noted for the larger *L* values; while in particular, the higher the *K* the larger the discrepancy, overall the agreement is again remarkable, especially for low-lying states.

These examples help to erase some of the mystery surrounding the origin of rotational motion in nuclei and the success of geometrical models. Rotational motion arises as a consequence of the approximate goodness of the *SU*(3) symmetry and the fact that the residual interaction, as complicated as it may appear in a non-*SU*(3)

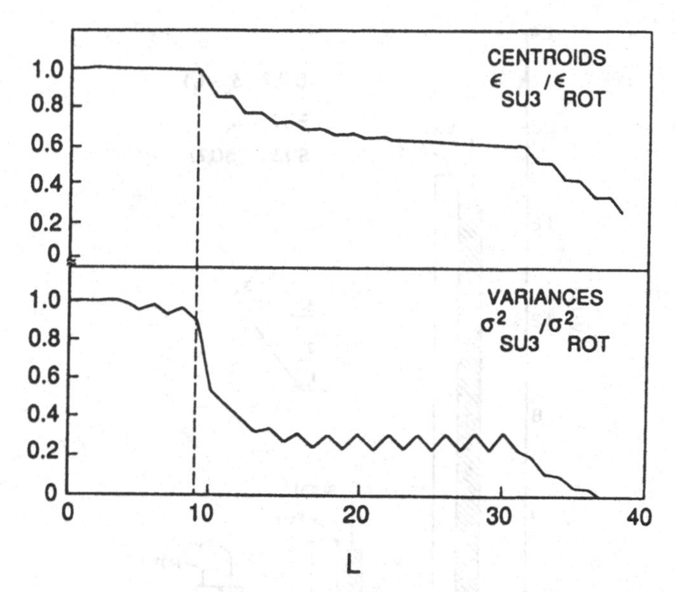

FIGURE 7.5. *Comparison of centroid and variance measures for the asymmetric rotor* ($\kappa = 0$) *and its algebraic image in the* $(\lambda, \mu) = (30, 8)$ *irrep of* SU(3). *Up to* $L = \min(\lambda, \mu) + 1 = 9$, *every state of the rotor has an algebraic counterpart. The fall off for* $L > 9$ *is a direct consequence of "missing" states in the algebraic theory. Note that for* $L = 37$ *and* $L = 38$ *the variance, which is a measure of the spread in eigenvalues, goes to zero (as it must) for there is but one of these states in the* (30,8) *representation. (From Leschber,* 1987.)

basis, behaves to a large extent like H_{SU3}. It remains to be shown just how the effective interaction gets renormalized into the H_{SU3} form, nonetheless, the fact that it does seems certain. In addition, many other issues remain unresolved, such as E2 transition strengths factors of 10 to even 1000 times greater than even an $SU(3)$ picture suggests. A recent surprise, again associated with rotations, is the incredible regularities found among rotational bands in neighboring and sometimes even rather dispersed (normal as well as superdeformed) nuclear species. Despite these matters however, it is clear that rotational bands appear as a consequence of the existence of $SU(3)$—and pseudo-$SU(3)$—symmetry in nuclei.

7.2.5. Shell-Model Operator for *K*-band Splitting

To close out this section on the quantum rotor and its shell-model realization, we want to use what we have learned from the H_{ROT}

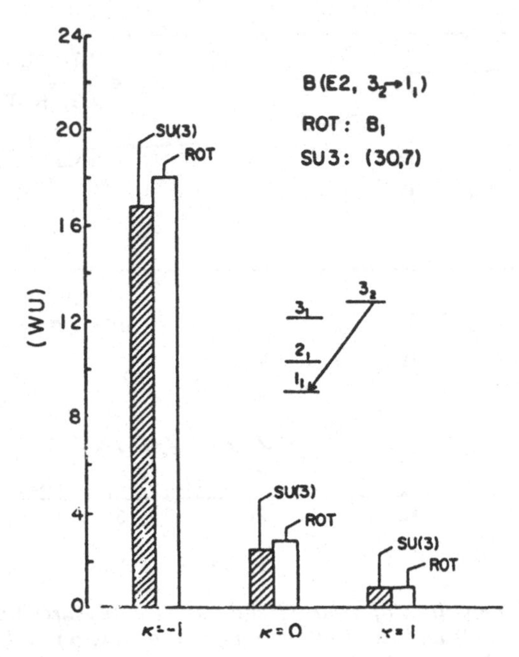

FIGURE 7.6. *Comparison of a representative B(E2) transition strength between two eigenstates of* H_{ROT} *and* H_{SU3} *for the* B_1 *symmetry type of* D_2*. Results are shown for* $\kappa = -1$ *(prolate),* $\kappa = 0$ *(asymmetric) and* $\kappa = +1$ *(oblate), where the asymmetry parameter* $\kappa = (2A_1 - A_3 - A_2)/(A_3 - A_2)$*. The parameters of* H_{SU3} *were determined from* H_{ROT} *through eq.* (7.9) *with the* λ_α *values determined through eq.* (7.13) *for the indicated* (λ,μ)*. (From Leschber,* 1987.)

$\leftrightarrow H_{SU3}$ mapping to consider *K*-band splitting—another prominent feature associated with rotational phenomena in nuclei. For example in ^{24}Mg, the 2_1^+ state lies at an excitation energy of 1.369 MeV above the ground state while the 2_2^+ state lies at 4.123 MeV (see Fig. 7.8). Since this is well below the second 0^+ state (6.432 MeV) it is natural to associate it with the ground state rotational sequence. Indeed, the simplest scheme makes the 2_1^+ level a member of the $K^\pi = 0^+$ ground band and the 2_2^+ state the band head of the second $K^\pi = 2^+$ band. On the other hand, the 2.754 MeV energy difference between the two 2^+ states is roughly twice the rotational $L(L + 1)$ excitation energy. Hence although we have identified them as members of the same rotational sequence, the 2^+ split is large. We can easily account for this with the simple Hamiltonian:

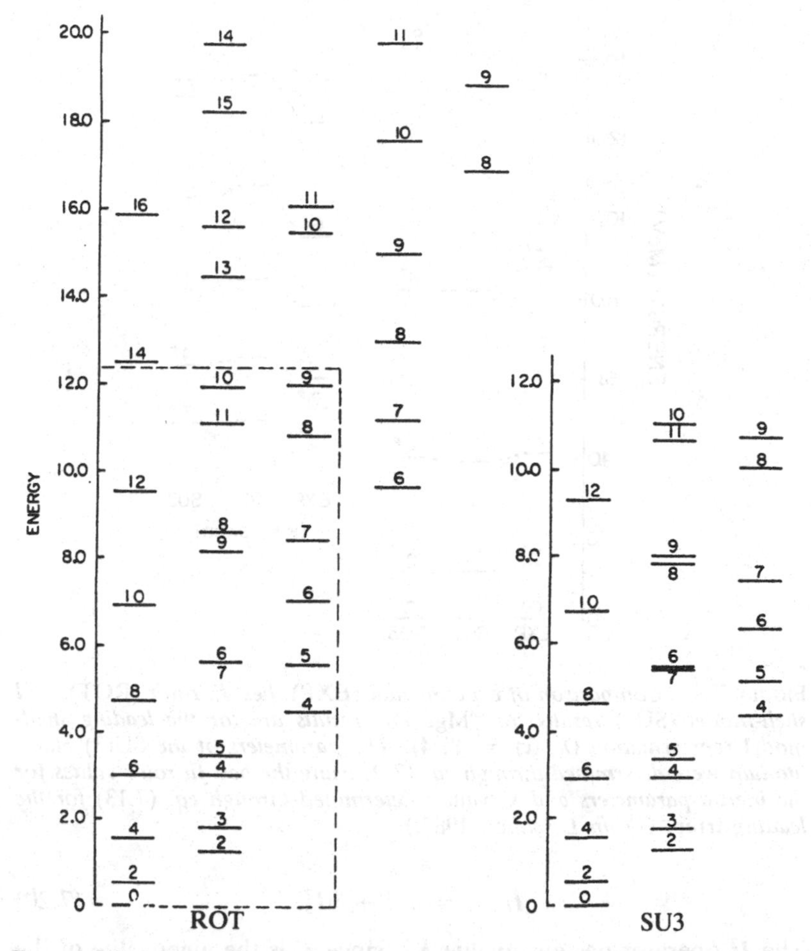

FIGURE 7.7. *Comparison of the excitation spectra of* H_{ROT} *with* H_{SU3} *for* $\kappa = 0$ *in the* $(\lambda, \mu) = (8, 4)$ irrep *of* SU(3). *Since* SU(3) *is a compact group its irreps are finite-dimensional. For the* $(\lambda, \mu) = (8, 4)$ *case this means that only rotor levels within the broken-boxed area have an* SU(3) *counterpart. Note that a state-by-state comparison shows excellent agreement for the lowest members in all three* K *bands, but this deteriorates as one goes to higher* L *values. The differences are larger and occur for lower* L *values the higher the* K. (*From Leschber,* 1987.)

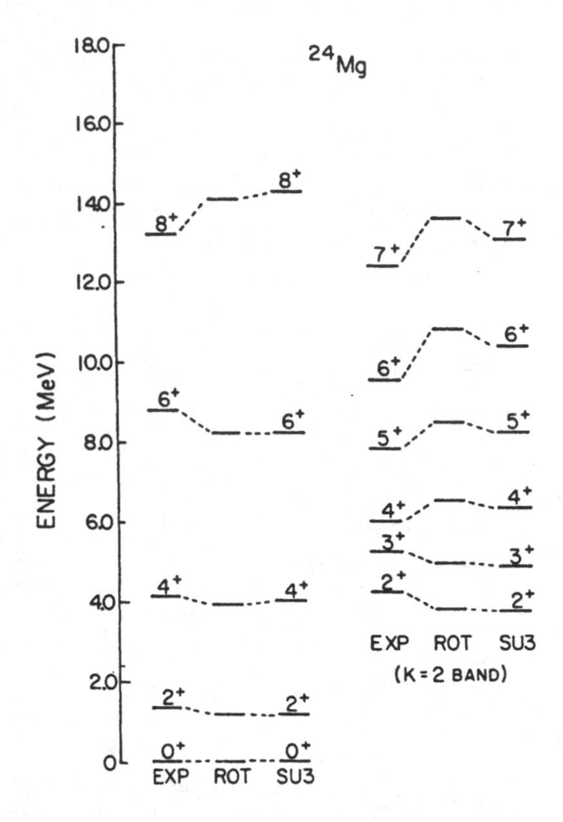

FIGURE 7.8. *Comparison of experimental* (EXP), *best-fit rotor* (ROT), *and shell-model* (SU3) *results for* ^{24}Mg. *The results are for the leading shell-model representation* (λ, μ) = (8, 4). *The parameters of the* SU(3) *Hamiltonian were determined through eq.* (7.9) *using the best-fit rotor values for the inertia parameters and* λ_α *values determined through eq.* (7.13) *for the leading* irrep. (*From Leschber,* 1987.)

$$H_{\mathrm{ROT}} = aI^2 + bI_3^2. \tag{7.20}$$

The I_3^2 operator has eigenvalue K^2, since K is the eigenvalue of the projection I_3 of I on the intrinsic symmetry axis of the system. By choosing a = 1.369/6 = 0.2282 MeV and b = 2.754/4 = 0.6885 MeV in eq. (7.20), we can reproduce the 2_1^+ and 2_2^+ energies of ^{24}Mg exactly. But this is not the whole story, because the two 4^+ levels are separated by only 1.887 MeV, nearly 1 MeV less than the separation of the 2^+ levels, though this simple form for H_{ROT} would make the separation of the two sets the same. The I_3^2 operator only shifts the K bands relative to one another; it does not otherwise affect the $L(L + 1)$ rotational pattern. Some improvement can be realized if we go to an asymmetric shape. For example, if rather than restricting $A_1 = A_2$ which specifies a prolate geometry ($\kappa = -1$), we adjust all three A_α's in H_{ROT} to get a best fit to the available data, including the 4^+ states as well as higher L's and even the easily calculated B(E2) transition strengths, improvement will occur. The results one gets when this is done for the ^{24}Mg is shown as H_{ROT} in Fig. 7.8 along with the corresponding $SU(3)$ shell-model results.

While we can find examples of rotors with large K-band splitting in the *ds*-shell (^{24}Mg being one of course), these nuclei have far too few valence nucleons to qualify as strongly deformed nuclei with well-developed rotational features. In heavier nuclei on the other hand, it is very common to find rotational spectra that follow H_{ROT} predictions through 10 and even more units of total angular momentum. An often cited example is ^{168}Er. If we use the same logic as we did in the ^{24}Mg case, then $a = 0.07980/6 = 0.01330$ MeV and $b = 0.74131/4 = 0.18533$ MeV for H_{ROT} of eq. (7.20). The b/a ratio is a measure of the size of the K-band splitting relative to rotational excitation energies. For ^{168}Er this ratio is 13.9 as compared to 3.02 for ^{24}Mg. On the other hand, the splitting of the 4^+ levels of ^{168}Er calls for $b = 0.73057/4 = 0.18264$ MeV which is only a 1.47% difference from the correspond 2^+ value. In the ^{24}Mg case the difference is 31.5%, so even though the K-band splitting is much greater than for ^{24}Mg, the ^{168}Er spectrum is much more rotational. It is common—among normally deformed nuclei—to find large K-band splitting when there is large deformation. Since superdeformed rotational sequences show almost no deviation from simple rotor predictions, it will be interesting to see if $K \neq 0$ bands exist in these systems, and if they do, the size of the band splitting.

While we can easily account for K-band splitting in a collective model framework, this feature has proven to be one of the most difficult to duplicate in shell-model calculations. A usual shell-model Hamiltonian has the form,

$$H = H_0 + C\sum_i l_i \cdot s_i + D\sum_i l_i^2 + \sum_{i<j} V_{ij}, \qquad (7.21)$$

where H_0 is the harmonic oscillator term, the C and D terms are one-body interactions chosen to reproduce the observed level ordering and shell structure, and the V_{ij} term is a residual two-body interaction. We have already seen that when C and D are taken to be zero and the two-body term is either the pairing operator P or the algebraic quadrupole-quadrupole interaction $Q^a \cdot Q^a$, the eigensolutions of this Hamiltonian can be respectively given analytically in a jj and LS → $SU(3)$ coupled basis. In particular, since $Q^a \cdot Q^a = 4C_2 - 3L^2$, the quadrupole-quadrupole form yields an $L(L + 1)$ spectrum. But this interaction does not distinguish between K-bands; the same L values in different bands are degenerate. For ^{24}Mg case this means the 2.754 MeV separation of the 2_1^+ and 2_2^+ states cannot be duplicated using only a $Q^a \cdot Q^a$ interaction. Indeed, even if we took C and D to be nonzero, and a linear combination of the two-body interactions P and $Q^a \cdot Q^a$, we would not be able to adequately reproduce the observed K-band splitting. The same applies to heavy deformed nuclei of the rare earth and actinide regions, although for these we need to use a pseudo-$SU(3)$ description, as we will explain in the next section.

There is, of course, a very simple solution to this dilemma. Since we know the shell-model image of H_{ROT}, we also know the shell-model image of the I_3^2 operator, which has eigenvalues K^2 and therefore generates K-band splitting, see eq. (7.7). If we denote this operator by K_{op}^2, then it has the following many-body shell-model form:

$$K_{\text{op}}^2 = (\lambda_1\lambda_2\lambda_3 L^2 + \lambda_3^2 X_3^a + \lambda_3 X_4^a)/(2\lambda_3^3 + \lambda_1\lambda_2\lambda_3),$$
$$\lambda_1 = \tfrac{1}{3}(-\lambda + \mu), \quad \lambda_2 = \tfrac{1}{3}(-\lambda - 2\mu - 3), \qquad (7.22)$$
$$\lambda_3 = \tfrac{1}{3}(2\lambda + \mu + 3).$$

Here, as with H_{SU3}, we have replaced the geometric X^cs with the algebraic X^as. Though this form appears to be simple, it is by no means a standard (1 + 2)-body interaction. Specifically, since X_3^a and X_4^a are cubic and quartic in the generators of $SU(3)$, they have 3-body and (3 + 4)-body parts, respectively, in addition to (0 + 1 + 2)-body terms. An important question that has not yet been answered is whether or not K_{op}^2 can be rewritten in (0 + 1 + 2)-body form. While some still resist accepting the necessity of including anything other than standard (0 + 1 + 2)-body interactions, it should be clear that the description is simpler when (3 + 4)-body terms are allowed. In this regard, it is important to note that K_{op}^2 is far from a general (3 + 4)-body operator, and it certainly cannot be ruled out as an appropriate effective interaction.

The eigenvalues of K_{op}^2 for the leading $(\lambda, \mu) = (30, 8)$ pseudo-

$SU(3)$ irrep for ^{168}Er are shown in Fig. 7.9. That these are not exactly 0, $2^2 = 4$, $4^2 = 16$, etc., is a direct result of the fact that $SU(3)$ is a compact group with finite dimensional irreps, while $T_5 \wedge SO(3)$ is noncompact and therefore has infinite dimensional irreps with K bands and L values extending out indefinitely. Nonetheless, as we discovered earlier, the agreement is excellent for low angular momentum values, progressively degenerating the higher the K and the larger the L. Actually, analytic results are available for matrix element of the K_{op}^2 in an Elliott basis, and these go over into the corresponding rotor result in the large (λ, μ) limit. Results similar to those given in Fig. 7.8 for ^{25}Mg are shown for ^{168}Er in Fig. 7.10. Clearly ^{168}Er is a better rotor than ^{24}Mg. This conclusion is borne out when comparisons of electromagnetic transitions are taken into account as well.

In conclusion, as a consequence of these developments—the

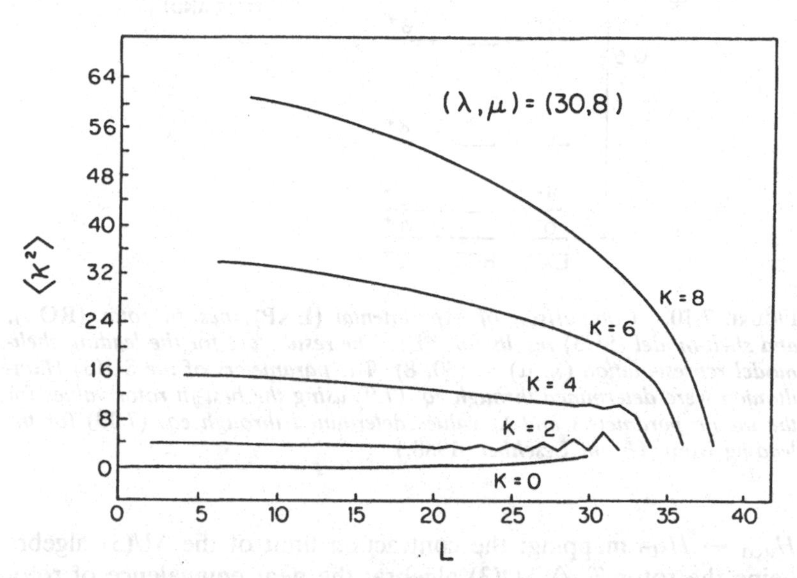

FIGURE 7.9. *Eigenvalue spectrum of the* K_{op}^2 *operator, eq.* (7.22), *in the* (30, 8) *irrep of* SU(3). *Note that whereas for members of the yrast* (K = 0) *band the eigenvalues are nearly zero, even for* L *values near the top of the band, for yrare* (K = 2) *and higher* (K = 4, 6, 8) *band members, there is a fall-off from the rotor* K^2 *values that increases with increasing* L *and is more pronounced the larger the* K. *The reason for this follows from the fact that* SU(3) *is a compact group with finite-dimensional irreps while the symmetry group of the rotor,* $T_5 \times SO(3)$, *is non-compact and therefore has infinite-dimensional representations. (From Naqvi and Draayer, 1990.)*

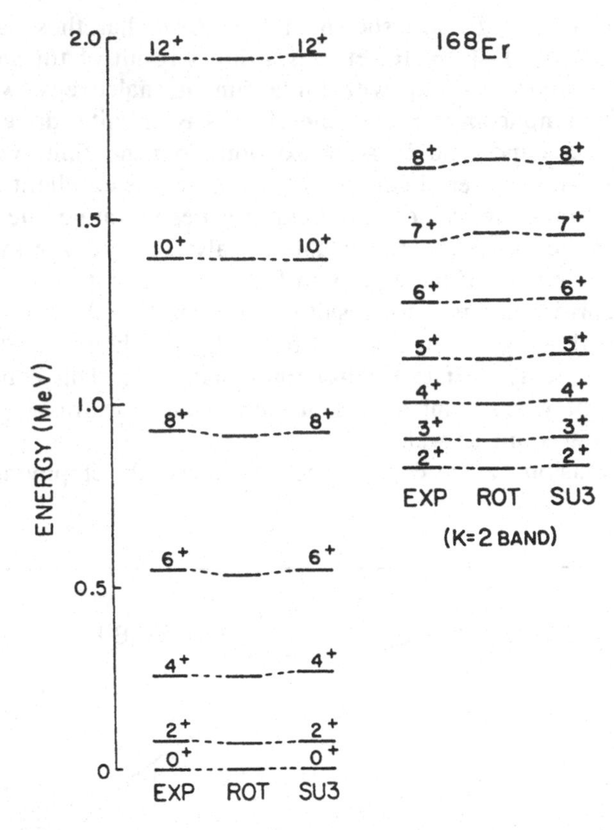

FIGURE 7.10. *Comparison of experimental* (EXP), *best-fit rotor* (ROT), *and shell-model* (SU3) *results for* ^{168}Er. *The results are for the leading shell-model representation* $(\lambda, \mu) = (30, 8)$. *The parameters of the* SU(3) *Hamiltonian were determined through eq.* (7.9) *using the best-fit rotor values for the inertia parameters and* λ_α *values determined through eq.* (7.13) *for the leading* irrep. (*From Leschber,* 1988.)

$H_{\text{ROT}} \leftrightarrow H_{SU3}$ mapping; the contraction limit of the $SU(3)$ algebra being the rotor $T_5 \wedge SO(3)$ algebra; the near equivalence of rotor and $SU(3)$ excitation spectra and B(E2) values; a shell-model form for an operator that generates K-band splitting; and so on—we claim that whenever rotor-like features arise in a finite-dimensional many-body system, one should look for an underlying $SU(3)$ symmetry. As we will see next, there is an $SU(3)$ for heavy nuclei that emerges as a result of good pseudo-spin symmetry.

7.3. PSEUDO-SPIN SYMMETRIES

The pseudo-spin concept follows directly and most naturally from the single-particle shell-model scheme independently introduced by Mayer as well as Haxel, Jensen and Suess. We will therefore begin with a short review of the single-particle shell-model scheme—the simplest of all microscopic theories. It accounts for the major shell closures and correctly predicts the ground state spin and magnetic moments of most odd-A nuclei. A convenient choice for the central field of this theory is the three-dimensional isotropic harmonic oscillator (H_0). This is usually augmented with one-body spin-orbit ($l \cdot s$) and orbit-orbit (l^2) interactions,

$$H = H_0 + Cl \cdot s + Dl^2. \tag{7.23}$$

The l^2 term ($D < 0$) pushes high angular momentum states down relative to those with lower l values, a feature that occurs automatically when a more realistic interaction like a Woods–Saxon form is used for the central potential. The phenomenological $l \cdot s$ term ($C < 0$)—coupling space and spin degrees of freedom—is required to achieve shell closures at the magic nucleon numbers, 2, 8, 20, 50, 82, 126 and 184. Unfortunately, the required value for C is so large that the spin-orbit interaction usually destroys the underlying $SU(3)$ symmetry of the isotropic oscillator for all but light ($A \lesssim 28$) nuclei, and renders it of little apparent value in attempts at unraveling the structure of heavy ($A \gtrsim 100$) systems. Specifically, for heavy nuclei the $j = n + \frac{1}{2}$ orbital of the n-th oscillator shell (which includes levels with $j = l \pm \frac{1}{2}$ and $l = n, n - 2, \ldots, 1$ or 0) is pushed down among the orbitals of the next lower shell. This yields new shells with normal parity $j = \frac{1}{2}, \frac{3}{2}, \ldots, n - \frac{1}{2}$ orbitals; plus a $j = n + \frac{3}{2}$ unique parity intruder from the shell above.

In this section we will show that this seemingly unfavorable scenario actually gives way to a much more favorable one for heavy nuclei, because $C \approx 4D$ or the Nilsson parameter $\mu = 2D/C \approx 0.5$. First of all, we will see that this leads to good pseudo-spin symmetry, because when $C \approx 4D$ (an empirical result deducible from single-particle energy systematics) the splitting generated by the $l \cdot s$ and l^2 interactions can be duplicated by a pseudo-oscillator Hamiltonian plus a pseudo l^2 interaction with (at most) a very small symmetry-breaking pseudo $l \cdot s$ term. We will then go beyond empirical evidence and show that the $C \approx 4D$ condition is actually consistent with relativistic mean-field predictions. In the case of the many-particle extension of the single-particle picture, we will discover that when the residual deformation-inducing quadrupole-quadrupole interaction dominates over the pseudo l^2 term—which must be the case when well-developed rotational bands are present—each pseudo-spin symmetry will have associated with it an yrast band that is dominated by its leading pseudo-$SU(3)$ irrep, the symmetry group of the pseudo-oscillator. We will also briefly consider the intruder's role in building up quadrupole collectivity, a subject only now being given careful consideration by algebraic structure theorists. To round out this section, we will give some typical results for characteristic phenomena such as backbending and electromagnetic transitions in rare earth and actinide nuclei. In a later section we will go beyond pseudo-$SU(3)$ to the pseudo-symplectic extension of the theory, which incorporates major shell mixing ($2n\hbar\omega$, $n = 1, 2, \ldots$) into the picture. This extension reinforces the goodness of the pseudo-$SU(3)$ picture. As a consequence of this synergism, strongly deformed configurations (such as those found in superdeformed bands) are expected to be simple when expressed in the framework of the pseudo-$SU(3)$ model and its pseudo-symplectic extension.

7.3.1. Spin-Orbit Doublets

Though historically the pseudo-spin concept has been introduced differently (see below), the quickest way to gain an appreciation for its importance and simplicity is by considering Fig. 7.11, in which the eigenvalues of the spherical ($\beta = 0$) single-particle Hamiltonian of eq. (7.23) are plotted as a function of the parameter $\mu = 2D/C$ which measures the relative strength of the l^2 and $l \cdot s$ terms. For the special value $\mu = 0.5$, the orbital pairs with $j = l + \frac{1}{2}$ and $j = (l + 2) - \frac{1}{2}$ are degenerate for all l values. Furthermore, notice that the splitting of these degenerate pairs follows an $\tilde{l}(\tilde{l} + 1)$ rule where $\tilde{l}$ is the average l of the pair, that is, $\tilde{l} = [l + (l + 2)]/2 = l + 1$. (Throughout, we will use a tilde (~) over a quantity to denote its pseudo-form.) This association is characterized as a special *normal* ↔ *pseudo* unitary transformation that makes this the degeneracy of pseudo-spin-orbit partners, $\tilde{\mathbf{j}} = \tilde{\boldsymbol{\ell}} + \tilde{\mathbf{s}}$ where $j = \tilde{j}$, $\tilde{l} = l + 1$ and $\tilde{s} = \frac{1}{2}$. This is shown schematically in Fig. 7.12. For the single-particle basis states this transformation has the form

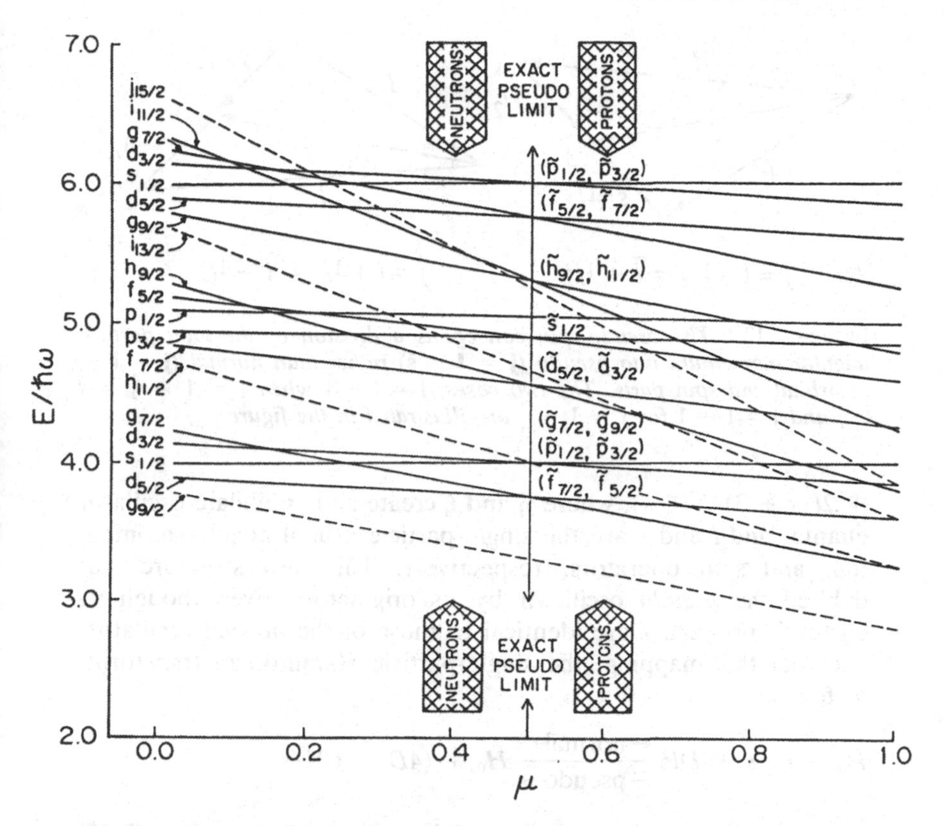

FIGURE 7.11. *Plot of the eigenvalues of the reduced single-particle Hamiltonian given by* $H/\hbar\omega = n - \kappa(2l\cdot s + \mu l^2)$, *where* $\mu = 2D/C$ *and* $\kappa = -C/2\hbar\omega$, *for the value* $\kappa = 0.05$ *and* $0.0 \leq \mu \leq 1.0$. *Notice that the* $j = (l+2)-\frac{1}{2}$ *and* $j = l+\frac{1}{2}$ *levels are degenerate for* $\mu = 0.5$. *(From Bahri, Draayer and Moszkowski, 1992.)*

$$|\tilde{n}(\tilde{l}, \tilde{s})j\tilde{m}\rangle = U_{njm,\tilde{n}j\tilde{m}}(l, \tilde{l})|n(l, s)jm\rangle, \qquad (7.24)$$

$$U_{njm,\tilde{n}j\tilde{m}}(l, \tilde{l}) = \delta_{n-1,\tilde{n}}\delta_{j,\tilde{j}}\delta_{m,\tilde{m}}\delta_{l\pm 1/2,\tilde{l}\mp 1/2}.$$

From the structure of $U_{njm,\tilde{n}j\tilde{m}}(l, \tilde{l})$, it should be clear that this unitary transformation amounts to a simple relabelling of the basis states with levels of the n-th shell (except $j = n + \frac{1}{2}$) associated with levels of the $\tilde{n}$-th shell of another oscillator, $\tilde{n} = n - 1$. A label-independent operator form which effects this special *normal* ↔ *pseudo* unitary transformation is known: $U_{njm,\tilde{n}j\tilde{m}}(l, \tilde{l}) = 2(\eta\cdot\xi$

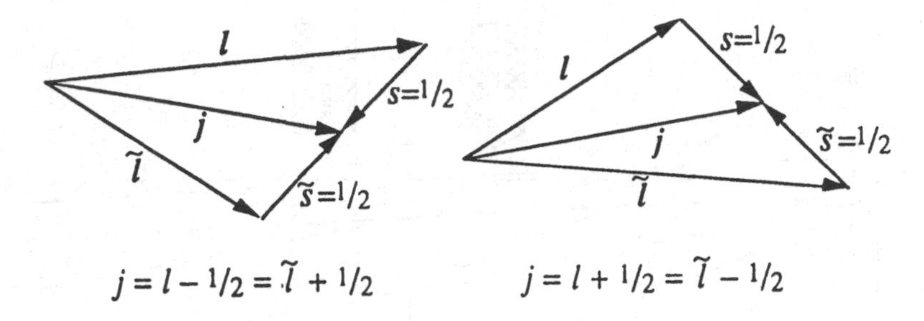

FIGURE 7.12. *The pseudo-spin concept is a division of the total particle angular momentum into pseudo* ($\mathbf{j} = \tilde{\mathbf{l}} + \tilde{\mathbf{s}}$) *rather than normal* ($\mathbf{j} = \mathbf{l} + \mathbf{s}$) *orbital and spin parts. The two cases,* $\tilde{l} = l + 1$ *when* $j = l + \frac{1}{2} = \tilde{l} - \frac{1}{2}$ *and* $\tilde{l} = l - 1$ *for* $j = l - \frac{1}{2}$ *are illustrated in the figure.*

$+ 2l\cdot s + 3)^{-1/2}(\xi\cdot s)$, where η and ξ create and annihilate oscillator quanta, and l and s are the single-particle orbital angular momentum and spin operators, respectively. This new structure was dubbed the *pseudo* oscillator by its originators, even though its algebraic properties are identical to those of the *normal* oscillator.

Under this mapping, the single-particle Hamiltonian transforms as follows:

$$H_0 + Cl\cdot s + Dl^2 \underset{\text{—pseudo}\rightarrow}{\overset{\leftarrow\text{normal—}}{\rule{3em}{0.4pt}}} \tilde{H}_0 + (4D - C)\tilde{l}\cdot\tilde{s} + D\tilde{l}^2 + (\hbar\omega + 2D - C). \quad (7.25)$$

Since the $(\hbar\omega + 2D - C)$ term is a constant, the pseudo form for the interaction, $\tilde{H} = \tilde{H}_0 + \tilde{C}\tilde{l}\cdot\tilde{s} + \tilde{D}\tilde{l}^2$, has the *same* excitation spectrum as the normal one ($H = H_0 + Cl\cdot s + Dl^2$) when $\hbar\tilde{\omega} = \hbar\omega$, $\tilde{C} = (4D - C)$ and $\tilde{D} = D$. This transformation is important, because for real ($A \gtrsim 100$) systems, $C \approx 4D$, so $\tilde{C} \approx 0$. As specifically indicated in the figure, $\mu_\nu \approx 0.4$ and $\mu_\pi \approx 0.6$ (ν for neutrons and π for protons); this places medium and heavy mass nuclei very close to the exact pseudo-spin limit ($\mu = 0.5$) of the theory. Indeed, the average μ value is almost exactly 0.5. The familiar single-particle shell-model Hamiltonian for medium and heavy nuclei can therefore be replaced by a less familiar, but equivalent, pseudo form which is inherently simpler due to its much smaller spin-orbit interaction strength.

Actually, an appreciation for the importance of the pseudo-spin picture was first gained in a different way, through consideration of diagrams like those given in Fig. 7.13 which show eigenvalues of the Nilsson Hamiltonian,

$$H = H_0 + Cl\cdot s + Dl^2 - \tfrac{1}{2}\chi(\kappa\beta)r^2 Y_{20}(\theta, \varphi), \qquad (7.26)$$

plotted as a function of the deformation β. This form is the single-particle Hamiltonian of eq. (7.23) augmented with a deformed (prolate) field. This type of single-particle Hamiltonian may be obtained by averaging over one of the Q^c operators in the $Q^c\cdot Q^c$ product: $Q^c\cdot Q^c \rightarrow Q^c\cdot\langle Q^c\rangle \rightarrow (\kappa\beta)r^2Y_{20}(\theta, \varphi)$, where $\kappa = 3k\sqrt{16\pi/5}$, and as before, $k^2 = (5/9\pi)(A\bar{r}^2)^2$. Clearly, eq. (7.26) reduces to the simpler eq. (7.23) expression in the limit $\chi \rightarrow 0$. The Nilsson levels are normally labeled with their asymptotic (cylindrical geometry) quantum numbers $\Omega[Nn_z\Lambda]$: N is the principal quantum number (denoted n elsewhere in this chapter and not to be confused with the number of bosons) and is therefore equal to the total number of oscillator quanta; $n_z = N, N - 1, \ldots$, is the number of oscillator quanta along the body-fixed symmetry axis

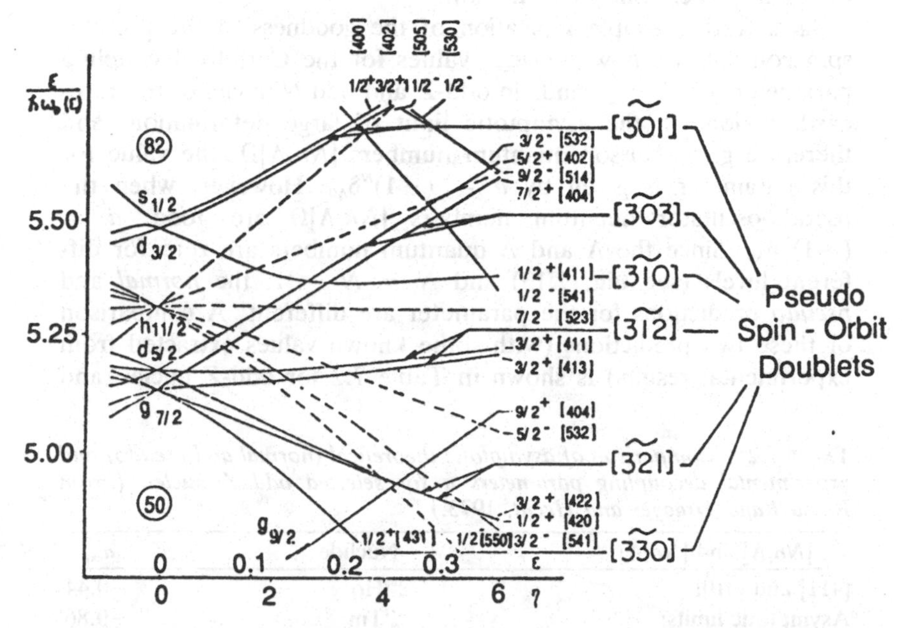

FIGURE 7.13. *Nilsson diagrams for the* $50 \le Z \le 82$ *shell with the single-particle levels labelled by* $\Omega[\mathrm{Nn_z\Lambda}]$ *and the corresponding pseudo-spin labels* $\Omega[\tilde{\mathrm{N}}\tilde{\mathrm{n}}_z\Lambda]$. *(From Ratna Raju, Draayer and Hecht,* 1973.)

(chosen to be the z-th direction); $\Lambda = \pm n_\perp, \pm(n_\perp - 2), \ldots, 1$ or 0 is the projection of the orbital angular momentum along the symmetry axis where $n_\perp = (N - n_z)$ is the number of quanta in the (x, y)-plane perpendicular to the symmetry axis; and $\boldsymbol{\Omega} = \boldsymbol{\Lambda} + \boldsymbol{\Sigma}$ where $\Sigma = \pm\frac{1}{2}$ is the projection of the spin on the body-fixed symmetry axis. In the pseudo-spin picture, these levels are relabeled as indicated by $\tilde{\Omega}[\tilde{N}\tilde{n}_z\tilde{\Lambda}]$ where $\tilde{N} = N - 1$, and the indexing of the levels is the same as for the normal scheme: $\tilde{n}_z = \tilde{N}, \tilde{N} - 1, \ldots; \tilde{\Lambda} = \pm\tilde{n}_\perp, \pm(\tilde{n}_\perp - 2), \ldots, 1$ or 0; and $\tilde{\boldsymbol{\Omega}} = \tilde{\boldsymbol{\Lambda}} + \tilde{\boldsymbol{\Sigma}} = \boldsymbol{\Omega}$ where $\tilde{\Sigma} = \pm\frac{1}{2}$ is the projection of the pseudo-spin on the body-fixed symmetry axis and $\tilde{\Omega} = \Omega$, because the total angular momentum and its projection are conserved under the *normal* $\leftrightarrow$ *pseudo* transformation. This relabeling of the orbitals yields pseudo-spin partners, $\tilde{\Omega} = \tilde{\Lambda} \pm \frac{1}{2}$ (identified with large $[\tilde{N}\tilde{n}_z\tilde{\Lambda}]$ labels on the right in the figure) which are nearly degenerate in energy. Notice that the near degeneracy of the pseudo-spin partners not only extends over the full range of β values (ϵ and η are alternate parameterizations of the deformation), but actually improves as a function of increasing deformation.

As a further simple indication of the goodness of the pseudo-spin concept, we now consider values for the Coriolis decoupling parameters of $K = \frac{1}{2}$ bands in odd-Z and odd-N nuclei of the rare-earth region. In the asymptotic limit of large deformation, and therefore good Nilsson quantum numbers $[Nn_z\Lambda]\Omega$, the value for this parameter is given by $a = (-1)^N\delta_{\Lambda 0}$. However, when the pseudo-oscillator quantum numbers $[\tilde{N}\tilde{n}_z\tilde{\Lambda}]\tilde{\Omega}$ are good, $a = (-1)^{\tilde{N}}\delta_{\tilde{\Lambda}0}$. Since the Λ and $\tilde{\Lambda}$ quantum numbers are zero for different levels (see Fig. 7.13) and $\tilde{N} = N - 1$, the *normal* and *pseudo* predictions for the parameter are different. A comparison of these two predictions (with some known values extracted from experimental results) is shown in Table 7.2 for odd-Z nuclei, and

TABLE 7.2. *Comparison of asymptotic theoretical (normal and pseudo) and experimental decoupling parameters a for selected odd Z nuclei. (From Ratna Raju, Draayer and Hecht,* 1973.*)*

$[Nn_z\Lambda]$ and $[\tilde{N}\tilde{n}_z\tilde{\Lambda}]$	Nuclide	a_{exp}
[411] and $[\tilde{3}\tilde{1}\tilde{0}]$	$^{165}_{67}$Ho	−0.44
Asymptotic limits:	$^{171}_{69}$Tm	−0.86
$a_{thy} = 0$ for good $[Nn_z\Lambda]$	$^{171}_{71}$Lu	−0.71
$a_{thy} = -1$ for good $[\tilde{N}\tilde{n}_z\tilde{\Lambda}]$	$^{177}_{71}$Lu	−0.91

Table 7.3 for odd-N rare earth nuclei. From these results, it is clear that the observed decoupling of $K = \frac{1}{2}$ bands falls much closer to the pseudo prediction than the normal Nilsson model result. Predictions similar to these can also be made for magnetic moments, and in most cases the experimental numbers fall closest to predictions based on the $[\tilde{N}\tilde{n}_z\tilde{\Lambda}]\tilde{\Omega}$ as compared with the $[Nn_z\Lambda]\Omega$ assignment of quantum numbers.

Before going on with a discussion of the many-particle pseudo-LS-coupling scheme, we need to emphasize an important feature of the *normal* ↔ *pseudo* unitary transformation we have introduced; namely, it carries H_0 into $\tilde{H}_0 + \hbar\omega$. This is an essential requirement—one that cannot be given up without losing the foundation for the pseudo-LS, pseudo-$SU(3)$, and pseudo-symplectic models that we come to next. We emphasize this point here, because a unitary operator form that transforms the combination $Cl \cdot s + Dl^2$ into $(4D - C)\tilde{l} \cdot \tilde{s} + D\tilde{l}^2 + (2D - C)$ has been identified: $U = \exp(i\pi h) = 2ih$ where $h = \bar{s} \cdot \hat{r}$ is the helicity. But this simple transformation—in contrast with our more complicated form, $U_{njm,\tilde{n}\tilde{j}\tilde{m}}(l, \tilde{l}) = 2(\eta \cdot \xi + 2l \cdot s + 3)^{-1/2}(\xi \cdot s)$—does not carry H_0 into $\tilde{H}_0 + \hbar\omega$. So, although the helicity operator is an attractive form because it transforms the one-body spin-orbit and orbit-orbit terms in H into their pseudo-spin counterparts, and furthermore, it leaves operators such as the $r^2Y_{20}(\theta, \varphi)$ part of the Nilsson Hamiltonian and the electric multipoles, which depend only on spatial coordinates, invariant, it does not commute with the kinetic energy operator, and consequently, does not affect the *normal* ↔ *pseudo* transformation we have identified. In particular, the helicity transformation does *not* lead to a realization of the pseudo-LS, pseudo-$SU(3)$, and pseudo-symplectic symmetries so important to a many-particle shell-model theory of medium and heavy mass nuclei.

7.3.2. Mean-Field Results

The pseudo-spin concept may be better understood by comparing an intuitive result for D with relativistic nuclear mean-field predictions for C. The origin of the l^2 term in H is in the flatness of the mean field in the interior region, as compared with the quadratic oscillator form ($V(r) = \frac{1}{2}M\omega^2r^2$). In the large mass limit ($A \to \infty$) the potential approaches that of a spherical well of finite depth. If this spherical well is replaced by one with an infinite depth, the single-particle energies are given by

TABLE 7.3. *Comparison of asymptotic theoretical (normal and pseudo) and experimental decoupling parameters a for selected odd* N *nuclei (from Ratna Raju, Draayer and Hecht,* 1973).

$[Nn_z\Lambda]$ and $[\tilde{N}\tilde{n}_z\tilde{\Lambda}]$	Nuclide	a_{exp}	$[Nn_z\Lambda]$ and $[\tilde{N}\tilde{n}_z\tilde{\Lambda}]$	Nuclide	a_{exp}
[521] and $[\bar{4}\bar{2}\bar{0}]$	$^{165}_{66}Dy$	0.58	[510] and $[\bar{4}\bar{1}\bar{1}]$	$^{177}_{70}Yb$	0.24
	$^{169}_{68}Er$	0.83		$^{179}_{72}Hf$	0.16
Asymptotic limits:	$^{173}_{72}Hf$	0.82	Asymptotic limits:	$^{181}_{72}Hf$	0.20
a_{thy} = 0 for good $[Nn_z\Lambda]$	$^{175}_{70}Yb$	0.75	a_{thy} = −1 for good $[Nn_z\Lambda]$	$^{183}_{74}W$	0.19
a_{thy} = +1 for good $[\tilde{N}\tilde{n}_z\tilde{\Lambda}]$	$^{179}_{74}W$	0.82	a_{thy} = 0 for good $[\tilde{N}\tilde{n}_z\tilde{\Lambda}]$	$^{185}_{76}Os$	0.02

$$E_{nl} = \frac{\hbar^2}{2MR^2} x_{nl}^2 \tag{7.27}$$

where M is the nucleon mass, R is the radius of the well, and the x_{nl} are zeros of spherical Bessel functions. These zeroes are approximately given by the result $x_{nl}^2 \approx [(n/2 + 1)\pi]^2 - l(l + 1)$. Table 7.4 illustrates the dependence of x_{nl} on l for the $n = 4$ case. The results show that the splitting follows an $l(l + 1)$ rule. Therefore, we have

$$D = \frac{-\hbar^2}{2MR^2}. \tag{7.28}$$

A more complete theory for D using the Klein–Gordon equation leads to the same conclusion when the kinetic energy is a small fraction of the nucleon mass.

Next, consider the strength of the spin-orbit coupling. Starting with the Dirac equation (with only the time component of the scalar and vector potentials taken into account) and using a non-relativistic reduction of the relativistic mean field theory, the spin-orbit interaction is given by

$$V_{ls} = \frac{\hbar^2}{2M} \frac{2}{r} \frac{d}{dr} \left(\frac{1}{1 - B\rho/\rho_0} \right) l \cdot s. \tag{7.29}$$

In this expression, ρ and ρ_0 are, respectively, the nucleon density at radius r and the nuclear matter density. The dimensionless quantity B in eq. (7.29) is related to the strength of the scalar and vector coupling constants. The spin-orbit strength C can be obtained from the average of C over the region inside radius R,

$$C = \frac{-\hbar^2}{2MR^2} \frac{6B}{1 - B}. \tag{7.30}$$

TABLE 7.4. *Zeros* (x_{nl}) *of spherical Bessel functions and* ($x_{n0}^2 - x_{nl}^2$) *compared with the simple* $l(l + 1)$ *approximation for the* $n = 4$ *case. (From Bahri, Draayer and Moszkowski, 1992.)*

n	l	x_{nl}/π	x_{nl}^2	$x_{n0}^2 - x_{nl}^2$	$l(l + 1)$
4	0	3.000	88.83	0.00	0
4	2	2.895	82.72	6.11	6
4	4	2.605	66.98	21.85	20

The fact that $d\rho/dr$ vanishes everywhere, except near the surface of the nucleus, was used in determining this result.

It follows from eqs. (7.28) and (7.30) that the ratio

$$\mu = \frac{2D}{C} = \frac{1 - B}{3B} \tag{7.31}$$

is independent of mass number. Furthermore, to obtain $\mu = 0.5$ requires $B = 0.4$. In the simplest version of the theory, $B = \frac{1}{2}(B_s + B_v)$ with its scalar ($i = s$) and vector ($i = v$) components given by $B_i = g_i^2\rho_0/\mu_i^2 Mc^2$ where μ_i and g_i, respectively, denote meson masses and coupling constants. Using this expression for B, the Nambu–Jona–Lasinio (NJL) model—which in its modern form starts with massless quarks and generates hadron masses out of the vacuum by spontaneous symmetry-breaking, and which has also been used to predict the coupling constants and masses appearing in a relativistic nuclear field theory—gives the result $\mu = 0.686$ shown in Table 7.5. As also shown in the table, results for the original Walecka model and a derivative coupling model due to Zimanyi and Moszkowski—which gives a more realistic equation of state for nuclear matter, which includes the effect of nucleon recoil, and when the original theory is extended to include exchange correlations—also yield reasonable results for μ.

7.3.3. Pseudo-LS Coupling

As we indicated above, the pseudo-scheme organizes the normal parity $j = \frac{1}{2}, \frac{3}{2}, \ldots, n - \frac{1}{2}$ levels of the n-th oscillator shell into

TABLE 7.5. *Comparison of $\mu = 2D/C$ values for various relativistic mean field theories. Exact pseudo-spin symmetry requires $\mu = 0.5$. Results given are for $\rho_0 = 0.16$ nucleons/fm^3 and a nuclear binding energy of -16 MeV. (From Bahri, Draayer and Moszkowski, 1992.)*

	B_s	B_v	B	μ
NJL[a]	0.339	0.316	0.327	0.686
Walecka[a]	0.487	0.368	0.427	0.447
Zimanyi[b]	0.252	0.088	0.344	0.635

[a] $B = \frac{1}{2}(B_s + B_v)$.

[b] $B = \frac{2}{3}B_s + 2\,B_v$ (including recoil and exchange effects).

a pseudo shell with $\tilde{n} = n - 1$. For example, the normal parity levels of the $n = 4$ shell are mapped onto the ($3\tilde{p}_{1/2}$, $3\tilde{p}_{3/2}$, $1\tilde{f}_{5/2}$, $1\tilde{f}_{7/2}$) orbitals of an $\tilde{n} = 3$ shell. In fact as we have already seen, this mapping of single-particle orbitals defines the pseudo-spin coupling scheme. A schematic diagram showing this correspondence for all single-particle levels of the oscillator is given in Fig. 7.14. To grasp its full significance, recall that the symmetry group for the usual many-particle generalization of the single-particle theory with particles distributed among the lowest available single-particle levels, is the unitary group $U(k\Omega)$ where $\Omega = (n + 1)(n + 2)/2$ is the spatial degeneracy of the n-th oscillator shell and $k = 2$ or 4 for a spin or spin-isospin formulation of the theory. The $U(\Omega) \otimes$

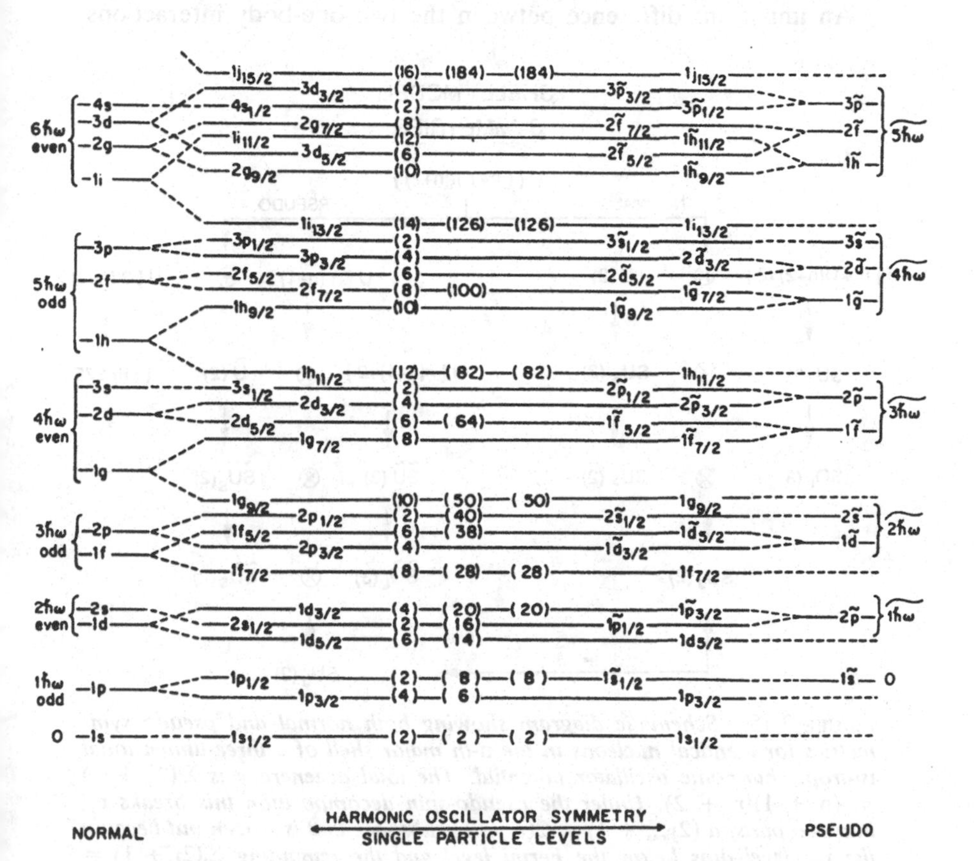

FIGURE 7.14. *Schematic diagram tracing the normal ↔ pseudo correspondence of spherical single-particle harmonic oscillator levels.*

$U(k)$ direct product subgroup of this $U(k\Omega)$ group separates the full (k by Ω)-dimensional space into its spatial (Ω) and spin (k = 2) or spin-isospin (k = 4) parts. Irreps of $U(\Omega)$, which are labeled by a Young pattern $[f] = [f_1, f_2, \ldots, f_\Omega]$, specify the space symmetry; irreps $[f^c] = [f_1^c, f_2^c, \ldots, f_k^c]$ of $U(k)$, which must be related to the $[f]$ of $U(\Omega)$ by row-column interchange in order to insure the overall antisymmetry in $U(k\Omega)$ required by the Pauli Exclusion Principle, label the complementary spin or spin-isospin symmetry. The reduction of the full model space into irreps of $U(\Omega) \otimes U(k)$ and its subgroups (normal coupling scheme) and $\tilde{U}(\Omega) \otimes \tilde{U}(k)$ and its subgroups (pseudo coupling scheme when the $j = n + \frac{1}{2}$ level frozen out) is shown for the $k = 2$ case in Fig. 7.15.

An important difference between the two one-body interactions

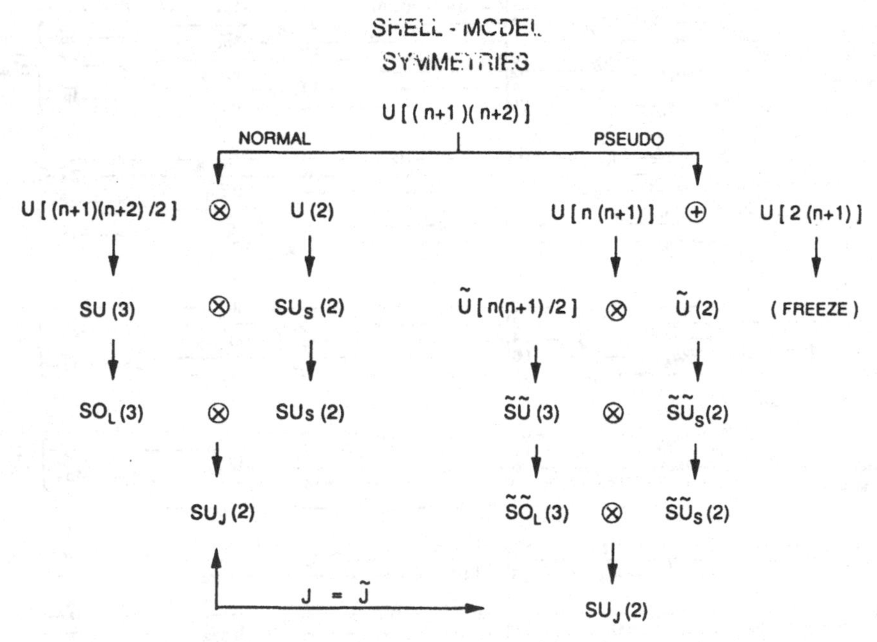

FIGURE 7.15. *Schematic diagram showing both normal and pseudo symmetries for identical nucleons in the* n*-th major shell of a three-dimensional isotropic harmonic oscillator potential. The total degeneracy is* $\Sigma_j(2j + 1) = (n + 1)(n + 2)$. *Under the pseudo-spin decomposition this breaks up into two parts, a* $(2j_{max} + 1) = 2(n + 1)$ *subspace that is frozen out because the* j_{max} *level dips below the Fermi level and the remaining* $\Sigma_j'(2j + 1) = (n + 1)(n + 2) - 2(n + 1) = n(n + 1)$ *valence space that can be mapped onto an oscillator shell of one less quanta,* $\tilde{n} = n - 1$.

is $l \cdot s$ couples different spatial symmetries, while l^2 does not. The $[f]$ and therefore $[f^c]$ are good quantum numbers when the strength of all terms like $l \cdot s$ that couple different spatial symmetries is small relative to the strength of others like l^2 which do not mix these symmetries. While this is not the case for the many-particle extension of the normal single-particle Hamiltonian ($H = H_0 + Cl \cdot s + Dl^2$), it is for its pseudo-counterpart ($\tilde{H} = \tilde{H}_0 + \tilde{C}\tilde{l} \cdot \tilde{s} + \tilde{D}\tilde{l}^2$). This feature is important, because it means we can partition the full model space into disjoint (pseudo) subspaces $\{[\tilde{f}]$ and $[\tilde{f}^c]$ of $\tilde{U}(\tilde{\Omega}) \otimes \tilde{U}(\tilde{k})$, respectively$\}$ that have much smaller dimensions than can be realized with the normal scheme. In addition (as is known to be the case for the surface-delta interaction, and as is shown below for the quadrupole-quadrupole interaction) if the residual two-body interaction is a pseudo-space scalar operator, it actually reinforces the goodness of the pseudo-space and pseudo-spin or pseudo-spin-isospin symmetries labelled by $[\tilde{f}]$ and $[\tilde{f}^c]$, respectively. We have already seen an example of this in Fig. 7.13: the near degeneracy of the pseudo-spin partners increases as a function of increasing deformation. The end result of partitioning the space via $\tilde{U}(\tilde{\Omega}) \otimes \tilde{U}(\tilde{k})$, rather than by $U(\Omega) \otimes U(k)$, is that we can deal with a subspace of the full space comprised of a collection of only a few pseudo-spatial symmetries.

For heavy nuclei, the valence protons and neutrons occupy different shells, so we must apply an identical particle ($\tilde{k} = 2$) formulation to each. Specifying the $[\tilde{f}]$ and $[\tilde{f}^c] = [\tilde{f}_1^c, \tilde{f}_2^c]$ labels is then equivalent to giving the total number of normal parity particles and their pseudo-spin: $\tilde{m} = \tilde{f}_1^c + \tilde{f}_2^c$ and $\tilde{S} = (\tilde{f}_1^c - \tilde{f}_2^c)/2$. We partition the normal parity $\tilde{m}$-particle space, with $\tilde{m} = \tilde{m}_\pi$ for protons and $\tilde{m} = \tilde{m}_\nu$ for neutrons, into subspaces with $\tilde{S} = 0, 1, 2, 3, \ldots, \tilde{S}_{max}$ for $\tilde{m}$ even and $\tilde{S} = \frac{1}{2}, \frac{3}{2}, \frac{5}{2}, \ldots, \tilde{S}_{max}$ for $\tilde{m}$ odd, where $\tilde{S}_{max}$ is the minimum of $\tilde{m}/2$ and $\tilde{\Omega} - \tilde{m}/2$. This means that $\Delta\tilde{S}$ (proton *and* neutron) is always an integer; also, there is a complementary set of spatial configura9tions for each $\tilde{S}$. To the extent pseudo-spin symmetry is good, we therefore expect to observe sets of states, such as rotational sequences, that differ in *total* angular momenta ($\mathbf{J} = \tilde{\mathbf{L}} + \tilde{\mathbf{S}}$) by integer (even-$A$ compared with even-A) or half-integer (odd-A with even-A) amounts. De-excitation gamma-ray spectra differing from one another by unit spin alignment—$\Delta E_i(J) = \Delta E_j(J + 1)$ for the i-th and j-th rotational sequences, where $\Delta E_\alpha(J) = E_\alpha(J + 1) - E_\alpha(J - 1)$—are therefore a natural consequence of good pseudo-spin symmetry.

It is important to understand that this alignment feature can be either proton or neutron in origin, or a combination of the two. This follows because the many-particle basis states are π-ν coupled configurations: $|\Psi^J\rangle = |[(\tilde{\alpha}_\pi \tilde{L}_\pi \tilde{\alpha}_\nu \tilde{L}_\nu)^{\tilde{L}} \times (\tilde{S}_\pi \tilde{S}_\nu)^{\tilde{S}}]^J\rangle$, where $\tilde{\alpha}_\kappa$ labels multiple occurrences of the $\tilde{L}_\kappa$ values, $\kappa = (\pi, \nu)$. The full symmetry group of this combined π-ν system is $[\tilde{U}_\pi(\tilde{\Omega}_\pi) \otimes \tilde{U}_\pi(2)] \otimes [\tilde{U}_\nu(\tilde{\Omega}_\nu) \otimes \tilde{U}_\nu(2)]$. This direct product structure can be reordered as for $|\Psi^J\rangle$ so the pseudo-space and pseudo-spin associations are made first, $[\tilde{U}_\pi(\tilde{\Omega}_\pi) \otimes \tilde{U}_\nu(\tilde{\Omega}_\nu)] \otimes [\tilde{U}_\pi(2) \otimes \tilde{U}_\nu(2)]$. In this expression, $\tilde{\Omega}_\kappa = (\tilde{n}_\kappa + 1)(\tilde{n}_\kappa + 2)/2$ is the pseudo-space degeneracy of the $\kappa = (\pi, \nu)$ subshell. Simple alignment is consistent with good total pseudo-spin symmetry, $\tilde{\mathbf{S}} = \tilde{\mathbf{S}}_\pi \times \tilde{\mathbf{S}}_\nu$, provided the π-ν interaction—like the π-π and ν-ν terms—conserves $\tilde{S}$. As is demonstrated below, the real quadrupole-quadrupole, $Q^\pi \cdot Q^\nu$, which only weakly (compared to the symmetry-preserving $\tilde{Q}^\pi \cdot \tilde{Q}^\nu$ interaction) couples configurations with different $\tilde{S}$ symmetry, is such an interaction.

7.3.4. Pseudo-*SU*(3) Scheme

We have seen that the importance of the $SU(3)$ model for light nuclei follows from the dominance of the collective quadrupole-quadrupole interaction ($Q^c \cdot Q^c$) over the one-body $l \cdot s$ and l^2 terms, as well as over all other two-body forms. Even though the spin-orbit interaction is relatively strong, yrast states of nuclei such as ^{20}Ne and ^{24}Mg are typically 60–80% pure leading representations of $SU(3)$. This comes about primarily because $Q^c \cdot Q^c$ conserves spatial symmetry; secondly, because $SU(3)$ is a subgroup of $U(\Omega)$ with second order invariant $C_2 = \frac{1}{4}(Q^a \cdot Q^a + 3L^2)$ where $Q^a \cdot Q^a$ is the single-shell $0\hbar\omega$ approximation for $Q^c \cdot Q^c$; and finally, because the expectation value of $Q^a \cdot Q^a$ is proportional to the square of the deformation, so each spatial symmetry $[f]$ is subdivided into (λ, μ) irreps of $SU(3)$ with the most deformed of these lying lowest and the least deformed highest. We now want to show that these same arguments apply for heavy deformed nuclei if the *normal* ↔ *pseudo* transformation is applied.

The amount and sharpness of the separation—first into irreps of $U(\Omega)$ and then by irreps of $SU(3)$—depends on the relative strength of the symmetry-preserving and symmetry-breaking interactions; and this, in turn, depends upon whether or not the available space supports strongly deformed configurations. As we have seen in the *ds*-shell case, systems such as ^{20}Ne and ^{24}Mg have leading (λ, μ)'s with relatively large deformation, so $Q^a \cdot Q^a$ overpowers the other interactions and yrast states have good $[f]$ and (λ, μ) quantum levels. We now argue that the pseudo-$SU(3)$ scheme, in which $\tilde{S}\tilde{U}(3)$ stands in the same relationship to $\tilde{U}(\Omega)$ as $SU(3)$ does to $U(\Omega)$ (see Fig. 7.15) provides a similar explanation for observed quadrupole collectivity in heavy deformed nuclei. There are, of course, some very important differences:

- the valence neutrons and protons for heavy nuclei of the rare earth and actinide regions occupy different major shells;
- the $Q^a \cdot Q^a$ interaction cannot be expressed solely in terms of quadratic invariants of $\tilde{S}\tilde{U}(3)$ and $\tilde{S}\tilde{O}(3)$;
- whereas the coefficent D of l^2 is positive for the *ds* shell, it is negative for rare earth and actinide nuclei.

Each of these points will now be considered.

We address the first of these three differences by noting that a very simple calculation can be carried out demonstrating that the collectivity of the system is not decreased, even when the protons and neutrons are in different shells, so long as the two species interact (albeit even weakly) through their quadrupole fields. Even when the valence protons and neutrons occupy different major shells and the separate two-body proton-proton ($V^{\pi\pi}$) and neutron-neutron ($V^{\nu\nu}$) interactions are dominated by pairing, a small $V^{\pi\nu} = Q^a_\pi \cdot Q^a_\pi$ interaction between the two suffices to drive the whole system towards the strong-coupled pseudo-$SU(3)$ limit of the theory. This is shown in Fig. 7.16 for the $[(fp)^{m_\pi=2}(gds)^{m_\nu=2}]$ case.

To consider the second matter, we simply note that the pseudo-spin scheme is an excellent starting point for a many-particle description of heavy nuclei, whether or not they are deformed. In particular, the $\tilde{l}\cdot\tilde{s}$ interaction term is weak relative to $\tilde{l}^2[\tilde{C}/\tilde{D} = (4D - C)/D \approx (C/D)/5]$ so the pseudo-spin symmetry is approximately good. Also, since the surface-delta interaction—which is known to be a good effective interaction for many applications—is a pseudo-spin scalar operator, the residual two-body interaction is not expected to change this picture very much, though it will induce some mixing among the different pseudo-space symmetries. As an example, although $Q^a \cdot Q^a$ is not an invariant of $\tilde{S}\tilde{U}(3)$, it is transformed under the *normal* ↔ *pseudo* mapping into its pseudo-counterpart plus small corrections,

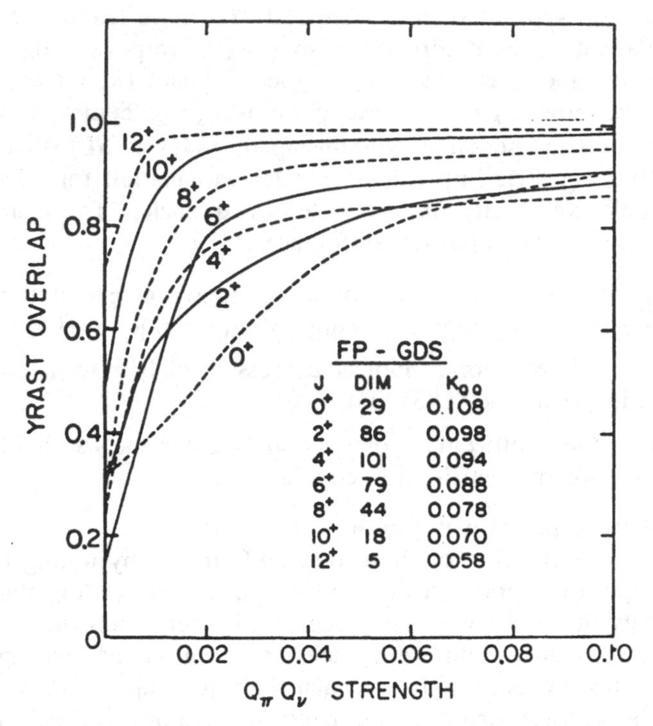

FIGURE 7.16. *Overlaps of calculated yrast-band states with the corresponding leading strong-coupled* SU(3) *basis states and fixed-*J *correlation coefficients for the configuration* $[(fp)^{m_\pi=2}(gds)^{m_\nu=2}]$. *The results are plotted as a function of the strength of the* $Q_\pi \cdot Q_\nu$ *interaction in the Hamiltonian* $H = H_{\pi\pi} + H_{\nu\nu} + 2\kappa Q_\pi \cdot Q_\nu$ *where* $H_{\pi\pi}$ *and* $H_{\nu\nu}$ *were taken to be pairing interactions. The results are normalized so a strength* $\kappa = 1$ *makes the norm of the* π-ν *interaction equal to that of the* π-π *and* ν-ν *interactions.* (*From Draayer, Weeks and Hecht,* 1982.)

$$Q^a \cdot Q^a = \kappa \tilde{Q}^a \cdot \tilde{Q}^a + \cdots . \tag{7.32}$$

Indeed, within the leading pseudo-space symmetry, the sum of all correction terms has been shown to induce less than a 1% change in calculated excitation energies and electromagnetic transition strengths.

To explore the third and final question concerning the relevance of the pseudo-$SU(3)$ scheme for strongly deformed nuclei with $D < 0$ (a condition responsible for helping to destroy $SU(3)$ in the *fp*-shell, for example), we consider the many-body problem with the Hamiltonian,

$$\begin{aligned} \tilde{H} &= \tilde{H}_0 + \tilde{C}\sum_i \tilde{l}\cdot\tilde{s} + \tilde{D}\sum_i \tilde{l}_i^2 - \tfrac{1}{2}\tilde{\chi}\tilde{Q}^a\cdot\tilde{Q}^a \\ &= \tilde{H}_0 + \tilde{C}\sum_i \tilde{l}\cdot\tilde{s} + \tilde{D}\sum_i \tilde{l}_i^2 - \tfrac{1}{2}\tilde{\chi}(4\tilde{C}_2 - 3\tilde{L}^2). \end{aligned} \tag{7.33}$$

The last form for $\tilde{H}$ follows because $\tilde{Q}^a\cdot\tilde{Q}^a = 4\tilde{C}_2 - 3\tilde{L}^2$ within a major shell of the oscillator. Since the $\tilde{C}_2$ and $\tilde{L}^2$ interactions are diagonal in an $\tilde{S}\tilde{U}(3)$ basis, they split but do not break the pseudo-oscillator symmetry; because $\tilde{\chi}$ is always positive, the $\tilde{S}\tilde{U}(3)$ representations with the largest eigenvalues for $\tilde{C}_2$—and hence the greatest intrinsic deformations ($\beta^2 \sim \tilde{\beta}^2 \sim \langle\tilde{Q}^a\cdot\tilde{Q}^a\rangle = \langle 4\tilde{C}_2\rangle$ in $J = \tilde{L} = 0$ states)—lie lowest. Intensities of calculated 0^+ states of this Hamiltonian, eq. (7.33), with $\tilde{C} = 0$ and $\tilde{D} < 0$ are plotted in Fig. 7.17 as a function of $\tilde{\chi}$ for the simple but representative case $(d\tilde{s})^4[\tilde{f}] = [4]$ with $(\tilde{\lambda}, \tilde{\mu})$'s $= (8, 0), (4, 2), (0, 4)$, and $(2, 0)$. This pseudo ^{20}Ne case (which differs from the real ^{20}Ne nucleus in that $\tilde{D} < 0$ rather than being positive and $\tilde{C} = 0$ instead of being nonzero) shows that under conditions very similar to those for pseudo-$SU(3)$ applications in rare earth and actinide nuclei, the symmetry-breaking *decreases* sharply as the strength of the deformation-inducing quadrupole-quadrupole interaction *increases*. A $\tilde{\chi}$ of 0.06-0.07 is realistic for both a normal and pseudo ^{20}Ne application of the theory. In the figure, the dashed curve is the $0^+_{1(8,0)}$ intensity when a spin-orbit interaction ($\tilde{C} < 0$) with one-tenth its normal strength is included in H. This is also an appropriate choice for pseudo-$SU(3)$ applications. Typically, calculated yrast states of heavy deformed nuclei are dominated to at least the 80% level by the leading $\tilde{S}\tilde{U}(3)$ symmetry.

From the results of this simple representative calculation, several important conclusions can be drawn. First of all, because $\tilde{C} \approx 0$ for nuclei of the rare earth and actinide regions, mixing among different pseudo-space symmetries is expected to be small. This means that the normal parity valence space can be truncated to a reasonable size (for example the $[\tilde{f}] = [4]$ symmetry in the pseudo ^{20}Ne case just considered) so simple yet realistic calculations can be carried out. Secondly, it is important to emphasize again that the breaking of the $\tilde{S}\tilde{U}(3)$ symmetry *decreases* very sharply as the

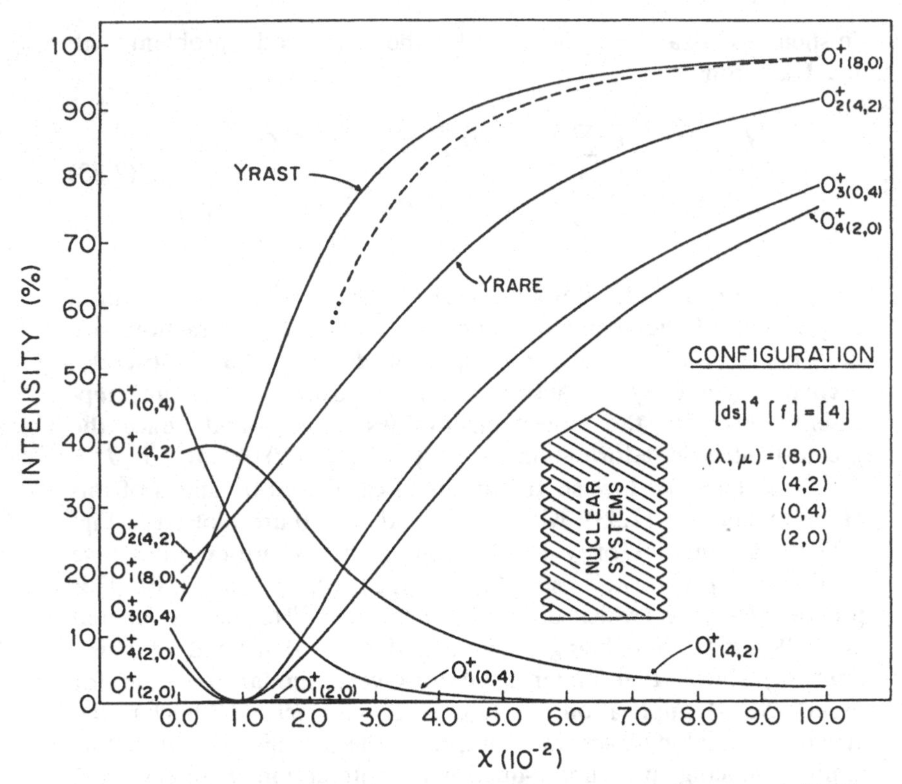

FIGURE 7.17. *Intensities of calculated* 0^+ *states of Hamiltonian eq.* (7.33) *with* $\tilde{C} = 0$ *and* $\tilde{D} < 0$ *are plotted as a function of* $\tilde{\chi}$ *for the representative pseudo* ^{20}Ne *case with* $(d\tilde{s})^4$, $[\tilde{f}] = [4]$ *and* $(\tilde{\lambda}, \tilde{\mu})$ *values* (8, 0), (4, 2), (0, 4), *and* (2, 0). *This differs from the normal* ^{20}Ne *case in that* $\tilde{D} < 0$ *instead of being positive and* $\tilde{C} = 0$ *instead of being nonzero. The results show that under conditions very similar to those that apply for pseudo*-SU(3) *applications in rare earth and actinide nuclei the symmetry breaking decreases sharply as the strength of the deformation inducing quadrupole-quadrupole interaction increases. A* $\tilde{\chi}$ *of* 0.06–0.07 *is realistic for both a normal and pseudo* ^{20}Ne *application of the theory. In the figure the dashed curve is the* $0^+_{1\,(8,\,0)}$ *intensity when a spin-orbit interaction* ($\tilde{C} < 0$) *with one-tenth its normal strength is included in* H. *This value is an appropriate choice for pseudo*-SU(3) *applications.* (*From Draayer,* 1990.)

strength of the deformation-inducing quadrupole-quadrupole interaction *increases*. More specifically, the spreading out of the various $\tilde{S}\tilde{U}(3)$ irreps within $[\tilde{f}]$ symmetries by $\tilde{Q}^a \cdot \tilde{Q}^a$ reduces the $\tilde{S}\tilde{U}(3)$ symmetry-breaking induced by the $\tilde{l}^2$ term. For example, this means that the normal parity contribution to yrast states of strongly deformed configurations should be dominated by the leading pseudo-$SU(3)$ symmetry.

It is also important for us to understand that the leading symmetry is the most deformed configuration available in the space under consideration. Going beyond normally deformed configurations to superdeformed, and perhaps even hyperdeformed structures, therefore implies shifting particles into higher-lying configurations. For instance, one candidate for a "superdeformed" band is the configuration in the pseudo ^{20}Ne example obtained by lifting two particles out of the p shell into the fp shell. This leads to $\tilde{S}\tilde{U}(3)$ irreps contained in the product $(8, 0) \times (0, 2) \times (6, 0)$. The leading irrep in this case is $(\tilde{\lambda}, \tilde{\mu}) = (14, 2)$. Since the square of the deformation is proportional to the expectation value of $\tilde{C}_2$ in $\tilde{L} = 0$ states, under the action of the *same* Hamiltonian this arrangement of nucleons will have a deformation that is nearly twice that of the leading (8,0) irrep: $\langle\tilde{C}_2(14, 2)\rangle/\langle\tilde{C}_2(8, 0)\rangle = 3.14$ which yields $\bar{\beta}(14, 2)/\bar{\beta}(8, 0) = 1.77$. Because the Hamiltonian does not change, we expect configurations like the one containing the (14, 2) to display even less representation mixing than the one containing the (8, 0), since for this arrangement of particles the dominance of the $\tilde{Q}^a \cdot \tilde{Q}^a$ term in the energy matrix will be even more pronounced.

7.3.5. Unique Parity Configurations

The pseudo-spin scheme considers the highest $j = n + \frac{1}{2}$ state associated with the n-th major shell of the oscillator to have defected out of the valence space—pushed down among occupied levels of the $(n - 1)$-th shell below by the strong spin-orbit interaction. Furthermore, the $j = n + \frac{3}{2}$ level from the $(n + 1)$-th shell immediately above is considered to intrude into the valence space—playing only a passive role in the dynamics of low-lying excitations, because it has the opposite parity to its new-found partners. Specifically, it is usually assumed that particles distributed in the unique parity intruder levels enter only as $J = 0$ coupled pairs for configurations below the backbending region, contributing binding energy to the system but nothing to the dynamics. Through the backbending region and beyond, however, alignment of the angular momentum of these pairs $(J \neq 0)$ sets in, and the intruder level can no longer be considered passive. We will now show some preliminary results which challenge this simple picture. In particular, the results suggest that the coupling of the intruder to its natural partners, even though these may be as much as a full major shell ($1\hbar\omega$) away, is strong and leads to a sizable contribution to the quadrupole moment and therefore E2 strengths of the many-particle system.

In Fig. 7.18, normalized expectation values of $Q^a \cdot Q^a$ in calculated ground states of the $(ds)^4$ system for Hamiltonian eq. (7.25) augmented with a $Q^a \cdot Q^a$ term ($H = H_0 + Cl \cdot s + Dl^2 - \frac{1}{2}\chi Q^a \cdot Q^a$) with $D = -0.2$ and C values as indicated are plotted as

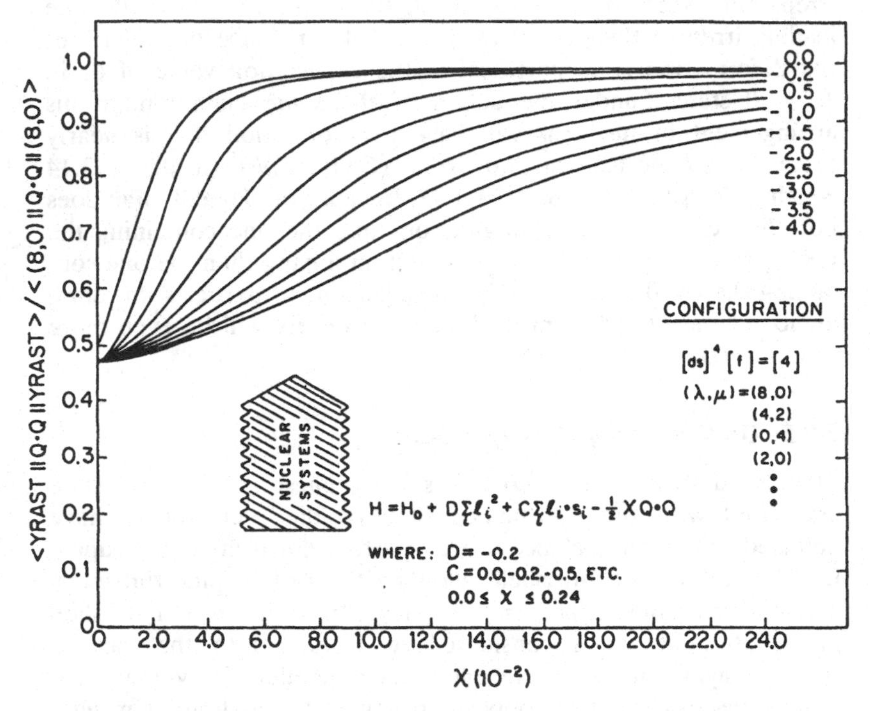

FIGURE 7.18. *Normalized expectation values of* Q·Q *in calculated ground states of the* $(ds)^4$ *system for Hamiltonian(4), with* D = −0.2 *and a spin-orbit term with* C *values as indicated, are plotted as a function of the strength* χ *of the quadrupole-quadrupole interaction. The normalization factor is the maximum eigenvalue of* Q·Q *in the* $(ds)^4$ *space which is just the expectation of* Q·Q *in the leading* (λ,μ) = (8,0) *irrep of* SU(3). *For the results to be representative of a pseudo* SU(3) *application,* C ≈ −2.5, *and as indicated,* $\chi \approx$ 0.06–0.07. *It follows from these results that even for a very strong spin-orbit splitting, the yrast state can achieve as much as 60–70% of its maximum total quadrupole collectivity.*

a function of the strength χ of the quadrupole-quadrupole interaction. We have chosen the normalization factor to be the maximum eigenvalue of $Q^a \cdot Q^a$ in the $(ds)^4$ space, that is, the expectation of $Q^a \cdot Q^a$ in the leading $(\lambda, \mu) = (8, 0)$ irrep of $SU(3)$. For these results to be representative of a pseudo-$SU(3)$ application, $C \approx -2.5$ and, as above, $\chi \approx 0.06 - 0.07$. It follows from this example that even for a very strong spin-orbit splitting, the yrast state can achieve as much as 60–70% of its maximum total quadrupole collectivity. For rare earth and actinide nuclei the number of particles in the intruder level is typically about $\frac{1}{3}$ of the total number of valence particles. Since the intruder level comes from the $(n + 1)$-th oscillator shell, a very rough estimate can be given for the ratio of the contribution to the quadrupole collectivity from particles in the unique parity intruder levels to the contribution from those in the normal parity orbitals: $\langle Q^a \cdot Q^a \rangle_{\text{unique}} / \langle Q^a \cdot Q^a \rangle_{\text{normal}} \approx 0.65 \times C_2 (m(n + 1)/3, 0) \div C_2(2mn/3, 0) \approx 0.65 \times [(n + 1)/(2n)]^2 \approx 0.25$. This argument assumes the $(\lambda, \mu) = (m(n + 1)/3, 0)$ and $(2mn/3, 0)$ irreps are representative of those respectively allowed by the exclusion principle for the unique and normal parity spaces. This suggests that particles in the unique parity orbitals can be expected to contribute to the quadrupole moment of a deformed system roughly in proportion to their number with a strength that is about half the strength with which the normal parity particles contribute. Since the number of particles in the unique parity space is typically not small, the intruder levels are important in determining quadrupole moments and E2 strengths.

7.3.6. Applications 1: Backbending, Forking, etc.

Backbending is the most characteristic non-rotational feature found in heavy deformed nuclei with rotational low-lying spectra. The term derives from the shape of the curves obtained when the system's effective moment of inertia, defined through the relation $E_I = \hbar^2/2\mathcal{I}[I(I + 1)]$, is plotted as a function of the angular frequency which is given via the relation $\mathcal{I}\omega = \hbar[I(I + 1)]^{1/2}$, that is,

$$\begin{aligned} 2\mathcal{I}/\hbar^2 &= (4I - 2)/\Delta E_\gamma \quad \text{versus} \\ (\hbar\omega)^2 &= (\Delta E_\gamma)^2/\{[I(I + 1)]^{1/2} - [(I - 2)(I - 1)]^{1/2}\}^2, \end{aligned} \tag{7.34}$$

where $\Delta E_\gamma = (E_I - E_{I-2})$. For a rigid rotor, the curve is a horizontal line since $\mathcal{I}$ is then a constant that is independent of the rotation rate. Typical spectra of strongly deformed nuclei show, however, that for low (high) rotational frequencies, $\mathcal{I}$ increases (decreases) slowly with increasing angular velocity, a centrifugal stretching (anti-stretching) phenomena, so the $2\mathcal{I}/\hbar^2$ versus $(\hbar\omega)^2$ curve is gently up (down) sloping. In the transition region between these two domains, $\mathcal{I}$ is frequently found to increase, sometimes sharply, and in some cases the rotational frequency actually decreases with increasing angular momentum. When this happens the $2\mathcal{I}/\hbar^2$ versus $(\hbar\omega)^2$ curve is S-shaped and the nucleus is said to backbend. Accompanying each backbend is a proportionate drop in the collective $B(\text{E2}; I \rightarrow I - 2)$ transition strengths.

There are two common explanations for backbending: it can be due to the alignment of a pair of either core or intruder level particles $(j^2)J = 0 \rightarrow (j^2)J = 2j]$, or it results from the crossing of the ground band by another with a larger moment of inertia. The first of these explanations simply states that it becomes energetically more favorable through the backbending region to add two units of angular momentum to the system by aligning a pair of particles rather than by increasing the rotational frequency of the core. Furthermore, the broken pair particles then participate in the rotation and increase the moment of inertia accordingly. The band-crossing mechanism is simply a statement that the nucleus can exist in different (orthogonal) intrinsic configurations with different inertia tensors and that this feature is maintained under rotation. A pseudo-$SU(3)$ scheme that takes into account more than a single $(\tilde{\lambda}, \tilde{\mu})$ irrep and $\nu \neq 0$ intruder configurations includes both of these features. The results of such a calculation for the $\tilde{f}\tilde{p}$-shell nucleus ^{126}Ba are shown in Fig. 7.19. The results of this microscopic pseudo-$SU(3)$ calculation show that the band-crossing and pair-alignment mechanisms are both operative and required to get the experimentally smooth S-shaped backbending curve.

A related phenomena called forking is found in some soft rotors. In this case, the ground state rotational band bifurcates or even trifurcates as energy increases. An example is ^{68}Ge, which has three 8^+ states feeding with roughly similar E2 strengths into the lone 6^+ of the ground state band. This is shown in Fig. 7.20, along with the results of a pseudo-$SU(3)$ calculation. Configurations of the type $(f_{5/2}, p_{3/2}, p_{1/2})^{n_n} \times (g_{9/2})^{n_u}$, in which the number of particles in the normal (n_n) plus unique (n_u) parity orbitals was restricted to 12, were included in the calculation. The normal parity space was

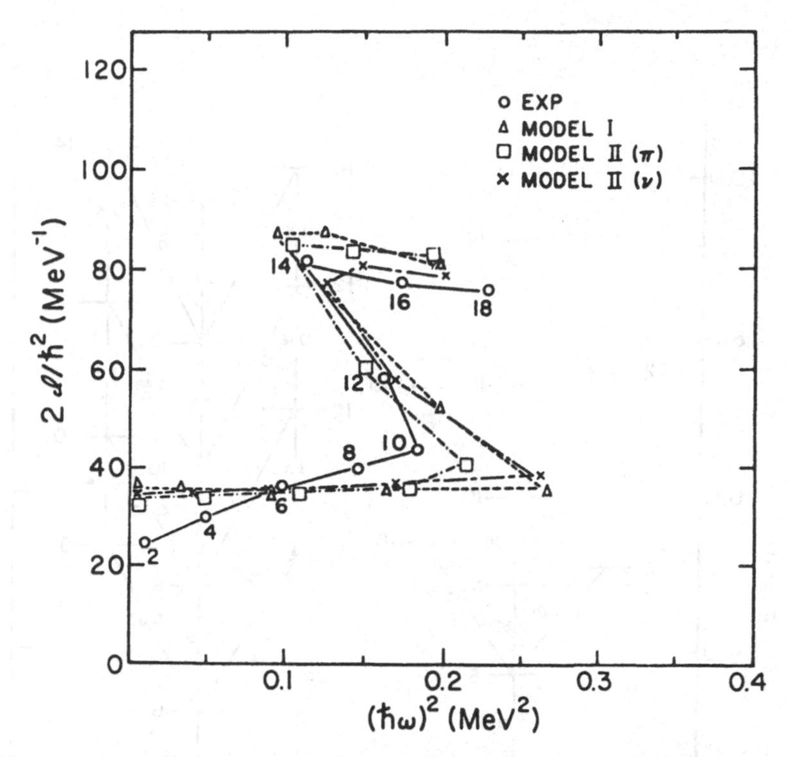

FIGURE 7.19. *Experimental and theoretical results showing backbending in* ^{126}Ba. *The Model* I *configuration is* $[(\tilde{f}\tilde{p})^{18}(\tilde{\lambda}, \tilde{\mu}) = (24, 0)T_N = 3; (h_{11/2})^8 T_U = 4]$ *where the* $h_{11/2}$ *particles were restricted to seniority* 0 *and* 2 *configurations. Models* II(π) *and* II(ν) *add to the Model I space a pair of protons* (π) *and neutrons* (ν) *scattered into the* $[(\tilde{f}\tilde{p})^{16}(\tilde{\lambda}, \tilde{\mu}) = (22, 4)T_N = 2; (h^{11/2})^{10}T_U = 5]$ and $[(\tilde{f}\tilde{p})^{16}(\tilde{\lambda}, \tilde{\mu}) = (20, 2)T^N = 4; (h_{11/2})^{10}T_U = 3]$ *configurations, respectively. Model* II(π) *results with proton alignment and scattering are closest to experiment.* (*From Draayer et al.*, 1981.)

mapped onto a pseudo-$(\tilde{d}_{5/2}\tilde{d}_{3/2}\tilde{s}_{1/2})$ space with the $[\tilde{f}]$ and $(\tilde{\lambda}, \tilde{\mu})$ configurations shown in Table 7.6 taken into account. The seniority of the unique parity configurations was restricted to $\nu = 0$ and 2 only. A modified surface-delta interaction was used for the interaction in the separate spaces, as well as for the interaction that scatters pairs of particles between the two spaces. The remaining interaction between particles in the two spaces was taken to be a quadrupole-quadrupole interaction, $Q_n \cdot Q_u$. As shown in the figure, the model successfully reproduces the main features of the spec-

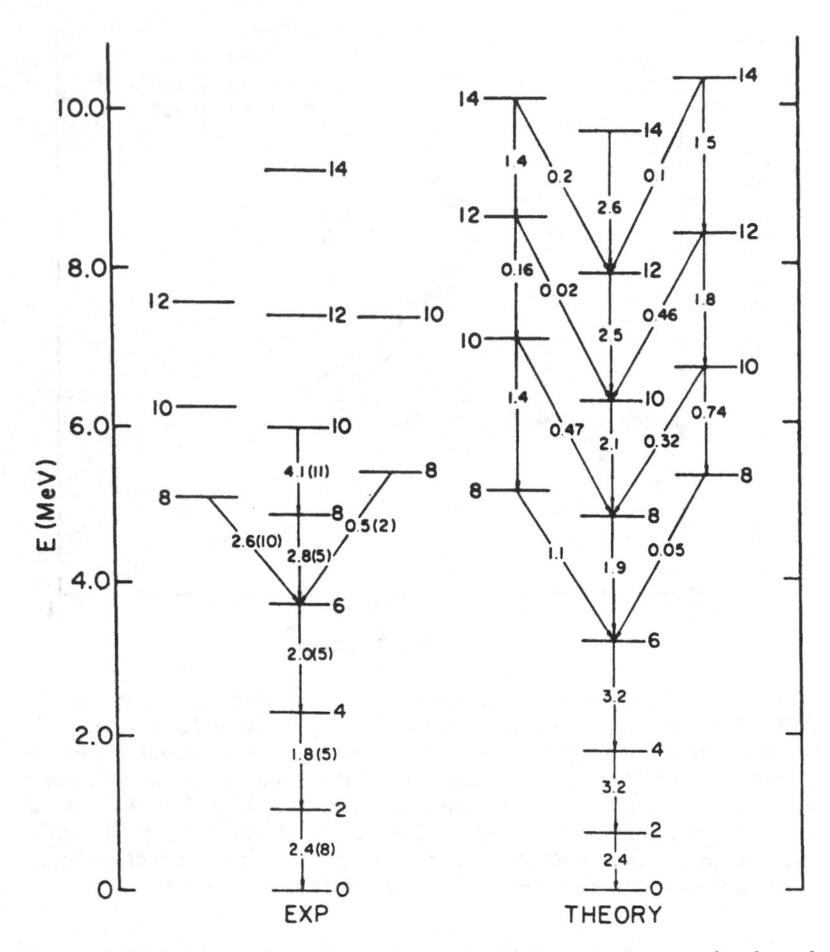

FIGURE 7.20. *Comparison of experimental and theoretical energy levels and* E2 *transition strengths in* ^{68}Ge. *The* B(E2) *values (units* = $10^{-2}e^2b^2$) *were determined without the use of effective charges* (e_π = 1, e_ν = 0). *The fact that the theoretical* B(E2) *values are larger than the corresponding experimental numbers is an indication that* ^{68}Ge *is less rotational than the model assumptions tend to make it. The sequence of states to the right of the* yrast *band involves the alignment of a pair of neutrons while the one on the left is an unaligned configuration similar to the* yrast *band. Proton alignment sets in at a higher excitation energy.* (*From Weeks, Han and Draayer,* 1981.)

TABLE 7.6. *Configuration symmetries included in the pseudo-SU(3) calculation* $[(f_{5/2}p_{3/2}p_{1/2}) \rightarrow (\tilde{d}_{5/2}\tilde{d}_{3/2}\tilde{s}_{1/2})]$ *of forking in* ^{68}Ge. (*From Weeks, Han and Draayer,* 1981.)

n_N	$[\tilde{f}]$	$(\tilde{\lambda}, \tilde{\mu})$	T_N	n_U	T_U
12	[4422]	(6, 6)	2	0	0
10	[442]	(10, 2)	1	2	1
10	[442]	(4, 8)	1	2	1
10	[4222]	(6, 2)	3	2	1
8	[44]	(8, 4)	0	4	2
8	[422]	(10, 0)	2	4	0
8	[422]	(10, 0)	2	4	2

trum. Because of space truncation, the theoretical results show stronger rotational features (enhanced E2 transition strengths and an $L(L + 1)$ excitation spectrum) than are found in nature. Nonetheless, the results show that the pseudo-$SU(3)$ scheme, even in this soft rotor region, works.

7.3.7. Applications 2: Electromagnetic Transitions

In setting about to carry out a shell-model calculation, two important choices must be made: the interaction Hamiltonian must be specified, and a many-particle basis selected. These choices can be closely interdependent, because an appropriate *effective* Hamiltonian must include parts that compensate for the excluded space. The field of effective interaction theory in nuclear physics deals with this difficult issue. Many years of work on this subject has taught us that the process is highly nonlinear and frequently divergent. The approaches used in almost all of the work done to date, however, did not exploit the now known synergistic relationships between interaction forms and model spaces that are so much a part of modern algebraic theories. Choosing an appropriate basis is like selecting the right coordinate system in classical physics—the wrong choice can actually render a solvable problem unsolvable, or solvable only with great difficulty. It is important to make the right choice.

We consider reproducing experimental results to be a necessary, but not sufficient condition; it is important that the Hamiltonian and basis space choices are reasonable and complementary. This can be insured, for example, by checking to make sure calculated results are stable under enlargement of the space, and by not allowing the Hamiltonian to change significantly as one moves through a series of neighboring nuclei. This is particularly important for AFM's, as we claim they are microscopic, and it is well-known that agreement can be forced in isolated cases by the *right* choice for the interaction parameters. Theories producing eigenstates that have small overlap with those determined in larger and more sophisticated calculations should be discounted. Regarding this matter, we need to emphasize that the pseudo-$SU(3)$ scheme yields stable solutions in heavy deformed nuclei because of the:

- validity of the pseudo-spin symmetry concept;
- complementary nature of the rotor and $SU(3)$ algebras;
- dominance of the quadrupole-quadrupole interaction.

These features are foundation elements undergirding the theory.

The general procedure for carrying out AFM calculations is the same, regardless of differences in detail that arise due to the fact that the symmetry groups that enter are model dependent. Things only simplify if the Hamiltonian and other operators of interest, such as those for electromagnetic transitions and particle transfer reactions, can be written down solely in terms of generators of the symmetry groups of the model, because analytic forms can then be given for their matrix elements. Indeed in this case—excluding the antisymmetry requirements—the AFM is like the IBM, as all observables are expressible in terms of generators of the highest group under consideration. This simplification can be forced, either by limiting the space to a single irrep of the largest symmetry group or by expanding operators of interest in terms of generators and nongenerator forms and discarding the latter. In most AFM applications, the cut is less sharp. The model space is restricted to a collection of irreps (rather than just one) of the largest symmetry group, and operators are expanded in terms of generator and nongenerator parts with at least the most important of the latter taken into account. Understanding what is large and important, as opposed to that which is small and unimportant, is necessary for all AFMs. Fortunately, quantitative measures for this can be defined, and in many cases, have been applied.

As an illustration of this procedure for operators, we now consider the case of M1 and E2 transitions in rare earth and actinide nuclei cast in the framework of the pseudo-$SU(3)$ picture. The electric (E) and magnetic (M) transition operators for a system of A nucleons in the long-wavelength approximation are given by

$$T_M^L(E) = b^L \sum_\sigma \sum_i e_\sigma r_\sigma^L(i) Y_{LM}(\hat{r}_\sigma(i)), \tag{7.35a}$$

$$T_M^L(M) = b^{L-1}\mu_N \sum_\sigma \sum_i \left\{ \left[g_\sigma^s \mathbf{s}_\sigma(i) + \frac{2g_\sigma^o}{L+1} \mathbf{l}_\sigma(i) \right] \cdot [\nabla_\sigma(i) r_\sigma^L(i) Y_{LM}(\hat{r}_\sigma(i))] \right\}, \tag{7.35b}$$

where σ is π or ν and $b \approx A^{1/6}$ (fm) is the harmonic oscillator size parameter, g^s and g^o are respectively in the spin and orbital g factors, e is the charge, and μ_N denotes the nuclear magneton. For M1 and E2 transition rates these expressions eq. (7.35) reduce to:

$$\begin{aligned} T_\mu^1(M) &= \sqrt{3/4\pi}\mu_N \sum_\sigma [g_\sigma^0 L_\mu^\sigma + g_\sigma^s S_\mu^\sigma], \\ T_\mu^2(E) &= \sqrt{5/16\pi}\mu_N b^2 \sum_\sigma e_\sigma Q_\mu^\sigma, \end{aligned} \tag{7.36}$$

where L_μ^σ, S_μ^σ and Q_μ^σ are the orbital angular momentum, spin and quadrupole operators.

Though the procedure for evaluating matrix elements of these operators is quite formidable, the required technology is available. For example, in Table 7.7 we give the tensorial expansion for $Q_\mu^a = \frac{1}{2}\sqrt{16\pi/5}\,[\Sigma_i(p_i^2 b^2 Y_{2\mu}(\hat{p}_i) + r_i^2 Y_{2\mu}(\hat{r}_i))]$; the interesting feature is that the $SU(3)$ generator part dominates. In terms of a normal $SU(3)$ expansion, Q_μ^a is a $[(\lambda_0, \mu_0)L_0, S_0] = [(1, 1)2, 0]$ tensor. When this operator is transformed into the pseudo scheme, the $[(\tilde{\lambda}_0, \tilde{\mu}_0)\tilde{L}_0\tilde{S}_0 = [(1, 1)2, 0]$ part still dominates but other tensors appear. In particular, because the *normal* ↔ *pseudo* transformation does not separately preserve the space-spin degrees of freedom, whenever Q_μ^a is expressed in terms of pseudo tensors it has $\tilde{S}_0 \neq 0$ parts. Nonetheless, the coefficient for each non-generator part is small relative to the coefficient that multiplies $\tilde{Q}_\mu^a$, the quadrupole generator $[(\tilde{\lambda}_0, \tilde{\mu}_0)\tilde{L}_0\tilde{S}_0 = [(1, 1)2, 0]$ of the pseudo-$SU(3)$ symmetry.

As another indication of the significance of the non-generator parts, in Table 7.8 we give two sets of results for B(E2) transition strengths in ^{168}Er, one for the complete algebraic quadrupole operator Q_μ^a and one for $\tilde{Q}_\mu^a$, which includes only the $[(\tilde{\lambda}_0, \tilde{\mu}_0)\tilde{L}_0\tilde{S}_0 =$

TABLE 7.7. *SU(3) tensor decomposition for the electric quadrupole operator,* $Q^a_\mu = \frac{1}{2}\sqrt{16\pi/5}\,[\Sigma_i(p_i^2 b^2 Y_{2\mu}(\hat{p}_i) + r_i^2 Y_{2\mu}(\hat{r}_i))]$. *The first set is the normal case with* Q^a a $(\lambda_0, \mu_0) = (1, 1)$ SU(3) *tensor only, while the second set is for* Q^a *rewritten in terms of tensors of the pseudo scheme, which involves all possible one-body* $(\tilde{\lambda}_0, \tilde{\mu}_0)$ *tensors:* (1, 1), (2, 2) . . . , (n, n) *for the* n-*th shell. The results for* Q^a *and* $\tilde{Q}^a$, *which dominate, are given in italics. All zero values are suppressed. (From Castaños, Draayer and Leschber,* 1987.*)*

(λ, μ_0)	κ_0	L_0	S_0	$n = 5$	$n = 4$	$n = 3$
Normal expansion						
(1, 1)	*1*	2	0	*28.98275*	*20.49390*	*13.41641*
$(\tilde{\lambda}_0, \tilde{\mu}_0)$	$\tilde{\kappa}_0$	$\tilde{L}_0$	$\tilde{S}_0$	$\tilde{n} = 5$	$\tilde{n} = 4$	$\tilde{n} = 3$
Pseudo Expansion						
(1, 1)	1	1	1	−0.95268	−0.76536	−0.58046
(1, 1)	*1*	*2*	*0*	*33.94606*	*24.44570*	*16.37716*
(2, 2)	1	2	0	−0.31804	−0.23130	−0.15032
(2, 2)	1	2	1	2.35558	1.73270	1.14296
(2, 2)	2	2	0	0.65178	0.47402	0.30808
(2, 2)	2	2	1	1.14940	0.84548	0.55770
(2, 2)	1	3	1	1.05264	0.78448	0.52754
(3, 3)	1	1	1	1.04658	−0.02848	−0.01416
(3, 3)	1	2	0	1.04658	0.65376	0.31786
(3, 3)	1	3	1	−0.11092	−0.06952	−0.03402
(3, 3)	2	3	1	0.16238	0.10176	0.04980
(4, 4)	1	2	0	−0.03012	−0.01392	
(4, 4)	1	2	1	0.34172	0.16024	
(4, 4)	2	2	0	0.05768	0.02666	
(4, 4)	2	2	1	0.17846	0.02666	
(4, 4)	1	3	1	0.18072	0.08522	
(5, 5)	1	1	1	−0.00186		
(5, 5)	1	2	0	0.09372		
(5, 5)	1	3	1	−0.00750		
(5, 5)	2	3	1	0.01052		

[(1, 1)2, 0] part of the full operator. The differences are small in all cases—for strong intra-band transitions as well as weak inter-band ones—which justifies replacing the complete operator Q^a_μ by its pseudo $\tilde{Q}^a_\mu$ counterpart, and dropping all non-generator terms. However, a word of caution is in order: because ^{168}Er is an even-even nucleus, the low-lying states are dominated by $S = 0$, and therefore $\tilde{S} = 0$ configurations; in odd-A nuclei, S is half integer so the effects of the $\tilde{S}_0 \neq 0$ tensors is more important. Nonetheless, the fact that the real quadrupole operator transforms into its

TABLE 7.8. *Comparison of selected* B(E2) *transition rates (units* = e^2b^2) *and quadrupole moments (units* = eb) *for states in* ^{168}Er. (*From Castaños, Draayer and Leschber,* 1987.)

I_i	I_f	A^*	B^*	I_i	I_f	A	B	I_i	I_f	A	B
Transition rates											
2_1	0_1	1.07	1.06	3_1	2_2	1.90	1.89	2_2	0_1	0.058	0.060
4_1	2_1	1.52	1.50	4_2	3_1	1.41	1.40	2_2	2_1	0.090	0.092
6_1	4_1	1.65	1.64	5_1	4_2	1.02	1.01	3_1	2_1	0.100	0.110
8_1	6_1	1.70	1.69	6_2	5_1	0.73	0.73	4_1	2_1	0.031	0.032
Quadrupole moments											
2_1	2_1	−2.09	−2.08	4_1	4_1	−2.65	−2.63				
2_2	2_2	2.09	2.08	4_2	4_2	−1.08	−1.08				

*Results for the real quarupole operator Q are given in column A while those for the pseudo quadrupole operator $\tilde{Q}$ are given in column B. The strength of $\tilde{Q}$ was set to reproduce the $(\lambda_0, \mu_0)L_0; S_0 = (1, 1)2; 0$ part of Q. The results demonstrate the near equivalence of $\tilde{Q}$ and Q.

pseudo counterpart with only small remaining terms is a clear example of what is required for an AFM to be a successful theory.

We should point out that in order to actually carry out calculations of the type referred above for ^{168}Er, one must be able to calculate many-particle matrix elements of the one-body operators given in eq. (7.35) as well as one and two-body Hamiltonian forms. This is no small task. If the Hamiltonian can be written in terms of invariants of the symmetry groups that one uses for specifying basis states, the problem simplifies. However, just as we have shown for operators, this will generally not be the case. For example, while $Q^a \cdot Q^a$ can be rewritten in terms of $SU(3)$ and $SO(3)$ invariants as $4C_2 - 3L^2$, the one-body orbit-orbit and spin-orbit interactions cannot be nor can the pairing be expressed in terms of simple operators. To include the effects of these important interactions, one must be able to calculate reduced matrix elements of one and two-body operator forms in the basis of the selected AFM. This is a solvable problem, so long as the coupling and recoupling coefficients of the relevant symmetry groups are available. For group chains that involve the $SU(3) \supset SO(3)$ reduction, the required technology is available. For others, like some of the higher symmetries that we will encounter in the next section on the FDSM, the required technologies are not yet available.

7.4. FERMION DYNAMICAL SYMMETRY MODEL

The Fermion Dynamical Symmetry Model (FDSM) is another shell-model theory which treats nucleons as fermions, thereby avoiding spurious boson degrees of freedom. It also exploits special group symmetries in an attempt to explain the observed properties of nuclei throughout the periodic table, especially the collective features of species with $A \gtrsim 100$. The original motivation for the FDSM grew out of an effort to provide a shell-model realization for the $U(5)$ vibrational, $O(6)$ γ-soft, and $SU(3)$ rotational limits of the IBM. This objective is realized by restricting the dynamics to a subspace of the full shell-model space spanned by the action of special $J = 0$ (S^+) and $J = 2$ (D^+_μ) fermion pair creation operators on the closed-shell vacuum. These pair creation operators and their annihilation operator counterparts (S and D_μ, respectively) are special because they form a subset of the set of all such pairs, and because they close (together with a complementary set of one-body operators, P^J_μ) under commutation. The FDSM Hamiltonian is built from scalar combinations of the pair creation and annihilation operators, and bilinear products of the one-body multipole operators: $[S^+S, (D^+ \times D)^0$, and $(P^J \times P^J)^0]$. Relaxing the restriction to S^+ and D^+_μ pairs renders the FDSM an alternative, complete shell-model scheme with features shared by other AFMs, in particular, the pseudo-$SU(3)$ scheme which also champions collective motion in medium and heavy mass deformed nuclei.

In this section we present a review of the FDSM, first giving background information concerning the structure of the model, then presenting selected results obtained in its application to various nuclear phenomena. The philosophy regarding the handling of defector and intruder levels in the FDSM is the same as used in the pseudo-LS and pseudo-$SU(3)$ models. The $j_{\max} = n + \frac{1}{2}$ level in both the proton and neutron subspaces is considered to have defected from the valence space to the shell below, and the intruder level with opposite (unique) parity and angular momentum $j = (n + 1) + \frac{1}{2} = n + \frac{3}{2}$—which penetrates into the valence space from the shell above—is assumed to be occupied with pairs coupled to a seniority zero (and hence spectroscopically inactive) configuration. As noted in Section 7.3, this assumption of spectroscopic inertness is reasonable only for configurations lying below the back-bending region, and then only if the quadrupole operator is renormalized appropriately. This restriction on the intruder level configurations can be relaxed for the FDSM in the same way as for the pseudo-$SU(3)$ scheme, but as in that case, it is a major extension that expands the scope and complexity of the theory considerably.

7.4.1. Favored Pair Concept

Although the success of the IBM stimulated a search for fermion realizations of the s and d boson operators, the closely related concept of favored pairs predates the IBM by nearly a decade. Specifically, in a valence space consisting of a set of orbitals $\{n(ls)j\}$, the identical-particle operator

$$A^+(JM) = \sum_{ll'} q^{(0)}(ll'J)[a_l^+ \times a_{l'}^+]^{JM;00} \tag{7.37}$$

where

$$q^{(0)}(ll'J) = \left[\frac{(2l + 1)(2l' + 1)}{2J + 1}\right]^{1/2} \langle l0l'0 \,|J0\rangle \qquad (7.38)$$

and

$$[a_l^+ \times a_{l'}^+]^{L=J,M_J,S=0,M_S=0}$$
$$= \sum_{mm'} \langle lml'm' |JM_J\rangle\langle \tfrac{1}{2}m_s\tfrac{1}{2} - m_s|00\rangle a^+_{lmm_s}a^+_{l'm'-m_s}, \qquad (7.39)$$

creates a favored pair state of relative angular momentum J with projection M. (For single-particle states with the same parity—as for the normal parity orbitals of a major shell—only even J values are allowed.) This superposition of two-particle states [eq. (7.37) with the $q^{(0)}(ll'J)$ weighting factor] is energetically favored by the surface-delta-interaction (SDI), which is known to be a remarkably good effective interaction in many regions of the periodic table. The favored pair concept includes quasi-spin symmetry (Section 7.1.4) as a special case: $S^+ = \sqrt{1/2}A^+(00)$, $S_- = \sqrt{1/2}A(00)$ and $S_0 = \frac{1}{2}[Q(00) - \Omega] = \frac{1}{2}[N_{op} - \Omega]$, where $\Omega = \Sigma(2l + 1)$ is the pair degeneracy number. The surface multipole operators $Q(JM)$ are one-body operator complements of the $A^+(JM)$ and $A(JM) = [A^+(JM)]^+$ operators:

$$Q(JM) = \sum_{ll'} (-1)^l q^{(0)}(ll'J)\sqrt{2}[a_l^+ \times a_{l'}]^{JM;00}$$
$$= \sum_i \sqrt{4\pi}Y_{JM}(\theta_i\phi_i). \qquad (7.40)$$

Clear evidence for the importance of such pairs can be seen in spectra of nuclei such as ${}^{40+n}_{20}\mathrm{Ca}_{20+n}$ and ${}^{90+n}_{40}\mathrm{Zr}_{50+n}$ where the configurations j^n ($j = \frac{7}{2}$ for Ca and $j = \frac{5}{2}$ for Zr) play a leading role.

One can construct a theory based on the concept of favored pairs by restricting the many-particle configurations to those which can be built from products of favored pair operators: $|\Psi_M^{N\alpha J}) = [A^+(J_N) \ldots \times [A^+(J_3) \times [A^+(J_2) \times A^+(J_1)]^{J_{21}}]^{J_{3(21)}} \ldots]_M^{J_{N(\ldots 3(21))}=J}|0\rangle$ where α stands for all intermediate coupling labels and the rounded ket $[|\Psi_M^{N\alpha J})]$ denotes an non-orthonormalized basis state. A logical choice for the Hamiltonian of such a system is $H = T + V$ where the interaction has a simple multipole form: $V = \Sigma_J G_J[\Sigma_M A^+(JM)A(JM)]$. Unfortunately, the favored pair operators do not form a closed algebra; that is, the commutator of an A^+ with an A includes more than just a Q, an A^+ with a Q is not just another A^+, and so on. This failure of the favored pairs to close under commutation means the action of the Hamiltonian ($H = T + V$) on a many-particle configuration built from products of favored pair operators [$|\Psi_M^{N\alpha J}\rangle$)] will carry one outside the favored pair subspace of the full model space. This is unlike the IBM situation in which the dynamics are completely contained within the space of s and d pairs, because the pairs form a $U(6)$ algebra which closes under commutation. Nonetheless, one can restrict the dynamics to the favored-pair subspace by simply setting all couplings to other than paired configurations to zero. An example testing this hypothesis is shown in Fig. 7.21 which makes a comparison of favored-pair results with complete calculations for the normal-$(g_{7/2}, d_{5/2}, d_{3/2}, s_{1/2})^4$ case considered to be a pseudo-$(\tilde{f}_{7/2}\tilde{f}_{5/2}\tilde{p}_{3/2}\tilde{p}_{1/2})^4$ space. Low-lying states of the favored pair approximation are in remarkably good agreement with the exact, full model space results. Of course, since the favored pairs are $\tilde{S} = 0$ objects, the $\tilde{S} = 1$ and $\tilde{S} = 2$ configurations of the full 4-particle space are not part of the favored pair picture, and are therefore not shown in the figure. Unfortunately, as shown by the figure, some of these excluded states lie low in the spectrum, raising doubts concerning the quality and reliability of the favored pair concept.

The principle undergirding the FDSM is that $J = 0$ (S^+) and $J = 2$ (D_μ^+) pairs built in another way, together with their annihilation operator counterparts (S and D_μ, respectively) and a complementary set of one-body operators (P_μ^J) form a closed algebra; this algebra defines a group structure that yields the IBM symmetries in specific limits. The key to the identification of a closed algebra is the introduction of pseudo-orbital and pseudo-spin operators with their vector sum equal to the total angular momentum: $\mathbf{j} = \mathbf{k} + \mathbf{i}$, where $\mathbf{k}$ and $\mathbf{i}$ are different from (and not to be confused with) the $\tilde{\boldsymbol{l}}$ and $\tilde{\boldsymbol{s}}$ of the pseudo-spin picture (Section 7.3). The name *pseudo* was adopted from the pseudo-spin scenario, as the concept of a separation of the total angular momentum into an orbital and spin part is the same in the two theories. In the usual fermion creation operator language, one can write

$$a_{jm}^+ = a_{(ki)jm}^+ = \sum_{m_k m_i} \langle k m_k i m_i | jm \rangle \, a_{k m_k i m_i}^+ \qquad (7.41)$$

where the (ki) subscript on a_{jm}^+ can be dropped because of the unique representation of the orbitals of a single major oscillator shell in terms of the k and i spin, as explicitly shown in Table 7.9.

In the spirit of the favored-pair picture, one can build generic

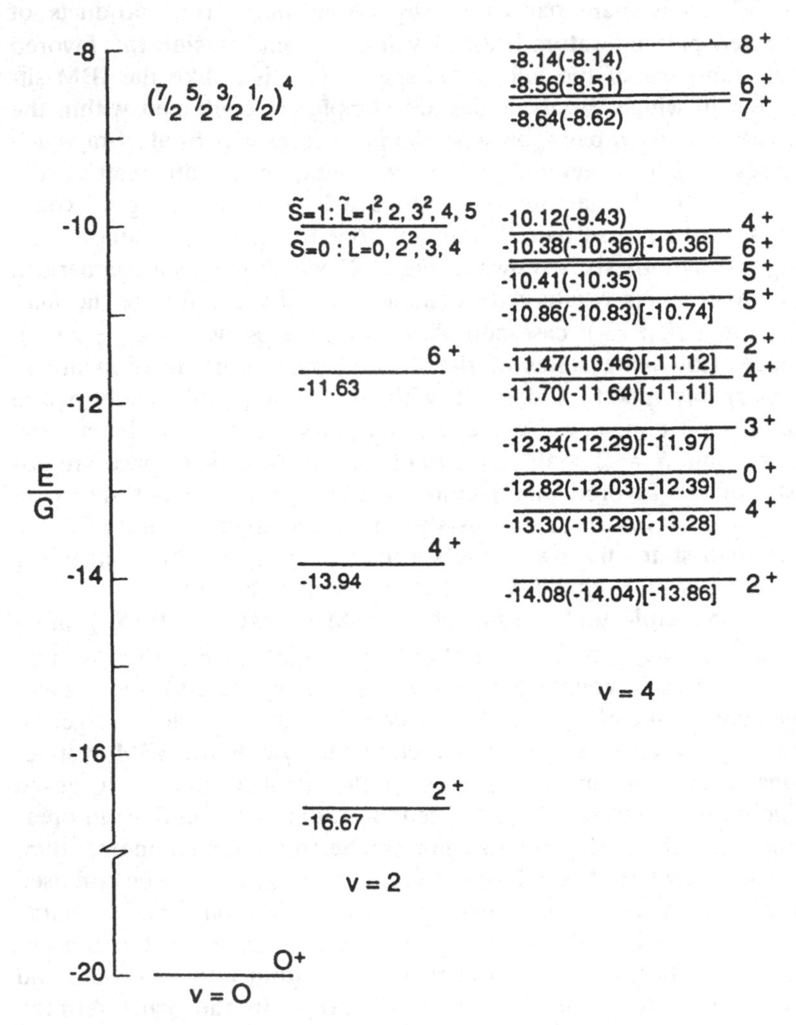

FIGURE 7.21. *A comparison of eigenvalue spectra for a favored-pair description and corresponding complete shell-model results for a normal* $(g_{7/2},d_{5/2},d_{3/2},s_{1/2})^4$ *configuration considered as a pseudo* $(\tilde{f}_{7/2}\tilde{f}_{5/2}\tilde{p}_{3/2}\tilde{p}_{1/2})^4$ *model space. Numbers below the levels are the exact eigenvalues of the Hamiltonian while those in parantheses are the eigenvalues in the truncated subspace of favored* v=4 *states constructed from favored* J ≠ 0 *pairs. Numbers in the square brackets are expectation values of the Hamiltonian in states obtained by seniority projection techniques from single states built from the lowest favored* J ≠ 0 *pairs. (From Hecht, McGrory and Draayer, 1972.)*

TABLE 7.9. *The single-particle levels of the shell-model as realized in terms of the pseudo-orbital* (k) *and pseudo-spin* (i) *labels of the Fermion Dynamical Symmetry Model:* **j** = **k** + **i** *within each shell or subshell with the unique parity intruder* (int) *level always identified as a* (k = 0, i = j) *orbital.* (*Based on Wu et al.*, 1986.)

#	1	2	3	4	5		6			7			8	
n k i	0 0 $\frac{1}{2}$	1 1 $\frac{1}{2}$	2 1 $\frac{3}{2}$	3 0 $\frac{7}{2}$	3 1 $\frac{3}{2}$	4 0 $\frac{9}{2}$	4 2 $\frac{3}{2}$	5 0 $\frac{11}{2}$	5 1 $\frac{1}{2}$	5 1 $\frac{7}{2}$	6 0 $\frac{13}{2}$	6 1 $\frac{3}{2}$	6 1 $\frac{9}{2}$	7 0 $\frac{15}{2}$
c o n f	$s_{1/2}$	$p_{1/2}$ $p_{3/2}$	$s_{1/2}$ $d_{3/2}$ $d_{5/2}$	$f_{7/2}$	$p_{1/2}$ $p_{3/2}$ $f_{5/2}$	$g_{9/2}$	$s_{1/2}$ $d_{3/2}$ $d_{5/2}$ $g_{7/2}$	$h_{11/2}$	$p_{1/2}$ $p_{3/2}$	$f_{5/2}$ $f_{7/2}$ $h_{9/2}$	$i_{13/2}$	$s_{1/2}$ $d_{3/2}$ $d_{5/2}$	$g_{7/2}$ $g_{9/2}$ $i_{11/2}$	$j_{15/2}$
s y m			G_6 G_8 G_3		G_6 G_8 G_3	i n t	G_8	i n t	G_6		i n t	G_6		i n t
Ω_0 Ω_1 Ω N	0 1 1 2	0 3 3 8	0 6 6 20	0 4 4 28	5 6 11 50		6 10 16 82		7 15 22 126			8 21 29 184		

$G_6 = (Sp_6^k \otimes SO_3^i) \otimes (SU_2 \otimes SO_2)$ (k-active)
$G_8 = (SO_8^k \otimes SO_3^i) \otimes (SU_2 \otimes SO_3)$ (i-active)
$G_3 = (SU_3^k \otimes SO_6^i) \otimes (SU_2 \otimes SO_3)$ (k-active or i-active)
$\Omega_0 = (2j_{int} + 1)/2$ (intruder level degeneracy)
$\Omega_1 = \Sigma(2k + 1)(2i + 1)/2$ (normal parity degeneracy)
$\Omega = \Omega_0 + \Omega_1$ (summed spatial degeneracy)
$N = 2\Sigma\Omega$ (total space-spin degeneracy)

pairs by coupling the pseudo-orbital and pseudo-spin fermion operators to total K and I as follows:

$$A^{+}\{[(k_1k_2)K(i_1i_2)I]JM\} = [a^{+}_{(k_1,i_1)} \times a^{+}_{(k_2,i_2)}]^{(KI)J}_{M}$$

$$= \sum_{j_1,j_2} \begin{bmatrix} k_1 & i_1 & j_1 \\ k_2 & i_2 & j_2 \\ K & I & J \end{bmatrix} [a^{+}_{(l_1s_1)j_1} \times a^{+}_{(l_2s_2)j_2}]^{J}_{M}.$$

The last form, which uses a 9-j coefficient to effect the $SU(2)$ recouplings, expresses the pair creation operators in terms of the usual $(ls)j$ coupling [not the $(ki)j$] scheme $(a^{+}_{(ls)jm} = a^{+}_{(kI)jm})$. The $A^{+}\{(k_1k_2)K(i_1i_2)I]JM\}$ span the full set of all possible fermion pair states, and a general two-body interaction can be written in terms of these operators:

$$V = \sum_{J} \sum_{[(k_1'k_2')K'(i_1'i_2')I'][(k_1k_2)K(i_1i_2)I]} C_J\{[(k_1'k_2')K'(i_1'i_2')I'] \times [k_1k_2)K(i_1i_2)I]\}$$
$$\times \sum_{M} A^{+}\{[(k_1'k_2')K'(i_1'i_2')I']JM\} A\{[(k_1k_2)K(i_1i_2)I]JM\}, \tag{7.43}$$

where $C_J\{[(k_1'k_2')K'(i_1'i_2')I'][(k_1k_2)K(i_1i_2)I]\}$ are constants that define the interaction and $A\{[k_1, k_2]K(i_1i_2)I]JM\} = (A^{+}\{[(k_1k_2)K(i_1i_2)I)JM\})^{+}$. In this unrestricted form, the $(k$-$i)$-coupling scheme is a complete shell-model theory. Special limits of the FDSM are obtained by placing restrictions on the k and i-space couplings. These restriction and their consequences are presented next.

7.4.2. Active and Inactive Spaces

The s and d operators of the IBM are considered to be fermion pairs coupled to total angular momentum 0 and 2, respectively. Within the framework of the $(k$-$i)$-scheme of the FDSM, pairs of this type can be realized in two distinct ways: by setting $K = 2$ and $I = 0$ when $k = 1$ (k-active); or with $K = 0$ and $I = 2$ when $i = \frac{3}{2}$ (i-active). If $\Omega_{ki} = (2k + 1)(2i + 1)/2$, the explicit form for the S and D pairs of the two FDSM limits are given by

$$S^+ \equiv A^{J=0+}_{M=0} \tag{7.44a}$$
$$= \sum_{k,i} \sqrt{\Omega_{ki}/2}\, A^+\{[(kk)0(ii)0]00\} \quad (k\text{-active or } i\text{-active}),$$

$$D^+_M \equiv A^{J=2+}_M = \sum_i \sqrt{\Omega_{1i}/2}\, A^+\{[(11)2(ii)0]2M\} \tag{7.44b}$$

$$(k\text{-active with } k = 1),$$

$$D^+_M \equiv A^{J=2+}_M = \sum_k \sqrt{\Omega_{k3/2}/2}\, A^+\,\{[(kk)0(\tfrac{3}{2}, \tfrac{3}{2})2]2M\} \tag{7.44c}$$

$$(i\text{-active with } i = \tfrac{3}{2}).$$

The complementary multipole operators are defined as follows:

$$P^0_0 = \sum_{k,i} \sqrt{\Omega_{ki}/2}[a^+_{ki} \times \tilde{a}_{ki}]^{00}_{00} \quad (k\text{-active or } i\text{-active}), \tag{7.45a}$$

$$P^J_M = \sum_i \sqrt{\Omega_{1i}/2}[a^+_{1i} \times \tilde{a}_{1i}]^{J0}_{M0} \quad (k\text{-active with } k=1), \tag{7.45b}$$

$$P^J_M = \sum_k \sqrt{\Omega_{k,3/2}/2}[a^+_{k,3/2} \times \tilde{a}_{k,3/2}]^{0J}_{0M} \quad (i\text{-active with } i=\tfrac{3}{2}). \tag{7.45c}$$

In eq. (7.50), $\tilde{a}_{kn,im} = (-1)^{k+n+i+m}a_{k,-n;i,-m}$ so the $\tilde{a}_{kn,im}$ are proper tensors under rotations. (A simple way to remember the phase factor in this relationship is to recall that the number operator is $n_{op} = \Sigma_n a^+_{kn,im}a_{kn,im} = \sqrt{(2k + 1)(2i + 1)}\; [a^+_{ki} \times \tilde{a}_{ki}]^{(00)0}$ where the $SU(2)$ coupling coefficients that enter are simply $\langle jm,\, j - m|00\rangle = (-1)^{j-m}/\sqrt{2j + 1}$.) The (k, i) sums are required when there is more than a single (k, i) value required to reproduce the single-particle shell-model orbitals [$i = \frac{1}{2}$ ($j = \frac{1}{2}$ and $\frac{3}{2}$) and $i = \frac{7}{2}$ ($j = \frac{5}{2}, \frac{7}{2}, \frac{9}{2}$) for the $n = 5$ shell and $i = \frac{3}{2}$ ($j = \frac{1}{2}, \frac{3}{2}$ and $\frac{5}{2}$) and $i = \frac{9}{2}$ ($j = \frac{7}{2}, \frac{9}{2}$ and $\frac{11}{2}$) for $n = 6$ case].

The total number of pair plus multipole operators in the k-active case ($k = 1$) is 21: 2 $J = 0$ pairs, (S^+, S); 10 $J = 2$ pairs, (D^+_μ and D_μ with projections $\mu = -2, -1, 0, +1, +2$); and $\Sigma_J(2J + 1) = 1 + 3 + 5 = 9$ multipoles (P^J_μ with $\mu = -J, -J + 1, \ldots, + J - 1, + J$ and $J = 0, 1$, and 2). These 21 operators close under commutation and generate a realization of the compact symplectic $Sp(6)$ group. This compact $Sp(6)$ group has two subgroups [$SU(3)$ and $SU(2) \otimes SO(3)$] that contain the exact symmetry $SO(3)$ angular momentum group as a subgroup. It can be shown that $SU(2) \otimes SO(3)$ and $SU(3)$ subgroups of $Sp(6)$ are formally equivalent to the $U(5)$ vibrational limit and $SU(3)$ rotational limit of the IBM, respectively. In the i-active limit ($i = \frac{3}{2}$), on the other hand, the total number of operators is 28: 2 $J = 0$ pairs, (S^+, S); the 10 $J = 2$ pairs, (D^+_μ and D_μ); and the $\Sigma_J(2J + 1) = 1 + 3 + 5 + 7 = 16$ multipoles, P^J_μ. Therefore in this case there are 28 operators, and these generate $SO(8)$ which has three subgroups [$SO(6)$, $SO(5) \otimes SU(2)$ and $SO(7)$] containing $SO(3)$. The $SO(5) \otimes SU(2)$ subgroup of $SO(8)$ can be shown to be formally equivalent to the $U(5)$ vibrational limit of the IBM, while the $SO(6)$ subgroup corresponds to the gamma-soft limit of the IBM theory. The $SO(7)$ subgroup of $SO(8)$ has no counterpart IBM counterpart. The relationships of these symmetry groups to one another—the dynamical symmetry group chains of the $Sp(6)$ and $SO(8)$ models of the FDSM—are shown in schematic form in Fig. 7.22.

The commutation relations of the A^{J+}_M, A^J_M and P^J_M operators are given by

$$[A^J_M, A^{J'+}_{M'}] = \Omega_N \delta_{JJ'} \delta_{MM'} - 2 \sum_{J''} (-1)^M K^{J''M''}_{J-MJ'M'} P^{J''}_{M''}, \qquad (7.46a)$$

$$[P^J_M, A^{J'+}_{M'}] = \sum_{J''} (-1)^M K^{J''M''}_{JMJ''M'} A^{J''+}_{M''}, \qquad (7.46b)$$

$$[P^J_M, P^{J'}_{M'}] = \frac{1}{2} \sum_{J''} [(-1)^{J''} - (-1)^{J+J'}] K^{J''M''}_{JMJ'M'} P^{J''}_{M''}, \qquad (7.46c)$$

where $\Omega_N = \Sigma(2j + 1)/2$ with the sum over the normal-parity levels, and

$$K^{J''M''}_{JM,J'M'} = (-1)^{2L} \sqrt{(2L + 1)(2J + 1)(2J' + 1)} \times \langle J'M', J'', M'' | JM \rangle \begin{Bmatrix} J & J' & J'' \\ L & L & L \end{Bmatrix} \qquad (7.47)$$

with $L = 1$ for the k-active case and $L = \frac{3}{2}$ for the i-active scenario. In. eq. (7.47), $\langle , | \rangle$ and $\{\ \}$ denote an $SU(2)$ Clebsch–Gordan and 6-j recoupling coefficient, respectively. The value for L in eq. (7.47) is the only difference in the commutation relations for the k-active and i-active cases, showing that the two schemes can be treated in a unified way.

The FDSM Hilbert space can be subdivided into subspaces characterized by a heritage quantum number u which counts the number of fermions *not* coupled to angular momentum $L = 0$ (S^+) or $L = 2$ (D^+_μ) pairs, see Fig. 7.23. The heritage quantum number is

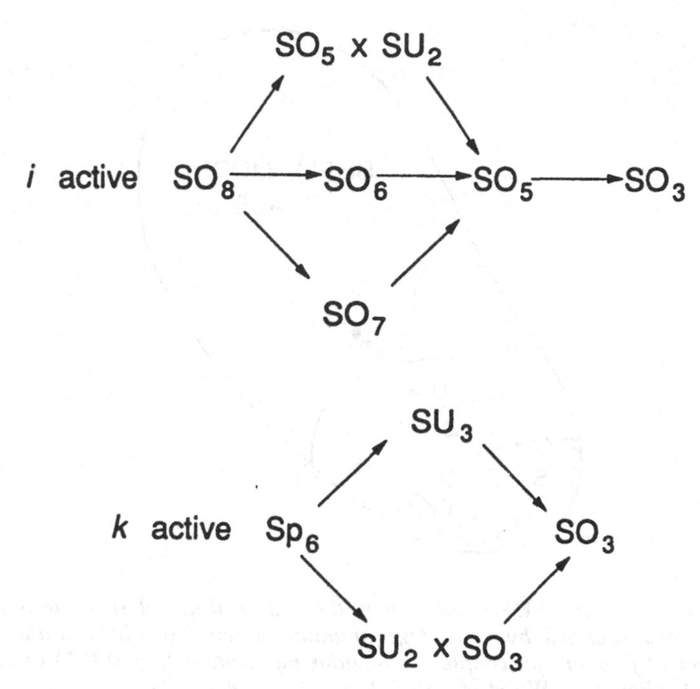

FIGURE 7.22. FDSM *dynamical symmetry subgroup chains of the* SO(8) *and* Sp(6) *symmetry limits of the theory.* (*Based on Wu et al.*, 1987.)

similar to the seniority quantum number v which counts the number of fermions *not* coupled to $L = 0$ (S^+) pairs. The m-fermion $u = 0$ space is spanned by $|\Psi^{N\alpha J}_M) = [A^{J_N^\dagger} \ldots \times [A^{J_3^\dagger} \times [A^{J_2^\dagger} \times A^{J_1^\dagger}]^{J_{21}}]^{J_{3(21)}} \ldots]^{J_{N(\ldots 3(21))}=J}_M|0\rangle$, where (as for the favored-pair case) α labels the intermediate couplings [in this case the J_α values of the A^+ operators are restricted to 0 and 2, see eq. (7.44)] and the rounded ket again denotes an non-orthonormalized basis state. A recursive build-up process can be used to construct the overlap matrices required to orthonormalize the $|\Psi^{N\alpha J}_M)$ basis states. It then follows that if n_S and n_D are, respectively, the number of $L = 0$ (S^+) and $L = 2$ (D^+_μ) pairs in an m-fermion system, $m = 2(n_S + n_D) + u = 2n_S + v$ where $v = 2n_D + u$ is the usual seniority quantum number including the number of particles not coupled to angular momentum zero pairs. Heritage non-zero configurations can be constructed by coupling the a^+_{jm} operators to the $|\Psi^{N\alpha J}_M)$, insuring at each stage that the added factors contain no $L = 0$ (S^+) or $L = 2$ (D^+_μ) pairs. The collection of basis states which includes

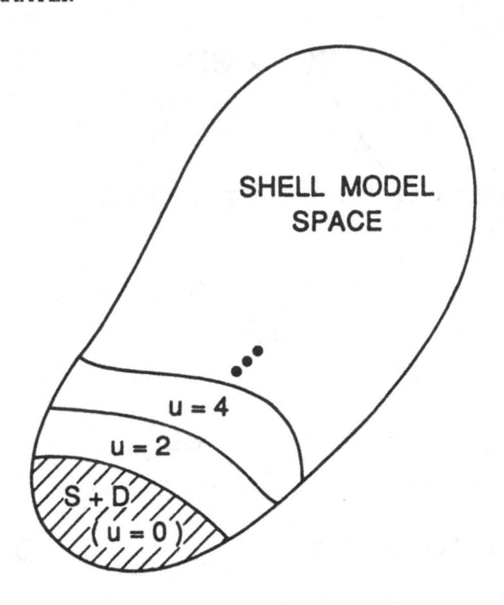

FIGURE 7.23. *The* FDSM *subdivides the full shell-model space into subspaces characterized by a heritage quantum number* μ *which counts the number of fermion* not *coupled to angular momentum* $L = 0$ (S^+) *or* $L = 2$ (D^+_μ). (*Based on Wu et al.*, 1987.)

all heritage values ($u = 0, 2, \ldots, m$ for m even and $u = 1, 3, \ldots, m$ for m odd) spans the full many-particle space; as with the IBM and the IBFM, only $u = 0$ (even-A) and $u = 1$ (odd-A) states are usually kept in the basis.

The FDSM Hamiltonian (restricted to heritage non-breaking interaction terms) is usually assumed to be scalar combinations of pair and multipole operator forms in the normal (N) and abnormal (A) parity spaces:

$$H_{\text{FDSM}} = \epsilon_A n_A + \epsilon_N n_N + H_A + H_N + H' \qquad (7.48)$$

where

$$H_A = \mathbf{G}_0\mathbf{S}^+\mathbf{S} + \mathbf{B}_0^A n_A(n_A - 1)/2, \qquad (7.49a)$$

$$H_N = G_0 S^+ S + G_2 D^+ \cdot D + \sum_J B_J P^J \cdot P^J, \qquad (7.49b)$$

$$H' = g_0(\mathbf{S}^+ S + S^+\mathbf{S}) + b_0 n_A n_N. \qquad (7.49c)$$

In eq. (7.48), ϵ_A is the single-particle energy of the intruder level and ϵ_N is the average single-particle energy of the normal parity orbitals $[\Sigma_j(2j + 1)\epsilon_j/\Sigma_j(2j + 1)]$; the n_A and n_N are operators that count the number of particles of each type. The H_A and H_N interactions act solely in the abnormal and normal parity spaces respectively, while H' is made up of a monopole term $(n_A n_N)$ with eigenvalues proportional to the product of the occupancies of the abnormal and normal parity orbitals and an operator that scatters $L = 0$ pairs $(\mathbf{S}^+\mathbf{S} + \mathbf{S}^+\mathbf{S})$ between the two spaces. The H_A term contributes an overall constant to the energy, when the number of particles in each space is fixed and the dynamics are restricted to seniority zero unique parity configurations. Fixing the number of particles of each type also means that H' only contributes to the binding energy of the system; the pair scattering option $(g_0 \neq 0)$ which carries one from a space with (m_A, m_N) particles to one with $(m_A \pm 2, m_N \mp 2)$ is blocked. (While the pair scattering term is important for describing backbending phenomena, it is not expected to play a major role in the dynamics of low-energy low-spin states below the backbending region. In the simplest version of the theory, it is customary to consider only the seniority zero intruder configuration and not the scattering of pairs between the abnormal and normal parity orbitals. With this restriction, the intruder level plays no role in the dynamics other than as a sink for m_A particles where the total number m is given by $m = m_A + m_N$. The ratio m_N/m is therefore the fraction of valence nucleons considered to be spectroscopically active.)

The most important term in the FDSM Hamiltonian describing low-energy, low-spin phenomena is H_N, which acts solely on the m_N particles in the normal parity part of the space. This term is responsible for determining the level structure and electromagnetic transition properties of low-lying states, and gives rise to the special (IBM-like) symmetries of the FDSM. In the next sub-section we will consider the $Sp(6)$ (k-active) and $SO(8)$ (i-active) limits of the FDSM Hamiltonian in somewhat greater detail. Before proceeding however, it is important to reiterate that without any additional restrictions, the (k-i)-scheme of the FDSM is a unitary transformation of the usual single-particle, shell-model orbitals—it is a complete shell-model theory that assumes an IBM-like form in certain of its limits. By including $u \neq 0$ basis states and adding heritage breaking terms to the Hamiltonian the full range of dynamics found in the complete shell-model problem is recovered.

7.4.3. Special Symmetries: SO(8) and Sp(6)

In the actinide region, the valence protons (π) and neutrons (ν) occupy k-active orbitals with $i = (\frac{1}{2}, \frac{7}{2})$ and $i = (\frac{3}{2}, \frac{9}{2})$, respectively, so the applicable FDSM symmetry is $Sp(6)$ for both nucleon types. The combined (π-ν) symmetry group is the direct product $Sp_\pi(6) \otimes Sp_\nu(6)$ structure. On the other hand, the valence protons of rare earth nuclei sit in an i-active ($k = 2$) model space so the $SO(8)$ FDSM symmetry applies. Therefore in the rare earth case, the (π-ν) symmetry group is the mixed-symmetry, direct-product $SO_\pi(8) \otimes Sp_\nu(6)$ structure. The subgroup scenarios for these direct product structures are identified in Table 7.10.

The FDSM group structure is very different from the IBM scenario in which the highest symmetry in every shell, whether proton, neutron, or combined proton and neutron, is $U(6)$. As a result, one can choose a combined $U(6)$ symmetry in the rare earth and actinide regions in the IBM case, such as $U_\pi(6) \otimes U_\nu(6) \supset U_{\pi+\nu}(6) = U(6)$ (with two-rowed Young diagrams irreducible representations allowed because the proton and neutron boson types are distinguishable) with its $U(5)$, $O(6)$ and $SU(3)$ limits. Or one may carry the (π-ν) direct-product couplings down to a lower level in the chain such as $SU_\pi(3) \otimes SU_\nu(3) \supset SU_{\pi+\nu}(3) = SU(3) \supset SO(3)$ or even $SU_\pi(3) \otimes SU_\nu(3) \supset SO_\pi(3) \otimes SO_\nu(3) \supset SO_{\pi+\nu}(3) = SO(3)$.

TABLE 7.10. FDSM *symmetries for heavy nuclei when the protons* (π) *and neutrons* (ν) *occupy neighboring shells. A maximally deformed product configuration via strong* $SU(3)$ *coupling can only be achieved in the actinide case:* $SU_\pi(3) \otimes SU_\nu(3) \supset SU(3) \supset SO(3)$.

FDSM *Symmetries for Heavy Nuclei*			
Nuclides	Protons	Neutrons	Coupled space
	$SO_\pi(8)$	$Sp_\nu(6)$	$SO_\pi(8) \otimes Sp_\nu(6)$
	$\downarrow$	$\downarrow$	$\downarrow$
rare earth	F_8^π	F_6^ν	$F_8^\pi \otimes F_6^\nu$
	$\downarrow$	$\downarrow$	$\downarrow$
	$SO_\pi(3)$	$SO_\nu(3)$	$SO_\pi(3) \otimes SO_\nu(3)$
	$Sp_\pi(6)$	$Sp_\nu(6)$	$Sp_\pi(6) \otimes Sp_\pi(6)$
	$\downarrow$	$\downarrow$	$\downarrow$
actinide	F_6^π	F_6^ν	$F_6^\pi \otimes F_6^\nu$
	$\downarrow$	$\downarrow$	$\downarrow$
	$SO_\pi(3)$	$SO_\nu(3)$	$SO_\pi(3) \otimes SO_\nu(3)$

$F_6 = SU(3)$ or $SU(2) \otimes SO(3)$ [which is formally similar to $U(5)$].
$F_8 = SO(6)$ or $SO(5) \otimes SU(2)$ [which is formally similar to $U(5)$] or $SO(7)$.

The FDSM symmetries can be combined at the highest level only in the actinide region: $Sp_\pi(6) \otimes Sp_\nu(6) \supset Sp_{\pi+\nu}(6) = Sp(6)$, etc. The direct-product structure is normally maintained down to the lowest level before effecting the proton and neutron coupling in the FDSM approach to the rare earth and actinide regions.

The symmetry preserving FDSM Hamiltonian, which applies in the k-active (rare earth neutron valence space and the actinide proton and neutron valence spaces) as well as the i-active (rare earth valence proton space) cases is given in eq. (7.48). As already discussed, the dynamics are generated by the H_N part of H_{FDSM}, given in eq. (7.49b), when the m valence particles are partitioned into a fixed (m_A, m_N) set. If terms proportional to the number of valence particles in the normal parity space are neglected, H_N involves respectively a total of 6 (G_0, G_2, and B_J where $J = 0, 1, 2,$ and 3) and 5 (G_0, G_2, and B_J where $J = 0, 1,$ and 2) parameters in the $SO(8)$ and $Sp(6)$ limits of the theory. However, since the sum of the number of $L = 0$ (S^+) and $L = 2$ (D^+) pairs in a heritage preserving system is fixed, the eigenvalues of S^+S and $D^+ \cdot D$ are related ($D^+ \cdot D + S^+S = N$, where N counts the number of pairs in the normal parity space) and as a consequence G_0 and G_2 are not independent parameters. Therefore, the number of independent parameters is, respectively, 5 and 4 for the $SO(8)$ and $Sp(6)$ limits of the theory. The number of independent second order group invariants in the $SO(8) \rightarrow SO(3)$ and $Sp(6) \rightarrow SO(3)$ chains [see Fig. 7.22 and Table 7.10] are precisely these numbers: $C_{SO(8)}$, $C_{SO(6)}$, $C_{SO(5)\otimes SU(2)}$, $C_{SO(7)}$, and $C_{SO(3)}$; and $C_{Sp(6)}$, $C_{SU(3)}$, $C_{SU(2)\otimes SO(3)}$, and $C_{SO(3)}$, respectively. This means the $SO(8)$ and $Sp(6)$ FDSM Hamiltonians can be represented faithfully by these sets of group invariants. Just as in the IBM case, one can then explore pure symmetry limits of the theory, two for $Sp(6)$, the $SU(3)$ and $SU(2) \otimes SO(3)$ branches, and three in the $SO(8)$ case, the $SO(8)$, $SO(5) \otimes SU(2)$ and $SO(7)$ paths.

The structure of even the simplest identical particle FDSM Hamiltonian is very rich. Since $Sp(6) \supset SU(3)$, $SO(8) \supset SO(6)$, and to the extent the $SO(8) \supset SO(5) \otimes SU(2)$ and $Sp(6) \supset SU(2) \otimes SO(3)$ chains are formally similar to the IBM $U(6) \supset U(5)$ limit, one certainly finds the $SU(3)$, $O(6)$ and $U(5)$ symmetries of the IBM in the FDSM scheme. Unlike the IBM case however, these three symmetries are not all found in the same higher group structure: $SU(3)$ and $U(5) \sim SU(2) \otimes SO(3)$ are contained in $Sp(6)$, while $O(6)$ and $U(5) \sim SO(5) \otimes SU(2)$ occur in $SO(8)$, where a tilde ($\sim$) is used to denote a formal correspondence. So while one can follow a transition between the $SU(3)$ and $U(5)$ limits within $Sp(6)$, and between the $SO(6)$ and $U(5)$ limits within $SO(8)$, one cannot study a transition between the $SU(3)$ and $O(6)$ limit of the theory, as these two are subgroups of different parent group structures. The FDSM provides a microscopy for studying the $U(5) \leftrightarrow SU(3)$ and $U(5) \leftrightarrow O(6)$ legs of the IBM triangle; it cannot be used to study the $O(6) \leftrightarrow SU(3)$ branch, as this is not a part of either the $SO(8) \rightarrow SO(3)$ or $Sp(6) \rightarrow SO(3)$ group chains. FDSM calculations are further complicated by the fact that each application requires the direct product coupling of a proton and neutron subspace with parameters for each determined separately, as well as additional ones associated with subspace couplings. Constants multiplying $n_\pi n_\nu$, the multipole-multipole $P^J_\pi \cdot P^J_\nu$ ($J = 0$, 1 and 2) interactions, and the pair scattering terms $\frac{1}{2}(D^+_\pi \cdot D_\nu + D^+_\nu \cdot D_\pi)$ and $\frac{1}{2}(S^+_\pi \cdot S_\nu + S^+_\nu \cdot S_\pi)$ serve as examples. Even if the multipole-multipole forms are left out, there are 13 parameters for actinides, 5 in the proton space and 5 in the neutron space, as well as 3 coupling the two. For rare earth nuclei the corresponding number is 14, as the applicable proton symmetry is $SO(8)$ rather than $Sp(6)$.

Since the FDSM has a microscopic foundation, it is a testable theory. Comparisons can be carried out on a variety of levels, for example:

- global predictions of nuclear systematics using special symmetry limits of the theory (some results given below);
- statistical measures for evaluating the probable goodness of the FDSM symmetries (as suggested in Section 7.1);
- expansions of common interactions such as pairing and $Q \cdot Q$ in terms of tensors of the (k-i)-coupling scheme;
- model-versus-model tests using, for example, available shell-model codes and techniques;
- detailed results for excitation spectra and electromagnetic transition rates of a specific nuclear species.

Only a few of these tests have been performed for the FDSM case, because only in special cases analytic solutions can be given solely in terms of the eigenvalues of the Casimir invariants of the relevant group chain. Some results from those investigations are given in the next section. To perform any of the other tests requires the development of shell-model technologies for the (k-i) coupling scheme, such as reduced matrix elements of the $Sp(6) \supset SU(3)$ coupling scheme. As for the symplectic scheme, the complementary $SU(3) \supset SO(3)$ coupling and recoupling coefficients are all available.

7.4.4. Applications: Global Predictions

The fermion underpinning of the FDSM theory means the fatal flaw of a boson picture, described in Section 7.1, can be avoided. This comes about because the $Sp(6) \supset SU(3)$ group structure dictates the set of $SU(3)$ irreps that can occur. In the IBM case the leading $SU(3)$ irrep is always $(\lambda, \mu) = (2N, 0)$ where N is the total number of bosons which is taken to be equal to half the number of valence particles. This is also the leading irrep in the FDSM case so long as $N < \Omega/3$, as illustrated in Table 7.11. For nuclei with $N > \Omega/3$ however, the $SU(3)$ irrep $(2N, 0)$ is spurious, because it is forbidden by the Pauli Principle. This constraint follows from the fact that Young diagrams representing allowed $Sp(6)$ configurations can have no more than $2\Omega/3$ columns, or equivalently, 3 rows. The factor of 3 dividing 2Ω enters because the full $(k\text{-}i)$-level degeneracy given by $2\Omega = \Sigma_i(2k + 1)(2i + 1)$ divides into $(2k + 1) = 3$ distin-

TABLE 7.11. *Irreps* (λ, μ) *of* $SU(3)$ *in the* $Sp(6)$ *irrep* $(\omega_1, \omega_2, \omega_3) = (5, 5, 5)$ *for heritage zero identical particles in a sdg shell with single-particle orbitals* $j = \frac{1}{2}, \frac{3}{2}, \frac{5}{2}, \frac{7}{2}$, *and* $\frac{9}{2}$. *The "missing"* $SU(3)$ *irreps (marked by* X*), such as the* (12, 0) *irrep for* $n = 12$ *particles, illustrate the dynamical Pauli effect. (From Hecht,* 1985.)

n	(λ, μ)							
30	(00)							
28	(02)							
26	(04)	(20)						
24	(06)	(22)	(00)					
22	(08)	(24)	(40)	(02)				
20	(0, 10)	(26)	(42)	(04)	(20)			
18	X	(28)	(44)	(60)	(06)	(22)	(00)	
16	X	X	(46)	(62)	(08)	(24)	(40)	(02)
14	X	X	(64)	(26)	(80)	(42)	(04)	(20)
12	X	(82)	(44)	(06)	(60)	(22)	(00)	
10	(10, 0)	(62)	(24)	(40)	(02)			
8	(80)	(42)	(04)	(20)				
6	(60)	(22)	(00)					
4	(40)	(02)						
2	(20)							
0	(00)							

guishable sets for $k = 1$, each of dimensionality $2\Omega/(2k + 1) = 2\Omega/3$. (In the usual shell-model picture there is a similar constraint—the full 2Ω-dimensional identical nucleon model space divides into allowed spin representations with a maximum of Ω columns, or equivalently, Young diagrams having no more than two rows.)

The $Sp(6) \supset SU(3)$ group plethysm for the heritage zero representation in a *sdg* shell is given in Table 7.11. The $SU(3)$ irreps missing in the center of the table, such as $(\lambda, \mu) = (12, 0)$ for $n = 12$ and the (14, 0) and (10, 2) irreps for $n = 14$, are absent because they represent fermion configurations that cannot be constructed. The constraint on allowed representations of $SU(3)$ is even more severe for heritage non-zero representations of $Sp(6)$. One cannot extend the $SU(3)$ collectivity by introducing $Sp(6)$ configurations with non-zero heritage quantum numbers: for $u = 2$, a second (8, 0) irrep can be built but an additional (10, 0) cannot be formed; for $u = 4$, an additional (6, 0) can be built but a third (8, 0) cannot be constructed; etc. The $SU(3)$ irreps given in Table 7.11 are the most collective FDSM configurations that can be built within a *sdg* shell-model space, all other $SU(3)$ representations (and there are many) are less collective than those given for the $u = 0$ case. This departure from the simple $(2N, 0)$ IBM rule has become known as the dynamical Paul effect. (Note that although here the allowed $SU(3)$ irreps follow from a knowledge of the $Sp(6) \supset SU(3)$ branching rules, the very simple argument given in Table 7.1 for the IBM also correctly identifies the most deformed spurious FDSM configurations. Also, the fatal flaw terminology originally referred to the FDSM, not the IBM, because the FDSM fails to yield $(\lambda, \mu) = (2N, 0)$ for the leading $SU(3)$ irrep when $N > \Omega/3$. Since the problem lies in the microscopic interpretation of the IBM (and not with the FDSM) the fatal flaw label, which carries a negative connotation, now normally refers to the IBM rather than the FDSM—a complete turnabout from the original use of the term!)

The dynamical Pauli effect cast in the language of $SU(3)$ representations means the leading irrep changes from one with $\mu = 0$ for $N < \Omega/3$ to one with $\mu \neq 0$ for $N > \Omega/3$, as shown in Table 7.11. (See Table 7.1 as well, and recall that $k = 2$ to take account of the spin degree-of-freedom.) The system with maximum prolate deformation (all quanta in the z-direction, none in the x-direction or y-direction) is found for $N = \Omega/3$; whenever $N > \Omega/3$, the excess quanta must be added in other than the z-direction and this drives the system towards an asymmetric shape. Because of this constraint the system realizes its maximum deformation when the valence shell is $\frac{1}{3}$ filled, dropping off from the maximum value by as much as 25% in the large Ω limit when the valence shell is $\frac{1}{2}$ filled. A plot of the deformation of an isotopic series of strongly deformed nuclei should reflect this change as the number of valence proton or neutron pairs crosses the $N = \Omega/3$ value. In rare earth and actinide nuclei the onset of this effect is postponed beyond the $N = \Omega/3$ value, due to the addition of pairs of particles to the unique parity orbitals. The transition from prolate to triaxial shapes with slightly less deformation is also expected to be smooth, not abrupt, due to vertical mixing with states at $\pm 2\hbar\omega$, $\pm 4\hbar\omega$, etc., which increase as one approaches the $N = \Omega/3$ value.

A plot of the deformation as a function of neutron number is

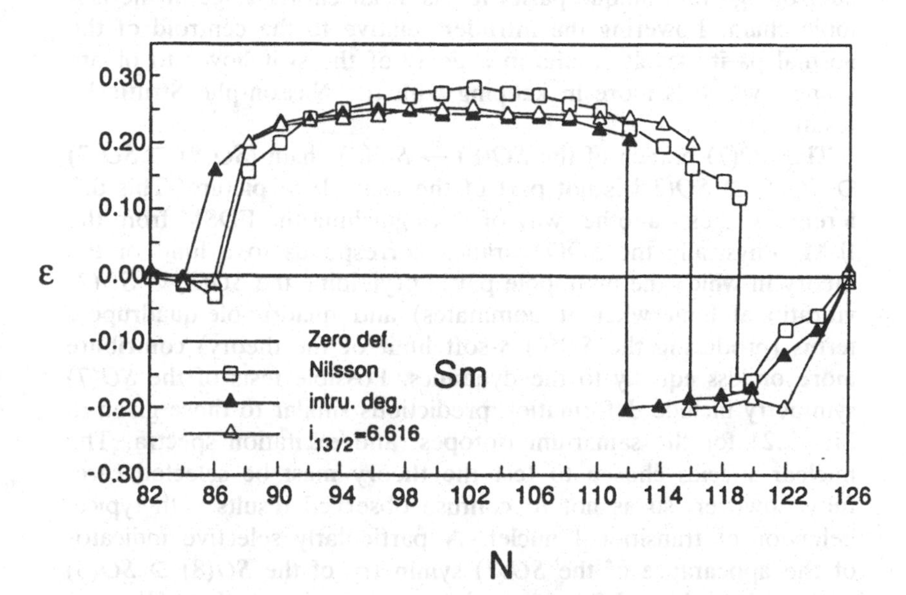

FIGURE 7.24. *Theoretical predictions for the deformation of samarium isotopes. The open squares are result of a Nilsson calculation which includes Strutinsky shell-corrections while the triangles label* FDSM *results: closed triangles for a scenario where the intruder level is filled as appropriate based on the Nilsson-Strutinsky results, and open triangles when the intruder level is lowered relative to its usual position. The latter has the effect of prolonging the switchover to oblate shapes because the intruder level absorbs a larger number of the nucleons earlier in the filling process. (From Feng, Guidry and Wu, 1992.)*

shown in Fig. 7.24 for isotopes of samarium. In this case $2\Omega = 30$, so the shell is $\frac{1}{3}$ filled when 10 particles occupy normal parity orbitals. Indeed, the results show a sharp rise in the deformation for neutron numbers 4–8 with saturation setting in for neutron numbers in the range 8–12. The uncertainty in these numbers reflects the fact that the $i_{13/2}$ unique-parity intruder level plays an important role, absorbing up to a total of 14 particles [126 − 82 = 44 = 30 (normal) + 14 (intruder)] as one adds neutrons. Particles placed in the unique-parity level also contribute to the deformation, but in a more gentle way, as the open square data (results for a Nilsson-plus-Strutinsky mean-field theory) shows. The triangular data points are for two different FDSM calculations: solid triangles for a normal filling of the intruder level, and open triangles for a scenario where the intruder orbital is artificially lowered so that more neutrons go into unique parity levels at an earlier stage in the isotopic chain. Lowering the intruder relative to the centroid of the normal parity levels results in a delay of the switchover to oblate shapes, which is more in keeping with the Nilsson-plus-Strutinsky results.

The $SO(7)$ branch of the $SO(8) \rightarrow SO(3)$ chain [$SO(8) \supset SO(7) \supset SO(5) \supset SO(3)$] is not part of the usual IBM picture. This difference suggests another way of distinguishing the FDSM from the IBM. Physically the $SO(7)$ branch corresponds to a limit of the theory in which the monopole pairing (yielding the $SO(5) \otimes SU(2)$ vibrational limit when it dominates) and quadrupole-quadrupole terms (producing the $SO(6)$ γ-soft limit of the theory) contribute more or less equally to the dynamics. Possible tests of the $SO(7)$ symmetry include deformation predictions similar to those given in Fig. 7.24 for the samarium isotopes, and excitation spectra. The nuclear species chosen to test the theory must be selected carefully however, so as not to confuse observed results with typical behavior of transitional nuclei. A particularly selective indicator of the appearance of the $SO(7)$ symmetry of the $SO(8) \supset SO(3)$ chain is the E2 strength ratio given by $R_0 = B(\mathrm{E2}{:}0_2^+ \rightarrow 2_1^+)/B(\mathrm{E2}{:}2_1^+ \rightarrow 0_1^+)$. Results for isotopes of ruthenium ($_{44}$Ru) and palladium ($_{46}$Pd) are shown in Fig. 7.25. The $SO(6)$ limit of the theory predicts a zero result for this ratio whereas the $SO(5) \otimes SU(2)$ limit yields a value that increases slowly from just above 2.0 to about 2.5 as particles are added to a system up through the mid-shell point. The $SO(7)$ limit, on the other hand, predicts values just below 1.0 at the beginning of the shell, and dropping off to

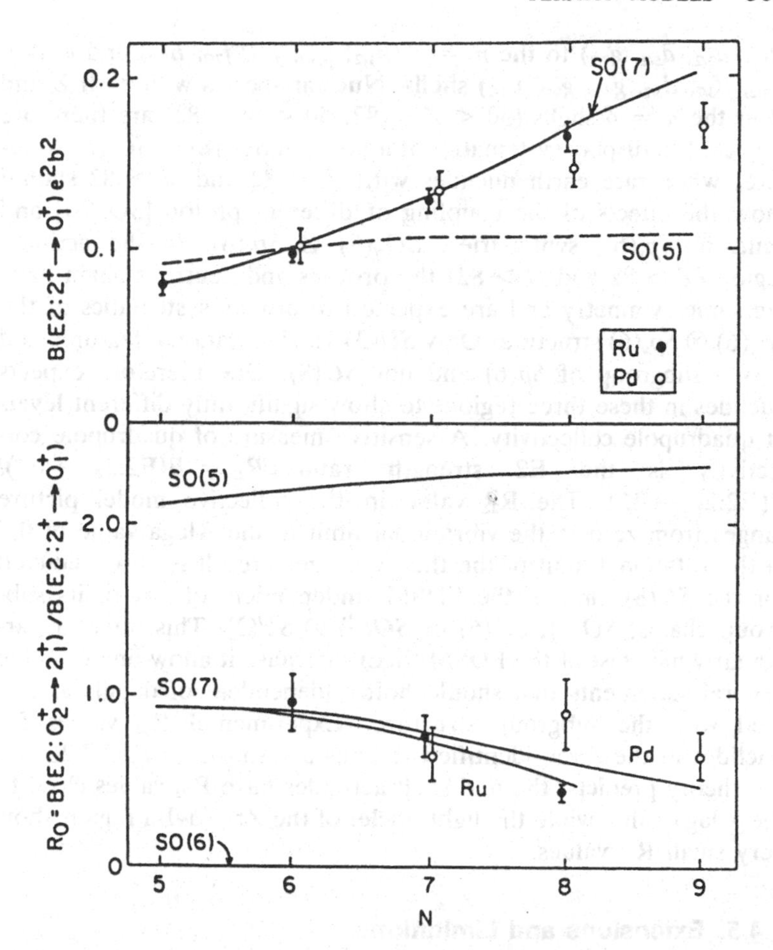

FIGURE 7.25. *Evidence for the* SO(8) ⊃ SO(7) ⊃ SO(5) ⊃ SO(3) FDSM *limit in the* Pd–Ru *region. The symmetry seems to account for structural variations in the* E2 *transition strength ratio* R_0 = B(E2:$0_2^+ \to 2_1^+$)/B(E2:$2_1^+ \to 0_1^+$) *that fall between the spherical vibrator* [SO(8) ⊃ SO(5) ⊗ SU(2) ⊃ SO(5) ⊃ SO(3)] *and γ-soft* [SO(8) ⊃ SO(6) ⊃ SO(5) ⊃ SO(3)] *limits of the theory.* (*From Casten et al.*, 1986.)

about 0.5 at the mid-shell point. The experimental numbers for Ru and Pd results are in good agreement with the *SO*(7) picture, and clearly distinguishable from the corresponding *SO*(5) ⊗ *SU*(2) and *SO*(6) limits of the *SO*(8)⊃ *SO*(3) chain.

Another unique feature of the FDSM is that the symmetry changes from *SO*(8) to *Sp*(6) as one moves from the $n = 4$ shell

($s_{1/2}$, $d_{3/2}$, $d_{5/2}$, $g_{7/2}$) to the $n = 5$ ($p_{1/2}$, $p_{3/2}$, $f_{5/2}$, $f_{7/2}$, $h_{9/2}$) and $n = 6$ ($s_{1/2}$, $d_{3/2}$, $d_{5/2}$, $g_{7/2}$, $g_{9/2}$, $i_{11/2}$) shells. Nuclear species with both Z and N in the $n = 4$ shells ($50 < Z < 82$, $50 < N < 82$) are therefore expected to display systematics of a coupled $SO_\pi(8) \otimes SO_\nu(8)$ structure, while rare earth nuclides with $Z < 82$ and $N > 82$ should show the effects of the coupling of different proton [$SO_\pi(8)$] and neutron [$Sp_\nu(6)$] symmetries, $SO_\pi(8) \otimes Sp_\nu(6)$. In the actinide region ($Z > 82$ and $N > 82$) the protons and neutrons again have the same symmetry and are expected to display systematics of the $Sp_\pi(6) \otimes Sp_\nu(6)$ structure. Only $SU(3)$ yields rotational features and it is a subgroup of $Sp(6)$ and not $SO(8)$. One therefore expects nuclides in these three regions to show significantly different levels of quadrupole collectivity. A sensitive measure of quadrupole collectivity is the E2 strength ratio $R_{22} = B(E2{:}2_2^+ \rightarrow 0_g^+)/B(E2{:}2_2^+ \rightarrow 2_g^+)$. The R_{22} value in the collective model picture ranges from zero in the vibrational limit to the Alaga value of 0.7 in the rotational limit of the theory. A zero result is also predicted for the $SO(8)$ limit of the FDSM, independent of any of its subgroup chains, $SO(7)$, $SO(6)$ or $SO(5) \otimes SU(2)$. This offers a particularly nice test of the FDSM theory because it allow one to make general statements that should hold independent of details associated with the subgroup structure. Experimental R_{22} values for nuclides in the three identified regions are shown in Fig. 7.26. As the theory predicts, the mid-shell actinides have R_{22} values close to the Alaga value while the light nuclei of the Zr–Cd–Ba region show very small R_{22} values.

7.4.5. Extensions and Limitations

The FDSM is, in principle, a complete shell-model theory. Its special symmetry limits are imbedded in the shell model, and it will be possible to explore and test their microscopic foundation once the required shell-model technologies are developed. The overall symmetry of a model space with identical particle shell degeneracy $2\Omega = \Sigma_j(2j+1)$ is $O(4\Omega)$. This orthogonal group in 4Ω dimensions is generated respectively by the $N = 2\Omega(2\Omega + 1)/2 + (2\Omega)^2 + 2\Omega(2\Omega N + 1)/2 = 2\Omega(4\Omega + 1)$ pair operators of types $a_\alpha^+ a_\beta^+$, $a_\alpha^+ a_\beta$ and $a_\alpha a_\beta$. The $SO(8)$ and $Sp(6)$ symmetry groups of the FDSM are subgroups of this $O(4\Omega)$ group, see eqs. (7.46) and (7.47). The full FDSM Hamiltonian eq. (7.48), including the H' part, eq. (7.49), that scatters pairs between the normal and intruder levels,

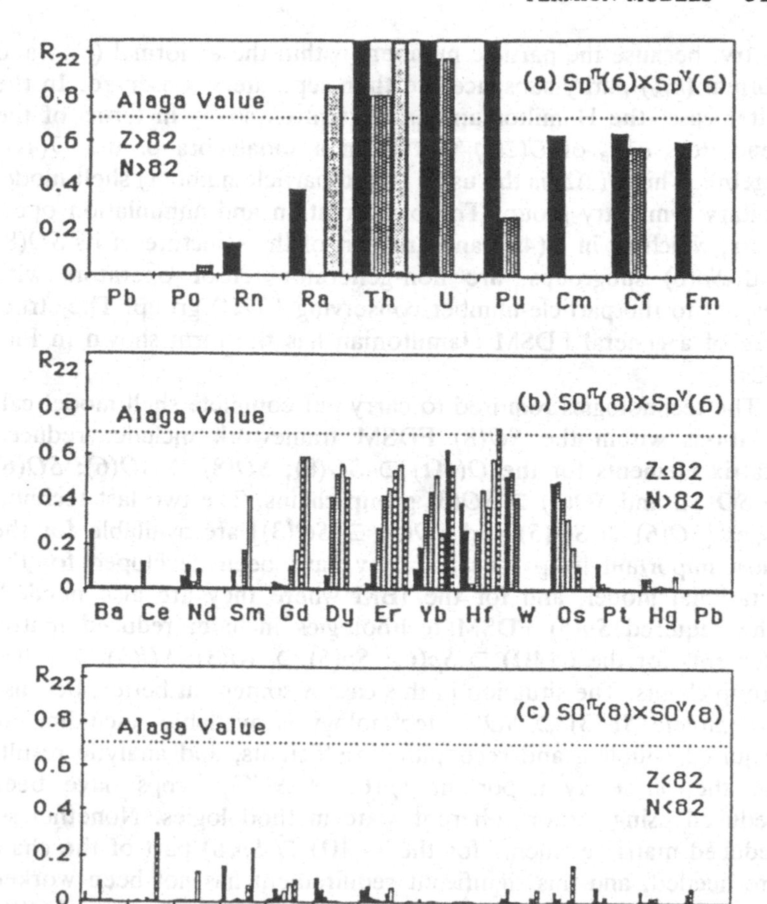

FIGURE 7.26. *The ratio* $R_{22} = B(E2{:}2_2^+ \to 0_g^+)/B(E2{:}2_2^+ \to 2_g^+)$ *are shown for* ~175 *even-even nuclei in three regions:* Z<82 *and* N<82 *where the* $SO^{\pi}(8) \times SO^{\nu}(8)$ FDSM *symmetry applies so a zero result for* R_{22} *is expected;* Z≤82 *and* N>82 *where the* $SO^{\pi}(8) \times Sp^{\nu}(6)$ FDSM *symmetry applies and accordingly results mid-way between zero and the rotor (Alaga) limit of* ~0.7 *are expected; and* Z>82 *and* N>82, *which is in the* $Sp^{\pi}(6) \times Sp^{\nu}(6)$ FDSM *symmetry range for strongly deformed nuclei and* R_{22} *predictions close to the rotor limit.* (*From Han et al.*, 1987.)

is built of generators of $O(4\Omega)$ and its matrix representation will therefore be block diagonal with no matrix elements coupling different $O(4\Omega)$ irreps. When the dynamics are restricted to a fixed-particle distribution (m_A, m_N), the Hamiltonian has a higher symmetry, because the particle numbers within the abnormal (m_A) and normal (m_N) parity subspace are then separately conserved. In the latter case, the Hamiltonian can be written solely in terms of the generators $a_\alpha^+ a_\beta$ of $U(2\Omega)$ that form a subalgebra of the $O(4\Omega)$ algebra. This $U(2\Omega)$ is the usual (fixed-particle number) shell-model unitary symmetry group. The pair creation and annihilation operators, which sit in $O(4\Omega)$ and are part of the structure of its $SO(8)$ and $Sp(6)$ subgroups, are non-generator tensor operators with respect to the particle-number conserving $U(2\Omega)$ group. The structure of a general FDSM Hamiltonian has the form shown in Fig. 7.27.

The technologies required to carry out complete shell-model calculations within the $SO(8)$ FDSM framework include: reduced matrix elements for the $O(4\Omega) \supset SO(8)$; $SO(8) \supset SO(6)$; $SO(6) \supset SO(5)$; and $SO(5) \supset SO(3)$ group chains. The two last technologies [$SO(6) \supset SO(5)$ and $SO(5) \supset SO(3)$] are available for the most important irreps because they have been developed for the rotational model, and for the IBM where they are also needed. The required $Sp(6)$ FDSM technologies include: reduced matrix elements for the $O(4\Omega) \supset Sp(6)$; $Sp(6) \supset SU(3)$; $SU(3) \supset SO(3)$ group chains. The situation in this case is somewhat better, because a complete $SU(3) \supset SO(3)$ technology is available, including all required coupling and recoupling coefficients, and analytic results for the physically important $Sp(6) \supset SU(3)$ irreps have been deduced using vector coherent state methodologies. Nonetheless, reduced matrix elements for the $O(4\Omega) \supset Sp(6)$ part of the chain are needed, and this significant requirement has not been worked out. Going beyond the simplest pure symmetry limits of the FDSM requires technologies that are currently not available; broader testing and applications of the FDSM await these technological developments.

When the single-particle orbitals making up the $2\Omega = \Sigma_j(2j + 1)$ sum form a regular series ($j = \frac{1}{2}, \frac{3}{2}, \frac{5}{2}, \ldots, n + \frac{1}{2}$), as for levels of n-th shell of the harmonic oscillator, the $U(2\Omega)$ subgroup of $O(4\Omega)$ has an $SU(3)$ subgroup. In the FDSM case, this $SU(3)$ subgroup of $U(2\Omega)$ is the pseudo-$SU(3)$ group introduced in Section 7.3.4. This follows because the handling of the intruder and defector levels from the shell-model oscillator picture is the same in the two theories. It is important to emphasize, however, that this is not the $SU(3)$ one encounters as a subgroup of $Sp(6)$. In the pseudo-$SU(3)$ case, each particle carries $(\tilde{n}, 0)$ quanta, so a pair

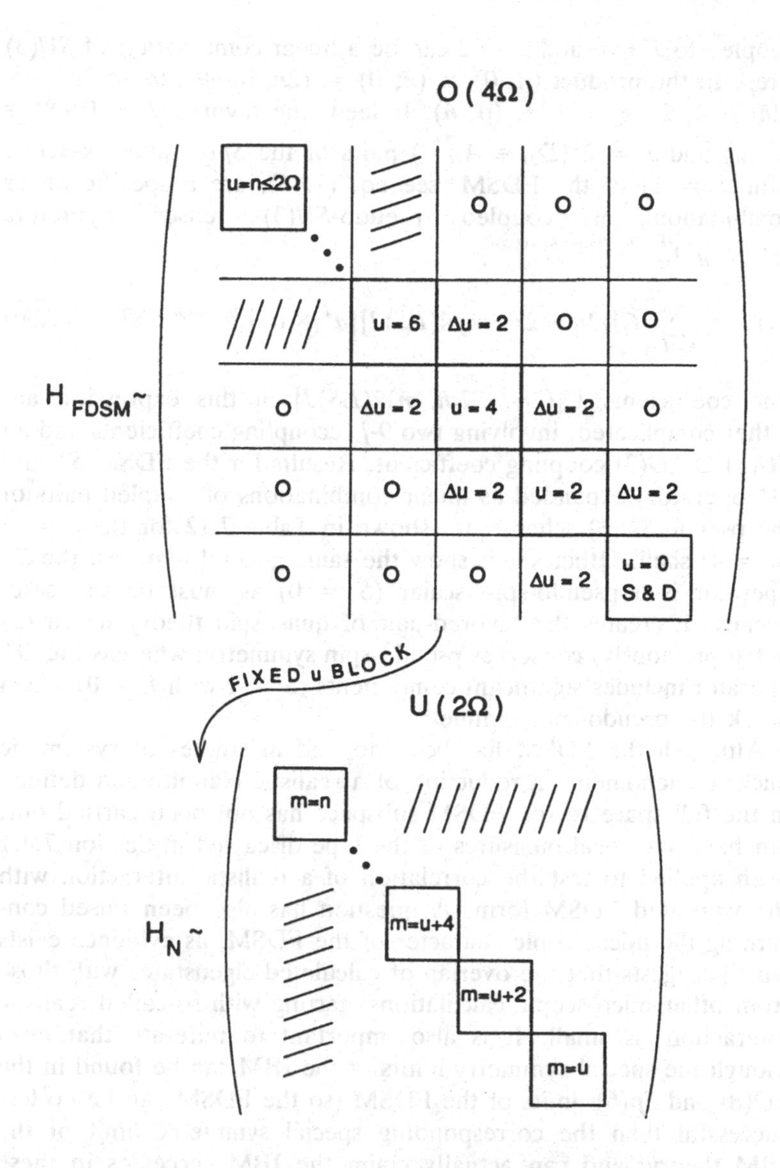

FIGURE 7.27. *The structure of the* FDSM *Hamiltonian eq.* (7.48) *matrix has the form shown in the diagram. An irreducible representation of* O(4Ω) *can be divided into subspaces according to the heritage quantum number which counts the number of fermions not coupled to angular momentum zero. If the Hamiltonian conserves the number of particles* (H′ = 0) *as well as the heritage, each* O(4Ω) *heritage block breaks up into* irreps *of* U(2Ω).

coupled to $J = 0$ and $J = 2$ can be a linear combination of $SU(3)$ irreps in the product $(\tilde{n}, 0) \times (\tilde{n}, 0) = (2\tilde{n}, 0) + (2\tilde{n} - 2, 1) + (2\tilde{n} - 4, 2) + \cdots + (0, \tilde{n})$. Indeed, the favored $J = 0$ ($S^+ \equiv A_{M=0}^{J=0^+}$) and $J = 2$ ($D_M^+ \equiv A_M^{J=2^+}$) pairs of the $Sp(6)$ limit (k-active with $k = 1$) of the FDSM, see eq. (7.44), are a specific linear combination of coupled pseudo-$SU(3)$ tensor operators $[a^+ \times a^+]_M^{(\tilde{\lambda}=2n-2m,\tilde{\mu}=m)\tilde{\kappa}(\tilde{L}\tilde{S})J}$:

$$A_M^{J^+} = \sum_{\tilde{m},\tilde{\kappa},\tilde{L},\tilde{S}} C[(2\tilde{n} - 2\tilde{m}, \tilde{m})\tilde{\kappa}(\tilde{L}\tilde{S})J][a^+ \times a^+]_M^{(2\tilde{n}-2\tilde{m},\tilde{m})\tilde{\kappa}(\tilde{L}\tilde{S})J}. \quad (7.50)$$

The coefficients $C[(2\tilde{n} - 2\tilde{m}, \tilde{m})\tilde{\kappa}(\tilde{L}\tilde{S})J]$ in this expression are rather complicated, involving two 9-j recoupling coefficients and an $SU(3) \supset SO(3)$ coupling coefficient. Results for the FDSM S^+ and D^+ operators expanded as linear combinations of coupled pairs of the pseudo-$SU(3)$ scheme are shown in Table 7.12 for the $n = 5$ ($\tilde{n} = 4$) shell. Other shells show the same general features: the S^+ operator is a pseudo-spin scalar ($\tilde{S} = 0$) as must be the case, because it creates the favored-pair of quasi-spin theory which (as noted previously) conserves pseudo-spin symmetry; whereas the D^+ operator includes significant components ($\tilde{S} = 1$ with $\tilde{L} \neq 0$) which break the pseudo-spin symmetry.

Although the FDSM has been applied in studies of systematic nuclear phenomena, a reduction of a realistic Hamiltonian defined in the full space to the FDSM subspace has not been carried out, nor have statistical measures of the type discussed in Section 7.4.1 been applied to test the correlation of a realistic interaction with the truncated FDSM form. A question has also been raised concerning the microscopic character of the FDSM, as evidence exists which suggests that the overlap of calculated eigenstates with those from other microscopic calculations starting with so-called realistic interactions is small. It is also important to reiterate that even though the special symmetry limits of the IBM can be found in the $SO(8)$ and $Sp(6)$ limits of the FDSM (so the FDSM can be no less successful than the corresponding special symmetry limit of the IBM theory, and can actually claim the IBM successes in these limits) the IBM does more, because it allows one to explore regions between the $O(6)$ and $SU(3)$ limits of the theory. The full range of IBM effective interaction possibilities is not part of the FDSM picture, since the $SO(6)$ and $SU(3)$ limits of the latter belong to

TABLE 7.12. *Expansion of the* S^+ *and* D^+ *pairs of the* Sp(6) *limit of the* FDSM *in terms of tensors of the pseudo*-SU(3) *scheme for the* n = 5 ($\tilde{n}$ = 4) *shell. Although* S^+ *is a pseudo-spin scalar operator* ($\tilde{S}$ = 0 *and pseudo-space symmetric*), D^+ *has large pseudo-spin vector* ($\tilde{S}$ = 1 *and pseudo-space antisymmetric*) *components. Numbers associated with* ($\tilde{L}\tilde{S}$) = (00) *and* ($\tilde{L}\tilde{S}$) = (20) *tensors, corresponding to* $\tilde{L}$ = 0 *and* $\tilde{L}$ = 2 *coupled pairs in the pseudo-spin scheme, are italicized.*

$(\tilde{\lambda}, \tilde{\mu})^*$	$\tilde{\kappa}$	$\tilde{L}$	$\tilde{S}$	$\tilde{n}=4$	
j orbitals		→		$\frac{1}{2}, \frac{3}{2}, \frac{5}{2}, \frac{7}{2}, \frac{9}{2}$	
FDSM pair		→		S^+	D^+
(8, 0)	1	0	0	*3.0000*	0
(8, 0)	1	2	0	0	*−0.3114*
(6, 1)	1	1	1	0	0.2400
(6, 1)	1	2	1	0	1.1436
(6, 1)	1	3	1	0	0.5237
(4, 2)	1	0	0	*−2.2361*	0
(4, 2)	1	2	0	0	*−1.5751*
(4, 2)	2	2	0	0	*1.4762*
(2, 3)	1	1	1	0	−0.4409
(2, 3)	1	2	1	0	0.5116
(2, 3)	1	3	1	0	−2.7969
(2, 3)	2	3	1	0	0.4671
(0, 4)	1	0	0	*1.0000*	0
(0, 4)	1	2	0	0	*0.3258*

*The FDSM S^+ and D^+ pair operators are (2, 0) tensors with respect to the $SU(3)$ subgroup of $Sp(6)$. In the pseudo-$SU(3)$ picture a nucleon carries $\tilde{n}$ quanta and transforms as a pure ($\tilde{n}$, 0) tensor, It follows from this that the S^+ and D^+ pairs of the FDSM scheme have pseudo-$SU(3)$ tensor character given by the product $(\tilde{n}, 0) \times (\tilde{n}, 0) = (2\tilde{n}, 0) + (2\tilde{n} - 2, 1) + (2\tilde{n} - 4, 2) + \cdots + (0, \tilde{n})$.

different primary symmetries, see Fig. 7.22 and discussion that follows.

It is important to reiterate that in principle the FDSM is a complete shell-model theory and as a consequence of this it can be used to address microscopic issues such as B(E2) transition strengths and particle transfer probabilities between specific initial and final nuclear states. This will require the development of a symmetry adapted code with options of the type currently available in existing shell-model programs. Because of its microscopic underpinnings, connection of the FDSM to other microscopic schemes can also be investigated—something that has already been done for the Nilsson model. The fact that the FDSM is an algebraic theory with more than one symmetry limit means issues related to masses of nuclear species off the line of stability can be explored. This opens up the possibility of its application in astrophysics. The FDSM, like its IBM forerunner, raises questions of fundamental importance to the field of nuclear physics and with this issues in related fields that depend on nuclear structure information for interpreting physical phenomena.

7.5. SYMPLECTIC SHELL MODEL

As stated in the introductory section to this chapter, understanding collective phenomena from a shell-model perspective has been a long-standing goal of nuclear structure physics. Although the IBM and AFM theories described in previous sections share this objective, they do not deal directly with the fact that reproducing observed E2 transition strengths in deformed nuclei (which even in light systems can be more than two orders of magnitude greater than simple single-particle estimates) requires extending the simple picture of single-shell valence spaces ($0\hbar\omega$ theory) to include configuration admixtures from higher-lying configurations ($n\hbar\omega$; n = 2, 4, etc.). A $0\hbar\omega$ theory cannot (without a large renormalization of the quadrupole operator) account for the observed E2 transition strengths. A simple argument can be given which demonstrates the inadequacy of a $0\hbar\omega$ shell-model theory for describing deformed phenomena. The $SU(3)$ coupling scheme organizes a model space according to the deformation of its many-particle configurations. This is illustrated in Fig. 7.28 which displays eigenvalues ϵ of $Q_0^a = 2N_z - N_x - N_y$ (decreasing vertically) and $2M_\Lambda$ of $2\Lambda_0 = N_x - N_y$ (increasing horizontally from left to right) for cartesian configurations of a given (λ, μ) symmetry. The maximum eigenvalue which Q_0^a can have is $\epsilon_{max} = 2\lambda + \mu$ where (λ, μ) is the leading m-particle $SU(3)$ representation. If all the valence nucleons could be put into the most deformed single-particle level—a scenario ruled out by the Exclusion Principle for $m > 4$ (spin-isospin) or $m > 2$ (spin)—one obtains $(\lambda, \mu) = (\eta m, 0)$ for that leading configuration where η is the oscillator shell number. The corresponding maximum eigenvalue for Q_0^a would then be $\epsilon_{max} = 2\eta m$; while this is usually larger than the allowed $2\lambda + \mu$ values, it places an upper limit on the deformation and shows that the E2 transition strengths ($\sim \langle Q^c \cdot Q^c \rangle$) can grow no stronger than quadratically with m in a $0\hbar\omega$ shell-model theory. The same conclusion follows directly from the fact that Q^a is a one-body operator. Observed E2 strengths show a

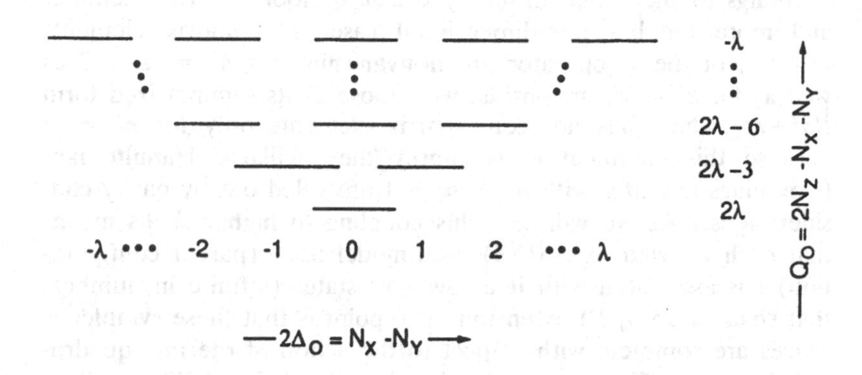

FIGURE 7.28. *Eigenvalues* ϵ *of* $Q_0^a = 2N_z - N_x - N_y$, *with the largest at the bottom of the pyramid, are plotted as a function of the eigenvalues of* $2M_\Lambda$ *of* $2\Lambda_0 = N_x - N_y$, *which vary horizontally, for intrinsic cartesian configurations of a* (λ, μ) = *(even,0) irreducible reprsentation of* SU(3). *The maximum eigenvalue that* Q_0^a *can have is* $\epsilon_{max} = 2\lambda + \mu$ *where* (λ, μ) *is the leading* m-particle SU(3) irrep. *The* N_α *are operators (denoted* n_α *previously) that count the* total *number of oscillator quanta in the* α*-th direction* (α=z,x,y).

growth with m as one moves away from a closed shell that is much stronger than this. This rapid buildup in strength can only come about through coherent admixtures from configurations with larger values for η; this means higher-lying, multi-$\hbar\omega$ configurations must be included from the outset if calculated eigenstates are to be considered realistic wavefunctions of the physical system.

The symplectic model, sometimes also called the microscopic collective model, is a shell-model scheme that extends the Elliott $SU(3)$ model to include mutliple $2\hbar\omega$ excitations of the monopole ($l = 0$) and quadrupole ($l = 2$) type. It can be used to describe both low-lying *and* giant resonance states of light ($A \lesssim 28$) nuclei that are strongly deformed. The multi-$\hbar\omega$ feature of the model means that the action of the real collective quadrupole operator, $Q^c = \sqrt{16\pi/5}\, r^2 Y_2(\hat{r})$, and not only its symmetrized intra-shell $0\hbar\omega$ preserving part, $Q^a = \sqrt{16\pi/5}\, \frac{1}{2}\, [r^2 Y_2(\hat{r}) + p^2 Y_2(\hat{p})]$, can be fully accommodated. The symmetry algebra of the scheme generates a realization of the non-compact (infinite dimensional irreps) symplectic group $Sp(3,R)$ which has the Elliott $SU(3)$ as its maximal compact subgroup. (More will be said about this structure below.) One can appreciate the fact that Q^c (in contrast with Q^a) involves couplings to shells that differ by $\pm 2\hbar\omega$ by looking at the simpler and more familiar one-dimensional case. The matrix elements $\langle n'|x^2|n\rangle$ of the x^2 operator are nonvanishing for $n' = n \pm 2$ as well as for $n' = n$, in contrast with those of its symmetrized form $\frac{1}{2}(x^2+p^2)$ which has non-zero matrix elements only for $n' = n$ because this combination is simply the oscillator Hamiltonian. (Couplings to states with $n' = n \pm 1$ are ruled out by parity considerations.) As we will see, this coupling to higher shells means that each allowed $0\hbar\omega$ $SU(3)$ shell-model irrep (parent configuration) has associated with it a tower of states (infinite in number) that span its $Sp(3, R)$ extension. The point is that these symplectic spaces are complete with respect to the action of the real quadrupole operator. This means intra-band and inter-band E2 transition strengths between low-lying, as well as giant resonance, states can be reproduced within the framework of the model without introducing proton and neutron effective charges. And as we will see below, it is also possible to define a symplectic extension for the pseudo-$SU(3)$ scheme—called the pseudo-symplectic model—which can be applied to the study of giant resonance and related phenomena in heavy ($A \gtrsim 100$) nuclei.

7.5.1. Tenets of the Theory

As indicated above, the symplectic model extends the $SU(3)$ picture by allowing for inter-shell $2\hbar\omega$ excitations of the monopole and quadrupole type. This feature is essential for modeling a system's quadrupole coherence. Specifically, if Q^c and Q^a denote the collective and algebraic forms for the quadrupole operator as defined above, then $Q^c = Q^a + \sqrt{6}/2\,(B_2^+ + B_2^-)$ where B_2^+ and B_2^- are pure $2\hbar\omega$ quadrupole raising and lowering operators, respectively. There are two types of raising (B_l^+) and lowering (B_l^-) operators in the symplectic scheme: the monopole ($l = 0$) and quadrupole ($l = 2$) parts of the cartesian tensor operator form, $Q^c_{\alpha,\beta} = (x_\alpha x_\beta - \frac{1}{3}\, r^2\delta_{\alpha,\beta})$. The 6 raising ($B^+_{lm}$) and 6 lowering ($B^-_{lm}$) operators ($l = 0$ and 2 with projections $m_l = -l, -l+1, \ldots, l-1, +l$), together with the number operator N (differing only by an additive constant from the oscillator Hamiltonian), as well as the 3 angular momentum operators L_α and the 5 quadrupole operators Q^a_μ, form a 21-dimensional set which closes under commutation of its members. This set of 21 operators generates a realization of the noncompact $Sp(3, R)$ (symplectic) group. Since this 21-dimensional Lie algebra contains the 8-dimensional $SU(3)$ subalgebra generated by the 3 angular momentum operators L_α and the 5 quadrupole operators Q^a_μ, $SU(3)$ is a subgroup of $Sp(3, R)$. Other subgroups, such as the symmetry group of the quantum rotor ($T_5 \wedge SO(3)$, see Section 7.2) generated by the 3 angular momentum operators L_α and the 5 quadrupole operators Q^c_μ, are also contained in the rich $Sp(3, R)$ group structure. However, $SU(3)$ is the largest (maximal) compact (finite dimensional irreps) subgroup of $Sp(3, R)$. The notation $Sp(6, R)$ is also used for this symplectic group structure. The use of a "6" rather than "3" is preferred by some, because there are 6 fundamental operators in the symplectic algebra—the 3 coordinates (x_α) and the 3 (conjugate) momenta (p_a). (The $Sp(6)$ of the FDSM is a compact realization of this symplectic group which also arises in a description of canonical transformations in classical mechanics, and in these contexts there is a similar choice.)

The structure of the $Sp(3, R)$ group and its relation to $SU(3)$ is shown in Fig. 7.29. Since $SU(3)$ is the largest compact subgroup of $Sp(3, R)$, it is useful to decompose the $Sp(3, R)$ generators into their $SU(3)$ tensor character. The number operator is an $SU(3)$ scalar $[(\lambda, \mu) = (0, 0)]$, while the L_α and Q^a_μ operators generate $SU(3)$ and therefore have the tensor character of the fundamental 8-dimensional $(\lambda, \mu) = (1, 1)$ irrep. The remaining 12 operators divide up into two sets of 6 each: the raising (B_l^+) operators which have the tensor character (2, 0) and the adjoint lowering (B_l^-) operators which have the conjugate (0, 2) symmetry. The B_l operators therefore have the same $SU(3)$ tensor character as the s ($l=0$) and d ($l=2$) boson creation and annihilation operators of the IBM. This feature (to which we will return later) comes about in both applications because the raising and lowering operators generate two-phonon excitations. It is important to emphasize, however, that the B_l operators of the symplectic model are *not* the same as the s and d boson operators of the IBM since the latter are defined solely within the $0\hbar\omega$ valence space while the former couple major oscillator shells differing by $\pm 2\hbar\omega$. [In the AFM picture, the addition of a pair of particles to a valence (η) shell adds $2\eta\hbar\omega$ of energy to the system (neglecting zero point motion which accounts for an

SYMPLECTIC EXTENSION

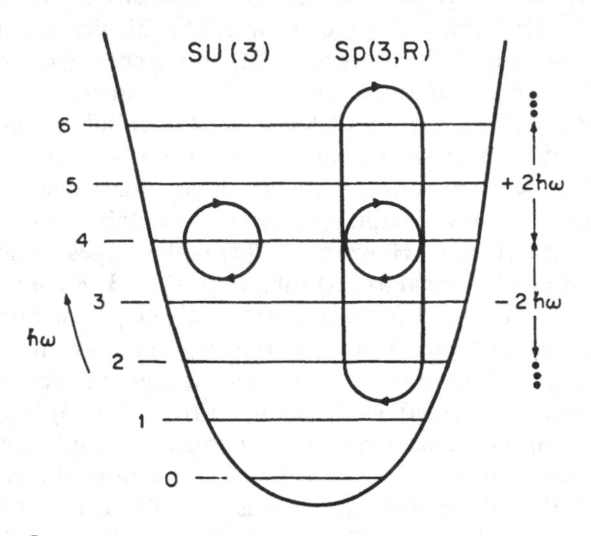

GENERATORS –

L – 3 A.M. } SU(3) } U(3) } Sp(3,R)

Q^a – 5 QUADRUPOLE

N – 1 NUMBER

B^+ – 6 +2ħω

B^- – 6 −2ħω

$$\ldots Q^c = Q^a + B^+ + B^-$$

FIGURE 7.29. *Symplectic extension of an* SU(3) *shell-model theory. The real quadrupole operator,* Q^c, *includes parts that couple to states in shells at* $\pm 2\hbar\omega$ *relative to the* $0\hbar\omega$ *valence space:* $Q^c = Q^a + \sqrt{6}/2\,(B_2^+ + B_2^-)$ *where the* 5 *algebraic quadrupole operators* Q^a, *which acts within the fixed* $n\hbar\omega$ *subspaces only, together with the* 3 *angular momentum operators* L, *generate the* 8-*dimensional* SU(3) *subalgebra of the full* 21 *dimensional* Sp(3, R) *symplectic algebra. The operator* N *counts the total number oscillator phonons while the* B_l^+ *and* B_l^- *operators, each with monopole* ($l = 0$) *and quadrupole* ($l = 2$) *parts are the* $2 \times 6 = 12$ *raising and lowering operators, respectively, of the symplectic extension of the* SU(3) *algebra. Because of the* B_l *operators,* irreps *of the symplectic group are infinite-dimensional.*

additional $3\hbar\omega$). This is to be compared with the IBM picture where the addition of a pair is effected by the s^+ or a d^+ operator which increases the system's energy by $2\hbar\omega_{IBM}$. Clearly, from this simple comparison it follows that $\hbar\omega_{IBM} \neq \hbar\omega_{AFM} \equiv \hbar\omega$.]

We now will explore the physics behind the B_l operators in greater depth, and ultimately discover why they can usually be replaced by simpler boson forms. First note that the B_l operators, as for Q^a and Q^c, are standard one-body objects which act in the same way on all particles. As a consequence, these operators do not change the particle permutation symmetry when they are applied to a many-particle wavefunction—the labelling of basis states by the irreducible representation $[1^m]$ of $U(k\Omega)$ as presented in Section 7.1 survives, even though the application of a B_l^+ operator to a $0\hbar\omega$ configuration implies the promotion of a particle into a higher-lying shell $2\hbar\omega$ up. The system has no way of knowing which of the particles gets promoted; all particles participate equally to the excitation process. In this regard it is important to emphasize that the B_l^+ operators generate very special (coherent) excited state configurations, and do not generate generic $2\hbar\omega$ shell-model states. For example, a $2\hbar\omega$ excitation generated by promoting two particles across a single major shell (rather than a single particle across two major shells) cannot be represented in terms of a B_l^+ operator acting on a $0\hbar\omega$ state; particle-hole (p-h) states of this type represent independent (excited but non-symplectic) shell-model configurations. Another example is a $4\hbar\omega$ configuration generated via a hexadecapole excitation operator—such a configuration cannot be represented as a double symplectic $(B_2^+ \times B_2^+)^4$ excitation $\{r^4Y_4(\hat{r}) \neq [r^2Y_2(\hat{r}) \times r^2Y_2(\hat{r})]^4\}$. Negative parity configurations must also to added to the symplectic picture, just as they are in the usual shell-model scheme. In general, any (non-spurious) excited (1p-1h, 2p-2h, 3p-3h, etc.) configuration that the B_l^- operator destroys is an independent shell-model configuration, however symplectic excitations can be superimposed on all such independent shell-model configurations in exactly the same way as they are added to $0\hbar\omega$ configurations.

Because the B_l operators act only on spatial coordinates, they are spin or spin-isospin scalar objects that preserve the $[f^c]$, as well as the $[1^m]$ symmetry label of a many-particle basis state. Since the spatial symmetry group $U(\Omega)$ and the spin or spin-isospin symmetry group $U(k)$ are complementary within antisymmetric irreps of $U(k\Omega) \supset U(\Omega) \otimes U(k)$, the $U(\Omega)$ and $U(k)$ irreps which enter into the construction of many-particle basis states are conjugates of one another. This means that the $[f]$ symmetry label is also good; the B_l operators are block-diagonal within irreducible representations of $U(\Omega)$ as well as $U(k)$. This property of the symplectic generators is extremely important—they do not involve spin or isospin degrees of freedom, and therefore do not couple different space-spin/space-spin-isospin symmetries of a standard LS/LST-coupled shell model. This feature means the symplectic scheme is actually a very simple and straightforward extension of the ordinary shell-model picture; everything one has learned about shell-model symmetries survives, and is expanded to include vertical (multi-$\hbar\omega$) excitations as required to accommodate large deformation and enhanced quadrupole transition strengths. Looking ahead, one can anticipate that this feature also insures the success of the pseudo symplectic model (which involves symplectic excitations superimposed on the pseudo-spin and pseudo-$SU(3)$ pictures)—quadrupole coherence built in via pseudo-symplectic operators will not destroy the underlying pseudo-spin symmetry.

We will now consider the structure of the many-particle basis states in greater detail (see Fig. 7.30). Each application of a $B_l^+(B_l^-)$ adds(subtracts) $2\hbar\omega$ of energy to(from) the system. It follows from this that the application of a B_l^- operator to *any* $0\hbar\omega$ basis state destroys it because these $0\hbar\omega$ states cannot give up $2\hbar\omega$ of energy without shifting a valence particle into an already occupied core configuration as this would violate the Exclusion Principle. The usual $0\hbar\omega$ shell-model states are therefore lowest-weight (LW) states with respect to the B_l^- operators of the symplectic algebra. As for the familiar angular momentum scenario, all the states belonging to a particular $0\hbar\omega$ $SU(3)$ irrep (labelled (λ_s, μ_s) in what follows) can all be generated from the unique state which is lowest weight with respect to the action of the generators of the $SU(3)$ subgroup of $Sp(3,R)$: $C_{ij}|(\lambda_s,\mu_s)LW\rangle = 0$, for all $i < j$. [Recall that in the $SU(2)$ (angular momentum) case, the J_- lowering operator destroys the lowest-weight state: $J_-|J, M=-J\rangle = 0$. The $2J$ other states (for a total of $2J + 1$) can be obtained from the lowest-weight one by repeated applications of the J_+ raising operator: $|J, M\rangle \sim (J_+)^{J+M}|J, -J\rangle$. In particular, the highest-weight state ($|J,M=J\rangle$) is given by $2J$ applications of the J_+ operator on the lowest-weight state. Furthermore, an additional application of the J_+ operator destroys the highest-weight state: $J_+|J,$

$M=J\rangle = 0$. Indeed, $(J_+)^{J+M}|J, -J\rangle \sim (J_-)^{J-M}|J, +J\rangle$, etc. This closure and LW-HW symmetry follows because $SU(2)$ is a compact group.] Applications of the $SU(3)$ raising operators, $C_{i,j}$ with $i > j$, to the lowest-weight state suffice to generate all states of the (λ_s, μ_s) irrep (equal in number to the dimensionality, $\dim(\lambda_s, \mu_s) = \frac{1}{2}(\lambda_s + \mu_s + 2)(\lambda_s + 1)(\mu_s + 1)$, of the representation) just as in the $SU(2)$ case. For example, the C_{21} operator shifts a phonon from the 1 direction to the 2 direction. These shifts conserve the number

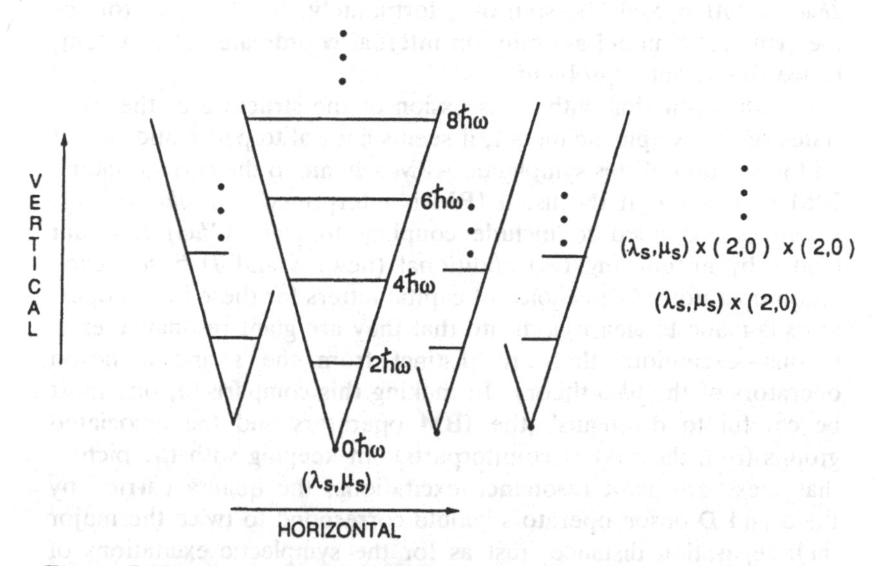

FIGURE 7.30. *Basis states of* Sp(3,R) *irreps. All states of an irrep (an infinite number) can be generated by an application of the raising generators to the* $0\hbar\omega$ SU(3) *lowest-weight state:* $|(\lambda_s, \mu_s)\text{LW}\rangle$, *which the lowering operators of* both SU(3) *and* Sp(3, R) *annihilate. The* SU(3) *raising operators acting on the lowest-weight state span the horizontal extension at* $0\hbar\omega$, *while the* B_i^+ *couple these states to those at excitation energy* $2\hbar\omega$ *that can be built via the coupling* $(\lambda_s, \mu_s) \times (2,0)$. *Another* B_i^+ *application pumps in an additional* $2\hbar\omega$ *of energy and carries one up to the* $4\hbar\omega$ *level, etc. For nuclear physics applications the lowest weight* SU(3) *irreps are* $0\hbar\omega$ *shell-model configurations for* 0p-0h *states, or more generally,* $n\hbar\omega$ *shell-model configurations for non-symplectic* np-nh *states.*

of oscillator quanta and are represented in Fig. 7.30 as horizontal displacements. As also indicated in the figure, each application of a B_l^+ operator adds $2\hbar\omega$ of energy to the system, but in this case [in contrast with the raising operators of $SU(2)$ and $SU(3)$] there is no cap on the number of allowed applications—the entire set of states (infinite in number because $Sp(3, R)$ is non-compact) belonging to a symplectic irrep can be generated from $|(\lambda_s, \mu_s)LW\rangle$, a single unique $0\hbar\omega$ configuration that is lowest weight—first with respect to the symplectic lowering operators, and second with respect to $SU(3)$. (As suggested above, this argument extends to excited non-symplectic shell-model configurations). It follows that $Sp(3, R)$ irreps can be identified by the (λ_s, μ_s) label of this lowest-weight state. A note of caution is in order, because an arbitrary $2\hbar\omega$ excitation could be spurious; fortunately, the B_l^+ operators of the symplectic model act only on internal coordinates of a system, hence this is not a problem.

Before continuing with a discussion of the structure of the basis states of the symplectic model, it seems natural to pause and reflect on the relation of this symplectic AFM scheme to the corresponding IBM realization. If the usual IBM is interpreted as a $0\hbar\omega$ theory, it can be extended to include coupling to giant $(2\hbar\omega)$ resonant modes by introducing two additional (new) S and D boson excitation operators. (The choice of capital letters for these boson operators is made to clearly indicate that they are giant resonance excitations—excitations that are distinct from the s and d boson operators of the $0\hbar\omega$ theory. In making this comparison, one must be careful to distinguish the IBM operators and the associated groups from their AFM counterparts). In keeping with the picture that these are giant resonance excitations, the quanta carried by the S and D boson operators should correspond to twice the major shell separation distance, just as for the symplectic excitations of the AFM case. If b_{lm}^- denotes the $s(l = 0)$ and $d(l = 2)$ boson operators, the $U_b(6)$ group of the usual $0\hbar\omega$ IBM theory is generated by the 36 distinct bilinear operator combinations $b_{l'm'}^+ b_{lm}^-$. Likewise, if B_{lm}^- denotes the giant $(2\hbar\omega)$ $(S(l = 0)$ and $D(l = 2)$ boson operators, then the $B_{l'm'}^+ B_{lm}^-$ also generate a $U_B(6)$ algebra. One approach for an extended IBM theory is to then consider the space spanned by representations of the direct product $U_B(6) \otimes U_b(6)$ group with the number of S and D bosons restricted: $N_B \lesssim (N_B)_{\max}$. (Basis states of the usual IBM theory $(|\Psi_{\text{IBM}}\rangle)$ are vacua of this picture: $B_{lm}^-|\Psi_{\text{IBM}}\rangle = 0$, etc.) However, the complete structure (in analogy with $Sp(3, R)$ of the AFM case) is obtained for $(N_B)_{max} \to \infty$ and leads to the non-compact $W_B(6) \otimes U_b(6)$ group where the $W_B(6)$ factor in this product is the Heisenberg–Weyl group generated by *all* powers and products of the new boson creation ($\overset{+}{B}_{lm}$) and annihilation ($\bar{B}_{lm}$) operators. The compact $U_B(6) \otimes U_b(6)$ structure is a subgroup of this larger non-compact $W_B(6) \otimes U_b(6)$ group. It describes the dynamics within a $2N_B\hbar\omega$ layer where the number of S and D bosons is fixed—it is not, however, the general case since inter-shell coupling (non-conservation of S and D bosons) is an integral part of the picture. In the $SU(3)$ limit of the IBM, this reduces to the $W_B(6) \otimes SU_b(3)$ group structure which is the IBM counterpart of $Sp(3,R)$ in the AFM theory.

A boson realization of the symplectic $B^{\pm}_{lm}$ generators (not to be confused with the like-labelled B^{+}_{lm} operators used in the preceding paragraph in identifying an IBM construction corresponding to the symplectic AFM scenario) can be introduced as well. This simplification leads to a more tractable version of the theory. In this case, a formal group theoretical expansion-contraction procedure such as the one presented in Section 7.2 for $SU(3)$ can be invoked to accomplish the transformation. Specifically, it can be shown that the generators of the $Sp(3, R)$ group transform as follows:

$$N \to N_s + 2N_b, \qquad L_m \to L^s_m + L^b_m,$$
$$Q^a_m \to Q^s_m + Q^b_m, \qquad B^{\pm}_{lm} \to \sqrt{(4/3)\, N_s}\; b^{\pm}_{lm}, \tag{7.51}$$

where $l = (0, 2)$ and the $b^{\pm}_{lm}$ operators satisfy the usual boson commutation relations. (The $b^{\pm}_{lm}$ boson (AFM) operators introduced in eq. (7.51) are different than (although labelled the same) and not to be confused with the $b^{\pm}_{lm}$ boson (IBM) operators used in the preceding paragraph.) The s and b indices (subscripts and superscripts) are used to distinguish the $0\hbar\omega$ shell-model and boson parts, respectively, of the N, L and Q^a operators. The N_s operator counts the total number of oscillator bosons in the $0\hbar\omega$ system, including the zero point energy, and is therefore a large number, while N_b counts the number of boson excitations, n for a $2n\hbar\omega$ theory. The 2 multiplying N_b in the expression for N enters, because each boson carries $2\hbar\omega$ units of energy. The L and Q operators have, in addition to their usual $0\hbar\omega$ shell-model parts, an added boson contribution which derives from the symplectic extension. It is the boson extension of the quadrupole operator that leads to the large enhancements observed in E2 transitions strengths. The contraction procedure replaces the symplectic algebra by the direct sum of a 6-dimensional Heisenberg–Weyl algebra and $su(3)$ [$sp(3,N) \rightarrow w(6) + su(3)$] while the group itself reduces to a semi-direct product structure, $Sp(3, R) \rightarrow W(6) \wedge SU(3)$. This structure is then similar (but not the same, as it involves a semi-direct product of $W(6)$ with $SU(3)$ rather than a direct product) to that suggested in the previous paragraph for a multi-$\hbar\omega$ extension of the IBM. The difference comes about, because even though the generators of $W(6)$ and $SU(3)$ close separately, the $W(6)$ and $SU(3)$ generator do not commute—symbolically: $[su(3), w(6)] \rightarrow w(6)$.

This group structure leads to a simple yet elegant scheme for labelling basis states of the contracted symplectic model,

$$\begin{array}{ccccccc} U_b(6) \wedge U_s(3) & \rightarrow & SU_b(3) \wedge SU_s(3) & \rightarrow & SU(3) & \rightarrow SO(3) & \rightarrow SO(2) \\ N_b \quad\quad N_s & & (\lambda_b\mu_b) \quad\quad (\lambda_s\mu_s) & \rho & (\lambda) \quad \kappa & L & M \end{array} \tag{7.52}$$

which explicitly shows how the usual $0\hbar\omega$ shell-model result is augmented with the boson extension. The $U_b(6) \rightarrow SU_b(3)$ reduction is the same as one encounters in the $U(6) \rightarrow SU(3)$ limit of the IBM. For example, the (λ_b,μ_b) irreps that can occur in a space with N_b bosons follows the rule for the decomposition of a symmetric $U(6)$ irrep $[n]$, built out of n products of $(2, 0)$ $SU(3)$ irreps, into a direct sum of $SU(3)$ representations:

$$\prod^{n} [\otimes (2, 0)] = \sum_{n_1 \geq n_2 \geq n_3 \geq 0,\ \text{all } n_i \text{ even}}^{n_1+n_2+n_3=2n} [\oplus (n_1 - n_2, n_2 - n_3)]. \tag{7.53}$$

Accordingly, the basis states for identical nucleon configurations of the symplectic model can be labelled as

$$|\Psi^{m[f]}_{\gamma JM_J}\rangle \equiv |\, m[f]\alpha[N_b(\lambda_b\mu_b)N_s(\lambda_s\mu_s)]\rho(\lambda\mu)\kappa LSJM_J\rangle \tag{7.54}$$

where we have replaced the K label used in Section 7.2 with a running index κ to denote an orthonormal set of bases states. The $\kappa = 1$ state corresponds to the normalized $K = 0$ or 1 configuration obtained when the minimum of (λ, μ) is even or odd, respectively, see eq. (7.17); the $\kappa = 2$ state refers to the normalized $K = 2$ or 3 configuration orthogonal to the $\kappa = 1$ solution; etc. If there are no boson ($N_b = 0$) excitations, this multi-$\hbar\omega$ scenario reduces to the usual $0\hbar\omega$ shell-model result.

To summarize, the structure of the set of states (infinite in number and quadrupole coherent) associated with every allowed ($0\hbar\omega$ or excited, but then independent and non-symplectic) $SU(3)$ shell-model configuration should now be clear. We refer to the collection of all $0\hbar\omega$ configurations as the $0\hbar\omega$ *horizontal* shell-model space and the towers built on each of the $SU(3)$ (λ_s, μ_s) irreps as the *vertical* extension of that irrep. Each vertical extension can be partitioned into horizontal slices with the states within each slice representable as a homogeneous polynomial function in the B^+_{lm} [symplectic → $Sp(3, R)$] or b^+_{lm} [boson approximation → $W(6)$] raising operators acting on a parent $0\hbar\omega$ (λ_s, μ_s) shell-model configuration. The degree of these homogeneous polynomials in the raising operators identifies the horizontal slice to which the basis state belongs; for example, it belongs to the $2n\hbar\omega$ slice of the (λ_s, μ_s) irrep if the polynomial acting on the (λ_s, μ_s) $0\hbar\omega$ configuration is of order n in the raising operators. Remember as well that in each (λ_s, μ_s) irrep there is one state which is also lowest-weight with respect to $SU(3)$. It follows that the full irrep space, its horizontal as well as its vertical extensions, can be generated by applying the raising operators of $SU(3)$ to span its horizontal extent, and the B^+_{lm} [$Sp(3, R)$] or b^+_{lm} [$W(6)$] raising operators to generate its vertical extension. (This representation of the model space as horizontal slices within vertical towers is illustrated in Fig. 7.31.) Because of this structure, when dealing with interactions (see below) it is useful to speak in terms of those that effect only horizontal mixing, such as pairing, and those like the quadrupole-quadrupole interaction which effect vertical mixing but do not break $SU(3)$ within the horizontal slices. Some interactions clearly generate horizontal *and* vertical mixing.

7.5.2. Symplectic Hamiltonian

An appropriate Hamiltonian for describing rotational phenomena within the symplectic model consists of the harmonic oscillator, which provides the background shell structure; the quadrupole-quadrupole interaction, $Q^c \cdot Q^c$; a residual interaction that should include (for example) the single-particle orbit-orbit and spin-orbit terms, as well as pairing and other interactions involving currents and a hexadecapole-hexadecapole form as needed to get M1 and E4 transition strengths correct, respectively; etc. However, most applications of the theory are much less ambitious than this, restricting the interaction to terms that can be expressed solely in terms of generators of the $Sp(3, R)$ algebra. The problem with proceeding to more realistic Hamiltonian forms is that codes for determining matrix elements of operators like pairing between basis

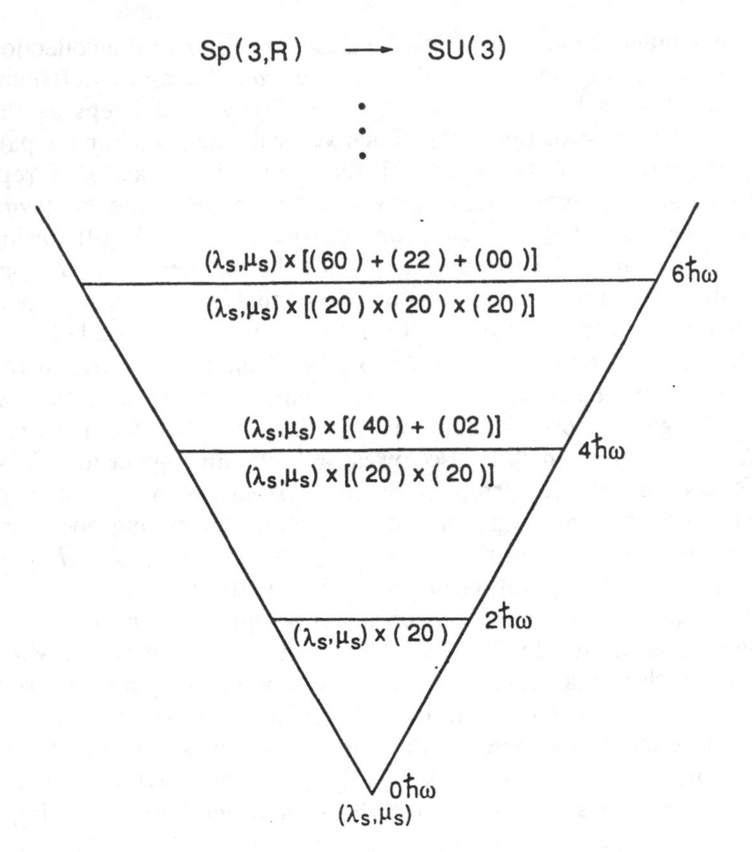

FIGURE 7.31. *A representation of the non-compact* Sp(3,R) *group is partitioned into* irreps *of its maximal compact* SU(3) *subgroup. The horizontal slices within the vertical tower at the* $2n\hbar\omega$ *level are all* SU(3) irreps *that arise in the product of the symmetric product of the* (20) irrep *with itself* n *times, which yield all* irreps $(n_1 - n_2, n_2 - n_3)$ *with the* n_i *even,* $n_1 \geq n_2 \geq n_3 \geq 0$, *and* $n_1 + n_2 + n_3 = 2n$, *and the lowest* $0\hbar\omega$ *shell-model* irrep (λ_s, μ_s) *which is unique and is therefore used to label the symplectic* irrep.

states of the $SU(3)$ scheme are not generally available—even for leading irreps of the $0\hbar\omega$ horizontal space. Despite this limitation, however, much can and has been learned about the buildup of quadrupole collectivity using the simpler operator forms.

There is another problem, because although observed E2 transition strengths in light nuclei (^{24}Mg, for example) can be reproduced with only the harmonic oscillator H_0 and $Q^c \cdot Q^c$ terms taken into account ($H = H_0 - \frac{1}{2}\chi\, Q^c \cdot Q^c$), the required value for χ is so large that the $Q^c \cdot Q^c$ interaction actually destroys the mean field of the oscillator established by H_0. To render $Q^c \cdot Q^c$ an appropriate residual interaction, it is necessary to first remove its contribution to the mean field. Once this is done, it becomes an acceptable residual interaction compatible with the structure established by H_0. An appropriate form is therefore

$$H = H_0 - \tfrac{1}{2}\chi[Q^c \cdot Q^c - (Q^c \cdot Q^c)_{\text{shell}}] + H_r, \tag{7.55}$$

where $Q^c \cdot Q^c$ has been replaced by $Q^c \cdot Q^c - (Q^c \cdot Q^c)_{\text{shell}}$ and $(Q^c \cdot Q^c)_{\text{shell}}$ is an operator that reproduces traces of $Q^c \cdot Q^c$ in all major shells of the oscillator. Simple analytic expressions can be given for the trace-equivalent forms of most operators. In particular, for $Q^c \cdot Q^c$ one finds that

$$(Q^c \cdot Q^c)^{\text{shell}} = \frac{65}{7}N^2 + \frac{250}{7}N + \frac{10}{3}NN_s + \frac{15}{4}C_2^s + 10N_s, \tag{7.56}$$

where N counts the total number of oscillator quanta [$N = N_s + 2N_b$, with N_s counting quanta up through the $0\hbar\omega$ level and N_b counting those associated with the $2\hbar\omega$ symplectic excitations, see eq. (7.51)], and C_2^s is the second order Casimir invariant of $SU(3)$ acting at the $0\hbar\omega$ level with eigenvalue $[(\lambda_s + \mu_s + 3)(\lambda_s + \mu_s) - \lambda_s\mu_s]$ in the (λ_s, μ_s) irrep.

The procedure one uses to obtain the boson form for Hamiltonian eq. (7.55) is straightforward. First of all, each operator is expressed in normal ordered form: after making the substitution $Q^c = Q^a + \sqrt{6}/2(B_2^- + B_2^-)$, the $2\hbar\omega$ raising generators (B_l^+) are arranged to be in leftmost positions, and the lowering generators (B_l^-) are shifted into rightmost positions. After this normal ordering of the raising and lowering operators, substitutions eq. (7.51) are made. This procedure yields the following boson form for the Hamiltonian:

$$\begin{aligned} H &= H_0 - \tfrac{1}{2}\chi[Q^c \cdot Q^c - (Q^c \cdot Q^c)_{\text{shell}}] + H_r \\ &\xrightarrow{\text{boson approx}} 2\hbar\omega N_b \\ &\quad + \chi\Bigg[\left(\frac{5}{3}N_s + \frac{75}{7}\right)N_b + \frac{25}{14}N_b^2 \\ &\quad - \tfrac{1}{2}Q^a \cdot Q^a - \sqrt{2N_s}(b^+ \cdot Q^a + Q^a \cdot b^-) \\ &\quad + 10\sqrt{N_s}(b_{00}^+ + b_{00}^-) - N_s(b^+ \cdot b^+ + b^- \cdot b^-) \\ &\quad - 2N_sN_d + \frac{5}{4}C_2(\lambda_s, \mu_s)\Bigg] + aL^2 + bX_3^a + cX_4^a, \end{aligned} \tag{7.57}$$

where C_2 is the second-order Casimir operator of the $SU(3)$ algebra, N_d counts the number of $l = 2$ bosons, the expansion of $(Q^c \cdot Q^c)_{\text{shell}}$ is included, and all additive constant terms have been dropped. A specific choice for H_r is included in this expression, namely, the interaction that we determined in Section 7.2 to be the image of the rotor—see eq. (7.11). In this form, matrix elements of the Hamiltonian [in the boson basis given by eqs. (7.52) and eq. (7.54)] can be easily evaluated and the matrix diagonalized.

An estimate for the strength χ of the quadrupole-quadrupole interaction can be determined by equating the difference in the expectation value of H in the 0^+ state of the $0\hbar\omega$ (λ_s, μ_s) irrep, and the so-called stretched $2\hbar\omega$ $(\lambda_s + 2, \mu_s)$ irrep to the excitation energy, $80A^{-1/3}$, of the giant monopole resonance:

$$\chi = \frac{A^{-1/3}}{4\lambda_s + 2\mu_s + 15/4 + N_s(\langle N_d \rangle - 5/6)}. \tag{7.58}$$

This result follows because in $L^\pi = 0^+$ states $Q^a \cdot Q^a = 4C_2$ and matrix elements of the X_3 and X_4 operators, like those of L^2, vanish. The only unknown in eq. (7.58) is the expectation value of N_d in the $2\hbar\omega$ space, and an analytic expression for it in terms of the (λ_s, μ_s) can be given:

$$\langle N_d \rangle = \frac{2\lambda_s(\lambda_s+2) + \mu_s(2\lambda_s+1)}{3(\lambda_s+1)(\lambda_s+\mu_s+2)}. \tag{7.59}$$

Though this estimate for χ has been shown to be quite good (see the results given below, for example) it can be improved by using it to calculate eigenfunctions of the Hamiltonian in eq. (7.57) with $a = b = c = 0$ in the 0^+ and 2^+ spaces and calculating the $2_1^+ \rightarrow 0_1^+$ reduced E2 transition probability. If the E2 strength is smaller (bigger) than the experimental result, a bigger (smaller) value of χ should be chosen. This step can be repeated until agreement with the experimental result is achieved. The assumption of a linear relation between E2 values and χ suffices in interpolating between results. The residual interaction can be ignored in this iterative process—even though it changes the excitation energy of the 2_1^+ state, it has virtually no effect on the calculated $2_1^+ \rightarrow 0_1^+$ transition strength.

Values for the parameters b and c of the residual interaction in eq. (7.57) can be determined from the splitting of the lowest (2_1^+) and first excited (2_2^+) $J = 2$ states. Specifically, in establishing a direct mapping between rotor and $SU(3)$ Hamiltonians, an operator expression was found for the square of the projection of the angular momentum on an intrinsic body-fixed symmetry axis, see eq. (7.22). In that expression, the λ_α's denote principal moments of inertia, and were related to the $SU(3)$ irrep labels λ and μ by requiring invariants of the rotor and $SU(3)$ to map onto one another. To the extent that the 2_2^+ state is a $K = 2$ band-head configuration, it follows that

$$b = \frac{\alpha\lambda_3}{2\lambda_3^2+\lambda_1\lambda_2} \quad \text{and} \quad c = \frac{\alpha}{2\lambda_3^2+\lambda_1\lambda_2} \qquad (7.60)$$
$$\text{where } \alpha = (E_{2_2^+} - E_{2_1^+})/4.$$

Note, in particular, that the b/c ratio is a constant given by $\lambda_3 = (2\lambda+\mu+3)/3$.

In a similar fashion, the parameter a that multiplies L^2 in eq. (7.57) can be given in terms of the experimental value for the inertia parameter $(\hbar^2/2\mathcal{I})$,

$$a = \hbar^2/2\mathcal{I} - \tfrac{3}{2}\chi + \frac{\alpha\lambda_1\lambda_2}{2\lambda_3^2+\lambda_1\lambda_2}. \qquad (7.61)$$

The second term in this expression reduces the parameter a from its $(\hbar^2/2\mathcal{I})$ value because an L^2 contribution also enters through $Q^a \cdot Q^a = 4C_2 - 3L^2$. The third term is required for the L^2 dependence in the K^2 operator. If the leading irrep has $\mu=0$, the theory predicts an absence of a $K=2$ band associated with the $K=0$ ground band. When this happens—for example, for ^{20}Ne of the *ds*-shell with $(\lambda, \mu) = (8,0)$ or as is shown below for the ^{238}U case with $(\lambda,\mu) = (54,0)$—it is simplest to set $\alpha = 0$ in both eqs. (7.60) and (7.61). The parameter, a, is then just the moment of inertia reduced by $\tfrac{3}{2}\chi$, the quadrupole-quadrupole contribution. In general, the parameter, a, is the least well-determined by these simple arguments. This is both understandable and fortunate: understandable as it involves a sum of three different terms and makes use of a replacement of $Q^c \cdot Q^c$ by $Q^a \cdot Q^a$; and fortunate, because the structure of the calculated eigenstates is independent of the value of the parameter, a.

7.5.3. Pseudo-Symplectic Scheme

As we discussed in Section 7.3, for heavy ($A \gtrsim 100$) nuclei there are two valence shells, one for protons (π) and one for neutrons (ν). Furthermore, each of these two spaces is comprised of a set of normal parity orbitals which may be identified as members of a pseudo shell ($\tilde{n}_\alpha = n_\alpha - 1$) and a unique parity intruder level with ($j_\alpha = n_\alpha + \frac{3}{2}$) from the shell above, where $\alpha = (\pi, \nu)$. If the normal (n) parity part of each space is partitioned into irreps of pseudo $SU(3)$, and a seniority scheme is used for the single j shell, unique (u) parity configurations, then the many-particle basis states can be given as angular-momentum-coupled products of four distinct and separate (proton and neutron) subspaces: $|\Psi_J\rangle = |\{[(J_{n_\pi} \times J_{n_\nu})J_n] \times [(J_{u_\pi} \times J_{u_\nu})J_u]\}J\rangle$. When the residual proton-neutron interaction is of the quadrupole-quadrupole type (regardless of the specific nature of the interaction in the separate subspaces) yrast states of strongly deformed even-even nuclei *below* the backbending region are well-represented by strong coupled pseudo $SU(3)$ basis states $[(\lambda_\pi, \mu_\pi) \times (\lambda_\pi, \mu_\pi) \rightarrow (\lambda, \mu)]$ and seniority zero-coupled ($J_{u_\pi} = 0$ and $J_{u_\nu} = 0$, so $J_u = J_{u_\pi} \times J_{u_\nu} = 0$) unique parity configurations. Of course, the structure of states *through and beyond* the backbending region will be more complicated. In particular, since they lie at an excitation energy that is comparable to the pairing gap, seniority two ($J_u = 2, 4, \ldots$) and higher unique parity configurations must be involved.

The pseudo symplectic scheme extends the pseudo $SU(3)$ picture by allowing for inter-shell $2\hbar\omega$ excitations of the monopole and quadrupole type among the normal parity orbitals, just as the symplectic scheme does for light ($A \lesssim 25$) nuclei. In this case, however, the underlying symmetry group is pseudo-$Sp(3,R)$ which contains pseudo-$SU(3)$ as a subgroup. The generators of the pseudo symplectic symmetry groups (one for the proton space and another for the neutrons) include the 3 angular momentum operators ($\tilde{L}_\alpha$'s) and 5 quadrupole operators ($\tilde{Q}^a_\alpha$'s)—which together generate an $\tilde{S}\tilde{U}_\alpha(3)$ subgroup—plus 6 raising $\tilde{B}^+_\alpha$ and 6 lowering $\tilde{B}^-_\alpha$ operators. In total, these 21 operators generate the non-compact $\tilde{S}\tilde{p}_\alpha(3,R)$ extension of the compact $\tilde{S}\tilde{U}_\alpha(3)$ group. Since multiple $2\hbar\omega$ excitations in the normal parity spaces are included, it would seem that unique parity excitations of this type should also be considered, as well as mixing between the two. But it can be argued as before that this is not necessary for states in even-even nuclei which lie below the backbending region, since the unique parity parts of these wavefunctions will be dominated by pairing correlations which differ from quadrupole correlations in that they do not require the involvement of higher shells for strength enhancement. Only for states immediately below and above the pairing gap (which coincides with the onset of backbending) or for odd-A nuclei are unique parity configurations expected to play an important role in the dynamics. However, we must point out that as in the $0\hbar\omega$ case, this assumption has not be thoroughly tested.

The pseudo symplectic model Hamiltonian is taken to be the sum of proton and neutron pseudo harmonic oscillator Hamiltonians, plus a real quadrupole-quadrupole interaction acting between the particles, including π-ν terms as well as the identical particle π-π and ν-ν interactions. And as indicated above contributions to the dynamics from protons and neutrons in the unique parity levels have been suppressed thus far. Also, to insure that the mean field of the oscillator is conserved under the quadrupole-quadrupole perturbation, as in the case of light nuclei, the $\tilde{Q}^c_\alpha \cdot \tilde{Q}^c_\alpha$ interactions are replaced by $\tilde{Q}^c_\alpha \cdot \tilde{Q}^c_\alpha - (\tilde{Q}^c_\alpha \cdot \tilde{Q}^c_\alpha)_{\text{shell}}$ where $(\tilde{Q}^c_\alpha \cdot \tilde{Q}^c_\alpha)_{\text{shell}}$ is an operator that reproduces single shell traces of the $\tilde{Q}^c_\alpha \cdot \tilde{Q}^c_\alpha$ operator in the α-th subspace. Thus we have

$$H = \tilde{H}_0 - \tfrac{1}{2}\chi[Q^c \cdot Q^c - (Q^c_\pi \cdot Q^c_\pi)_{\text{shell}} - (Q^c_\nu \cdot Q^c_\nu)_{\text{shell}}] + H_r, \quad (7.62)$$

where

$$\tilde{H}_0 = \tilde{H}_{0\pi} + \tilde{H}_{0\nu} \quad \text{and} \quad Q^c = Q^c_\pi + Q^c_\nu. \quad (7.63)$$

As before, the superscript c in these equations denotes the collective quadrupole operators as opposed to the symmetrized algebraic ones of a $0\hbar\omega$ pseudo-$SU(3)$ scheme. An appropriate residual interaction, H_r, is also included and in applications has been chosen to be a rotor-like form (see eqs. (7.57) and (7.65), below).

We can express the mass quadrupole operators Q^c_π and Q^c_ν in terms of generators of the normal (not pseudo) symplectic algebras: $Q^c_\alpha = Q^a_\alpha + \sqrt{6}/2[B^+_{2\alpha} + B^-_{2\alpha}]$. And as for the pseudo-$SU(3)$ case, these operators ($O = Q^a, B^+, B^-$) can all be expanded in terms of pseudo-$SU(3)$ tensors as indicated in eq. (7.32) and Table 7.7:

$$O = \kappa\tilde{O} + \cdots, \quad (7.64)$$

where $\tilde{O}$ has the same tensorial character as O and the ellipses represents other tensors which arise because the normal $\leftrightarrow$ pseudo-transformation is not symmetry preserving. Once again, the constant κ in this expansion is always greater than unity, ranging from ~1.4 for $O = Q^a(\tilde{n} = 2)$ to ~1.1 for $O = B^+_{2m}(\tilde{n} = 6)$. An average

κ value for actinide nuclei is 1.14. As can be seen for the B^+ operators in Table 7.13 (see also Table 7.7), the other terms in the series have a different tensorial character, and their expansion coefficients are usually less than ten percent of the leading term. In

TABLE 7.13. *Tensor expansion coefficient for the $2\hbar\omega$ raising operators, B^+_{ln} with $l = 0$ and 2 of the* Sp(3, R) *symplectic algebra. The $2\hbar\omega$ lowering operators, B^-_{ln}, are the hermitian adjoints of the raising generators. Corresponding normal and pseudo forms are italicized. (From Castaños et al.,* 1991.)

(λ_0, μ_0)	κ_0	L_0	S_0	$n = 4$
B^+_{00}				
Normal				
(2, 0)	*1*	*0*	*0*	*16.73320*
Pseudo				
(2, 0)	*1*	*0*	*0*	*19.89008*
(3, 1)	1	1	1	1.95141
(4, 2)	1	0	0	0.69593
(5, 3)	1	1	1	0.29740
(6, 4)	1	0	0	0.06295
B^+_{2m}				
Normal				
(2, 0)	*1*	2	*0*	*16.73320*
Pseudo				
(2, 0)	*1*	*2*	*0*	*18.54388*
(2, 0)	1	2	1	2.41666
(3, 1)	1	1	1	0.65306
(3, 1)	1	2	0	0.84002
(3, 1)	1	2	1	0.82559
(3, 1)	1	3	1	0.35250
(4, 2)	1	2	0	0.37102
(4, 2)	1	2	1	0.35373
(4, 2)	2	2	0	−0.08262
(4, 2)	2	2	1	0.03246
(4, 2)	1	3	1	0.16523
(5, 3)	1	1	1	0.09171
(5, 3)	1	2	0	0.12822
(5, 3)	1	2	1	0.12796
(5, 3)	1	3	1	0.06454
(5, 3)	2	3	1	−0.01430
(6, 4)	1	2	0	0.03749
(6, 4)	1	2	1	0.03438
(6, 4)	2	2	0	−0.00518
(6, 4)	2	2	1	0.00607
(6, 4)	1	3	1	0.01584

pseudo-$SU(3)$ applications, these terms have been found to have no noticeable effect on the spectrum and yield less than a one percent change in calculated electromagnetic transition rates. In applications that employ the symplectic extension, it is therefore customary to drop the additional tensors.

Neglecting higher order terms, the Hamiltonian eq. (7.62) can be rewritten as

$$H = \hbar\omega\tilde{N} - \tfrac{1}{2}\tilde{x}[\tilde{Q}^c\cdot\tilde{Q}^c - (\tilde{Q}^c_\pi\cdot\tilde{Q}^c_\pi)_{\text{shell}} - (\tilde{Q}^c_\nu\cdot\tilde{Q}^c_\nu)_{\text{shell}}] \\ + a\tilde{L}^2 + b\tilde{X}^a_3 + c\tilde{X}^a_4. \tag{7.65}$$

In this expression, the average value for the κ introduced in eq. (7.64) is absorbed into the $\tilde{x}$, and the last three terms are an explicit form for H_r (as in eq. (7.57), these can be used to fix the K-band splitting and moment of inertia of the system). Note that $\tilde{L}$ is the total pseudo angular momentum operator while the $\tilde{X}^a_3$ and $\tilde{X}^a_4$ terms are third and fourth order rotational scalars built out of the algebraic quadrupole (Q^a) operators: $\tilde{X}^a_3 \sim [\tilde{L}x\tilde{Q}^a x\tilde{L}]$ and $\tilde{X}^a_4 \sim [(\tilde{L}x\tilde{Q}^a)^1 \times (\tilde{Q}^a x\tilde{L})^1]$—see eq. (7.6). As in the pseudo-$SU(3)$ case, this special form for the residual interaction means the moment of inertia and band splitting of the low-lying states can be adjusted without otherwise affecting the dynamics. In particular, note that matrix elements of this H_r vanish in 0^+ states. In analogy with eq. (7.57) for the normal symplectic case, one can write down a boson expansion form for eq. (7.65). In this case the eigenvalue of the $\tilde{N}_s$ operator is much larger, and accordingly the boson approximation much better, because one is closer to the contraction limit of the theory. And as for light nuclei, an estimate for the strength $\tilde{x}$ of the quadrupole-quadrupole interaction can be made by equating the difference in the expectation value of H in the 0^+ state of the $0\hbar\omega$ $(\tilde{\lambda}, \tilde{\mu})$ irrep and the stretched $(\tilde{\lambda} + 2, \tilde{\mu})$ pseudo-$SU(3)$ irrep at $2\hbar\omega$ to the excitation energy, $80A^{-1/3}$, of the giant monopole resonance [see eq. (7.58)]. And as before, this estimate for $\tilde{x}$ can be improved upon by using it to calculate eigenfunctions of the Hamiltonian, eq. (7.65), with $a=b=c=0$ in the 0^+ and 2^+ spaces and calculating the $2^+_1 \rightarrow 0^+_1$ reduced E2 transition probability. In fact, all the expressions and procedures given for determining the parameters $\tilde{x}$, a, b, and c in the normal symplectic case apply without change to the pseudo symplectic situation. For example, if the E2 strength is smaller (larger) than the experimental result a larger (smaller) value of $\tilde{x}$ should be chosen.

7.5.4. Giant Resonant Modes

To illustrate the symplectic theory, we will now give some calculated results for low-lying E2 transition and giant resonant mode strengths in ^{24}Mg $[\lambda_s, \mu_s) = (8,4)]$ based on the symplectic extension of the Elliott $SU(3)$ model, and for the strongly deformed heavy nuclei ^{168}Er $[(\tilde{\lambda}_s, \tilde{\mu}_2) = (30,8)$ and ^{238}U $[(\tilde{\lambda}_s, \tilde{\mu}_s) = (54,0)]$ using the pseudo symplectic scheme. The Hamiltonians used in these shell-model studies were the symplectic version of the $Q^c \cdot Q^c$ interaction and its boson approximation for the ^{24}Mg case (see. eq. (7.55) and eq. (7.57), respectively) and the pseudo symplectic form, eq. (7.65), for ^{168}Er and ^{238}U. To insure that the major shell structure survived the perturbation of the quadrupole-quadrupole interaction, the trace-equivalent form $Q^c \cdot Q^c - (Q^c \cdot Q^c)_{\text{shell}}$ was used rather than just $Q^c \cdot Q^c$, and the residual interaction chosen was the $0\hbar\omega$ rotor form specified in eqs. (7.57) and (7.65). It is important to reiterate that the objective of the symplectic model is the reproduction of the measured E2 transition strengths, and quadrupole moments of the low-lying collective states. Missing from the analysis are questions regarding other types of states, such as those based on pairing correlations. Specifically, the calculations recorded here—and others that have been done so far—are single symplectic irrep analyses. The various algebraic and algorithmic technologies that will allow for mixed-irrep calculations are still under development.

In Fig. 7.32, the excitation spectrum of ^{24}Mg is plotted as a function of χ for $0 \leq \chi \leq 0.042$ which extends just beyond the $\chi = 0.0415$ value required to reproduce the observed $B(\text{E2}, 2_1^+ \to 0_1^+)$ strength. This is roughly half the estimate for χ that one gets from eqs. (7.58) and (7.59) with $A = 24$, $(\lambda_s, \mu_s) = (8,4)$, and $N_s = 62.5$. The reason for this discrepancy in the predicted and actual value for χ is that with $A = 24$ one is quite far from the asymptotic limit of the theory where the boson approximation, which was used to derive the approximation for χ, works well. In the calculation, $\hbar\omega$ was set at 12.6 MeV, which is its self-consistent value $(45A^{-1/3} - 25A^{-2/3})$; and the a, b, and c parameters of the rotor form for the residual interaction H_r were assigned best fit values for this χ, namely 0.141, 0.0424, and 0.00554, respectively. An interesting number is the ratio $b/c = 7.66$. For a prolate rotor based on the (8, 4) irrep of $SU(3)$, b/c is given by $\lambda_3 = (2\lambda + \mu + 3)/3 = 7.67$, since this is the b/c value that is required for the X_3^a and X_4^a oper-

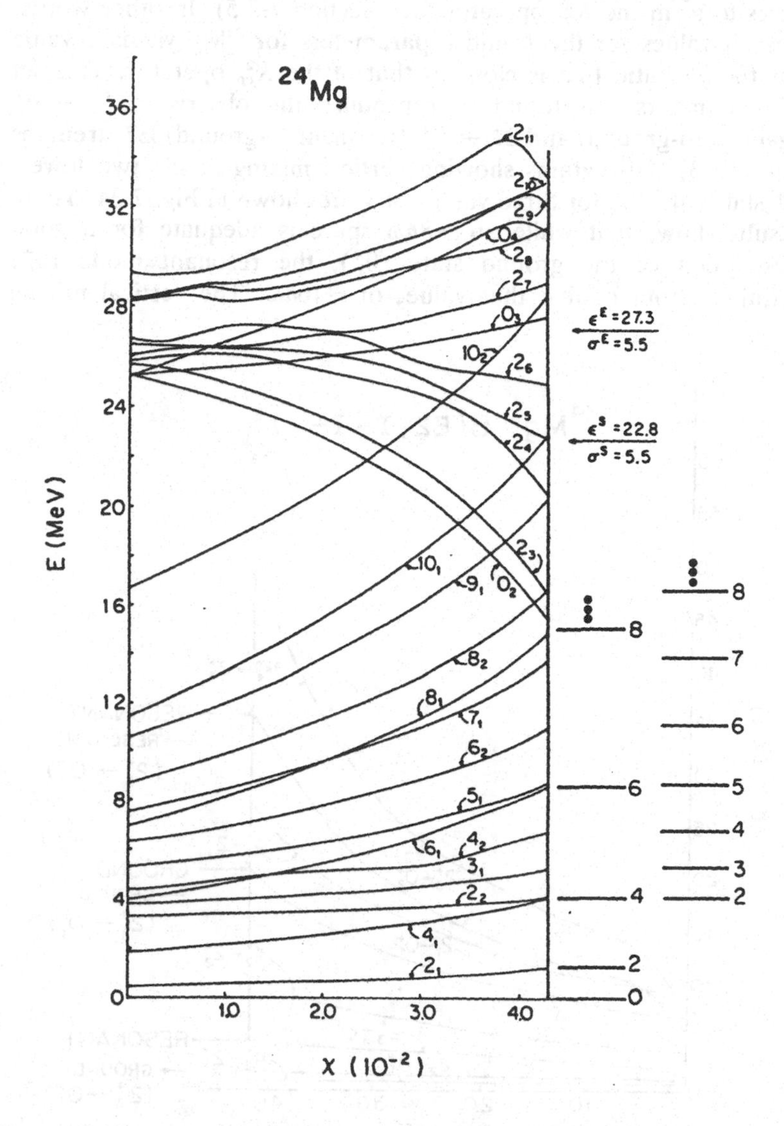

FIGURE 7.32. *Calculated eigenenergies for* ^{24}Mg *plotted as a function of the observed* E2 *strengths can be reproduced with* χ = 0.415 *and* no *effective charge. Calculated results for the low-lying spectrum and the centroid and width of the resonant mode* (2_3) *are indicated on the far right.* (From Draayer, 1990.)

ators to form the $K_{\rm op}^2$ operator (see Section 7.2.5). In other words, best-fit values for the b and c parameters for ^{24}Mg yields a value for the b/c ratio that is close to that of the $K_{\rm op}^2$ operator. This set of parameters was found to reproduce the observed $2_1^+ \rightarrow 0_1^+$ (gamma-to-ground) and $2_3^+ \rightarrow 0_1^+$ (resonant-to-ground) E2 strengths (Fig. 7.33). Histograms showing vertical mixing in the two lowest 0^+ states of ^{24}Mg for three values of χ are shown in Fig. 7.34. These results show, that while an $\sim 8\hbar\omega$ space is adequate for a good description of the ground state (0_1^+), the resonant mode (0_2^+) requires about double that value, or $\sim 16\hbar\omega$. The vertical mixing

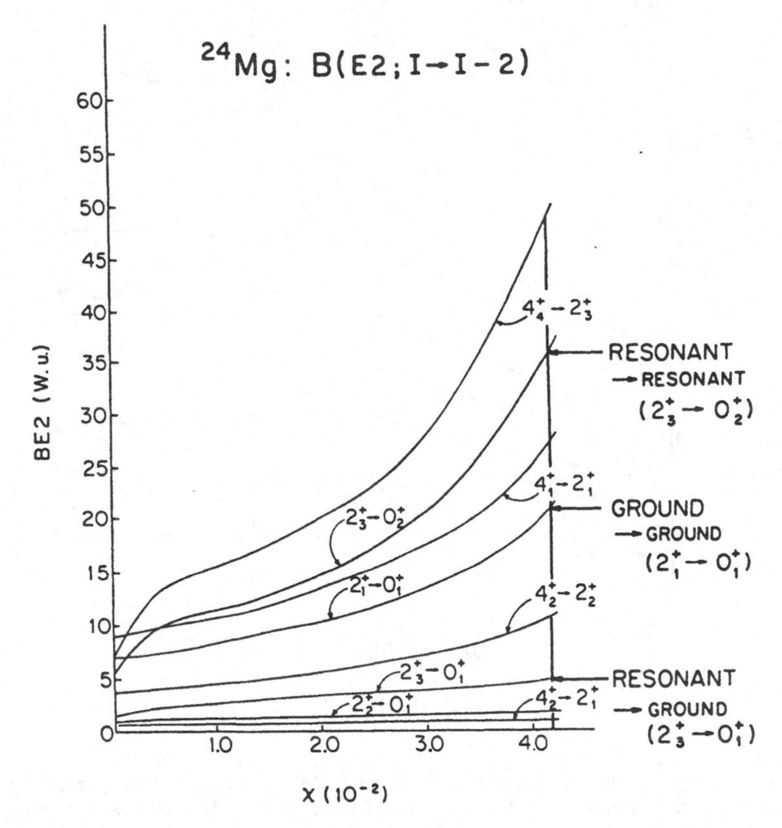

FIGURE 7.33. *Calculated E2 transition strengths for ^{24}Mg plotted as a function of the strength χ of the real quadrupole-quadrupole interaction. The observed strengths, intraband and interband, are reproduced with χ = 0.415 and no effective charge. (From Draayer, 1990.)*

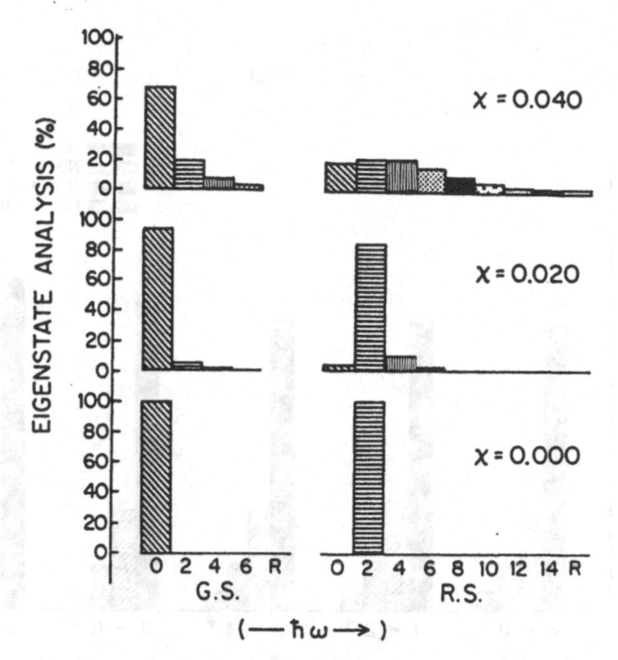

FIGURE 7.34. *Intensity analysis of the calculated* 0_1 *ground state* (G.S.) *and* 0_2 *resonant state* (R.S.) *eigenstates for* ^{24}Mg *are shown for three values of the strength* χ *of the real quadrupole-quadrupole interaction. For* $\chi = 0.415$, *which is the value required to reproduce observed* E2 *transition strengths without an effective charge, the* $0\hbar\omega$ *component of the ground state is slightly less than about* 70% *while the resonant state extends out to* $16\hbar\omega$. (*From Draayer,* 1990.)

accounts for approximately 25% of the total ground state wavefunction intensity, and is the origin of the enhanced E2 rates. The intensity distribution across the major shells for members of the ground state band is shown in Fig. 7.35. Note that the resonant mode (which peaks at $2\hbar\omega$ even as the simplest picture suggests) is strongly distributed across many shells. An interesting number is the ratio of the expectation value of $Q^c \cdot Q^c$ in the resonant mode and ground state: $b_r/b_g \approx 1.4$. This means that the resonant mode configuration has a deformation representative of the 3:2 axis ratio found for some superdeformed structures.

Similar calculations have been carried out using the full group theoretical hardware introduced above for selected rare earth and actinide nuclei. For these nuclei, one is much closer to the asymptotic (boson) limit of the theory, and therefore the estimate for χ

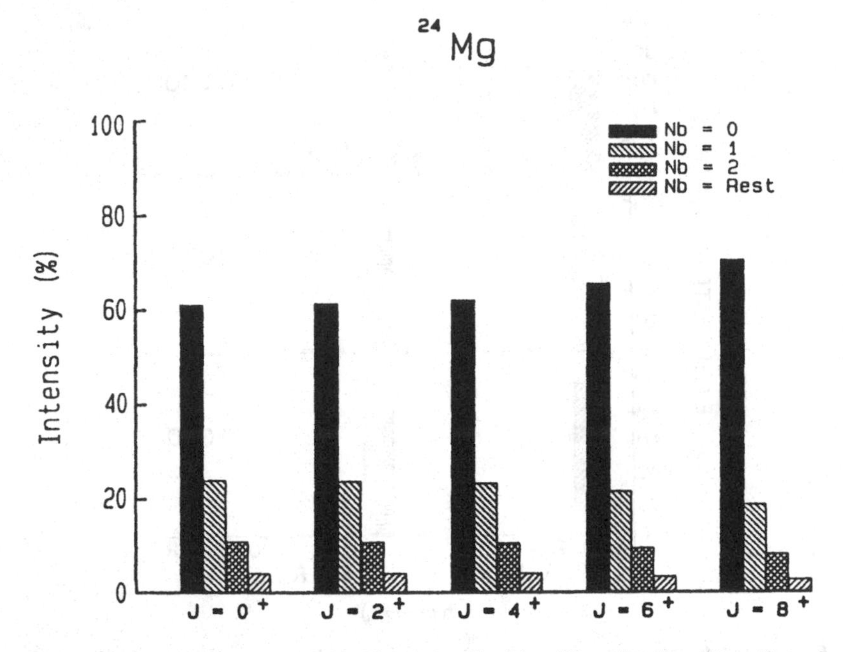

FIGURE 7.35. *Intensities of the* $0\hbar\omega$ ($N_b=0$), $2\hbar\omega$ ($N_b=1$), $4\hbar\omega$ ($N_b=2$) *and higher shells (Rest) in members of the ground state band in* ^{24}Mg. *The results show that the sum of the* $n\hbar\omega$ *contributions,* $n\neq 0$, *decreases from about* 40% *for the* $J^\pi = 0^+$ *state to less around* 30% *for the* $J^\pi = 8^+$ *state.*

is typically within 10–20% of the final best-fit value. A comparison of experimental and theoretical B(E2) rates for ^{168}Er is given in Fig. 7.36. Whereas no effective charges are need for *ds*-shell nuclei, for rare earth and actinide applications where the intruder levels are not counted, the E2 rates need to be rescaled using the factor

$$f^2 = \left[\left(\frac{Z_n}{A_n}\right)\left(\frac{ZA^{2/3}}{Z_nA_n^{2/3}}\right)\right]^2 = \left(\frac{Z}{A_n}\right)^2\left(\frac{A}{A_n}\right)^{4/3}. \tag{7.66}$$

In this expression, the subscript n on Z and A refers to nucleons that are distributed in the normal parity part of the full space whereas the unsubscripted Z and A count total numbers. As the notation suggests, the need for such a factor follows from the fact that contributions to the E2 strength from the intruder levels have been neglected. (This particular form can be deduced from the collective model expression for Q^c_π. For a uniform charge distribution

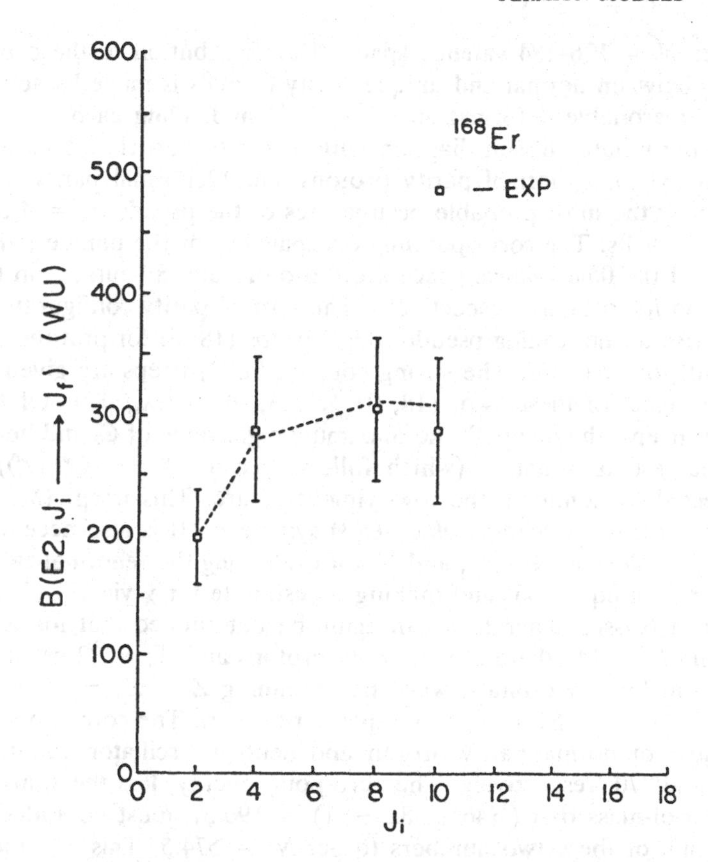

FIGURE 7.36. *Comparison of calculated and experimental* E2 *rates for* ^{168}Er. *The theoretical results (-) are the* Sp(3,R) *calculations generated using a boson approximation of the pseudo symplectic model. (From Draayer,* 1990.)

Q^c_π scales as $ZA^{2/3}$, where Z measures the charge and the $A^{2/3}$ factor comes from the quadratic radial dependence of Q^c and integration out to the nuclear surface, $R_0 = r_0A^{1/3}$. In addition, if the neutron and proton distributions track one another, $Q^c_\pi = (Z/A)\ Q^c$.) In the pseudo-$SU(3)$ picture, $Z \rightarrow Z_n$ and $A \rightarrow A_n$ where Z_n and A_n are the equivalent of Z and A for the normal parity space. The f^2 renormalizes the B(E2) rates accordingly. For isospin zero nuclei ($A_n = A$, $Z_n = Z = A/2$) this yields $f^2 = \frac{1}{4} = (Z/A)^2$ as it should.

Now we will consider the ^{238}U case in slightly greater detail. It has 10 valence protons in the Z = 82–126 shell, and 20 neutrons in the $N = 126$–184 valence space. The distribution of these particles between normal and unique parity orbitals is made by selecting a reasonable deformation ($\beta \approx 0.25$) and filling each level of the appropriate Nilsson diagram with a pair of particles. This procedure yields 6 normal parity protons and 12 normal parity neutrons for the most probable occupancies of the pseudo $\tilde{n}_\pi = 4$ and $\tilde{n}_\nu = 5$ shells. The corresponding occupancies for the unique parity parts of the $0\hbar\omega$ valence space are 4 protons and 8 neutrons in the $i_{13/2}$ and $j_{15/2}$ orbitals, respectively. The normal parity configurations give rise to the leading pseudo-$SU(3)$ irreps (18, 0) for protons and (36, 0) for neutrons. The strong coupled $SU(3)$ irreps are given by the product of these two, $(18, 0) \times (36, 0) \rightarrow \{(\tilde{\lambda}_s, \tilde{\mu}_s)\}$. Of this set of irreps, the one with the maximum eigenvalue of C_2 and hence the largest deformation (which follows because $\beta^2 \sim \langle Q^a \cdot Q^a \rangle$) is expected to dominate the low-lying structure. This irrep, $(\tilde{\lambda}_s, \tilde{\mu}_s) = (54, 0)$, is the leading $0\hbar\omega$ $SU(3)$ symmetry. It is also necessary to have values for Z_n, A_n, and $\tilde{N}_s$ for evaluating the renormalization factor f^2 in eq. (7.66) and making an estimate for $\tilde{\chi}$ via eq. (7.58). From a Nilsson diagram it can again be determined that for $Z = 92$ and $N = 146$ there are $Z_n = 46$ protons and $A_n = 82$ neutrons in normal-parity orbitals, with the remaining $Z - Z_n = 46 = Z_u$ and $N - N_n = 64 = N_u$ in unique-parity ones. The corresponding number of normal-parity proton and neutron oscillator quanta is 114 and 270, respectively. The zero-point energy, less the spurious center-of-mass part $(\frac{3}{2}(46 + 82 - 1) = 190.5)$, must be added to the sum of these two numbers to get $\tilde{N}_s = 574.5$. This is a much larger number than for ^{24}Mg ($N_s = 62.5$) and hence the ^{238}U application is much closer to the asymptotic boson limit of the theory. Consequently, the estimate for $\tilde{\chi}$ based on eq. (7.58) is expected to be much closer to the final value than in the ^{24}Mg case.

Applying these considerations to ^{238}U yields a value of 0.00138(MeV) for $\tilde{\chi}$. With this $\tilde{\chi}$, the calculated $2_1^+ \rightarrow 0_1^+$ transition rate is 2.48 e^2b^2, which is nearly within the error bars of the adopted experimental number, 2.42 ± 0.04. Since the leading pseudo-$SU(3)$ irrep for ^{238}U is (54, 0), it is analogous to ^{20}Ne in the ds-shell in that the theory predicts the absence of a $K^\pi = 2^+$ (gamma) band. In this application, the b and c parameters were therefore set to zero and the predicted value for a is then 0.00541. A best fit to the experimental energy spectrum and E2 transition strengths for the $L^\pi = 0^+$–12^+ states was obtained with $\tilde{\chi}p = 0.00135$ (MeV) and $a = 0.00465$. In the previous calculation a full

$20\hbar\omega$ symplectic basis was used, with dimensionalities ranging from 489 for the $L^\pi = 0^+$ space up to 4069 for $L^\pi = 12^+$. While the best fit value for $\tilde{\chi}$ is within 5% of the inital estimate, the final value for the parameter a is about 15% below the simplest prediction. This larger difference for parameter a can be attributed to the fact that $Q^c \cdot Q^c$ is more effective in generating rotations than $Q^a \cdot Q^a$ (which replaced it in generating the simple analytic prediction) suggests. Rather than showing a near perfect fit to the experimental spectrum, only calculated intensities of the ground (0_1^+) and resonant (0_2^+) states are given in Fig. 7.37 for the final best-fit $\tilde{\chi}$ value. Recall that for $\tilde{\chi} = 0.0$, the ground and giant monopole states are $0\hbar\omega$ [$(\tilde{\lambda}, \tilde{\mu}) = (54, 0)$] and $2\hbar\omega$ [$\tilde{\lambda}, \tilde{\mu}) = (56, 0)$] excitations, respectively.

An important feature is the amount of vertical mixing required to reproduce the observed E2 transition strengths. The $(\tilde{\lambda}, \tilde{\mu}) = (54,0)$ $0\hbar\omega$ configuration makes up 81.0% of the intensity profile of the ground state wavefunction, with slightly over a 15% contribution from the $2\hbar\omega$ space [11.3% $(\tilde{\lambda}, \tilde{\mu}) = (56, 0)$ and 4.0% $(\tilde{\lambda}, \tilde{\mu}) = (52, 2)$] and under 5% from the $4\hbar\omega$ space [1.4% for $(\tilde{\lambda}, \tilde{\mu}) = (58, 0)$, 1.3% for $(\tilde{\lambda},\tilde{\mu}) = (54,2)$ and 0.4% for $(\tilde{\lambda}, \tilde{\mu}) = (50,4)$], etc. On the other hand, the resonant mode has significant multi-$\hbar\omega$ admixtures extending out to $12\hbar\omega$, but in agreement with the simplest picture, the dominant contribution is $(\tilde{\lambda}, \tilde{\mu}) = (56, 0)$ with 45.5% intensity. The $(\tilde{\lambda}, \tilde{\mu}) = (52, 2)$ irrep adds an additional

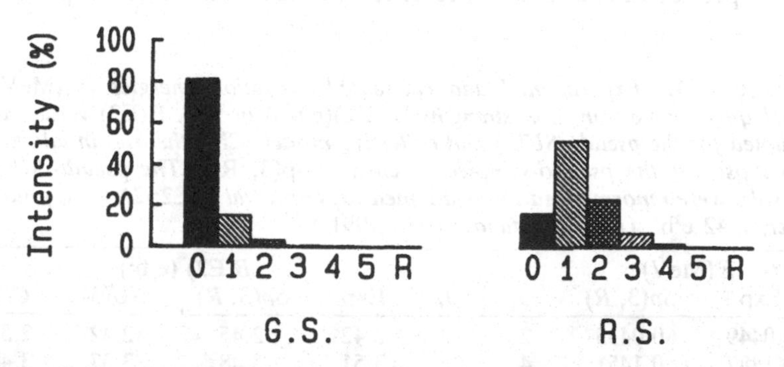

FIGURE 7.37. *Calculated intensities for the ground state* ($0_1^+ \rightarrow$ G.S.) *and resonant mode* ($0_2^+ \rightarrow$ R.S.) *of* ^{238}U *using the pseudo symplectic Hamiltonian in a full* $20\hbar\omega$ *single-irrep symplectic basis with the best-fit value for the strength of the* $Q^c \cdot Q^c$ interaction, $\tilde{\chi} = 0.00135$ (MeV). *The abscissa labels the number of* $2\hbar\omega$ *excitations.* (*From Castaños et al.*, 1991.)

5.1% for a total of 50.6% in the $2\hbar\omega$ space. The $0\hbar\omega$ and $4\hbar\omega$ spaces contribute 16.8% and 23.6%, respectively, to the structure of the resonant mode wavefunction. At each level, $\tilde{n} = 1, 2, \ldots$, the stretched $SU(3)$ irrep, $(\tilde{\lambda} + 2\tilde{n}, \tilde{\mu})$, was found to be the most important. Detailed results for excitation energies and electric quadrupole transition strengths are given in Table 7.14. Results for the pseudo-$SU(3)$ scheme and the collective model are included in the table for comparison. Note that the pseudo-$SU(3)$ $B(E2)$ strengths saturate at $J_i = 6$, while the collective model results show no such trend. The pseudo symplectic $B(E2)$ results, on the other hand, show saturation near the experimentally observed spins of $J_i = 12$–16.

Calculations of the type described above for ^{24}Mg and ^{238}U, while numerically intensive and not as simple as IBM calculations which give many more analytic results, are not overly difficult to perform, and can yield considerable insight into the structure of these nuclei. Furthermore, it should be clear that the problem of doing a many $(\tilde{\lambda}_s, \tilde{\mu}_s)$ calculation (the direction in which the theory is moving) maps onto a parallel processing strategy, a new and emerging computer technology. Short of a multi-(λ_s, μ_s) calculation, however, recent applications of the theory have been to a study of electron scattering form factors and double β-decay. The first of these applications is important because measured cross-sections can be used to study the flow of nuclear matter—determining whether nuclei rotate rigidly or irrotationally, a question requiring critical review as it probes fundamental differences between the macroscopic and

TABLE 7.14. *Experimental and calculated excitation energies* [E_i(MeV)] *and quadrupole transition strengths* [B(E2)(e^2b^2)] *in* ^{238}U. B(E2) *values are quoted for the pseudo-SU*(3) *and collective model* (CM) *theories in addition to those for the pseudo-symplectic scheme* [Sp(3, R)]. *The pseudo-SU*(3) *results were renormalized to the adopted experimental* B(E2: $2_1^+ \rightarrow 0_1^+$) *number*, 2.42 e^2b^2. (*From Castaños et al.*, 1991.)

E_i(MeV)				B(E2) (e^2b^2)			
Exp	*Sp*(3, *R*)	J_i	J_f	Exp	*Sp*(3, *R*)	*SU*(3)	*CM*
0.0449	0.0435	2	0	2.42	2.45	2.42	2.34
0.1487	0.1451	4	2	3.51	3.48	3.37	3.40
0.3072	0.3048	4	2	3.87	3.81	3.53	3.85
0.5178	0.5225	8	6	3.57	3.96	3.42	4.16
0.7757	0.7982	10	8	4.21	4.02	3.14	4.43
1.0765	1.1320	12	10	4.33	4.03	2.73	4.68

microscopic pictures of nuclear structure. Specifically, transverse form factors extracted from (e, e') data are sensitive to nuclear currents, and therefore the microscopic mechanisms responsible for observed rotational phenomena. High momentum transfer data is particularly sensitive to the multi-$\hbar\omega$ mixing that must be present. Applications of the theory to double β-decay have significance regarding the mass of the neutrino, and hence are of interest in high-energy and astrophysics. These go well beyond simple static electromagnetic measures such as E2 and M1 strengths which almost any theory (including a rotor-plus-particle model) seems to be able to reproduce so long as one is willing to accept the notion of effective charges.

7.5.5. Lesson for the Future

Bosons are easier to handle than fermions because the Pauli Principle constraint can be set aside. From a practical perspective, this means one can work with symmetric group representations, and for these the required group → subgroup structures (plethysms) are usually rather simple, so analytic results can often be given. To appreciate this point, one can compare the $U(6)$ symmetry group of the IBM and its symmetric representations ($[m/2]$ where m is the number of valence particles) with the much larger $U(k\Omega)$ group of the AFM approach and its complex antisymmetric irreps ($[1^m]$ where $\Omega = \Sigma(2l + 1)$ is the spatial degeneracy and k is 2 or 4 for a spin or spin-isospin version of the theory). Setting aside complications associated with antisymmetric forms frees one to focus on the physics of the problem, rather than becoming engulfed in sophisticated group methodologies. There is a price to be paid however, because with the boson ansatz comes the possibility of an under representation of the model space (missing configurations for small m values in the vibration limit of the theory) as well as an over representation of the model space (spurious configurations for large m values in its rotational limit). Attempts to correct for these deficiencies renders the boson approach as complex as a fermion scheme, and therefore compromises its appeal as a simple scheme understandable by all for studying nuclear structure phenomena.

The symplectic extension of the Elliott model, which has been the topic of this section, gives further insight into the complementary nature of the boson and fermion approaches. In particular, we have seen that the symplectic raising and lowering operators can be replaced by boson creation and annihilation operator forms when the occupancy of the excited configurations is respectively low or high. In this case the substitution is easily understood, since the promotion of a particle into an empty shell (or out of a filled one) is free of Pauli blocking effects. One does not always have to rely on intuitive arguments however, because sometimes a formal group contraction-expansion procedure can be applied to investigate the transition from a fermionic to a bosonic picture. As a specific example, we noted that the $SU(3)$ algebra contracts to $T_5 \wedge SU(3)$. From a purely mathematical perspective this allows one to understand the complementarity of the $SU(3)$ shell-model and collective model pictures, because the contraction procedure establishes a relationship between the group algebras of the systems.

Typical representations of $SU(3)$ encountered in applications of the IBM and AFM theories are rather far from the contraction limit of the theory. Nonetheless, the fact that mid-shell nuclei display rotational features is well documented. Indeed, results for excitation energies and transition rates show that $SU(3)$ results duplicate those of the rotor even for large L values and small representations—far from the contraction limit of the theory which guarantees this type of behavior. The lesson to be learned is a simple one: while it is important to have established a contraction limit of the theory guaranteeing rotational results, rotational motion may actually appear in non-contraction limit scenarios. This is certainly the case for the $0\hbar\omega$ Elliott and pseudo-$SU(3)$ AFM theories, and also appears to be true for the boson realization of the symplectic scheme, a $2n\hbar\omega$ ($n \neq 0$) theory. Whether this extends to the even smaller non-spurious representations of the IBM and FDSM theories is not entirely clear, though the successes of these models certainly suggest that it does.

ACKNOWLEDGEMENTS

There are several people who helped considerably in the preparation of this chapter: I am particularly grateful to Charolette Curtis for suggestions regarding style and for proofreading the text; the support of Beverly Rodrigues, an outstanding and trusted assistant, was essential; and I would also like to acknowledge the encouragement I received from Hortensia Valdes, a co-worker and personal friend. The work was supported in part by a grant from the U.S. National Science Foundation.

BIBLIOGRAPHY

SECTION 7.1 INTRODUCTION

Historical References

M. G. Mayer, *Phys. Rev.* **75**, 1969 (1949); **78**, 16 (1950).

O. Haxel, J.H.D. Jensen and H. E. Suess, *Phys. Rev.* **75**, 1766 (1949); *Z. Physik* **128**, 295 (1950).

A. Bohr and B. Mottelson, *Dan. Vid. Selsk. Mat. Fys. Medd.* **27** (1953).

S. G. Nilsson, *Dan. Vid. Selsk. Mat. Fys. Medd.* **29** (1955).

Background Reviews

J. B. French, *Proceedings of the International School of Physics,* Enrico Fermi, Course 36, ed. C. Bloch (Academic Press, New York, 1966).

M. Harvey, *Advances in Nuclear Physics,* eds. M. Baranger and E. Vogt (Plenum Press, New York, 1968), Vol. 1, p. 67.

K. T. Hecht, *Annual Review of Nuclear Science* (Academic Press, New York, 1973) Vol. 23, p. 123.

SECTION 7.2 QUANTUM ROTOR AND THE SU(3) MODEL

SU(3) and the Shell-Model

J. P. Elliott, *Proc. Roy. Soc. London* **A245**, 128, 562 (1958).

M. Moshinsky, *Rev. Mod. Phys.* **34**, 813 (1962).

J. P. Elliott and M. Harvey, *Proc. Roy. Soc.* **A272**, 557 (1963).

J. P. Elliott and C. E. Wilsdon, *Proc. Roy. Soc.* **A302**, 509 (1968).

SU(3) ⊃ R(3) Group Algebra

G. Racah in *Group Theoretical Concepts and Methods in Elementary Particle Physics,* ed. F. Gürsey (Gordon and Breach, New York, 1964) 31.

J. D. Vergados, *Nucl. Phys.* **A111**, 681 (1968).

J. P. Draayer and Y. Akiyama, *J. Math. Phys.* **14**, 1904 (1973).

B. R. Judd, W. Miller, Jr., J. Patera and P. Winternitz, *J. Math. Phys.* **15**, 1787 (1974).

Rotor ↔ SU(3) Mapping

Y. Leschber, *Hadronic J. Supp.* **3**, 95 (1987).

O. Castaños, J. P. Draayer and Y. Leschber, *Z. Phys.* **A329**, 33 (1988).

H. A. Naqvi and J. P. Draayer, *Nucl. Phys.* **A536**, 297 (1992).

SECTION 7.3 PSEUDOSPIN SYMMETRIES

Spin-Orbit Doublet Picture

A. Arima, M. Harvey and K. Shimizu, *Phys. Lett.* **B30**, 517 (1969).

K. T. Hecht and A. Adler, *Nucl. Phys.* **A137**, 129 (1969).

Theoretical Underpinnings

A. Bohr, I. Hamamoto and B. R. Mottelson, *Phys. Scr.* **26**, 267 (1982).

O. Castaños, M. Moshinsky and C. Quesne, *Phys. Lett.* **B277**, 238 (1992).

C. Bahri, J. P. Draayer and S. A. Moszkowski, *Phys. Rev. Lett.* **68**, 2133 (1992).

Mean-Field References

Y. Nambu and G. Jona-Lasinio, *Phys. Rev.* **122**, 345 (1961); **124**, 246 (1961).

B. D. Serot and J. D. Walecka, in *Advances in Nuclear Physics*, ed. J. W. Negele and E. Vogt (Plenum Press, New York, 1986), Vol. 16.

J. Zimanyi and S. A. Moszkowski, *Phys. Rev. C* **42**, 1416 (1990).

Pseudo-SU(3) Extension

R. D. Ratna Raju, J. P. Draayer and K. T. Hecht, *Nucl. Phys.* **A202**, 433 (1973).

O. Castaños, J. P. Draayer and Y. Leschber, *Ann. Phys.* **180**, 290 (1987).

J. P. Draayer, in *AIP Conference Proceedings: Capture Gamma-Ray Spectroscopy, Pacific Grove, CA 1990*, ed. Richard W. Hoff (American Institute of Physics, New York, 1991), p. 30.

Miscellaneous Applications

R. D. Ratna Raju, K. T. Hecht, B. D. Chang and J. P. Draayer, *Phys. Rev. C* **20**, 2397 (1979).

J. P. Draayer, C. S. Han, K. J. Weeks and K. T. Hecht, *Nucl. Phys.* **A365**, 127 (1981).

K. J. Weeks, C. S. Han and J. P. Draayer, *Nucl. Phys.* **A371**, 19 (1981).

J. P. Draayer, K. J. Weeks and K. T. Hecht, *Nucl. Phys.* **A381**, 1 (1982).

K. J. Weeks and J. P. Draayer, *Nucl. Phys.* **A393**, 69 (1983).

O. Castaños and J. P. Draayer, *Nucl. Phys.* **A491**, 349 (1989).

P. Rochford and J. P. Draayer, *Ann. Phys.* **214**, 341 (1992).

Corroborating References

W. Nazarewicz, P. J. Twin, P. Fallon and J. D. Garrett, *Phys. Rev. Lett.* **64**, 1654 (1990).

F. S. Stephens *et al.*, *Phys. Rev. Lett.* **64**, 2623 (1990); **65**, 301 (1990).

J. Dudek, B. Herskind, W. Nazarewicz, Z. Szymanski and T. R. Werner, *Phys. Rev. C* **38** (1988) 940.

SECTION 7.4 FERMION DYNAMICAL SYMMETRY MODEL

Theoretical Underpinnings

K. T. Hecht, J. B. McGrory and J. P. Draayer, *Nucl. Phys.* **A197**, 369 (1972).

J. N. Ginocchio, *Ann. Phys.* **126**, 234 (1980).

C.-L. Wu, D. H. Feng, X.-G. Chen, J.-Q. Chen and M. W. Guidry, *Phys. Lett.* **168B**, 313 (1986).

C.-L. Wu, D. H. Feng, X.-G. Chen, J.-Q Chen and M. W. Guidry, *Phys. Rev. C* **36**, 1157 (1987).

X.-L. Han, M. W. Guidry, D. H. Feng and C.-L. Wu, *Phys. Lett.* **192B**, 253 (1987).

Some Limits of the Theory

R. F. Casten, C.-L. Wu, D. H. Feng, J. N. Ginocchio and X.-L. Han, *Phys. Rev. Lett.* **56**, 2578 (1986).

J.-Q. Chen, X.-G. Chen, D. H. Feng, C.-L. Wu, J. N. Ginocchio and M. W. Guidry, *Phys. Rev. C* **40**, 2844 (1989).

Miscellaneous Applications

C.-L. Wu, X.-L. Han, Z.-P. Li, M. Guidry and D. H. Feng, *Phys. Lett.* **194B**, 447 (1987).

W.-M. Zhang, C. C. Martens, D. H. Feng and J. M. Yuan, *Phys. Rev. Lett.* **61**, 2167 (1988).

C.-L. Wu, D. H. Feng and M. W. Guidry, *Phys. Rev. Lett.* **66**, 1377 (1991).

Tests of the Theory

K. T. Hecht, in *Proceedings of the VIII Oaxtepec Symposium on Nuclear Physics, Oaxtepec, Morelos, México, January 8–10, 1985,* p 165.

P. Halse, *Phys. Rev. C* **39**, 1104 (1989).

J.-Q. Chen, D. H. Feng, M. W. Guidry and C.-L. Wu, *Phys. Rev. C* **44**, 559 (1991).

SECTION 7.5 SYMPLECTIC SHELL MODEL

Theoretical Underpinnings

H. Ui, *Progr. Theoret. Phys.* **44**, 153 (1970).

D. J. Rowe, *Rep. Progr. Phys.* **48**, 1419 (1985).

Pseudo-Symplectic Scheme

C. Bahri, J. P. Draayer, O. Castaños, and G. Rosensteel, *Phys. Lett.* **B234**, 430 (1990).

O. Castaños, P. O. Hess, J. P. Draayer, and P. Rochford, *Nucl. Phys.* **A524**, 469 (1991).

Miscellaneous Applications

G. Rosensteel, J. P. Draayer and K. J. Weeks, *Nucl. Phys.* **A419**, 1 (1984).

J. P. Draayer, *Nucl. Phys.* **A520**, 259c (1990).

C. Bahri, J. P. Draayer, O. Castaños and G. Rosensteel, *Phys. Lett.* **B234**, 430 (1990).

J. P. Draayer, S. C. Park, and O. Castaños, *Phys. Rev. Lett.* **62**, 20 (1989).

Index

As this book is a coherent whole devoted to a focused range of topics, many ideas and concepts appear repeatedly throughout. This index therefore makes no attempt to cite every occurrence of each rubric. The most important and extensive discussions are listed. For some topics, such as the IBM, s,d bosons, deformed rotors, quadrupole interactions, etc. that pervade the entire text, no specific listings are given. Also, many specific group labels are not separately indexed. For technical terms, such as Lie Algebra, Casimir operator, and the like, often only the first, or defining, occurrence is given. The list of specific nuclei or regions of nuclei is also not complete: only those most important in the history of these models are identified here.

Printed in the United States
by Baker & Taylor Publisher Services

Printed in the United States
by Baker & Taylor Publisher Services